CU00900750

The Millennium Atlas of Butterflies in Britain and Ireland

The Millennium Atlas of Butterflies in Britain and Ireland

Jim Asher, Martin Warren, Richard Fox, Paul Harding,
Gail Jeffcoate and Stephen Jeffcoate

With assistance from Nick Greatorex-Davies and Estella Roberts

Butterfly Conservation

Biological Records Centre

The Dublin Naturalists' Field Club

OXFORD
UNIVERSITY PRESS

Great Clarendon Street, Oxford OX2 6DP

Oxford University Press is a department of the University of Oxford
It furthers the University's aim of excellence in research, scholarship,
and education by publishing worldwide in

Oxford New York

Athens Auckland Bangkok Bogotá Buenos Aires Calcutta
Cape Town Chennai Dar es Salaam Delhi Florence Hong Kong Istanbul
Karachi Kuala Lumpur Madrid Melbourne Mexico City Mumbai
Nairobi Paris São Paulo Singapore Taipei Tokyo Toronto Warsaw

with associated companies in Berlin Ibadan

Oxford is a registered trade mark of Oxford University Press
in the UK and in certain other countries

Published in the United States
by Oxford University Press Inc., New York

First published 2001

A catalogue record for this book is available from the British Library

Library of Congress Cataloging in Publication Data

Millennium atlas of butterflies in Britain and Ireland / Jim Asher .. [et al.]
Includes bibliographical references
1. Butterflies—Great Britain. 2. Butterflies—Ireland. 3. Butterflies—Great
Britain—Geographical distribution—Maps. 4. Butterflies—Ireland—Geographical
distribution—Maps. I. Asher, Jim.

QL555.G7 M56 2001 595.78′9′0941—dc21 00-050160

ISBN 0 19 850565 5 (alk. paper)

Typeset by Footnote Graphics, Warminster, Wilts

Printed in Great Britain
on acid-free paper by
Butler & Tanner Ltd, Frome, Somerset

Dedicated to the many recorders who made it all possible

The Esmée Fairbairn Charitable Trust

Contents

Foreword *viii*
Preface *ix*
Acknowledgements *xi*

1 Background *1*
Butterflies for the New Millennium *1*
The importance of butterflies in Britain and Ireland *1*
Butterfly recording in Britain and Ireland *3*
Why the Butterflies for the New Millennium project was initiated *7*
About the atlas *10*

2 Butterfly habitats *11*
Geology, soils, and climate *11*
Habitats and management *12*
Major habitats and key features *19*

3 Recording and data collection *30*
Principles and standards for butterfly recording *30*
Organizing the survey—Butterfly*Net* *32*
BNM recording media *33*
Improving coverage *34*
Data collation *36*
Data verification and validation *36*
Bias and constraints *37*
Collation of pre-1995 data *38*
The Butterfly Monitoring Scheme *38*

4 Interpreting the data *40*
Mapping the data *40*
Recording effort *43*
Species richness *44*
Interpretation of Butterfly Monitoring Scheme data *45*
The timing of butterfly life cycle stages *47*
Comparison with historic data sets *48*

5 Species accounts *50*
Guide to content and layout *50*
Resident and regular migrant species
Hesperiidae: the skippers *52*
Papilionidae: the swallowtails *84*
Pieridae: the whites *88*
Lycaenidae: the hairstreaks, coppers, and blues *116*
Nymphalidae: the nymphalids and fritillaries *184*
Nymphalidae: the browns *244*
Rare migrant and extinct species *288*

6 The pattern and cause of change *324*

Habitat specialists and wider countryside species *324*
Distribution change since the 1970–82 survey *325*
Changes in distributions since the nineteenth century *332*
Changes in local distribution and abundance *333*
Causes of decline *336*
Causes of range expansion *343*

7 Conserving butterflies in the new millennium *346*

Butterflies in crisis *346*
Developing a strategic approach to conservation *347*
Threats to butterflies of conservation concern *350*
Conserving butterflies in the wider countryside *355*
Recommendations for future action *362*
Conclusions *365*

Appendices

1 Vernacular and scientific names of butterfly species *366*
2 Vernacular and scientific names of plants and other organisms *368*
3 Local butterfly atlases *373*
4 Butterflies for the New Millennium Instructions for recorders *377*
5 Standard recording forms *380*
6 UK monthly climate summaries (1995–9) *385*
7 Useful contact addresses *387*
8 Butterfly phenograms *390*
9 Vice-counties of Britain and Ireland *399*

References *401*

Glossary and abbreviations *421*

Index *425*

Photo and artwork credits *432*

Foreword

Human beings have been changing the landscape of Britain and Ireland for thousands of years. During the Mesolithic Period, they felled great areas of the forests that once covered most of these islands. By the end of the first millennium AD, lowland agriculture, fields, hedges, wetland drainage, coppiced woodland, and rural settlements were all well established. Butterflies, like other insects, while continuing to flourish in the wilder, uncultivated regions, also managed to colonize these new man-made habitats.

But now, at the beginning of a new millennium, humanity has encroached upon and degraded so much of the land that the countryside is in crisis. Intensive agriculture, habitat destruction, pollution, and urban and inter-urban developments have all contributed to this wholesale rural degradation. The result has been a great loss of wildlife in both numbers and variety. Butterflies, once widespread, have been seriously affected. Some species have retreated from the wider countryside into specialized fragmented habitats. Others have become extinct. Many of these losses have occurred within the last few decades. They are still continuing unabated.

The beginning of a new millennium should surely be the spur for us to stem and reverse this tide. This Millennium Atlas Project has caught the imagination of many thousands of individuals who have contributed a great deal of time and expertise in assembling the one and a half million records on which this book is based. But butterflies are not just beautiful creatures to observe. They also have a wider value. Like miners' canaries they can give warnings of environmental dangers.

This Atlas is now the baseline of our knowledge of butterfly distribution in Britain and Ireland. Future generations will certainly find it invaluable. Whether they will regard it as a measure of their own conservation successes or a sad record of lost glories, only time will tell.

Sir David Attenborough CH, FRS
President of Butterfly Conservation

Preface

Butterflies have a special place in our experience and vision of the countryside and are amongst the most beautiful groups of creatures. Their popular appeal, together with the small number of native species and the ease with which most can be identified, accounts for the long history of butterfly recording in Britain and Ireland. There are also good scientific reasons for recording butterflies: they respond rapidly to change and are thus good indicators of the quality of our countryside and the environment in general.

The second half of the twentieth century saw dramatic changes in the countryside of Britain and Ireland. Our native wildlife is increasingly threatened as habitats are damaged or destroyed, in particular by agricultural intensification, forestry, buildings, and roads. Added to these are the insidious threats from changes in the global climate caused by human activities with largely unknown, but potentially adverse, impacts on our biodiversity.

Understanding these threats and attempting to reverse them through practical conservation and research requires up to date, reliable information. Species and their habitats cannot be conserved unless we know where they are, what requirements they have and why they are declining. We also need to target the scarce resources available for nature conservation on the most threatened species and habitats. Such priorities cannot be determined without data on the distribution and abundance of species.

Many changes in distribution were noted in the previous Atlas of Butterflies in Britain and Ireland, published in 1984. Subsequently, interest in butterfly recording and the data that it generates has continued to grow. The Butterflies for the New Millennium project was launched in 1995 to co-ordinate this recording effort in a more comprehensive and systematic way than ever before. The project, organized by Butterfly Conservation and the Biological Records Centre, has created a network of butterfly recorders, spanning all parts of Britain, Ireland, the Isle of Man, and the Channel Islands. The network collates records from a wide range of sources and provides detailed, accurate, and ongoing information on the status of our butterflies. This atlas marks the successful completion of the first five years of the project and highlights the changing fortunes of the species that inhabit our islands at the beginning of the new millennium. In addition, the book summarizes the wealth of recent information provided by ecological research and conservation management experience and draws heavily upon the dataset of the Butterfly Monitoring Scheme. The implications for the conservation of our native butterflies are the central theme throughout and the book culminates in a synthesis of the challenges being faced and the means by which they may be overcome.

This atlas would not have been possible without the enthusiastic participation and support of many organizations. Local partnerships involving Butterfly Conservation branches, the Dublin Naturalists' Field Club, county wildlife trusts, local biological records centres, and many others have been the cornerstones of the survey. Special thanks are due to the individuals and organizations who have acted as co-ordinators for local areas. National organizations, including the statutory nature conservation agencies, other government departments and conservation charities have also lent their support. Many private landowners have permitted butterfly surveys to take place on their land and their kind co-operation is much appreciated. We are also indebted to our sponsors the Esmée Fairbairn Charitable Trust, the Vincent Wildlife Trust, ICI, The Heritage Council, and the Joint Nature Conservation Committee.

This book is dedicated to the many thousands of individuals who have walked the countryside across the length and breadth of these islands recording butterflies between 1995 and 1999. The authors owe these recorders a huge debt of gratitude and offer them our most sincere thanks. We only hope that this book does justice to their expectations in its presentation and content.

Jim Asher
Martin Warren
Richard Fox
Paul Harding
Gail Jeffcoate
Stephen Jeffcoate
March 2000

Acknowledgements

Central project co-ordination

STEERING GROUP:

Jim Asher, Richard Fox, Nick Greatorex-Davies, Paul Harding, Gail Jeffcoate, Stephen Jeffcoate, and Martin Warren.

STEERING GROUP (Satellite Members):

Trevor Boyd, Paul Kirkland, David Nash, Roy Neeve, Gary Roberts, Roger Smith, Richard Sutcliffe, Rob Whitehead, and Mike Williams.

Butterfly Conservation staff (especially Estella Roberts), Conservation Committee, and Roger Sutton

Biological Records Centre staff (especially Henry Arnold)

The Dublin Naturalists' Field Club

Centre for Ecology and Hydrology (Monks Wood)

Joint Nature Conservation Committee

Major sponsors and supporters

The Vincent Wildlife Trust

The Esmée Fairbairn Charitable Trust

ICI

The Heritage Council

The Ernest Kleinwort Charitable Trust

The H.D.H. Wills 1965 Charitable Trust

The following organizations provided records for the survey or supported the project in other ways: British Trust for Ornithology, British Trust for Conservation Volunteers, Butterfly Monitoring Scheme, Countryside Council for Wales, English Nature, Environment and Heritage Service, Farming and Wildlife Advisory Group, Ministry of Defence, National Trust, Royal Society for the Protection of Birds, Scottish Natural Heritage, and The Wildlife Trusts.

Production of the Atlas

Reviewers

We are extremely grateful to the following colleagues who reviewed chapters and species accounts: Bob Aldwell, Rich Austin, Nigel Bourn, Nick Bowles, Trevor Boyd, Paul Brakefield, Tom Brereton, Lincoln Brower, Simon Bryant, Caroline Bulman, Sue Clarke, Martyn Davies, Jack Dempster, Roger Dennis, Harry Eales, Sam Ellis, John Feltwell, Adrian Fowles, David Gutiérrez, Marney Hall, Jane Hill, Paul Hillis, John Holloway, Martin Honey, Christina James, Jenny Joy, Paul Kirkland, Roger and Margaret Long, Jamie McDonald, Tim Melling, Dorian Moss, David Nash, Brian Nelson, Matthew Oates, Ernie Pollard, Colin Pratt, Andrew Pullin, Neil Ravenscroft, Deborah Sazer, Tim Shreeve, Rob Souter, Odette

Sutcliffe, Richard Sutcliffe, Chris Thomas, Jeremy Thomas, Mike Tucker, Andrea Turner, Mike Williams, and Ken Willmott.

Advisors

The authors would also like to thank Philip Buckley, Mike Freeman, Tristan Lafranchis, Guy Meredith, Robert Pyle, David Roy, Michael Skelton, Bill Smyllie, Tim Sparks, Chris van Swaay, and Mark Tunmore for providing useful information and advice.

Photographs and illustrations (see pp. 432–3)

Jim Asher, Alan Barnes, Butterfly Conservation slide library, Centre for Ecology and Hydrology (Monks Wood), Dutch Butterfly Conservation, English Nature (Peter Wakeley), Harley Books, John Haywood, Joan Heath, Tony Hoare, Gail Jeffcoate, Stephen Jeffcoate, Roger Key, Richard Lewington, Jeremy Thomas, Robert Thompson, Martin Warren, Kars Veling (Dutch Butterfly Conservation), and Ken Willmott.

Publicity and fundraising

Gary Roberts (Archmain Ltd.), Sue Daniells (Butterfly Conservation).

Local project co-ordination (Butterfly*Net*)

Many thousands of people have taken part in the project and the authors regret that it is impossible to acknowledge them all individually. Instead, for each area that collated records the current local co-ordinator is listed (in bold) together with previous co-ordinators and other individuals who assisted in the co-ordination process (in brackets). Organizations whose staff and members provided records and support are then acknowledged.

England

Avon

CO-ORDINATOR: **Sarah Myles** (with Philippa Burrell).

Bristol Regional Environmental Records Centre, Avon Butterfly Project, West Country Branch of Butterfly Conservation, Gloucestershire Branch of Butterfly Conservation, Avon Wildlife Trust, City of Bristol Museum and Art Gallery, English Nature (Somerset Team).

Bedfordshire

CO-ORDINATOR: **Charles Baker** (with Michael Healy).

Bedfordshire Natural History Society, Bedfordshire and Northamptonshire Branch of Butterfly Conservation.

Berkshire, Buckinghamshire, and Oxfordshire

CO-ORDINATOR: **Jim Asher** (with Mike Wilkins, Roger Kemp, Grahame Hawker, Mark Calway, and David White).

Upper Thames Branch of Butterfly Conservation, Berkshire, Buckinghamshire and Oxfordshire Wildlife Trust, Oxfordshire Biological Records Centre, Buckinghamshire Environmental Records Centre, Buckinghamshire Invertebrate Group, Milton Keynes Natural History Society, Banbury Ornithological Society, West Oxfordshire Field Club.

Cambridgeshire and Essex

CO-ORDINATOR: **Val Perrin** (with Iris Newbery and Barry Dickerson).

Cambridgeshire and Essex Branch of Butterfly Conservation, Writtle College, Essex Moth Group, National Trust (Wicken Fen), Huntingdonshire Moth and Butterfly Group, Huntingdonshire Fauna and Flora Society.

Cheshire

CO-ORDINATOR: **Barry Shaw**.

Cheshire and Peak District Branch of Butterfly Conservation, Cheshire and Wirral Ornithological Society, Lyme Natural History Group.

Cornwall and the Isles of Scilly

CO-ORDINATOR: **John Worth** (with Adrian Spalding and Colin French).

Cornwall Branch of Butterfly Conservation, the former Cornwall Biological Records Centre, Caradon Field and Natural History Club, North Cornwall Field and Natural History Club.

Cumbria

CO-ORDINATOR: **Stephen Hewitt** (with Bill Kydd and Geoff Naylor).

Tullie House Museum, North of England Branch of Butterfly Conservation, Cumbria Naturalists' Union, English Nature (Cumbria Team), Cumbria Wildlife Trust.

Derbyshire

CO-ORDINATOR: **Ken Orpe** (with Roy Frost and Fred Harrison).

East Midlands Branch of Butterfly Conservation, Cheshire and Peak District Branch of Butterfly Conservation, Derbyshire and Nottinghamshire Entomological Society, Derbyshire Wildlife Trust, Derbyshire Ornithological Society, Derbyshire County Council (Ranger Service), English Nature (Peak District and Derbyshire Team), Derby City Council (Derby Biological Records Centre).

Devon

CO-ORDINATOR: **Roger Bristow** (with Simon Mitchell).

Devon Branch of Butterfly Conservation, Dartmoor National Park, Devon Wildlife Trust, Devonshire Association, Royal Albert Memorial Museum.

Dorset

CO-ORDINATOR: **Bill Shreeves** (with Bernard Franklin, David Jeffers, Carolyn Steele, and Brian Dicker).

Dorset Branch of Butterfly Conservation, Dorset Environmental Records Centre, Dorset Wildlife Trust and Urban Wildlife Project, Butterfly Conservation UK Conservation Office, Herpetological Conservation Trust, University of Leeds (Department of Biology).

Gloucestershire

CO-ORDINATOR: **Guy Meredith**.

Gloucestershire Branch of Butterfly Conservation, West Midlands Branch of Butterfly Conservation, Gloucestershire Naturalists' Society, Gloucestershire

Wildlife Trust, Gloucestershire Environmental Data Unit, Bristol Regional Environmental Records Centre.

Greater Manchester

CO-ORDINATOR: **Peter Hardy** (with Philip Kinder, Richard Greenwood, and Stephen Hind).

Lyme Natural History Group, Wigan Field Club, Bolton Museum and Art Gallery, Oldham Local Interest Museum.

Hampshire and the Isle of Wight

CO-ORDINATOR: **John Taverner** (with David Green, Alison Harper, Ken Bailey, Ian Small, and Andy Barker).

Hampshire and Isle of Wight Branch of Butterfly Conservation, Isle of Wight Natural History Society, English Nature (Hampshire and Isle of Wight Team), Hampshire Wildlife Trust, Hampshire County Council.

Hertfordshire and North London

CO-ORDINATOR: **John Murray** (with Michael Healy).

Hertfordshire and Middlesex Branch of Butterfly Conservation, London Natural History Society, Hertfordshire Natural History Society, Hertfordshire Biological Records Centre.

Kent and South-east London

CO-ORDINATOR: **John Maddocks**.

Kent Branch of Butterfly Conservation, Amateur Entomologists' Society, English Nature (Kent Team), Dungeness Bird Observatory, Kent Field Club, Kent Ornithological Society, Kent Wildlife Trust, London Wildlife Trust, North Kent Wildlife Preservation Society, Orpington Field Club, Royal Society for the Protection of Birds (Kent Reserves), Sandwich Bay Bird Observatory, White Cliffs Countryside Project, Whitstable Natural History Society, Corporation of London.

Lancashire and Merseyside

CO-ORDINATOR: **Laura Sivell** (with Peter North, Bob Murphy, and Simon Hayhow).

Lancashire Branch of Butterfly Conservation, Museum of Lancashire, English Nature (North West Team and Cumbria Team), Wildfowl and Wetlands Trust (Martin Mere), Lancashire Wildlife Trust, North Lancashire Naturalists, Council for the Protection of Rural England, North West Water.

Leicestershire and Rutland

CO-ORDINATOR: **Adrian Russell** (with Maurice Paul).

East Midlands Branch of Butterfly Conservation, Leicestershire Lepidoptera Recording Scheme, Leicestershire Entomological Society, Leicester Literary and Philosophical Society (Natural History Section), Rutland Natural History Society, Loughborough Naturalists' Club, Hinckley and District Natural History Society, Market Bosworth and District Natural History Society, Market Harborough Natural History Society, Leicestershire and Rutland Wildlife Trust, Hinckley and

Bosworth Borough Council, Leicestershire County Council (Holly Hayes Environmental Resources Centre).

Lincolnshire

CO-ORDINATOR: **Mark Tyszka** (with **Allan Binding**, Joseph Duddington, and Rex Johnson).

Lincolnshire Branch of Butterfly Conservation, Lincolnshire Naturalists' Union, Lincolnshire Wildlife Trust, English Nature (Humber to Pennines Team and East Midlands Team), Lincolnshire Bird Club, Lincoln City Council, Lincolnshire County Council.

Norfolk

CO-ORDINATOR: **Brian McIlwrath** (with Mike Hall, Francis Farrow, Fiona Joliffe, Barry and Bridget Dummell).

Norfolk Branch of Butterfly Conservation, Norfolk Wildlife Trust, Wymondham Nature Group, Norfolk Museum Services (Natural History Department).

Northamptonshire

CO-ORDINATOR: **Douglas Goddard**.

Bedfordshire and Northamptonshire Branch of Butterfly Conservation, Wildlife Trust for Bedfordshire, Cambridgeshire, Northamptonshire and Peterborough.

Northumberland and County Durham

CO-ORDINATOR: **Ian Waller** (with Harry Eales and Hew Ellis).

North of England Branch of Butterfly Conservation, Durham Wildlife Trust, Northern Naturalists' Union, Natural History Society of Northumbria, Durham Bird Club, Northumberland and Tyneside Bird Club, Northumberland Wildlife Trust, Tyne and Wear Museums, Sir James Knott Trust.

Nottinghamshire

CO-ORDINATOR: **Michael Walker**.

East Midlands Branch of Butterfly Conservation, Nottinghamshire Biological Records Centre, Nottinghamshire and Derbyshire Entomological Society.

Somerset

CO-ORDINATOR: **Roger Sutton** (with Tony Liebert, Paul Williams, and Jerry Board).

Somerset Environmental Records Centre, Somerset Butterfly Group, West Country Branch of Butterfly Conservation, Mid-Somerset Naturalists' Society, Exmoor Natural History Society, Somerset Archaeological and Natural History Society, English Nature (Somerset Team).

Suffolk

CO-ORDINATOR: **Richard Stewart**.

Suffolk Branch of Butterfly Conservation, Royal Society for the Protection of Birds (Suffolk Reserves), Suffolk Wildlife Trust, Forestry Commission (East Anglia District Office), English Nature (Suffolk Team), Suffolk Naturalists' Society, National Trust

(Dunwich Heath and Orfordness), Suffolk Biological Records Centre, Ipswich and District Natural History Society.

Surrey and South-west London

CO-ORDINATOR: **Gail Jeffcoate** (with Bill Gerrard and Mike Thurner).

Surrey and South-west London Branch of Butterfly Conservation, Corporation of London, Surrey Wildlife Trust, Surrey Wildlife Atlas Project, Surrey Bird Club, Downlands Countryside Management Project, Lower Mole Countryside Management Project, Wildfowl and Wetlands Trust (Barn Elms), Richmond Park Wildlife Group, Haslemere Natural History Society.

Sussex

CO-ORDINATOR: **Joyce Gay**.

Sussex Branch of Butterfly Conservation, West Sussex County Council, East Sussex County Council, National Trust (South Region), English Nature (Sussex and Surrey Team), South East Water PLC, Sussex Wildlife Trust, Sussex Downs Conservation Board, Chichester Harbour Conservancy, Conservators of Ashdown Forest, Royal Society for the Protection of Birds (Pulborough Brooks and Forewood), Forestry Commission, Worthing Natural History Society.

Warwickshire

CO-ORDINATOR: **Keith Warmington** (with Mike Slater, Neil Thompson, and Phil Parr).

Warwickshire Branch of Butterfly Conservation, West Midlands Branch of Butterfly Conservation, Warwickshire Museum Biological Records Centre, Coventry Museum, EcoRecord, Defence Munitions (Kineton), and Royal Engineers (Long Marston).

West Midlands (Herefordshire, Shropshire, Staffordshire, and Worcestershire)

CO-ORDINATOR: **Andy Nicholls** (with Richard Southwell, Jean Armstrong, John Brayford, David Jackson, Patrick Taylor, Ian Duncan, Christine and **Jim Chance**, Steve Harper, and Gunter Petters).

West Midlands Branch of Butterfly Conservation, Warwickshire Branch of Butterfly Conservation, National Trust (Shropshire and Staffordshire Wardens), Shropshire Wildlife Trust, Shropshire County Council, Staffordshire Wildlife Trust, Worcester Wildlife Trust.

Wiltshire

CO-ORDINATOR: **Mike Fuller**.

Wiltshire Branch of Butterfly Conservation, Wiltshire Butterfly Mapping Scheme (1982–94) Recorders, Ministry of Defence (Salisbury Plain Training Area, Boscombe Down, and Porton Down Conservation Groups), Wiltshire Wildlife Trust, Wiltshire and Swindon Biological Records Centre, Bentley Wood Trustees, Salisbury and District Natural History Society, Hampshire Branch of Butterfly Conservation, Dorset Branch of Butterfly Conservation.

Yorkshire

OVERALL CO-ORDINATORS: **Howard Frost** and **Philip Winter**.

Yorkshire Branch of Butterfly Conservation, Yorkshire Naturalists' Union.

South-east Yorkshire (Vice-county 61)

CO-ORDINATOR: **Sean Clough** (with Howard Frost, Christine Frost, Barry Spence, Peter Crowther, David Woodmansey, Adrian Johnson, Craig Ralston, John Killingbeck, Jack Whitehead, and John Sanderson).

Filey Bridge Ornithological Group, Yorkshire Wildlife Trust (Spurn National Nature Reserve and Filey Dams Nature Reserve), Yorkshire Water PLC (Tophill Low Nature Reserve).

North-east Yorkshire (Vice-county 62)

CO-ORDINATOR: **Peter Waterton** (with Steve Kirtley, Robert Parks, and Peter Robinson).

Tees Valley Wildlife Trust, Yorkshire Wildlife Trust, Cleveland Naturalists' Field Club, Harrogate and District Naturalists' Society, Forest Enterprise (North York Moors Forest District), Scarborough Field Naturalists' Society.

South-west Yorkshire (Vice-county 63)

CO-ORDINATOR: **Roy Bedford** (with Bill Smyllie and Susan Stead).

Rotherham Biological Records Centre, Wintersett Wildlife Group, Bradford Urban Wildlife Group, Yorkshire Wildlife Trust.

Mid-west Yorkshire (Vice-county 64)

CO-ORDINATOR: **Terence Whitaker** (with David Blakeley, Susan Stead, and Mike Barnham).

Malham Tarn Field Centre, Royal Society for the Protection of Birds (Fairburn Ings).

North-west Yorkshire (Vice-county 65)

CO-ORDINATOR: **Derek Parkinson** (with Arnold Robson).

The Darlington and Stockton Times, Catterick Conservation Group (Catterick Army Training Area).

Isle of Man

CO-ORDINATOR: **Kate Hawkins** (with Gordon Craine and John Lamb).

Manx National Heritage, Manx Bird Atlas, Isle of Man Branch of Butterfly Conservation, Isle of Man Biological Records Centre, Manx Wildlife Trust, Isle of Man Government (Wildlife Office of Department of Agriculture, Fisheries, and Forestry).

Wales

OVERALL CO-ORDINATOR (North Wales): **Rob Whitehead**

OVERALL CO-ORDINATOR (South Wales): **Neil Jones**

North Wales and South Wales Branches of Butterfly Conservation.

Anglesey and Caernarvonshire

CO-ORDINATOR: **Doug Murray** (with **Alan Wagstaff** and Paul Whalley).

Royal Society for the Protection of Birds (Isle of Anglesey Senior Warden), Countryside Section Conwy County Borough Council, North Wales Wildlife Trust.

Brecknockshire

CO-ORDINATOR: **Chris Dyson** (with Jane Goodwin and John Clarkson).
Brecknock Wildlife Trust, Countryside Council for Wales (East Area Office).

Carmarthenshire

CO-ORDINATOR: **Steve Lucas**.
Wildlife Trust West Wales, Countryside Council for Wales (Llandeilo–West Area Office).

Ceredigion

CO-ORDINATOR: **Lin Gander**.
Wildlife Trust West Wales.

Denbighshire

CO-ORDINATOR: **Rob Whitehead**.
North Wales Branch of Butterfly Conservation, North Wales Wildlife Trust, Dr Jean Green (BSBI Recorder for Denbighshire), Countryside Council for Wales (North West Area and North East Area Offices), Forest Enterprise (Llanrwst Forest District), Royal Society for the Protection of Birds, Leeds University (Department of Biology).

Flintshire

CO-ORDINATOR: **Lawrence Rawsthorne**.
Dyserth and District Field Club.

Merionethshire

CO-ORDINATOR: **Andrew Graham** (with Chris Hall).
Royal Society for the Protection of Birds (Mawddach Valley Reserve), Snowdonia National Park Authority, Countryside Council for Wales (North West Area Office), North Wales Wildlife Trust.

Mid and South Glamorgan

CO-ORDINATOR: **Rob Nottage** (with Richard Smith).
South Wales Branch of Butterfly Conservation, Cardiff County Council, Glamorgan Wildlife Trust, Countryside Council for Wales (South Wales Area Office), Glamorgan Moth Recording Group, Rhondda Cynon Taff County Borough Council, Bridgend County Borough Council.

Monmouthshire

CO-ORDINATOR: **Martin Anthoney**.
South Wales Branch of Butterfly Conservation, Gwent Wildlife Trust, WING (Wildlife in Newport Group).

Montgomeryshire

CO-ORDINATOR: **Simon Spencer** (with Michael Edmonds and Dennis Goodbody).
Montgomeryshire Wildlife Trust, Countryside Council for Wales (East Area Office).

Pembrokeshire

CO-ORDINATOR: **Bob Haycock** and **Annie Poole**.

Butterfly Conservation, Countryside Council for Wales (Pembrokeshire–West Area Office), National Trust (Pembrokeshire), Pembrokeshire Coast National Park Authority, Royal Society for the Protection of Birds (Pembrokeshire), Wildlife Trust West Wales, Ministry of Defence (South Pembrokeshire Ranges Recording and Advisory Group), Pembrokeshire County Wildlife Recorders.

Radnorshire

CO-ORDINATOR: **Joyce Gay**.

Radnorshire Wildlife Trust, Countryside Council for Wales (East Area Office), Dwr Cymru–Welsh Water (Elan Valley Rangers).

West Glamorgan

CO-ORDINATOR : **Barry Stewart**.

South Wales Branch of Butterfly Conservation, Glamorgan Moth Recording Group, Countryside Council for Wales (South Wales Area Office), Gower Ornithological Society, Glamorgan Wildlife Trust.

Scotland

OVERALL CO-ORDINATOR: **Richard Sutcliffe**.

East Scotland

CO-ORDINATOR: **Chris Stamp** (with Eleanor Stafford).

East Scotland Branch of Butterfly Conservation, Fife Nature Biological Records Centre, Perth Museum Biological Records Centre, Scottish Wildlife Trust, Scottish Borders Biological Record Centre, Dundee Natural History Museum, Central Area Recording System for the Environment.

Fife

CO-ORDINATOR: **Anne-Marie Smout**.

Fife Nature, Fife Nature Butterfly Interest Group.

Highland and Western Isles

CO-ORDINATOR: **David Barbour** (with Stephen Moran, Paul Hackett, Jimmy Stewart, and John Trevor).

Highland Branch of Butterfly Conservation, Highland Biological Recording Group, Scottish Natural Heritage (North Areas Offices), Curracag (Western Isles Natural History Society).

Lothian

CO-ORDINATOR: **Bob Saville**.

Lothian Wildlife Information Centre, Scottish Wildlife Trust.

Orkney

CO-ORDINATOR: **Sydney Gauld**.

Orkney Field Club, Royal Society for the Protection of Birds (Orkney), Scottish Natural Heritage (North Areas–Orkney Office).

Perth and Kinross

CO-ORDINATOR: **Mark Simmons**.

Perth Museum Biological Records Centre.

Renfrewshire

CO-ORDINATOR: **Shona Allan** (with David Mellor and Douglas Breingan).

Renfrewshire Biological Records Centre (Paisley Museum and Art Gallery), Paisley Natural History Society, Clyde Muirshiel Regional Park.

Shetland

CO-ORDINATOR: **Mike Pennington**.

Shetland Entomological Group.

South-west Scotland

CO-ORDINATOR: **Richard Sutcliffe** (with James Black, Bill Davidson, Susan and Keith Futter, Graham Irving, Jessie McKay, Fraser Simpson, and the late M. Neal Rankin).

Glasgow and South-west Scotland Branch of Butterfly Conservation, Glasgow City Council (Land Services and Glasgow Museums Biological Records Centre), North Lanarkshire Council (Conservation and Greening), South Lanarkshire Ranger Service, Mugdock Country Park, Wildfowl and Wetlands Trust (Caerlaverock), National Trust for Scotland (Brodick, Culzean, Rockcliffe, Threave), Scottish Natural Heritage (West Areas Offices), Islay Field Studies Centre, the Ayrshire Bird Report.

Northern Ireland

CO-ORDINATOR: **Trevor Boyd**.

Northern Ireland Branch of Butterfly Conservation, Centre for Environmental Data and Recording, National Trust (Murlough National Nature Reserve), Ulster Wildlife Trust, Belfast Naturalists' Field Club, Royal Society for the Protection of Birds (Northern Ireland).

Channel Islands

Guernsey, Sark, and Alderney

CO-ORDINATOR: **Rich Austin**.

La Société Guernesiaise, La Société Sercquiaise, The Alderney Society.

Jersey

CO-ORDINATOR : **Margaret Long** (with Rich Austin).

Entomology and Ornithology Sections, Société Jersiaise.

Republic of Ireland

Butterflies for the New Millennium–Ireland

CO-ORDINATOR: **David Nash** (with Mary Willis, Bob Aldwell, Ken Bond, and Michael Salter).

The Dublin Naturalists' Field Club, The Heritage Council.

1 Background

Butterflies for the New Millennium

The Butterflies for the New Millennium (BNM) project has completed a comprehensive re-survey of the distribution of butterflies in Britain (including the Isle of Man and the Channel Islands) and Ireland, between 1995 and 1999. BNM consists of two independent but closely inter-related projects. In Great Britain (England, Scotland, and Wales), Northern Ireland, the Isle of Man and the Channel Islands, the project has been organized by Butterfly Conservation in association with the national Biological Records Centre (BRC). It was launched at a meeting held at BRC on 25 May 1995. In the Republic of Ireland, the 'Butterflies for the New Millennium–Ireland' project has been run by the Dublin Naturalists' Field Club, and was launched formally at a public meeting in Dublin on 25 April 1998. The BNM project has received support and encouragement from many sources, for which Butterfly Conservation, BRC, and DNFC are most grateful. However, the most important contribution was, without doubt, that of the many thousands of volunteers who recorded butterflies for the project.

This chapter gives the background to the project and puts BNM in context with earlier work to record and document the butterfly fauna of Britain and Ireland. It also describes the atlas itself.

The BNM organizations

Butterfly Conservation, previously known under its official name, the British Butterfly Conservation Society, was founded in 1968 and is dedicated to saving wild butterflies, moths, and their habitats. It is Europe's largest voluntary organization devoted to insect conservation.

Biological Records Centre (BRC) has been based at Monks Wood near Huntingdon since it was formed in 1964. BRC is funded by the Centre for Ecology and Hydrology, and the Joint Nature Conservation Committee, as part of the National Biodiversity Network, to collate, manage and disseminate information about the occurrence of plants and animals in the UK.

The Dublin Naturalists' Field Club (DNFC) was founded in 1886 and is the leading natural history society in the Republic of Ireland for the survey and recording of plants and animals.[1] Generations of its members have been actively involved with most of the major survey and atlas projects co-ordinated by BRC.

The importance of butterflies in Britain and Ireland

The present-day butterfly fauna of Britain and Ireland is the product of the natural development of thousands of generations since butterflies re-colonized these islands following the last glaciation. The fauna has survived modifications to almost all the terrestrial environment that have been brought about by humans over many millennia. These modifications have taken place alongside significant natural

[1] For further details of the history of the DNFC see DNFC (1986).

fluctuations in the climate, including long periods that were wetter, drier, warmer, or colder than at present.[2] The associated changes of land use by humans, determined by varying economic and social conditions, have favoured different species at particular times. For example, heath and grassland species almost certainly increased at the expense of woodland ones as native woodlands were cleared. These same species later decreased as their habitats declined in the face of the growth of the human population and accompanying changes in land use.

The number of species in Britain and Ireland

In a European or global context, there is nothing remarkable about the butterfly fauna of these islands. Approximately 560 species of butterflies occur in Europe, but only 59 species in Britain and 28 in Ireland are historically resident.[3] A further three (the Clouded Yellow, Red Admiral, and Painted Lady) are regular migrants that breed in these islands. Several more species occur as rare migrants or vagrants. No species is endemic to Britain or Ireland. Details of the species that have been recorded regularly in Britain and Ireland are given in Chapter 5. More than 30 additional species have been found in Britain and Ireland at some time in the past, but most have been classified as accidental introductions or have occurred only a few times as vagrants.[4] Many authors have distinguished subspecies, races, or forms of species within these islands.[4] The taxonomic status of these subdivisions is often the subject of dispute; therefore we have chosen to use the term *race* to include all such subdivisions below the level of species.

The number of butterfly species occurring in these islands may seem trivial in terms of the many thousands of species known worldwide, but it is one of the most intensively studied butterfly faunas in the world. During the eighteenth and nineteenth centuries, entomologists worked hard to discover the butterfly fauna of Britain and Ireland. Thus, unlike almost every other group of invertebrates, the list of resident and breeding butterflies was not added to during the twentieth century.[5] The good knowledge of butterflies and the ease with which they can be recognized have meant that butterflies have become increasingly important in assessing our natural heritage. This accessibility is not the only reason for regarding butterflies as 'important' for nature conservation or environmental assessment. Knowledge of the distribution and ecological requirements of butterflies is now being used, and butterflies are increasingly seen as indicators of the quality of habitats.

Ranges, changes and specialization

At the present time, only 16 species can be considered to occur throughout most of Britain and Ireland. The remaining resident species have a discernible boundary to their range within these islands, although in many cases these boundaries have changed with time. Of the 59 resident species in Britain, more than 70% have a geographic range boundary within Britain. Thirty-one species that occur in Britain do not occur in Ireland, and nearly 40% of the Irish species have a range boundary within Ireland. These edges of range are important in a wider context, being the north-western limits of the range of the species in Europe. Topography, climate, geology, soils, and land use contribute to the availability of suitable habitats and foodplants, and therefore to the range of individual species of butterflies; these factors are discussed in subsequent chapters.

The geographical distributions of butterflies have responded to changes in land

[2] e.g. Thomas (1993).

[3] Resident species are defined as naturally occurring species that are known to have bred regularly within the last 200 years, but some resident species are now extinct. The Large Chequered Skipper, which occurs only in the Channel Islands, is excluded from these totals.

[4] Emmet and Heath (1989).

[5] Excluding species that have been accidentally or intentionally introduced (for example the Geranium Bronze), the most recent addition was the Essex Skipper. This was distinguished as separate from the Small Skipper, and therefore new to Britain, in 1889. The Pearl-bordered Fritillary was not discovered in Ireland until 1922.

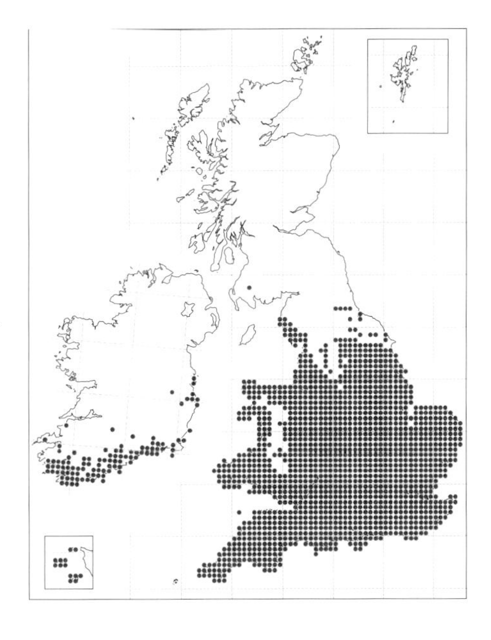

The distribution of the Gatekeeper shows distinct boundaries in both Britain and Ireland.

use and climate, but as with other animals and plants at the edge of their range, some species have become specialized to environmental conditions that are far from ideal. Such species may have become restricted to narrowly defined habitats or have developed population structures or reproductive strategies not usually found in their main range (e.g. the Swallowtail, Large Blue, and Heath Fritillary). Since so many butterfly species in Britain and Ireland are near the edge of their European range and have adapted accordingly, they are vulnerable to even slight changes in their environment. This vulnerability, together with their short life spans and limited mobility, makes such species particularly valuable as indicators of environmental changes. Populations near the edge of the range of a species often show greater fluctuations in abundance than are observed nearer the centre of the range, possibly leading to more frequent extinctions and colonizations at the local scale.[6]

Whilst many butterfly species are able to breed successfully in a range of habitats, others are more specialized. To help in the interpretations of species distribution, it is useful to distinguish the following two groups of species. **Wider countryside species** are those that are habitat generalists or use habitats that are very widespread, and **habitat specialist species** are those that are restricted to scarce or very 'narrow' habitats. These terms are explained in more detail in Chapter 6.

Butterfly recording in Britain and Ireland

Early history

Artists and writers have used butterflies as symbols of beauty and fragility for centuries and we know that they have attracted the attention of naturalists in Britain and Ireland since at least the late 1500s. Butterfly collecting was a popular hobby from the late seventeenth century,[7] but it was generally the preserve of the privileged few. It was not until the second part of the nineteenth century that increasing affluence and access to the countryside, by rail and bicycle, allowed the study of nature to come within the reach of the less-privileged many. Lepidoptera were probably the most studied group of insects throughout the seventeenth and eighteenth centuries.[8] The proliferation of books and journal articles from the early 1800s onwards, suggests that butterflies, in particular, have been the subject of considerable popular interest for some 200 years.

The first distribution lists and maps

The earliest attempt to produce a summary of the distribution of butterflies of Britain and Ireland was published by H. J. Fust in 1868,[9] using the provinces and sub-provinces devised by the botanist H. C. Watson. Much local and national

[6] Thomas *et al.* (1994).

[7] Marren (1998).

[8] Lisney (1960).

[9] Fust (1868).

E. B. Ford (1901–88) was the first person to publish distribution maps of butterflies in Britain and Ireland, in 1945.

information on the occurrence of butterflies was published throughout the nineteenth century and the first part of the twentieth century. The next attempt at a national overview was collated in 1945 by E. B. Ford, the leading professional specialist in Lepidoptera of his generation in Britain. He mapped the overall range of 32 species in his *New Naturalist* volume on butterflies.[10] He chose species to demonstrate various features such as the origins of our butterfly fauna, restricted distributions, and increases in ranges. However, he used whatever information was available to him from existing sources; this was incomplete, patchy, and, to some extent, biased.

Botanists give a lead

The collation of mainly *ad hoc* information about the occurrence of species at a national scale was the standard method by which baselines of information were compiled for most of our flora and fauna until at least the 1960s. A notable exception was the work of the Irish botanist R. L. Praeger. Between 1895 and 1900, he walked the mountains, bogs, lanes, and fields of half of Ireland systematically to record the flora, in a representative range of habitats in each biological vice-county, for his *Irish Topographical Botany* published in 1901.[11] It was also botanists who pioneered the use of newly published Ordnance Survey maps of Britain, which included the National Grid for the first time, for recording and mapping the distribution of species. This method was first used for the *Atlas of the British Flora* project that was proposed by A. R. Clapham in 1950.[12]

The Biological Records Centre's butterfly recording scheme (1967–82)

The database from the *Atlas of the British Flora* formed the nucleus of the Biological Records Centre (BRC) when it was established in 1964. In 1967, BRC launched the Lepidoptera Distribution Maps Scheme (co-ordinated by John Heath) covering the butterflies and larger moths. This project ran until Heath retired in January 1982.[13] During the following year BRC compiled the detailed computerized database for butterflies from record cards collated by Heath's project, in preparation for the 1984 atlas.[14] The text of the 1984 atlas was written by John Heath and two colleagues at the Institute of Terrestrial Ecology, Ernie Pollard and Jeremy Thomas. The data set for the 1984 atlas has been used in preparing this atlas and the records from 1970–82 provide the baseline against which the BNM data have been compared. The data set also contained records from before 1970, including some from the nineteenth century.

John Heath (1922–87) ran the BRC Lepidoptera Distribution Maps Scheme from 1967–82 and was Head of BRC from 1979–82. As an enthusiastic 'amateur' lepidopterist he initiated collaborative work between Butterfly Conservation and BRC after his retirement in 1982. This portrait shows John entering data at BRC in the 1970s using 80-column punched cards, the available technology at that time.

[10] Ford (1945).

[11] Vice-counties are sub-divisions of the traditional counties of Britain and Ireland (Appendix 9). They were devised originally by botanists for recording the occurrence of plants, but have been used by naturalists for many other groups. The system for Britain was devised by H. C. Watson in 1852 (Dandy 1969) and that for Ireland by R. L. Praeger (Praeger 1901; Webb 1980).

[12] Perring and Walters (1962).

[13] Heath *et al.* (1984).

[14] Harding and Greene (1984).

Left
Ernie Pollard was a senior ecologist with the Nature Conservancy and Institute of Terrestrial Ecology at Monks Wood from 1963–89. He developed the Butterfly Monitoring Scheme and organized it from 1976–87.

Right
Jeremy Thomas has carried out pioneering work on butterfly ecology, throughout his career with the Nature Conservancy and the Institute of Terrestrial Ecology (now the Centre for Ecology and Hydrology). He is a Vice-President of Butterfly Conservation. This photograph shows Jeremy searching for Brown Hairstreak eggs during surveys for the 1984 Atlas.

Butterfly Conservation and BRC collaborate on recording

In 1983, John Heath acted as an intermediary between BRC and John Tatham and Roger Sutton of Butterfly Conservation to organize a new national recording scheme for butterflies, on behalf of BRC. One of the main aims of the new project was to collate records of 36 Target Species at a site level and to pass on any reports of threats to sites or populations to local Butterfly Conservation branches and other organizations. This approach was an early example of the way in which recording schemes using volunteers are capable of development to serve a wider range of objectives than mere 'species mapping'. This Target Species project was generally poorly supported, mainly because, after the publication of the 1984 atlas, many recorders became involved in local atlas projects (see below). Data from the Target Species project span the period 1983–94 and were computerized at BRC for use with the Butterflies for the New Millennium data set.

The Moths and Butterflies of Great Britain and Ireland (Volume 7, Part 1)

The volume that covers butterflies in the Harley Books series was in preparation when John Heath, the main editor, died in July 1987. His co-editor, Maitland Emmet, and the publishers completed the volume.[15] BRC supplied the basic maps for this volume (essentially those from the 1984 atlas), but the publishers contacted some county and regional specialists for additional 10 km square records and corrections to update these maps. Subsequently, the publishers deposited at BRC the original annotated maps and other related papers received from their contacts, but records from this source are incomplete and have not been used in this Atlas project.

The Butterflies of Ireland: a field guide

In 1992, a field guide written and illustrated by Norman Hickin was published posthumously under the editorship of Tim Lavery.[16] The book included 10 km square distribution maps, which the publisher's acknowledgement states were produced by Lavery. Although Lavery also wrote the introduction to the book, there is no mention of how or where the data for these maps were acquired, compiled, or stored. In a brief section on recording, passing reference is made to the existence of updated maps in *The Moths and Butterflies of Great Britain and Ireland* volume (above). The maps in the two publications appear to be almost identical. Data may exist for the maps in this Irish field guide, and from subsequent recording co-ordinated through 'The Irish Butterfly and Moth Society' that is mentioned in the guide.

[15] Emmet and Heath (1989).

[16] Hickin (1992).

However, neither the Dublin Naturalists' Field Club nor Butterfly Conservation has been able to trace the whereabouts of any data.

Local butterfly atlases

Local atlases have developed in scale and complexity since the early 1980s. Prior to this, the preparation of county or regional lists of species had been fashionable for at least a century.[17] One of the earliest regional atlases, for Scotland, provided a model for the depth and breadth of biological, ecological, and historical detail that could be included in a 'local' atlas.[18] Interest in butterflies and enthusiasm for recording their distribution and abundance grew rapidly from the early 1980s and much of this energy was channelled into recording for local atlases. Several reviews of these local butterfly atlases have been published[19] and an up-to-date list is included in Appendix 3. Some local atlases mapped species at the 10 km square scale, but the majority have mapped species using either 2 km or 1 km squares. Several more recent authors have been able to emulate the example of the Scottish atlas, although covering much smaller areas. The activity of butterfly enthusiasts, working on locally-based projects such as atlases, provided an essential starting point for the establishment of Butterfly*Net* (see Chapter 3) and formed a basis for the BNM project. Indeed, a few local organizers had already compiled detailed computerized data sets for their atlases.

Butterfly Monitoring Scheme

The national Butterfly Monitoring Scheme (BMS) is organized by the Centre for Ecology and Hydrology (Monks Wood) with financial support from the Joint Nature Conservation Committee. It was set up in 1976 and operates at over 100 sites throughout the UK. The objectives of the BMS and methods for monitoring and analysing the resultant data are summarized in Chapter 3. Many scientific papers have been published from the BMS and a major summary of the results was published in 1993.[20] Some of the long-term results from the BMS are used throughout this atlas, in particular the graphs of annual abundance of individual species. Summary data from the sites in the BMS have been incorporated in the BNM data set for distribution mapping.

[17] Chalmers-Hunt (1989).

[18] Thomson (1980).

[19] Corke and Harding (1990); Dennis *et al.* (1999).

[20] Pollard and Yates (1993*a*).

[21] Hall (1981).

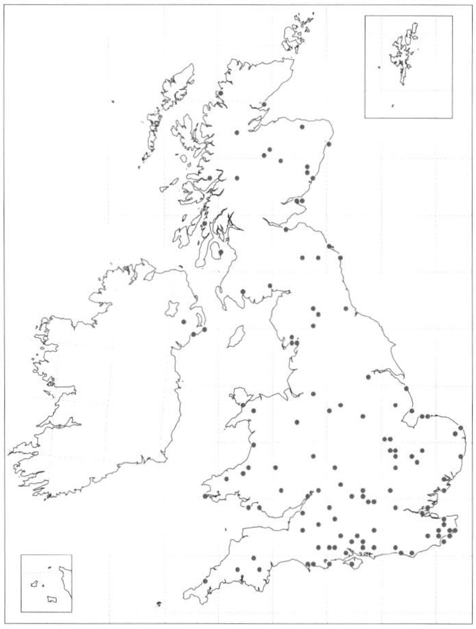

The location of Butterfly Monitoring Scheme sites from 1995–9, including sites in the Environmental Change Network.

Butterfly monitoring based on a transect method

The methods for assessing butterfly abundance using transect walks, developed for the national BMS, were described in a simple handbook published in 1981.[21] These methods have been adopted by many individuals and groups for monitoring at least 500 additional site transects in recent years. Although the results of some of these transects are collated, analysed, and published at a local level, much of the information is not readily accessible. Butterfly Conservation is establishing a network to co-ordinate transect monitoring of a large number of sites.[22]

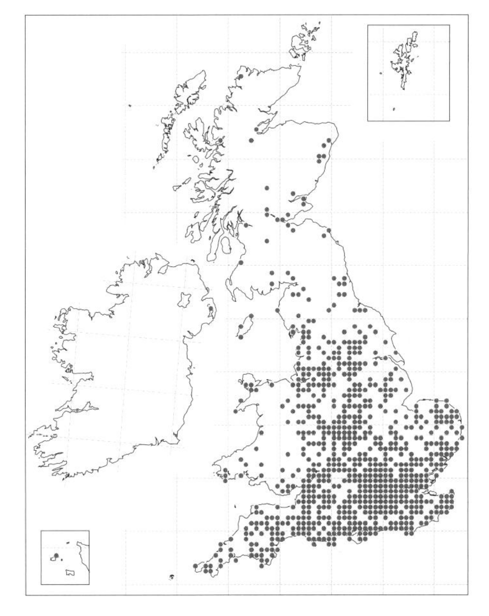

The locations of gardens providing records for the Butterfly Conservation national garden butterfly survey.

Garden butterfly surveys

There have been several local surveys of butterflies in gardens and two national surveys undertaken by young people. Most of these surveys are summarized in a paper by Margaret Vickery, which also describes the Butterfly Conservation national garden butterfly survey that was launched in 1991.[23] This survey has involved over 1000 gardens and the records have been included in the BNM data set after thorough verification.

Species surveys, action plans and research

Individual species have been the subject of many special surveys, usually to establish the national or regional status of rare or threatened species, and research has been conducted on numerous species. Much of the literature from this work has been reviewed elsewhere.[24] Butterfly Conservation has published 25 detailed Species Action Plans as a contribution to the UK Biodiversity Action Plan process, drawing on much new work in the 1990s (see Chapter 7). The authors of the species accounts in this atlas have drawn heavily on these published and unpublished sources.

Comparison of butterfly recording with other taxonomic groups

The BNM project is the second major survey of butterflies in Britain and Ireland. Only two other groups have been the subject of more than one thorough survey of this type. Breeding birds were surveyed in 1968–72 and 1988–91 in projects led by the British Trust for Ornithology.[25] These projects collated records from only the years of the two surveys, but some earlier records were summarized in a historical atlas covering the period 1875–1900.[26] Records of flowering plants were collated by the Botanical Society of the British Isles (BSBI) from 1954–62 in the *Atlas of the British Flora*.[27] The resultant maps distinguished records only as 'prior to 1930' or '1930 onwards', although the majority of records were from the 1950s. The Atlas 2000 project for vascular plants, also led by BSBI, collated plant records since 1962, of which the majority are from the period 1987–99.[28] The national societies and recording schemes that cover many other groups of plants and animals are still at the stage of having completed a first 'national' scale collation of all or mainly recent records, or are actively working towards that stage.

Why the Butterflies for the New Millennium project was initiated

The changing environment

Throughout the twentieth century, human activities led to rapid and largely detrimental changes to the landscape, to habitats, and to species. A desire to identify

[22] Brereton and Goodhand (2000).

[23] Vickery (1995).

[24] Emmet and Heath (1989); Harding and Green (1991); Heath *et al.* (1984).

[25] Gibbons *et al.* (1993); Sharrock (1976).

[26] Holloway (1996).

[27] Perring and Sell (1968); Perring and Walters (1962).

[28] Preston *et al.* (in press).

and measure, and then to limit or reverse these changes, led to the formation of numerous voluntary and statutory conservation organizations in both Britain and Ireland and to partnerships between them.[29] The first comprehensive survey of butterflies, organized by BRC from 1967–82, showed marked changes in the range of many species since 1940.[30] During the 1980s and early 1990s, surveys and research on individual species, the BMS and several local atlases, reinforced the observation that there was a general decline in both range and abundance, particularly of habitat specialist species. At the same time, evidence was accumulating that some of the wider countryside species were increasing in abundance and extending their geographical range. Clearly, butterfly distribution and abundance were undergoing significant changes. In the early 1990s, Butterfly Conservation and BRC identified the need to re-survey the butterfly fauna of Britain and Ireland and the idea of the BNM was born in an internal report.[31]

Planning BNM

After several years of discussion and planning, the BNM project was launched formally in 1995 jointly by Butterfly Conservation and BRC. Any project of the scale and complexity of BNM must address the needs for the information that it will generate. Several different requirements helped to shape the BNM project:

- to inform and support statutory measures in nature conservation;
- to inform practical action for the conservation of species and habitats;
- to obtain data for comparison with similar data sets for other fauna and flora and with other environmental data;
- to develop better methods and to encourage more people to record butterflies effectively;
- to make appropriate information about butterflies, their habitats and their conservation readily available.

Butterfly Conservation and BRC agreed that these five needs were sufficiently urgent and challenging to proceed with BNM, without attempting a more complex and statistically robust project. Given the absence of co-ordination of butterfly recording before BNM started, the recording methods are a considerable advance on those of the 1970–82 survey (see Chapter 3). It proved difficult to obtain funding for the co-ordination of BNM and it took time to develop the network to assemble data that would fulfil even these five basic needs. There is now a growing demand for more rigorously collected data to allow detailed statistical analyses based on structured and repeatable sampling methods, but there is no evidence that funding for such a project would have been forthcoming in the mid 1990s.

Protecting butterflies and their habitats

When the original atlas survey was initiated in 1967, no butterflies were protected under legislation in Britain or Ireland. Also, at that time, few sites of importance for butterflies were protected as nature reserves or even, in Britain, as Sites of Special Scientific Interest (SSSI). No comparable site designations existed in Northern Ireland or the Republic of Ireland at that time. Thus much has already been achieved in the conservation of butterflies and their habitats. In 1999, six butterfly species were protected in British legislation (seven in Northern Ireland) and two

[29] Sheail (1998).

[30] Heath *et al.* (1984).

[31] Asher (1992); Harding *et al.* (1995).

species were protected under European or other international legislation and conventions. Butterflies now feature in the designation descriptions of many SSSI in Britain and the equivalent Areas of Special Scientific Interest in Northern Ireland. Also, Butterfly Conservation has acquired over 25 nature reserves primarily for their butterfly fauna. Butterflies are increasingly recognized as being of importance in determining policies and practice in nature conservation and many are listed as priorities under the UK Biodiversity Action Plan (see Chapter 7).

The role of Butterfly Conservation and its members

In developing its role as the UK's foremost organization concerned with Lepidoptera conservation, Butterfly Conservation has a particular need for authoritative information on butterflies. This information is required to establish priorities for positive action to protect species and their habitats, and to promote these priorities to statutory organizations and to other voluntary organizations in the UK, the Republic of Ireland, and elsewhere in Europe. The BNM project offered Butterfly Conservation an opportunity, not only to obtain up-to-date information, but also to develop the role of its members and others, in supplying information in a dispersed, but centrally co-ordinated system.

The role of the Biological Records Centre

Working with a wide range of voluntary groups, mainly through national specialist societies and recording schemes, BRC has been promoting the collection of data by skilled amateurs for more than 35 years. BRC is responsible for collating, managing, quality assurance, and dissemination of national (UK) data on species, for use in nature conservation, biogeographic and environmental research, and in providing public information.

The role of statutory conservation organizations

The statutory nature conservation organizations (JNCC and the individual country agencies)[32] require up-to-date information at site, county, country, and national levels to carry out their statutory functions of advising government and conserving important sites and threatened species. In the Republic of Ireland, the BNM project has been run by DNFC, a voluntary society, with some financial support from the Heritage Council, a statutory governmental agency.[33] Nature conservation is overseen in the Isle of Man and the individual Channel Islands by departments within the respective administrations. Appendix 7 includes the addresses of relevant organizations.

BNM and the future

BNM has taken a more structured approach to the collection of data than was the case with the survey summarized in the 1984 atlas. It has benefited from several years of planning and shared experience from both detailed local butterfly surveys and national surveys of other groups. Nevertheless, alternative methods could have been used that might have reduced some of the inevitable bias, and which might have provided better data for assessing local abundance and measuring short-term regional and national changes. These and other improvements are being examined for the next phase of butterfly recording, which will build on the current survey and recording network.

[32] Countryside Council for Wales, English Nature, Environment and Heritage Service (Northern Ireland) and Scottish Natural Heritage.

[33] In the Republic of Ireland the statutory organisation with responsibility for nature conservation is Dúchas: The Heritage Service, which operates the National Parks and Wildlife department.

About the atlas

The atlas summarizes the results of the Butterflies for the New Millennium project from 1995 to 1999. The main results of the survey provide the core of the atlas, which consists of distribution maps and detailed accounts of each species (Chapter 5). In addition, BNM has drawn on many sources, for example from recent research and monitoring, to compare the results of this survey with other information, and to put them into a wider context. In Chapter 2, the habitats of butterflies, and their importance, are described and illustrated. Chapter 3 gives details of the main types of recording and data collection for the atlas and in Chapter 4 we explain how the data have been used in the maps and species accounts. Following the maps and species accounts (Chapter 5), changes in range and abundance, and their probable causes, are described in Chapter 6. Finally, in Chapter 7, the future conservation of butterflies and their habitats in Britain and Ireland is considered.

This atlas does not mark an end to butterfly surveys or research in Britain and Ireland, or of discussion of the conservation of species and their habitats. It is the starting point, with a baseline of detailed data, for fresh ideas, better methods, and increased activity in the recording, monitoring, research, and conservation of butterflies in the new millennium.

2 Butterfly habitats

The distribution of butterflies in Britain and Ireland is influenced by a wide variety of environmental factors including the topography, the underlying geology and soils (and their direct influence on the flora), the pattern of rainfall, and the temperature in different regions. This physical environment has been shaped by the activities of humans for over 10 000 years to create the pattern of habitats and land use that we see today.[1] This chapter outlines some of the factors influencing the main butterfly habitats found in Britain and Ireland.

Geology, soils, and climate

A fundamental characteristic of Britain and Ireland is the great variety of geology, soils, and climate that occur in such a small area. The land is composed of sedimentary rocks such as limestones and sandstones, metamorphic rocks, and igneous rocks such as granites. Over the last few million years, the land has been tilted, exposing rocks of early geological eras to the north-west and of more recent eras to the south-east. In broad terms, Britain can be divided into two categories: 'Lowland Britain' in the south and east being typified by low rolling hills and large plains and valleys; and 'Upland Britain' in the north and west typified by mountains and tracts of higher land interspersed with narrow valleys. In Ireland, there is a central lowland region surrounded by higher land, with ranges of low mountains mainly in the east and west. These topographical differences are reflected in the underlying geology with the harder, often acidic rocks in upland areas, and softer, sometimes calcareous rocks in lowland areas (see maps on pp. 14–15).

Much of Britain and Ireland is covered by soils that have not been derived simply from the progressive weathering of the underlying rocks (p. 16). In many areas the soils have been formed from deposits of glacial clays, silts, and sands that were eroded during the Ice Ages from rocks tens or even hundreds of miles away. Thus the gently rolling chalk hills of parts of west Norfolk are covered with sandy soils that form the Breckland heaths. In areas of high rainfall or poor drainage, peats have formed, both on hills and in valleys and river basins. For example, the 'solid' geology of central Ireland is mainly Carboniferous limestone, but this is overlain by huge areas of peat bogs and glacial drift. Soils can vary greatly within short distances, leading to markedly different flora and fauna.

The climate of Britain and Ireland is also variable, influenced partly by latitude and altitude, and partly by the position of the islands off the coast of mainland Europe where they are surrounded by warming ocean currents (the Gulf Stream) and are the first landfall of north Atlantic weather systems. The northern regions tend to be colder and wetter, but the pattern varies throughout the year (p. 17). In winter, the south and west have a relatively mild climate with some coastal areas experiencing no frosts, whereas central inland areas experience greater temperature extremes. In summer, the hottest areas are in the south-east, especially in the Thames basin around London. The wettest parts tend to be in the mountainous areas in northern Britain and Ireland, and in the west of Britain, which bear the brunt of the predominantly westerly air current off the Atlantic.

[1] The land use history of Britain and Ireland is described by Cabot (1999), Rackham (1986), and Stamp (1955).

Sunshine levels follow a different pattern, with the sunniest places being around the coast, especially along the south coast which receives an average of over 5 hours of sunshine per day, rising to over 7 hours in July. The least sunny areas are the highlands of Scotland and the north and west of Ireland (p. 17).

Habitats and management

Some butterfly species are highly mobile and move freely around the countryside, breeding in small pockets of suitable vegetation scattered over a wide area. However, many species form discrete colonies in certain types of habitat that provide the larval foodplants and other resources they need to complete their life cycle. The main broad habitat types used by butterflies in Britain and Ireland are described below, with an indication of the characteristic species that may be found if local conditions are suitable. The review is not intended to be comprehensive and there are many additional habitats that may be used by one or more species. These specific habitats are described in greater detail within each of the species accounts and in other books.[2]

Key features for butterflies

Within each broad habitat type, every resident butterfly species requires several particular features in order to complete its life cycle successfully. For example, each species not only requires the presence of particular larval foodplants, but generally needs these to be growing in exactly the right conditions to allow successful development from egg, through larva and pupa. Most adult butterflies also require food resources such as flowers for nectar or the honeydew produced by aphids. Many butterflies prefer sheltered positions, either to establish territories or for basking, or to provide the warmth needed for egg and larval development. For these reasons, many butterflies are far more restricted than are their larval foodplants. In the following sections, the chief features and factors influencing habitat suitability for butterflies are shown in the column alongside each photograph.

The importance of current and historical management

The vegetation of Britain and Ireland has been altered by human activity for thousands of years, resulting in a heavily modified or cultural landscape (see map on p. 18). The great variety of habitats we see today is the product of many generations of management: including livestock grazing, burning, cultivation, woodland management, and development. Without this influence much of the countryside would revert to woodland. Most habitats in Britain and Ireland are described as semi-natural, that is containing mostly native plants and some natural features, but substantially altered by human activity. For example woodlands have been coppiced, managed as wood-pasture, or felled and replanted; whereas grasslands have been cut for hay or grazed by livestock such as sheep and cattle. Only a few truly natural and unmodified habitats still exist, usually where human exploitation is impractical, for example inland cliffs, mountain-tops, or steep or inaccessible areas around the coast.

Over the past two millennia, the patterns of grazing and arable farming have altered, reflecting changes in human settlement and economic activity, ranging from the rise and fall of the Roman Empire, the Dark Ages, invasions and wars, to the effects of diseases and agrarian reforms. In Britain, the original cover of native

[2] Descriptions of the British vegetation are given by Rodwell (1991*a*,*b*, 1992, 1995, 1999) and Tansley (1939); and the Irish vegetation by White and Doyle (1982). A comprehensive list of habitats used by butterflies is given in Appendix 2 of Dennis (1992*a*).

broad-leaved woodland has been reduced to about 2% of the land area, although a further 6% is now planted (mainly coniferous or mixed) woodland. In Ireland these figures are halved.[3] During the twentieth century, the extent of other semi-natural habitats was greatly reduced by modern farming methods. About 20% of Britain is now regularly cultivated as arable land and semi-natural grasslands now cover only 2% of lowland areas.[4] In Ireland, only 6% is currently arable but over 40% was cultivated prior to the great famine in the 1840s.[5] A large proportion of land is now intensively managed 'improved' grassland, usually planted with highly productive grasses such as Perennial Rye-grass (*c.* 28% in Britain and 61% in Ireland). These and other changes have had a profound effect on butterflies and are discussed in Chapters 5 and 6.

Most of the semi-natural habitats used by butterflies rely on the continuation of regular management in order to prevent them from changing radically or reverting to scrub and woodland (a process known as succession). Local populations can rise and fall dramatically if management is changed and extinction may occur if adverse conditions persist, even for just a few critical months. The role of traditional and long-established regimes in conserving butterflies is discussed in Chapter 7.

[3] Peterken (1981).
[4] Barr *et al.* (1993).
[5] Cabot (1999).

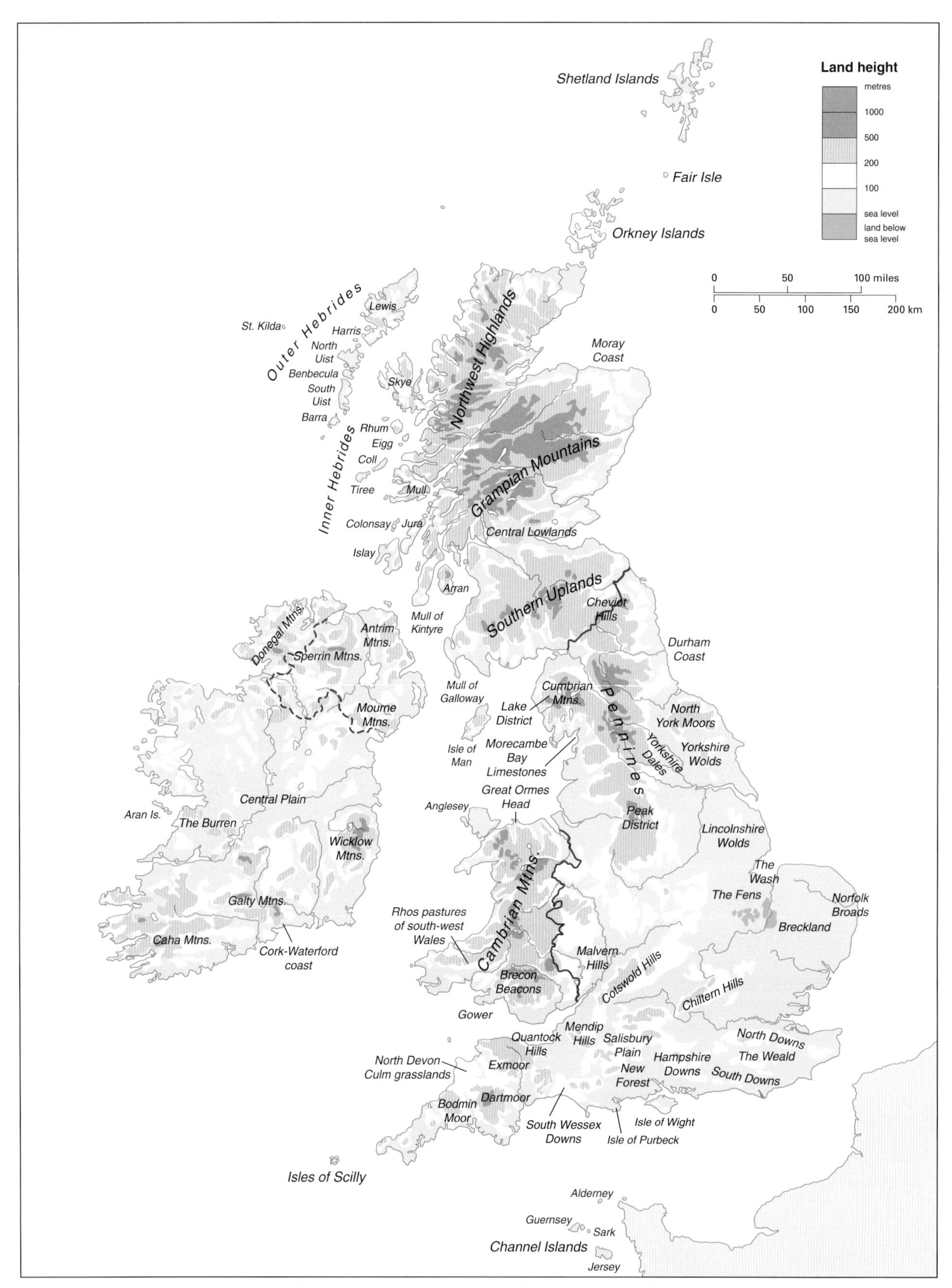

Relief and main physical features, including important butterfly areas referred to in the text.

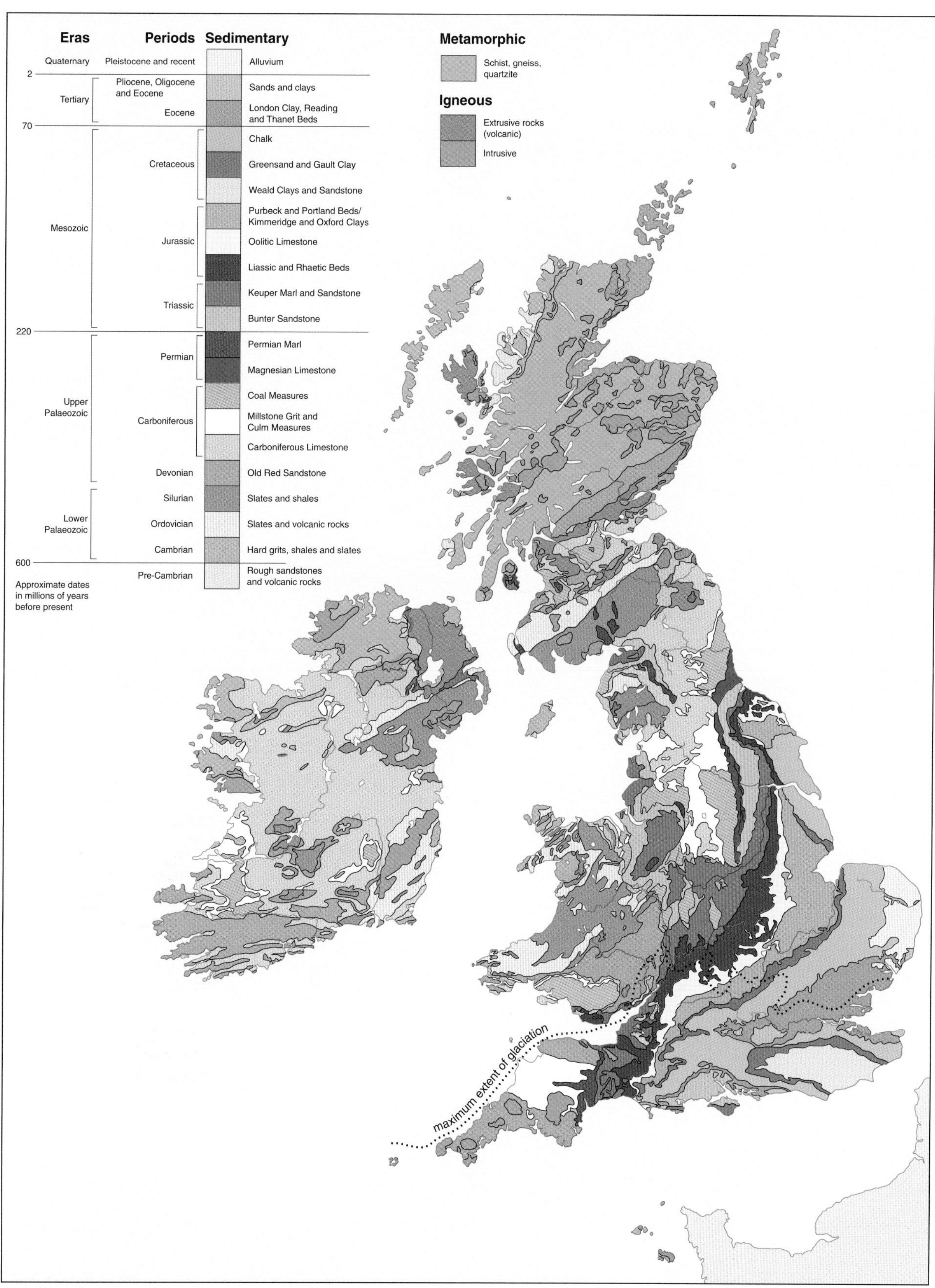

Solid geology.

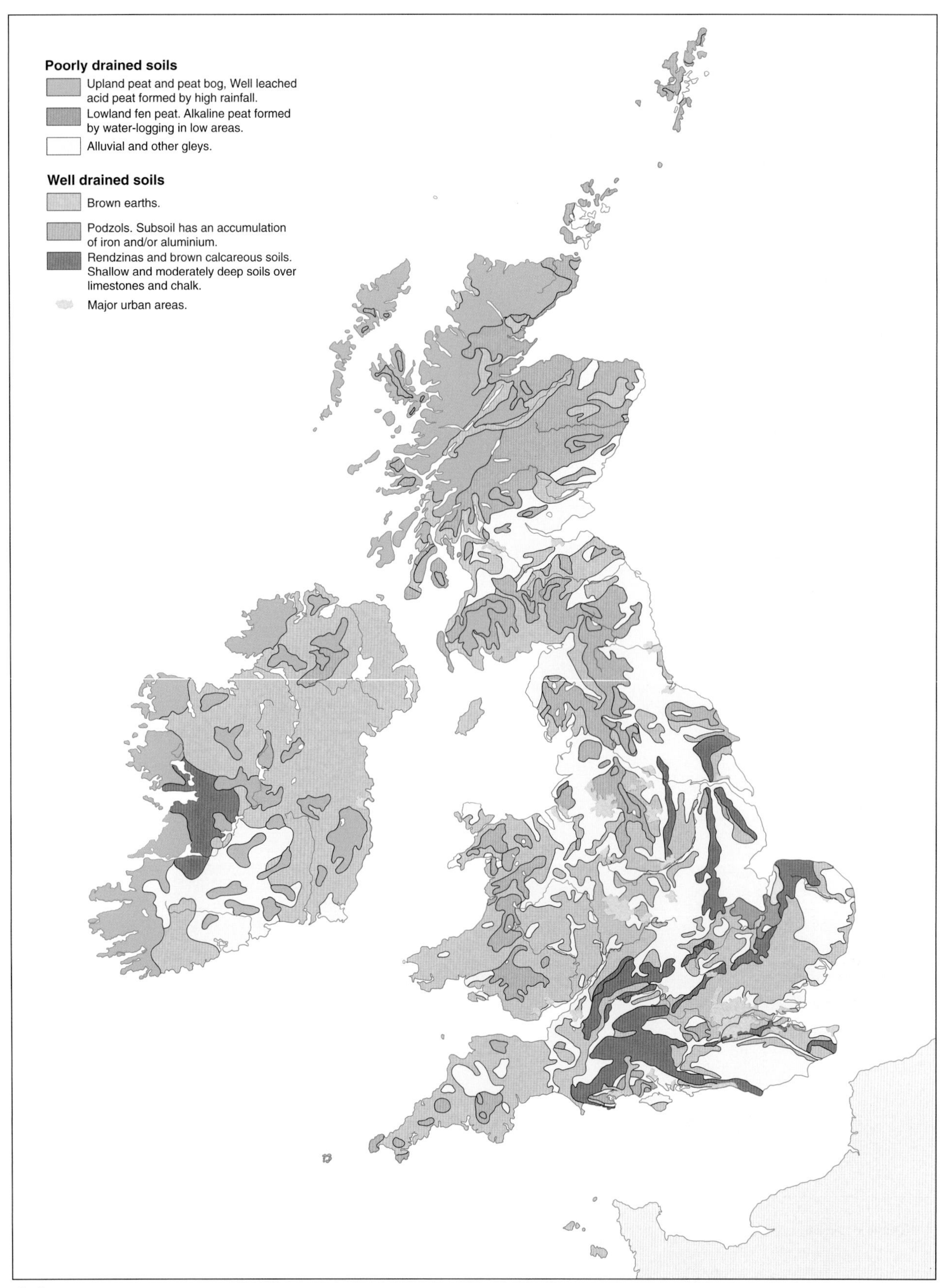
Poorly drained soils
Upland peat and peat bog, Well leached acid peat formed by high rainfall.
Lowland fen peat. Alkaline peat formed by water-logging in low areas.
Alluvial and other gleys.
Well drained soils
Brown earths.
Podzols. Subsoil has an accumulation of iron and/or aluminium.
Rendzinas and brown calcareous soils. Shallow and moderately deep soils over limestones and chalk.
Major urban areas.

Soils.

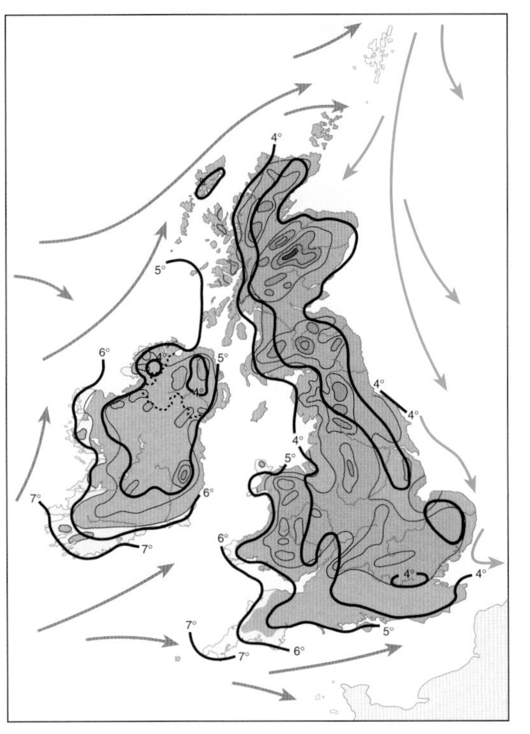

January mean temperature.

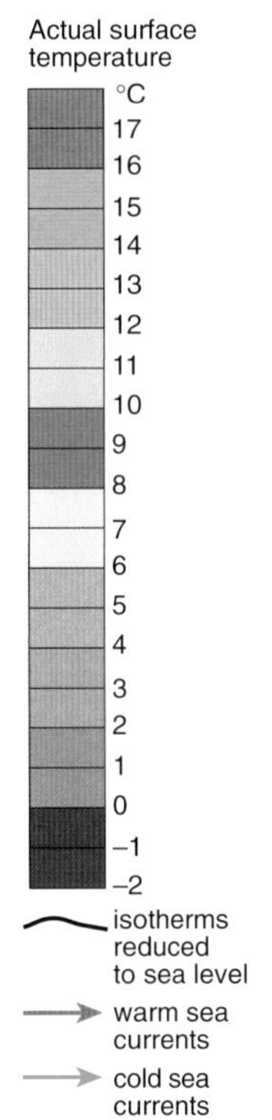

July mean temperature.

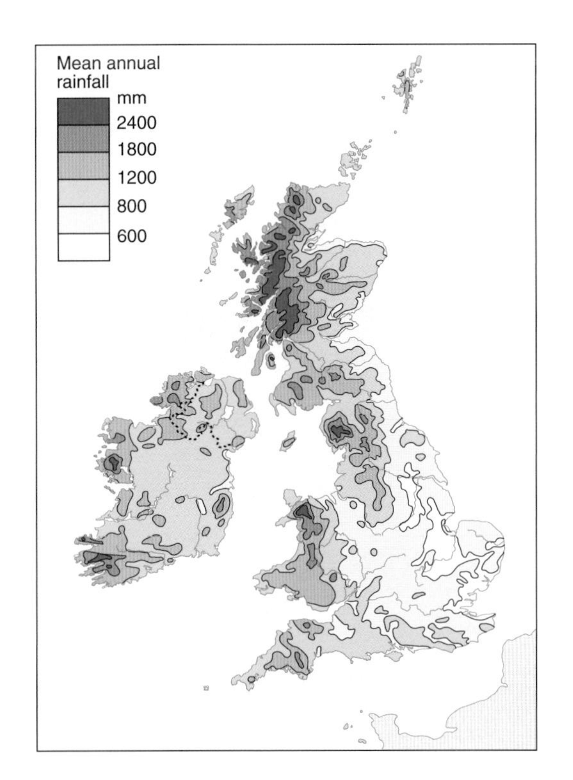

Mean annual rainfall.

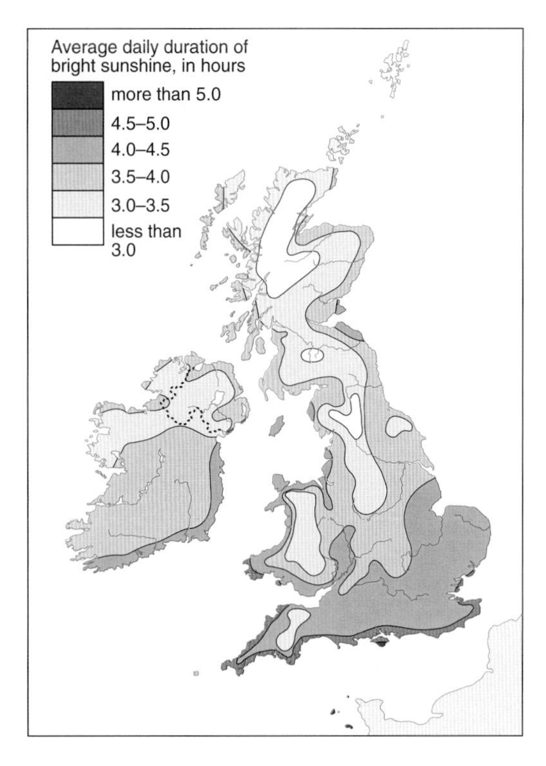

Average daily sunshine, in hours.

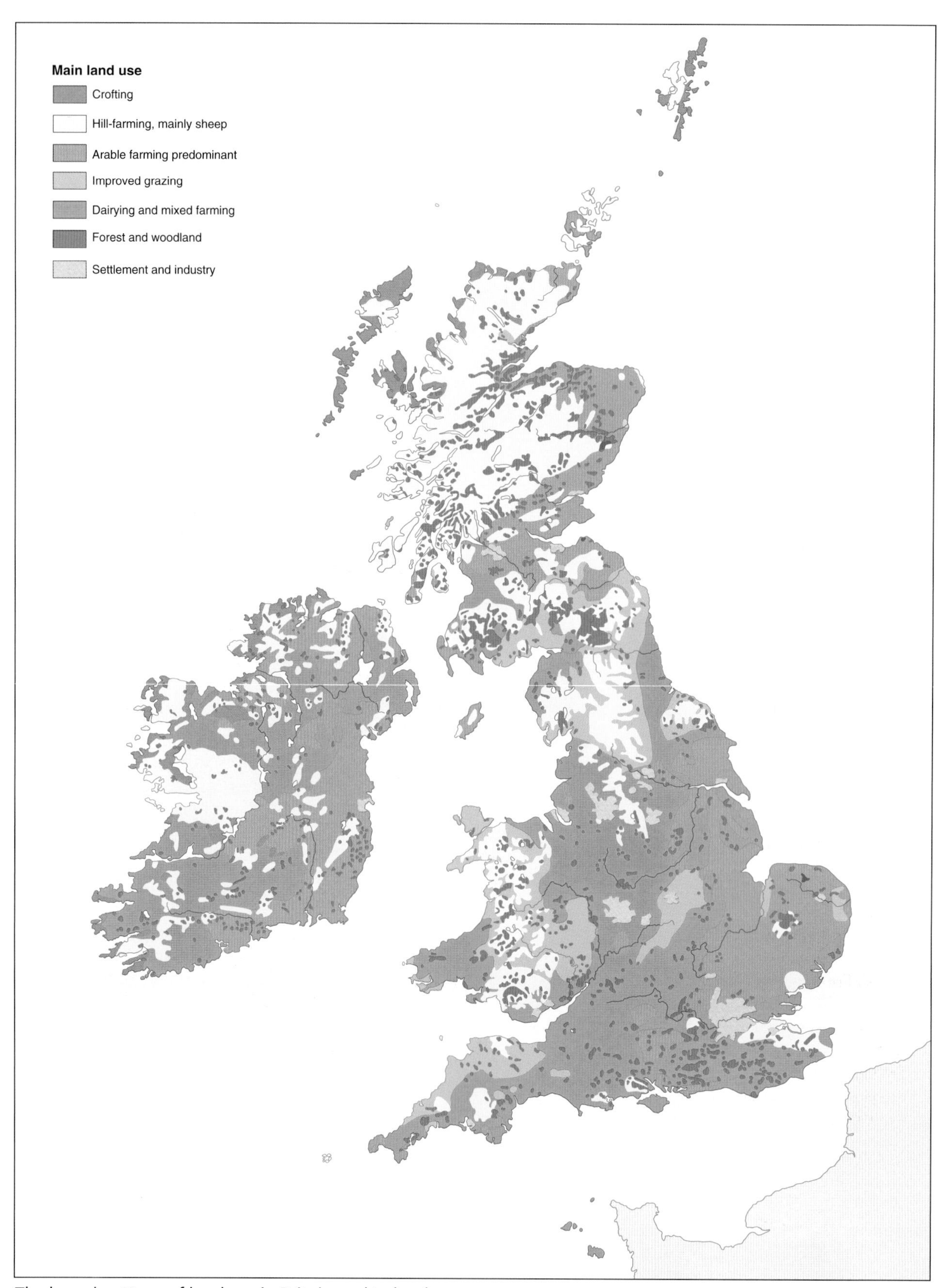

The broad pattern of land use in Britain and Ireland.

Major habitats and key features

GRASSLAND

Chalk and limestone grassland

KEY FEATURES FOR BUTTERFLIES

- Semi-natural grasslands on calcareous soils, limestone pavement, and eskers
- Thin, dry, and nutrient-poor soils
- Wide variety of alkaline-tolerant plants, including many larval foodplants such as Common Bird's-foot-trefoil, Horse-shoe Vetch, Kidney Vetch, Cowslip, and grasses
- Presence of warm, south-facing slopes
- Height and structure of sward
- Timing and intensity of grazing and/or cutting
- Type of livestock or other grazing animals such as rabbits
- Presence of scrub patches to provide shelter

CHARACTERISTIC BUTTERFLIES

Small Skipper
Lulworth Skipper*
Silver-spotted Skipper*
Dingy Skipper
Grizzled Skipper
Green Hairstreak
Small Blue
Brown Argus
Northern Brown Argus
Common Blue
Chalkhill Blue*
Adonis Blue*
Duke of Burgundy
Dark Green Fritillary
Marsh Fritillary
Wall
Marbled White
Meadow Brown
Small Heath
(*species using this habitat exclusively in Britain and Ireland)

Chalk grassland, Sussex, England.

Wet meadow, Co. Fermanagh, Northern Ireland.

Damp grassland and meadows

KEY FEATURES FOR BUTTERFLIES

- Semi-natural grasslands on moist soils
- Larval foodplants such as Cuckooflower, Devil's-bit Scabious, Marsh Violet, and grasses
- Height and structure of sward
- Intensity and timing of grazing and/or cutting
- Type of livestock

CHARACTERISTIC BUTTERFLIES

Small Skipper
Green-veined White
Orange-tip
Common Blue
Small Pearl-bordered Fritillary
Marsh Fritillary
Ringlet

Acidic grassland and bracken/grass mixtures

CHARACTERISTIC BUTTERFLIES

Small Skipper
Essex Skipper
Small Copper
Small Pearl-bordered Fritillary
Pearl-bordered Fritillary
High Brown Fritillary
Dark Green Fritillary
Wall
Grayling
Small Heath

Acidic grassland and bracken, Dartmoor, England.

KEY FEATURES FOR BUTTERFLIES

- Semi-natural grasslands on acidic soils
- Presence of warm, dead bracken litter
- Presence of warm, south-facing slopes
- Presence and type of livestock and other grazing animals (e.g. deer)
- Presence of scrub patches
- Larval foodplants such as Common Dog-violet, sorrels, and grasses

Coastal grassland and dunes

KEY FEATURES FOR BUTTERFLIES

- Thin, sandy and/or eroding soils
- Presence of natural grassland on cliffs, under-cliffs, and machair
- Larval foodplants such as Kidney Vetch, Common Dog-violet, and grasses
- Sunny climate

CHARACTERISTIC BUTTERFLIES

Small Blue
Common Blue
Dark Green Fritillary
Glanville Fritillary
Wall
Grayling
Meadow Brown
Small Heath

Coastal grassland, Isle of Wight, England.

Silage harvesting, Aberdeenshire, Scotland.

Improved grassland

KEY FEATURES FOR BUTTERFLIES

- Limited range of plants and usually no larval foodplants
- Intensive chemical inputs and/or regular cultivation

CHARACTERISTIC BUTTERFLIES

None (apart from mobile species in transit)

WOODLAND AND SCRUB

Mature broad-leaved woodland

CHARACTERISTIC BUTTERFLIES

Brimstone
Purple Hairstreak
White-letter Hairstreak
Black Hairstreak
White Admiral
Purple Emperor
Silver-washed Fritillary
Speckled Wood

KEY FEATURES FOR BUTTERFLIES

- Specific trees, shrubs, and ground flora for larval foodplants (e.g. oaks, willows, elms, buckthorns, Blackthorn, Holly, Honeysuckle, Common Dog-violet)
- Sunny rides, clearings, and wood edges
- Woodland structure and management
- History of management and length of continuous woodland cover (ancient woodlands more than 400 years old are generally richer)
- Abundance and type of grazers and browsers (e.g. rabbits and deer)

Broad-leaved woodland, Hampshire, England.

Clearings and young woodland

KEY FEATURES FOR BUTTERFLIES

- Sparse vegetation and warm, sunny conditions
- Larval foodplants such as Common Dog-violet, Primrose, and grasses
- Presence of leaf litter
- History of management and length of continuous woodland cover

CHARACTERISTIC BUTTERFLIES

Duke of Burgundy
Small Pearl-bordered Fritillary
Pearl-bordered Fritillary
High Brown Fritillary
Dark Green Fritillary
Heath Fritillary

Coppice woodland, Suffolk, England.

Woodland ride, Cambridgeshire, England

Rides and glades

CHARACTERISTIC BUTTERFLIES

Chequered Skipper
Large Skipper
Dingy Skipper
Grizzled Skipper
Wood White
Green Hairstreak
Common Blue
Comma
Small Pearl-bordered Fritillary
Pearl-bordered Fritillary
Dark Green Fritillary
Scotch Argus
Ringlet
+ other grassland species

KEY FEATURES FOR BUTTERFLIES

- Warm, sheltered and sunny conditions
- Varied structure (short and tall grassland, scrub, and wood edge)
- Larval foodplants such as Common Dog-violet, Common Bird's-foot-trefoil, Wild Strawberry, Primrose, vetches, and grasses
- Frequency of cutting and disturbance of vegetation
- History of management

Conifer plantation, Argyll, Scotland

Conifer plantations

CHARACTERISTIC BUTTERFLIES

Can contain many woodland species if rides and glades are present, or if planted on an ancient woodland site with a varied ground flora

KEY FEATURES FOR BUTTERFLIES

- Interior of plantations too dark and shady for butterflies
- Possible breeding habitats confined to rides and clearings
- Overall structure and management of plantations
- Presence of broad-leaved trees and shrubs in ride edges, especially larval foodplants such as oaks, willows, elms, buckthorns, Blackthorn, Holly, and Honeysuckle
- Previous land use

Scrub

KEY FEATURES FOR BUTTERFLIES

- Specific shrubs for larval foodplants such as Blackthorn, gorses, Holly, buckthorns, and elms
- Size, age, and density of shrubs
- Structure of clearings and edges

CHARACTERISTIC BUTTERFLIES

Brimstone
Green Hairstreak
White-letter Hairstreak
Brown Hairstreak
Black Hairstreak
Holly Blue
Duke of Burgundy
Dark Green Fritillary

Scrub habitat, Co. Armagh, Northern Ireland.

HEDGES, VERGES, AND FARMLAND

Hedges and arable field margins

CHARACTERISTIC BUTTERFLIES

Small Skipper
Large Skipper
Brimstone
Large White
Small White
Green-veined White
Orange-tip
Brown Hairstreak
Holly Blue
Small Tortoiseshell
Peacock
Gatekeeper
Meadow Brown
Ringlet

KEY FEATURES FOR BUTTERFLIES

- Size and age of hedge
- Presence of larval foodplant shrubs such as Blackthorn, Holly, buckthorns, and elms
- Presence of permanent grass edge with larval foodplants such as Garlic Mustard, nettles, and grasses
- Level of chemical spray drift and presence of unsprayed field margin
- Type and frequency of hedge cutting
- Context in the landscape (i.e. density of hedgerows and presence of nearby woodland or other semi-natural habitats)
- Sheltered edges and intersections

Arable field margin and adjacent hedgerow, Sussex, England.

Arable crops

KEY FEATURES FOR BUTTERFLIES

- Regular cultivation and intensive chemical inputs usually eliminate all larval foodplants
- Some crops such as Oil-seed Rape and clovers may provide nectar sources
- Brassica crops can provide breeding habitats for Large and Small Whites
- Non-rotational set-aside fields may provide breeding habitats for several species including Small and Essex Skippers, Small Copper, Brown Argus, Common Blue, Marbled White, and Meadow Brown.

CHARACTERISTIC BUTTERFLIES

See above

Arable fields, Dorset, England.

Roadside verge flora, Herefordshire, England.

Road verges

KEY FEATURES FOR BUTTERFLIES

- Larval foodplants, especially grasses
- Verge width
- Type and frequency of cutting
- Presence of adjacent hedgerow and shelter
- Possible use of herbicides or growth retardants

CHARACTERISTIC BUTTERFLIES

Small Skipper
Essex Skipper
Brimstone
Green-veined White
Orange-tip
Small Tortoiseshell
Peacock
Marbled White
Gatekeeper
Meadow Brown

Disused railway line, Dorset, England.

Railway lines

KEY FEATURES FOR BUTTERFLIES

- Warm, sheltered and sunny conditions
- Presence of short, sparse vegetation and scrub
- Variety of larval foodplants
- Frequency of cutting and disturbance of vegetation
- Current use or disuse

CHARACTERISTIC BUTTERFLIES

Small Skipper
Large Skipper
Dingy Skipper
Grizzled Skipper
Wood White
Green Hairstreak
Small Blue
Common Blue
Wall
Marbled White
Ringlet
+ other grassland species

UPLAND HABITATS

Mountains and upland grassland (above 300 m)

KEY FEATURES FOR BUTTERFLIES

- Montane climate
- Larval foodplants such as Mat-grass and other grasses
- Presence of nectar sources and shelter
- Presence of livestock and wild grazing animals

CHARACTERISTIC BUTTERFLIES

Green-veined White
Mountain Ringlet*
Small Heath

(*species using this habitat exclusively in Britain and Ireland)

Mountains and upland grassland, Highland, Scotland.

Moorland, Highland, Scotland.

Moorland

KEY FEATURES FOR BUTTERFLIES

- Damp moorland conditions with damp grassland
- Larval foodplants (e.g. Marsh Violet and grasses)
- Grazing levels by livestock and deer
- Frequency of burning

CHARACTERISTIC BUTTERFLIES

Green-veined White
Green Hairstreak
Small Pearl-bordered Fritillary
Dark Green Fritillary
Scotch Argus
Small Heath
Large Heath

Lowland heathland, Dorset, England.

Lowland heathland

CHARACTERISTIC BUTTERFLIES

Green Hairstreak
Small Copper
Silver-studded Blue
Grayling
Small Heath

KEY FEATURES FOR BUTTERFLIES

- Presence of warm, dry soils with bare patches and areas of wet heathland
- Larval foodplants such as heathers and fine grasses
- Presence of livestock and other grazing animals
- Frequency of burning

WETLAND HABITATS

Fenland

KEY FEATURES FOR BUTTERFLIES

- Waterlogged soils with fen vegetation such as reeds and sedges
- Larval foodplants such as Milk-parsley and Water Dock
- Frequency of cutting or level of grazing
- Presence of carr woodland with buckthorns

CHARACTERISTIC BUTTERFLIES

Brimstone
Swallowtail*
Large Copper*
(*species using this habitat exclusively in Britain and Ireland)

Fenland, Norfolk, England.

Bog habitat, Co. Tyrone, Northern Ireland.

Bogs

CHARACTERISTIC BUTTERFLIES

Small Pearl-bordered Fritillary
Large Heath

KEY FEATURES FOR BUTTERFLIES

- Waterlogged, peaty soils
- Larval foodplants such as Hare's-tail Cottongrass and Marsh Violet
- Level of grazing
- Frequency of burning
- Extent of peat cutting

URBAN AND POST-INDUSTRIAL HABITATS

Gardens, parks and churchyards

CHARACTERISTIC BUTTERFLIES

Large White
Small White
Green-veined White
Orange-tip
Purple Hairstreak
Small Copper
Holly Blue
Red Admiral
Small Tortoiseshell
Peacock
Speckled Wood
Meadow Brown

KEY FEATURES FOR BUTTERFLIES

- Abundant nectar sources
- Sunny, sheltered conditions
- Larval foodplants such as brassicas, nettles, Holly, Ivy, and grasses

Garden habitat, Butterfly Conservation's Gold Medal garden at the Chelsea Flower Show (1998).

Disused quarries

CHARACTERISTIC BUTTERFLIES

Small Skipper
Dingy Skipper
Grizzled Skipper
Small Blue
Brown Argus
Northern Brown Argus
Common Blue
Wall
Marbled White
Grayling
Meadow Brown
Small Heath

Disused quarry, Co. Durham, England.

KEY FEATURES FOR BUTTERFLIES

- Thin, dry soil with bare patches
- Larval foodplants such as Common Bird's-foot-trefoil, Kidney Vetch, and grasses
- Presence of warm, south-facing slopes
- Presence of scrub and shelter

Urban and industrial derelict land

CHARACTERISTIC BUTTERFLIES

Small Skipper
Large Skipper
Dingy Skipper
Grizzled Skipper
Brown Argus
Common Blue
Small Tortoiseshell
Wall
Meadow Brown
Small Heath
+ other grassland species

Urban habitat, Hereford, England.

KEY FEATURES FOR BUTTERFLIES

- Varied structure (short and tall often weedy vegetation, with bare ground and scrub)
- Variety of larval foodplants and nectar sources

3 Recording and data collection

This chapter covers the structure and organization of the Butterflies for the New Millennium (BNM) project, starting with the standards set for butterfly recording during the survey period. It also outlines the other main source of butterfly data used in this book, the Butterfly Monitoring Scheme (BMS).

Principles and standards for butterfly recording

The basic unit of information for the atlas is a recorded sighting of one or more butterflies at a specified location. Prior to the initiation of the BNM survey, the Biological Records Centre (BRC) and Butterfly Conservation drew up the minimum requirements for a record, if it was to be useful and relevant to conservation and research.[1] These new standards were an advance over previous national schemes, in which there had been no consistent use of precise grid references and no assessment of the number of butterflies of each species seen.

Species name

Species must be correctly identified and named. Most butterflies in Britain and Ireland can be unambiguously identified with modest experience, aided by a field guide.[2] Long-established vernacular names for butterflies are normally used in Britain and Ireland, and the English names are used in this book. Species are recognized internationally by their scientific names, but even these are subject to change. For example, in the taxonomic revision adopted for this atlas[3] the scientific name of the Purple Hairstreak has changed from *Quercusia quercus* to *Neozephyrus quercus*. A checklist of the vernacular and scientific names used in this book is given in Appendix 1.

Grid reference

Precise grid references enable sightings to be plotted on maps at different levels of resolution and provide a means of revisiting sites to check on the continued presence

[1] Asher (1992); Harding *et al.* (1995).

[2] Recommended guides include Lewington (1999), Thomas (1989), Tolman and Lewington (1997), and Whalley and Lewington (1996). Porter (1997) provides a guide to larvae. In addition the Field Studies Council have produced an AIDGAP leaflet on butterflies.

[3] Karsholt and Razowski (1996).

The minimum requirements for a BNM record.

Recorded information	*Example*	*Need*
Species name	Meadow Brown	Accurate identification
Grid reference	SU990338	To locate the record accurately and precisely on a map
Location	Oaken Wood (Butterfly Conservation reserve)	To describe the location and to verify the grid reference
Date	14 July 1997	To place the record in time, so that phenology and long-term distribution changes can be analysed. May also help to verify the species identification
Recorder name	Ian Cunningham	To attribute the record and allow checking of queries
Number seen	code: D (30–99 individuals seen)	To assess the local status of species and distinguish vagrants from colonies, for example. May be useful in assessing the conservation importance of a site

of species. They are also essential to inform local development and conservation planning and for scientific research purposes. BNM recorders were encouraged to provide records with a precision of a 100 m grid square.

Across Britain and the Isle of Man, the Ordnance Survey (OS) National Grid reference system is used. A grid reference is described by a two-letter code designating a 100 km square, followed by numbers referring to the eastings (numbers running eastwards across a map) and then to the northings (numbers running northwards). A two-figure grid reference, such as NN61, defines a 10 km square within the 100 km square NN, whereas a six-figure one (e.g. NN603172) refers to one 100 m square within the 10 km square (see Appendix 4 for further information).

The National Grid system adopted jointly by the Ordnance Survey of Northern Ireland and the Ordnance Survey of Ireland is similar, except that a single letter designates each 100 km square. Both national grids are centrally aligned to north and appear at an angle to one another because of the curvature of the earth. The Channel Islands are mapped on yet another grid, a Universal Transverse Mercator projection.

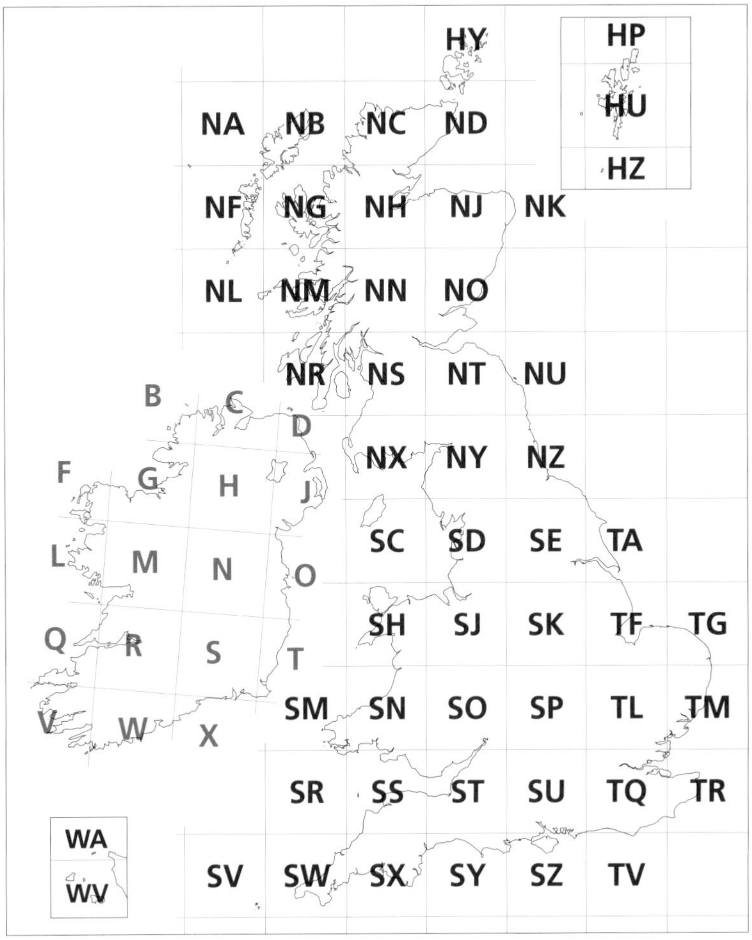

The survey area showing 100 km grid squares and letter codes.

Location or site name

A familiar name for the location provides context for the sighting and a method of checking the accuracy of grid references. Recorders were encouraged to use location names that appear on Ordnance Survey maps.

Date of sighting

The date of the sighting is needed:

- to assess changes in distribution over time. Consequently, the submission of records designated only within a range of years was discouraged.
- to provide information on the flight period or the timing of the immature stages of the butterfly. The effects of factors such as weather patterns, climate, latitude, and altitude can then be assessed.
- to provide an additional verification check on species identification (e.g. adult Orange-tips are unlikely in September).

Recorder name

The name of the recorder establishes the source of the record in case further information or confirmation is required for significant or unusual records. Recorders retain the Intellectual Property Right of their records and so the association of their name with the data is also important for proper acknowledgement.

Number seen

Many distribution maps depict only the recorded presence of species, irrespective of the numbers seen. This can give a misleading impression, as no distinction is made between a solitary individual and one or many separate populations within a square. This leads to an overestimation of distribution, which is most likely to affect the more mobile butterfly species, and is of particular concern if they are scarce or declining.

An indication of the number of butterflies seen is also important for conservation and research purposes, although it must be interpreted with care. The number seen is not necessarily correlated with abundance, but may depend on the recorder's skill, the time spent searching, and the weather. Information on these latter variables was not collected systematically and was considered to be outside the scope of this project.[4]

Within the BNM survey, standard abundance codes were used and the number of individuals of each species seen on a visit was allocated to one of five categories: A for a single individual; B: 2–9; C: 10–29; D: 30–99; and E: 100 or more.

Non-adult stages

Life cycle stages other than the adult butterfly provide direct evidence of breeding and were also recorded. This allowed some species to be recorded outside the (often short) flight period. In addition, paired (copulating) butterflies were recorded.

Other information

The potential value of other optional information associated with a record was also recognized, including the time of day, weather conditions, habitat description (based on CORINE biotope codes[5]), and the time spent searching the location.

Organizing the survey—Butterfly*Net*

At the start of BNM, over 200 local butterfly recording schemes were operating in Britain and Ireland and it was unnecessary and inappropriate to introduce a new national-scale recording scheme. Instead, local schemes were encouraged to submit data sets that met the minimum standards, for central collation. A three-tier structure of recorders, local co-ordinators, and a central database was set up. This had the advantages that recorders could send records to one place and existing local schemes could continue to meet their own objectives. However, it did present the challenge of contacting and co-ordinating many different organizations. Furthermore, in some areas there was no existing local co-ordination of butterfly records. It took three years, starting in 1993, to build up a fully co-ordinated network to cover the UK. This was successfully extended to the rest of the survey area by 1998.[6]

The network, named Butterfly*Net*, is made up of local co-ordinators with

[4] More systematic methods have been developed and used to assess abundance for intensive surveys of rare or threatened species. For examples see Brereton *et al.* (1999), Thomas (1983*a*), and Warren (1993*a*).

[5] See Glossary for explanation.

[6] Fox (1998*a*).

responsibility for specific geographical areas. The co-ordinators represent a wide range of organizations including Butterfly Conservation branches, local biological records centres, county wildlife trusts, and natural history societies. The area covered by each co-ordinator varies from a vice-county or county (typical in England and Wales), through regions (typical in Scotland) to whole provinces or countries (e.g. the Channel Islands, Isle of Man, Northern Ireland, and the Republic of Ireland). Their roles are similar, irrespective of the size of area covered, and each acts as the focal point for recording in their area, receiving, checking, and computerizing records before transferring them to the central database. Co-ordinators also provide feedback to recorders, generate publicity, and forge co-operation with organizations involved in recording, nature conservation, and land management in their area.

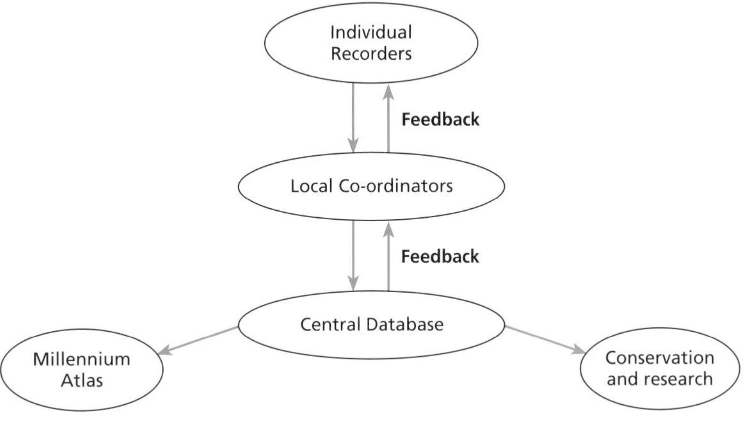

The structure and flow of butterfly records and feedback in Butterfly*Net*.

Central co-ordination was necessary to create and maintain Butterfly*Net* and a project officer was employed to support the local co-ordinators. Feedback was provided to the co-ordinators at annual meetings and national meetings were also conducted in Ireland, Scotland, and Wales to promote the project.[7]

Some records were received directly from national organizations, including the Countryside Council for Wales, English Nature, Scottish Natural Heritage, the Royal Society for the Protection of Birds, the National Trust, and the Ministry of Defence. Those records not in electronic form were re-routed to the appropriate local co-ordinator for verification. Electronic data were incorporated directly into the central database. Verified data from other Butterfly Conservation schemes (e.g. the Garden Butterfly Survey[8] and the Pearl-bordered Fritillary survey[9]) and summary information from the BMS, were also incorporated.

In Ireland, the six counties of Northern Ireland were covered by the local branch of Butterfly Conservation, working in partnership with the Centre for Environmental Data and Recording in Belfast. In the Republic of Ireland, after initial problems in finding an appropriate organization, the Dublin Naturalists' Field Club (DNFC) took on the co-ordination of recording across the country, with support from The Heritage Council.

BNM recording media

A simple instruction sheet (see Appendix 4) was produced to guide recorders and this was adopted by many local co-ordinators. It recommended making visits in good weather to a variety of sites representative of different habitats, three or four times during the butterfly season (e.g. in early May, mid-June, mid-July, and the second half of August). This provided an opportunity to record all species present. Two recording forms were designed around the recording standards (see Appendix 5). One form was intended for repeated visits to a site, such as a regularly walked footpath, nature reserve, or garden and the second was designed for casual sightings from a number of sites. Regional versions of these forms were produced, each with a

[7] Proceedings documents were produced for most of these meetings and are available from Butterfly Conservation (Harding 1995, 1996; Fox 1997*a*,*b*, 1999*a*, 2000).

[8] Vickery (1995).

[9] Brereton (1998).

species list appropriate to that area, so as to simplify use and reduce errors, both when recording and when computerizing data.

Improving coverage

Although the aim was to achieve complete coverage across Britain and Ireland at a high level of precision, resources for recording differed greatly throughout the survey area. There were larger numbers of recorders in the more densely populated areas and it was possible to achieve good coverage at the 2 km square level. In contrast, parts of northern and western Britain and Ireland are more sparsely populated and a similar level of coverage even at the 5 km square level could not be achieved in only five years. Strenuous efforts were made to improve the coverage in under-recorded areas in the last two years of the survey.

Information on coverage was fed back locally to recorders, identifying areas that were under-recorded. Various approaches were developed, which proved valuable in stimulating recording effort:

- Circulation of coverage maps at national and local scales, showing recorded squares and the number of species recorded in each.
- Circulation of provisional distribution maps for individual species, emphasizing gaps in recording coverage.
- Use of distribution maps combining recent and historical records to highlight areas where a species was previously recorded but for which there were no recent records.
- Publication of lists of local places (e.g. villages, nature reserves) with no records.
- Allocation of specific grid squares to individual recorders.
- Organization of field meetings to survey under-recorded areas.

Although such initiatives improved coverage in many areas, it became clear after three years of the project that some areas were still inadequately recorded, as is shown by the coverage maps of the Green-veined White.

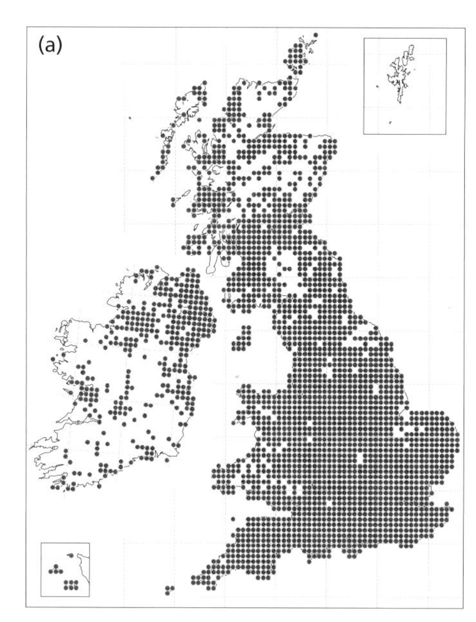

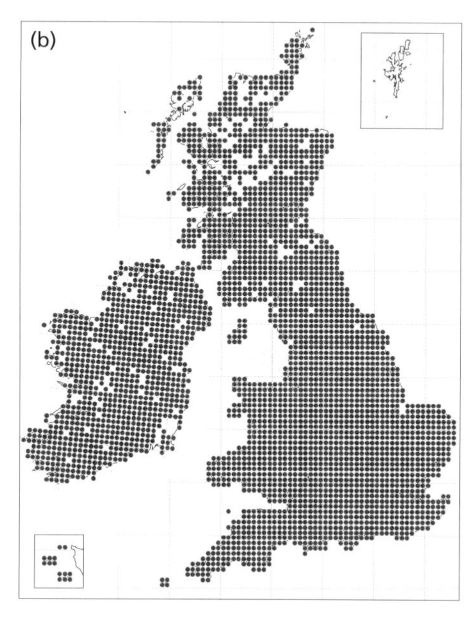

The Green-veined White is considered to be common and widespread, yet after three years of survey work the distribution map (Map a: 1995–7) showed substantial gaps in coverage. Many were filled in 1998–9 using targeted recording, to give the final distribution (Map b: 1995–9).

Filling the gaps—targeted recording

In 1998, a targeted recording scheme was organized in close liaison with local co-ordinators. Recorders were given details of areas in need of survey and the locations of previous records of key species. Over 100 volunteers took part in this scheme during 1998 and 1999. As a result, records were obtained from over 1500 10 km squares, many of which had not been visited during the earlier years of the survey. Targeted recording played an important role in ensuring adequate coverage of many under-recorded areas and species.

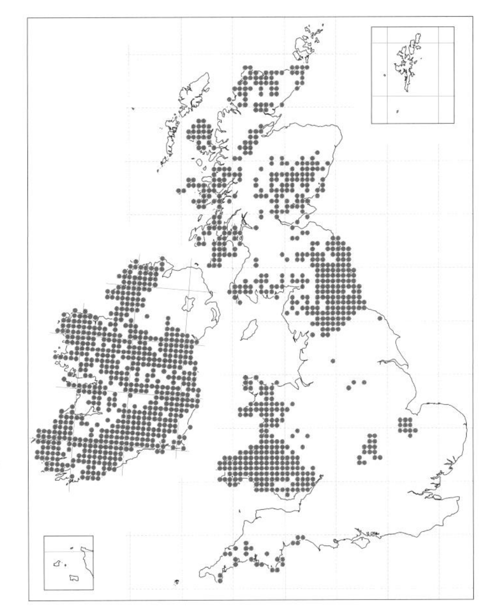

The 10 km squares visited by volunteers working on the BNM targeted recording scheme, 1998-9.

Publicity

Recording was stimulated by local and national publicity campaigns. In addition to the BNM meetings, a twice-yearly newsletter was circulated and an Internet web site was constructed. Articles in Butterfly Conservation News,[10] other conservation press, and the national media were aimed at audiences with an interest in wildlife.[11] Recorders of other taxonomic groups were also asked to help through articles in appropriate publications[12] and through direct approaches (e.g. to bird clubs).

Postcard surveys

A novel method of stimulating recording using postcards was developed for the Orange-tip in Scotland in 1997.[13] Such was its success, the scheme was repeated in

[10] e.g. Asher (1995); Fox (1999*b*).

[11] Fox (1998*b*, 1999*c*).

[12] Fox (1999*d*).

[13] Sutcliffe and Kirkland (1998).

The unmistakable male Orange-tip was used in a successful postcard recording scheme in Scotland, Wales, and the Republic of Ireland. The reverse side of the card had a pro-forma for details of sightings and a return address.

subsequent years and extended to Wales and the Republic of Ireland in 1999. In addition, postcard surveys for the Mountain Ringlet and the Scotch Argus were organized in the Lake District (northern England) and Scotland. The cards were distributed in public places such as libraries, visitor centres, tourist information offices, and youth hostels, and proved successful in motivating people with no previous experience of biological recording to send in records.

Data collation

Most local co-ordinators entered records onto a computerized database. Many used the *Levana* software developed for butterfly recording,[14] but others used the comprehensive RECORDER system.[15] A few used non-standard formats based on other database or spreadsheet software. After computer input, data for each year were submitted to the central database either by disk or e-mail. Updates of data from preceding years, arising from late records, were sent in a similar way.

In virtually all cases, any problems of format posed by the use of different software were solved (normally via a tab-separated text format) and the data were converted into the *Levana* system used for the central database. This fulfilled a goal of the project that records should be entered onto computer only once, after which all data transfers should be electronic.

Data verification and validation

A key feature of any recording scheme is the system by which the accuracy of the data is checked. The methods that were applied fall into the general categories of verification (confirming that a record is accurate) and validation (confirming that a record has been correctly transcribed, for example on to a computer). Verification and validation are time consuming but essential. All reasonable efforts have been made to ensure the veracity of each record, but Butterfly Conservation, BRC, and DNFC cannot be held responsible for errors of fact in, or interpretation of, the data.

It was important that verification took place locally, where there was expert knowledge of local butterfly populations and sites. The regional structure of Butterfly*Net* made this straightforward and the majority of BNM records have been first verified and then validated by the local co-ordinators.

Further checks on the data, once submitted, were carried out centrally. They were plotted on a national map and the project team investigated records falling outside the local co-ordinator's area. Some of these were genuine records from outside the area, whereas in other cases grid reference errors were identified, confirmed with the local co-ordinator, and corrected. Provisional distribution maps

[14] *Levana* Butterfly Recording System © Specklewood Software (1992–9). User manual V1.2 Sept 1996.

[15] RECORDER V3.21b © JNCC (1995). User manual V3.21b, JNCC (September 1995).

Verification of records.

Data attribute	*Verified against*
Species identification	Knowledge of local fauna, date, grid reference, site name, and level of experience of the recorder
Grid reference	Site name and expected distribution of species
Date	Expected life cycle phenology

were also sent out. These provided feedback on progress, enabled co-ordinators to check for anomalies, and resulted in further errors being identified, documented, and corrected.

Bias and constraints

BNM was designed and conducted in order to achieve as complete geographical coverage as possible. Even with an estimated 10 000 recorders it was not easy to survey the whole of Britain and Ireland at the desired level of precision and in a short time period. The project relied largely upon volunteers submitting records in a non-systematic way and it is important to identify the limitations of such an approach[16] and consider them when interpreting the data.

Recorders

Most participants in the survey were already experienced in butterfly recording. However, the BNM project encouraged people to begin recording for the first time and this may have introduced a few errors in species identification or grid references. Local co-ordinators worked throughout the survey to eliminate errors and ensure that a high level of accuracy was achieved.

A geographical bias of recording effort exists within the database, due to the uneven distribution of recorders. There is also local-scale bias because recorders are often drawn to sites with high diversity or abundance of butterflies and they are more likely to visit sites close to their homes.[17] Land with limited or no public access is less likely to be recorded, as is land at higher altitudes.[18] However, the focus on attaining complete coverage led to increased recording of the wider countryside and reduced such bias in the BNM survey.

Bias in the data can also be caused by time-related factors. The amount of time spent recording a particular site or grid square and the number of visits made were not standardized, although guidance was given to recorders. In addition, there is a seasonal bias that particularly affects single-brooded species with early flight periods (e.g. the Grizzled Skipper). In the spring and early summer, fewer butterfly recorders are active and the weather is often less conducive to recording adult butterflies. Finally, there was an increase in recording effort towards the end of the survey (nearly 40% more 10 km squares were visited in 1999 than in 1995), as a result of targeted recording initiatives.

Weather

Climatic factors influence the abundance of many butterfly species (p. 343)[19] and their distribution.[20] Furthermore, the weather affects the opportunities for observers to record butterflies. It is possible that short-term climatic trends or extreme weather events within the survey period biased the results. A short survey time period would be expected to increase this bias.

UK climate summaries[21] show that the period 1995–9 was warmer, drier, and more sunny than the 1951–80 average. Only 1996 was cooler than average and 1998 and 1999 were wetter. Although the early part of the year was generally unfavourable for butterfly recording, 1999 was the warmest year in the Central England Temperature series since records began in 1659. Each of the BNM survey years had more sunshine than the long-term average. The years, therefore, appear to have been

[16] Dennis *et al.* (1999); Rich (1998).

[17] Dennis and Thomas (2000)

[18] Dennis and Hardy (1999).

[19] Pollard (1988); Pollard and Yates (1993*a*).

[20] Hill *et al.* (1999, in press); Parmesan *et al.* (1999); Pollard (1979); and Chapters 5 and 6.

[21] Supplied by M. Hulme, Climate Research Unit, UEA, Norwich on www.cru.uea.ac.uk/~mikeh/datasets/uk/month_uk

generally favourable for butterflies. Monthly anomaly tables (Appendix 6) provide a useful guide to assessing good or poor periods for particular species.

The BMS collated indices have also been used to assess the effects of weather during the BNM survey. Weather is probably the most important single factor that affects population levels of the indexed species from year to year.[22] The analysis,[23] which ranked years in which species had annual indices in the upper or lower quartile, indicated that 1995–9 was a favourable period for butterflies on the monitored sites. Three consecutive years (1995–7) were placed in the top seven BMS years and even the poorest recent year (1998) was ranked sixteenth (of 24).

Survey time period

The BNM survey period involved a compromise. There had to be sufficient time to achieve adequate coverage, but the value of the data for conservation is diminished by a long survey period. Most butterfly distributions are in a state of flux, so a short 'snapshot' of the current situation will generally provide more meaningful comparison with previous and future situations. In this respect, the five-year BNM period represents a considerable improvement over the 13-year survey period of the previous atlas.

Certain species appear to undergo cycles of abundance and distribution. A bias might be expected for species whose cycles are of similar duration to the survey period (e.g. the Holly Blue). The distribution of such species must be interpreted in relation to the phase of the population cycle.

Recording absence

The distribution maps in this atlas show only the recorded presence of species. No distinction has been made between squares with negative results for a particular species and those that have not been visited. In order to state that a species is not present, rigorous standardization of recording effort is necessary. The lack of this in BNM prevents the use of negative records (i.e. records of species absence) and they were not collated in the central database.

Collation of pre-1995 data

In addition to the BNM survey, earlier records have been collated in the central database. The BRC database (spanning the period 1800–1982) was made available, as were data from the BMS (1976–99). Locally collated data sets covering a variety of date periods were also submitted through Butterfly*Net*. However, many of the butterfly records made between the publication of the 1984 Atlas and the start of BNM had not been computerized. Therefore, BRC carried out the computerization of all record cards submitted as part of the Butterfly Conservation/BRC Target Species Project. This contributed a further 75 000 records to the central database.

The Butterfly Monitoring Scheme

Considerable use has been made in this atlas of data from the Butterfly Monitoring Scheme. At each BMS site a fixed route, or transect, is walked every week, from 1 April until 29 September. An important aspect is that recording conditions are standardized in relation to the weather and time of day.[24] All butterflies seen within 5 m in front or to the sides of the recorder are counted and summed to give weekly

[22] Pollard and Yates (1993*a*).

[23] T. Sparks and N. Greatorex-Davies (personal communication).

[24] Hall (1981); Pollard *et al.* (1986); Pollard and Yates (1993*a*). The BMS methods specify that the temperature must be above 13 °C providing that there is more than 60% sunshine, or above 17 °C in overcast weather. The wind speed should not normally exceed Force 5 on the Beaufort scale, and the walk should be started after 10.45 and completed before 15.45 (BST).

Key differences between recording for the BNM project and monitoring by the BMS.

	Recording method	*Abundance data*	*Geographical representation*
Butterflies for the New Millennium (BNM)	Non-standard recording conditions. Other life cycle stages recorded	Limited numerical interpretation possible	All areas, including the wider countryside, although still some bias towards special sites.
Butterfly Monitoring Scheme (BMS)	Standardized transect counts of adult butterflies	Considerable numerical interpretation possible	Mainly special sites

totals for each species. These totals are then combined to give an annual index for each species. Where a species has several broods, a separate index for each generation is produced, providing that the broods are distinguishable. For species with overlapping broods (e.g. the Speckled Wood) a single annual index is calculated.

There are important differences between transect monitoring and recording for a distribution atlas. BMS records are gathered systematically and over many years, using a standardized method and the data provide information on trends in relative abundance. The BMS sites (p. 6) tend to be of high conservation value and most are designated nature reserves or Sites of Special Scientific Interest or both. Consequently, the data may not be representative of the countryside as a whole.

BNM, on the other hand, recorded butterflies on a wider basis, extending into areas with no protected sites or special wildlife significance. It therefore helps to demonstrate the effects of change in the wider countryside. However, the task of covering a large area meant that much recording had to be opportunistic and was not standardized in relation to time spent searching and weather conditions. The two data sets thus complement each other and both are used in the species accounts. It is in the context of the methodology and the biases outlined in this chapter, that the database has been analysed and interpreted to produce the results presented in this atlas.

4 Interpreting the data

Butterflies for the New Millennium has been the most comprehensive survey in the history of butterfly recording. Both the quantity and quality of the data collected and the 10 000 or so participants have far exceeded previous work and the database provides a rich source of information for future conservation and research.

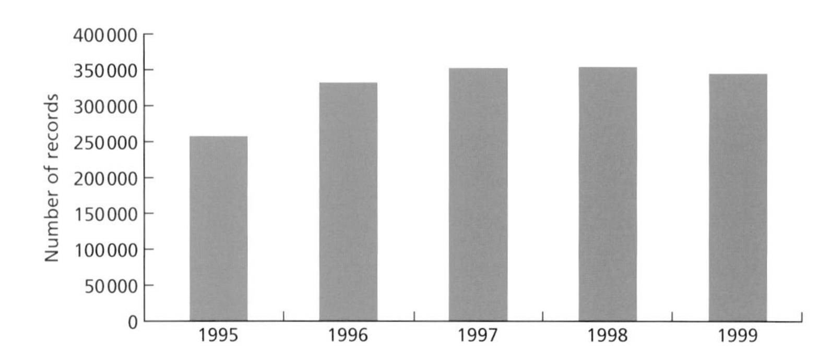

The number of records entered in the BNM database for each year of the survey.

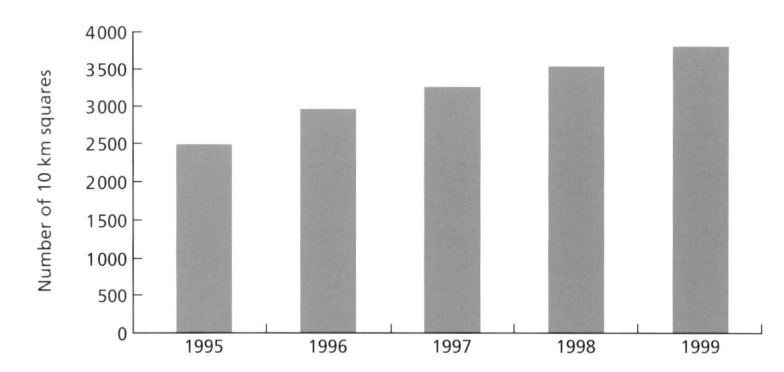

The cumulative coverage of 10 km squares during the BNM survey.

Mapping the data

The principal aim of this atlas is to present a summary of the records from the BNM project as distribution maps. Using a 10 km grid square standard format, these maps enable comparison between species. Where information is needed at a more detailed scale, supplementary maps at a finer resolution have been prepared. The geographical area covered by the survey made it necessary to use inset boxes for the Channel Islands and Shetland to fit easily on to a page. The blue lines overlaid on Britain (including the Isle of Man) show the Ordnance Survey 100 km gridlines. Those across Ireland are the equivalent lines of the Irish National Grid and the Universal Transverse Mercator grid is used across the Channel Islands (p. 31).

The main map for each species depicts the BNM distribution (1995–9) and historical data. Records from 1970–82 are represented by open circles and those from before 1970 by crosses. These historical data indicate only the recorded presence of a species. BNM records are superimposed upon historical ones, so that only declines in distribution can be represented on the standard species maps. Subsidiary maps with different symbols are used to show expansion of distribution, where relevant.

Choice of mapping resolution

Dot maps of flora and fauna distribution across Britain and Ireland have been plotted typically at 10 km square resolution, and butterflies have been no exception.[1] There are three main reasons for this choice of scale. Firstly, the maps must be legible and 1 km or 2 km square dots are too small to be distinguished on a map covering the whole of Britain and Ireland in a standard sized book. Secondly, many national surveys have been able to achieve good coverage for mapping only at coarse levels of resolution (e.g. 10 km squares). Finally, the same mapping resolution is used for ease

[1] e.g. Emmet and Heath (1989); Heath *et al.* (1984); Hickin (1992); Thomson (1980).

of comparison with previous data sets. Thus 10 km square maps are used in this atlas, even though the survey has achieved extensive coverage at a much finer resolution.

The BNM project set out to collect detailed records; nearly 94% of BNM records have a resolution of at least 1 km, compared to only 51% of 1970–82 records, which resulted from a project with more limited objectives. However, the choice of scale has serious implications for the analysis of the distribution change of a species on maps that only show its recorded presence (or the absence of records). 10 km grid squares may contain many colonies or individuals of a particular species. The extinction of many colonies and massive declines in abundance may occur in such squares without being registered on the map. The fact that colonies of many species (particularly habitat specialists) occur together, rather than being spread throughout the countryside, further exacerbates this problem, which is of more than theoretical concern.

Summary statistics of the BNM survey (1995–9) and 1970–82 data set.*

	Overall	*Britain and Isle of Man*	*Ireland*	*Channel Islands*
BNM:				
Total number of butterfly records	1 617 071	1 548 935	50 483	17 653
Coverage of 10 km squares	98.7%	98.9%	98.2%	100%
1970–82:				
Total number of butterfly records	137 426	124 978	12 340	108
Coverage of 10 km squares	90.4%	92.4%	85.6%	46.7%

* Heath *et al.* (1984)

The proportion of records at different levels of precision in the BNM and 1970–82 survey data sets. Records that have grid references referring only to a 10 km or 5 km square cannot be represented on a distribution map at finer levels of resolution. Precise records can be plotted on maps at many different levels of resolution.

Resolution	*BNM survey (1995–9)*	*BRC survey (1970–82)*
100 m square	61.5%	26.2%
1 km square	31.9%	24.9%
2 km square	5.5%	0%
5 km square	0.4%	0%
10 km square	0.7%	48.9%

In several studies, butterfly distribution data collected at fine levels of resolution have been compared over time at different scales.[2] Coarse grids were found to underestimate rates of decline in distribution for all but the rarest species. It was in recognition of this problem that the BNM survey aimed to achieve satisfactory coverage at much finer resolution than those used previously. Although much of the detail cannot be shown in this atlas, there is considerable potential for the BNM data set to be analysed at finer levels of resolution in the future.

Interpreting the number of butterflies seen

Previous atlases of butterflies in Britain and Ireland have shown only the simple presence of species. In this survey the number of butterflies seen was incorporated in

[2] León-Cortés *et al.* (1999, 2000); Thomas and Abery (1995); and see p. 336.

The interpretation of numbers of butterflies seen on the distribution maps in the species accounts.

Mapping symbol	*Category*	*Recorded sighting*
• (small, light)	1 sighting	Maximum of one individual recorded on any one visit
• (medium)	2–9 max seen	Minimum of two and a maximum of nine individuals seen on any one visit; or egg, larva, or pupa recorded; or records of single individuals from two or more 2 km squares within the 10 km square in any one year
● (large, dark)	10+ max seen	10 or more individuals seen on any one visit

the minimum recording standards. Where there was no indication of the numbers seen, a conservative approach was adopted and records were interpreted as being of single individuals.

These data give an indication of the status of the species (resident or vagrant) within each grid square. Different coloured symbols have been used to show the maximum number of individuals recorded on any single visit to a site within the relevant square. This method is appropriate to the BNM data set, which includes records from multiple visits to some sites, repeated year on year, through to casual sightings from a recorder passing once through a site. By considering the maximum seen on any one visit, the results of repeat visits to a site do not bias the interpretation, although the probability of sighting a species is increased by making repeated visits (p. 44). Records of non-adult life cycle stages were placed in the '2–9 max seen' category, as they demonstrate breeding. Information on the numbers of eggs, larvae, or pupae seen was not collated in the database.

A few local data sets contained no information on numbers seen, although they had good coverage at a 2 km square resolution. If the simple representation is used for numbers seen, these areas would appear to be anomalous on the distribution maps because all records would be counted as single individuals. To provide a more realistic picture, symbols were upgraded from the '1 sighting' category to the '2–9 max seen' category where the species was recorded in two or more separate 2 km squares within a particular 10 km square in any one year. This was carried out across the complete data set (with the exception of rare migrant species), to avoid introducing any additional bias, but it does not affect the appearance or the interpretation of the maps elsewhere to any great extent.

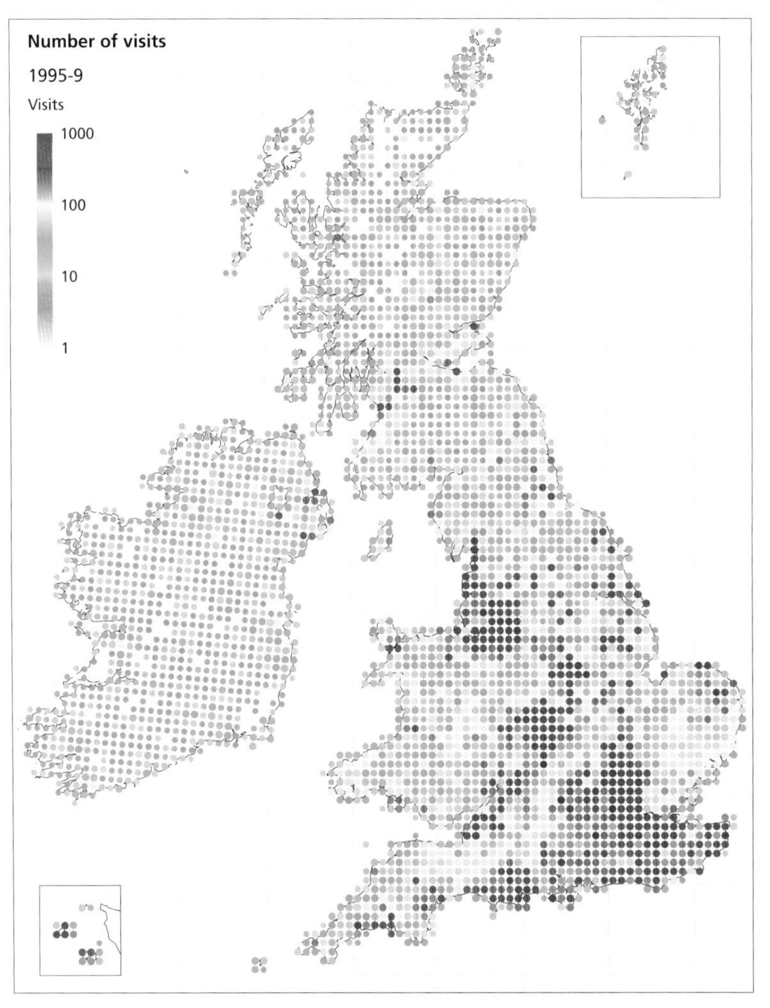

Recording effort map showing the minimum number of visits to each 10 km square during the 1995–9 survey period. Summary records provided for a site over a whole year (e.g. from a transect walk), have been counted as a single visit.

This representation of the number of individuals seen is simplified and allowances must be made in any interpretation of the results based solely on the maps (as opposed to using the underlying data). The recorded number of butterflies seen is not representative of the relative or actual abundance, because recording effort was not standardized. The categories and symbols were selected to indicate only the general status of species within a square and to separate single sightings from what would appear to be colonies. In well-recorded parts of Britain and Ireland the '1 sighting' symbol suggests a vagrant, but little can be assumed if the square was visited only once. The '2–9 max seen' symbol may represent evidence of breeding colonies (i.e. records of non-adult stages) and residency, but interpretation is limited. The '10+ max seen' category is particularly useful and for most species it indicates breeding populations (particularly for species that form discrete colonies). The system, therefore, represents an improvement over maps showing presence of species only.

Recording effort

An assessment of recording effort is important for the interpretation of survey results. The distribution of recording effort in the BNM survey can be represented by maps showing the number of recorder visits to each 10 km square and the proportion of 2 km squares visited within each 10 km square. The latter indicates the level of recording coverage within each 10 km square, highlighting areas that have been surveyed at a detailed scale.

There were high levels of recording effort in south-east England, across large parts of the West Midlands, and in the north-west. In Scotland and Wales, the under-recorded uplands contrast with better coverage in low-lying areas. The maps also show variation between different local recording schemes. Areas with established recording schemes were able to generate large data sets (e.g. Cheshire, Suffolk, and Hertfordshire/Middlesex). In other areas, schemes had to be set up and recorders recruited specifically for the BNM project. These faced a greater challenge to achieve good coverage and, inevitably, generated fewer records. This was also the case in the Republic of Ireland where a co-ordinated scheme was in operation for only the last two years of the survey, although some records from the preceding years were available.

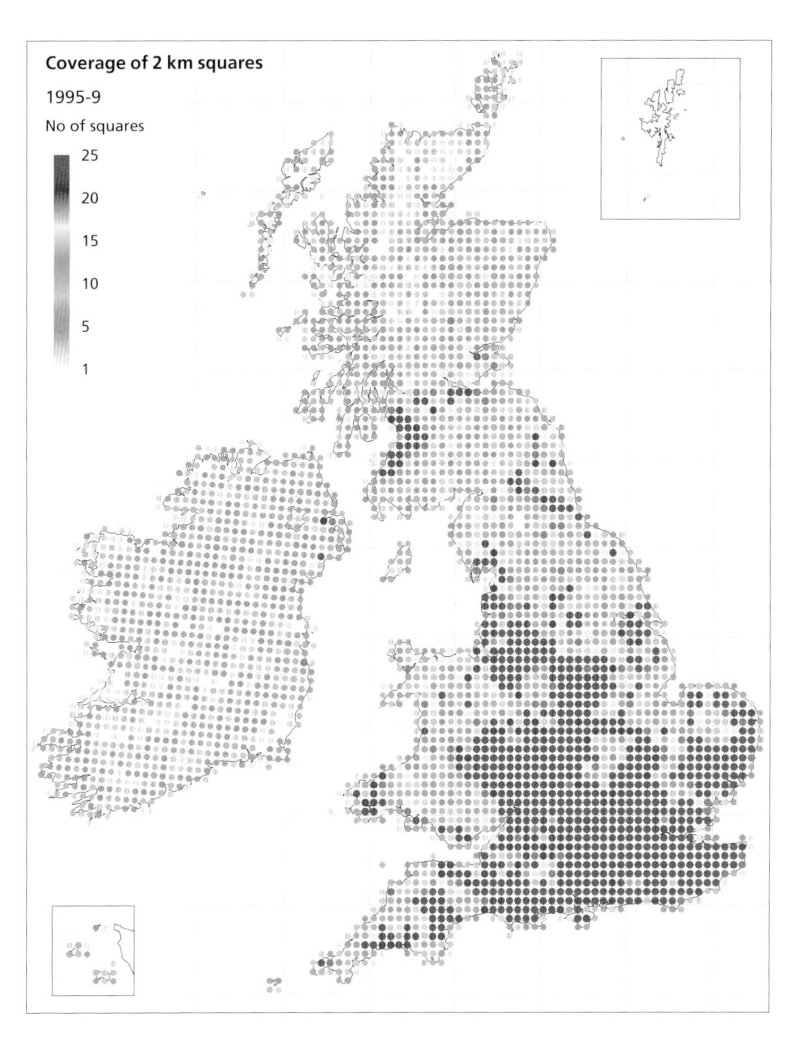

Recording effort map showing the proportion of 2 km squares recorded within each 10 km square. Records supplied at only 5 km or 10 km square resolution have not been included. Thus some squares that have been visited remain blank on this map, most notably in Shetland.

The maps also highlight 'honey-pot' areas, known to have relatively rich habitats and a good range of species, and which attract butterfly recorders (e.g. Morecambe Bay in north-west England, parts of the Dorset coast, and the Burren, in western Ireland).

Recording effort varied during the year. For example, fewer visits were made during April, May, and June than later in the summer (July and August). The completeness of the recorded distributions of predominantly single-brooded species, with flight periods in this early part of the season (e.g. the Dingy Skipper, Grizzled Skipper, and Green Hairstreak) is likely to have been adversely affected by this bias in recording effort. It is less likely to have affected rarer species such as the

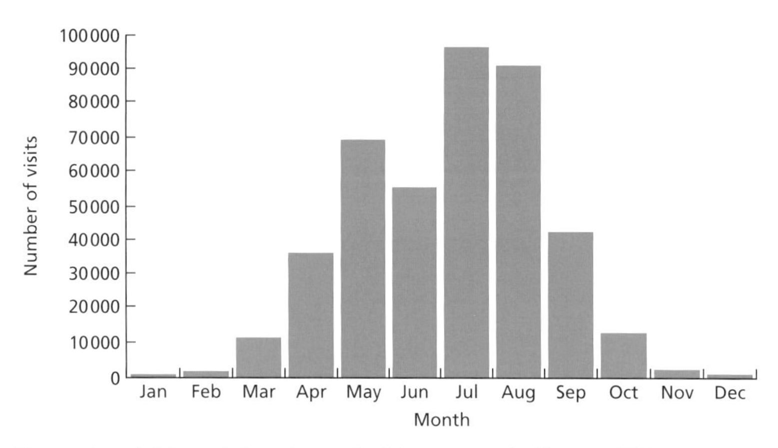

The number of visits made in each month of the year over the five-year BNM period.

The number of species recorded in each 10 km square in the BNM survey.

Chequered Skipper and the Pearl-bordered Fritillary, where special efforts were made to ensure good coverage.

Variation in recording effort was inevitable and a full analysis is limited by the absence of data relating to visits when no species were recorded. The bias caused by uneven effort in recording should be borne in mind when using the maps of distribution and species richness. Where relevant, the species accounts include an assessment of the effect of recording effort on the distribution map.

Species richness

More species occur in the south and east than in the north and west. This is mainly a reflection of the patterns of climate across Britain and Ireland,[3] although the number of species recorded may also be under-estimated in some areas through lower recording effort. Analysis of a local data set for Greater Manchester[4] has shown that the number of visits to a grid square is a very significant factor in determining the recorded presence of many individual species and the estimated species richness.[5]

Re-analysis of the BNM data at a 5 km square resolution also shows the biodiversity gradient from the south and east to the north and west and reveals both the 'hot-spots' and the less species-rich areas. In the south of Britain, geographical features with many butterfly species are emphasized, including the North and South Downs, the Chilterns, the Cotswolds, Salisbury Plain, the Isle of Purbeck, and Dartmoor (see p. 14 for locations). In northern Britain and much of Ireland, it is more difficult to establish genuine differences in the levels of species richness due to the confounding effects of recording effort. Nevertheless, some areas with high numbers of resident species are clearly

[3] Dennis (1992*b*, 1993); Dennis and Williams (1986); Turner *et al.* (1987).

[4] Hardy (1998).

[5] Dennis (2000); Dennis *et al.* (1999); Dennis and Hardy (1999); Hardy and Dennis (1999).

The number of species recorded in each 5 km square in the BNM survey.

revealed, such as around Morecambe Bay, in Argyll, and the Burren. Other apparent 'hot-spots' (e.g. around Belfast, Dublin, Edinburgh, and Glasgow) may be due to increased recording in the countryside surrounding conurbations, relative to outlying areas. The distribution of habitat specialist species (overleaf) presents the clearest indication of the key areas, and further emphasizes the importance of southern England for butterflies (see also pp. 358 and 359).

Interpretation of Butterfly Monitoring Scheme data

The Butterfly Monitoring Scheme (BMS) provides a wealth of information that can aid the interpretation of the status of species. The annual abundance indices from over 100 sites in the scheme are combined to produce collated indices of abundance

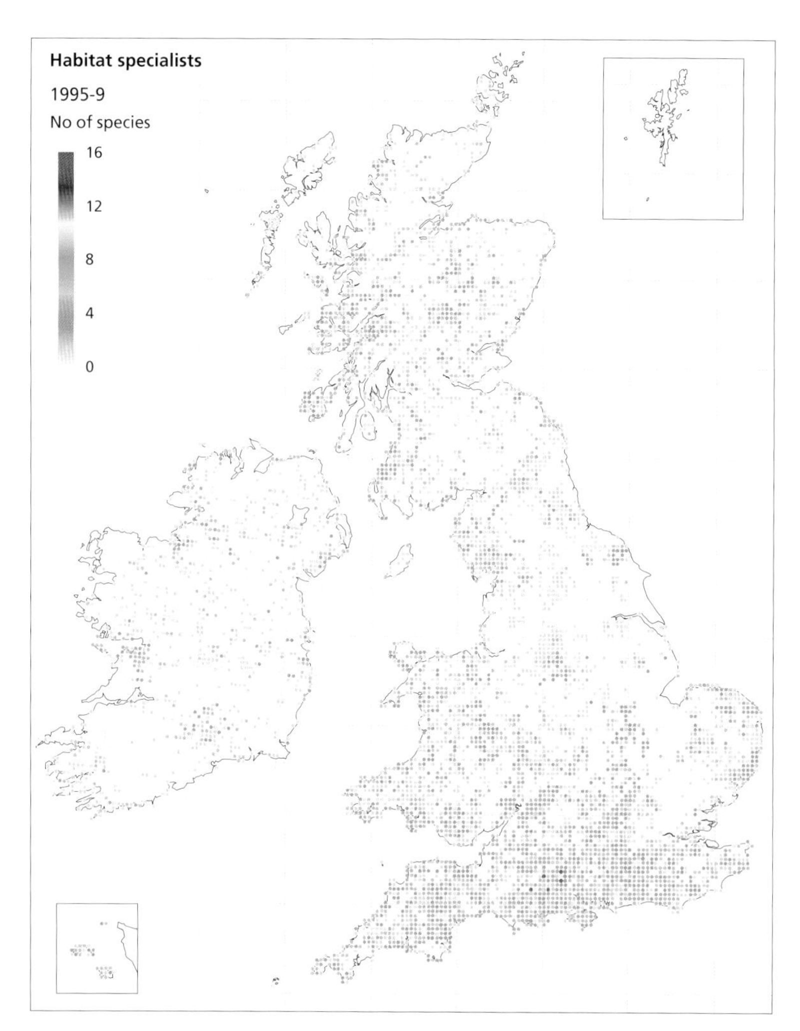

The number of habitat specialist species recorded in each 5 km square in the BNM survey.

for more than 30 species.[6] These are not unduly influenced by changes at individual sites. Therefore, wider influences on butterfly abundance (most notably the weather) can be separated from local, site-specific effects (e.g. habitat management). Annual collated indices can be compared for a sequence of years to assess trends in species abundance over time and long-term trends (such as those associated with climate change) can be distinguished from fluctuations caused, for example, by short-term weather events. In addition, the BMS data set has provided valuable information about butterfly phenology, voltinism, migration, and local distribution changes.

It has been emphasized previously that care should be exercised in the use of BMS data,[7] and this caution should be borne in mind in comparisons with the BNM findings. Some species that have undergone recent expansion of distribution have also shown a significant increase in abundance as measured by BMS collated indices (e.g. the Comma and Speckled Wood).[8] Several species show contracting distributions and decreases in abundance (e.g. the Pearl-bordered Fritillary and Wall). However, some other species (e.g. the Orange-tip and Gatekeeper) have not shown clear changes in BMS abundance during the recent period of range expansion. The factors relating change in distribution to local abundance are not fully understood.

The BMS sites were not selected randomly and they have an uneven geographical distribution in Britain and Northern Ireland. Many of the sites are managed for nature conservation, sometimes with specific efforts being made to conserve the key species that are being monitored. This can result in increased abundance on individual sites, against a background of decline in the overall collated index and recorded distribution of a particular species. Therefore, conditions on most BMS sites are not representative of the wider countryside. Some rare species are not sufficiently represented in the BMS to allow for the calculation of collated indices of abundance and those that are presented are less reliable than those for common and widespread butterflies.[9]

The value of the BMS is that data have been gathered over many years and the longer the scheme continues, the easier it will be to identify trends. The BNM survey has been conducted over a five-year period and the longer data set of the BMS helps to place it in context. Although there are underlying differences between BMS and

[6] Moss and Pollard (1993).
[7] Pollard and Yates (1993*a*).
[8] Pollard *et al.* (1995).
[9] Pollard and Yates (1993*a*).

BNM data, the combination of the two data sets provides a more powerful insight into the status of butterflies than would be possible with either in isolation.

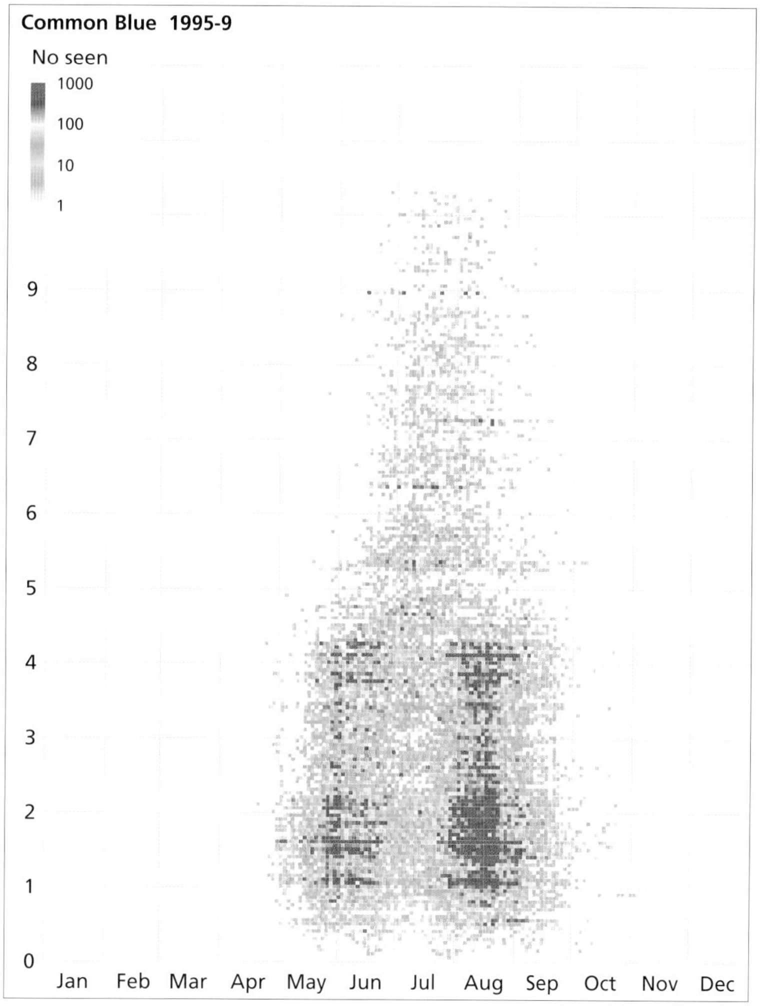

The timing of butterfly life cycle stages

This atlas shows the distributions of species in Britain and Ireland and the variations in these distributions with time. The BNM data set contains other information, including the dates of sightings, which can be analysed to provide information on flight periods. In contrast to the BMS, the BNM data were not gathered systematically through the butterfly season. Nevertheless, the large volume and extensive coverage of BNM records enables a useful analysis of the timing of life cycle stages (known as phenology).

As an example, the information relating to the flight period of the Common Blue derived from the five-year BNM survey is shown. The grid reference northing (equivalent to latitude) of each Common Blue record was plotted against the date of the sighting. In this plot (termed a **phenogram**) the horizontal lines are at the 100 km grid intervals of the National Grid of Britain. The vertical lines denote months. The colour scale represents the number of adult butterflies recorded on each date.[10] The clear picture that emerges is the transition from two flight periods in the south to a single flight period in the north. The zone of transition is at a latitude corresponding to Yorkshire. There is also a slight trend showing that the flight period(s) occur at later dates towards the north. Insufficient data were available to be able to prepare a similar analysis for Ireland, where the phenology pattern may be different.

Phenogram for the Common Blue in Britain and the Isle of Man showing the transition from double to single flight periods.

The latitude at which the Common Blue changes from a double to a single generation each year might be expected to shift in response to climatic factors. Global warming might result in the latitude of this transition zone shifting further north and phenograms of data from different year ranges might show whether this is occurring. The 1970–82 survey does not contain enough data to plot a similar phenogram, but the BNM data set provides a baseline against which such comparisons can be made in the future. Appendix 8 and some of the species accounts give further examples of phenograms from the BNM data set.

[10] For each record a mid-range value of the abundance code is taken.

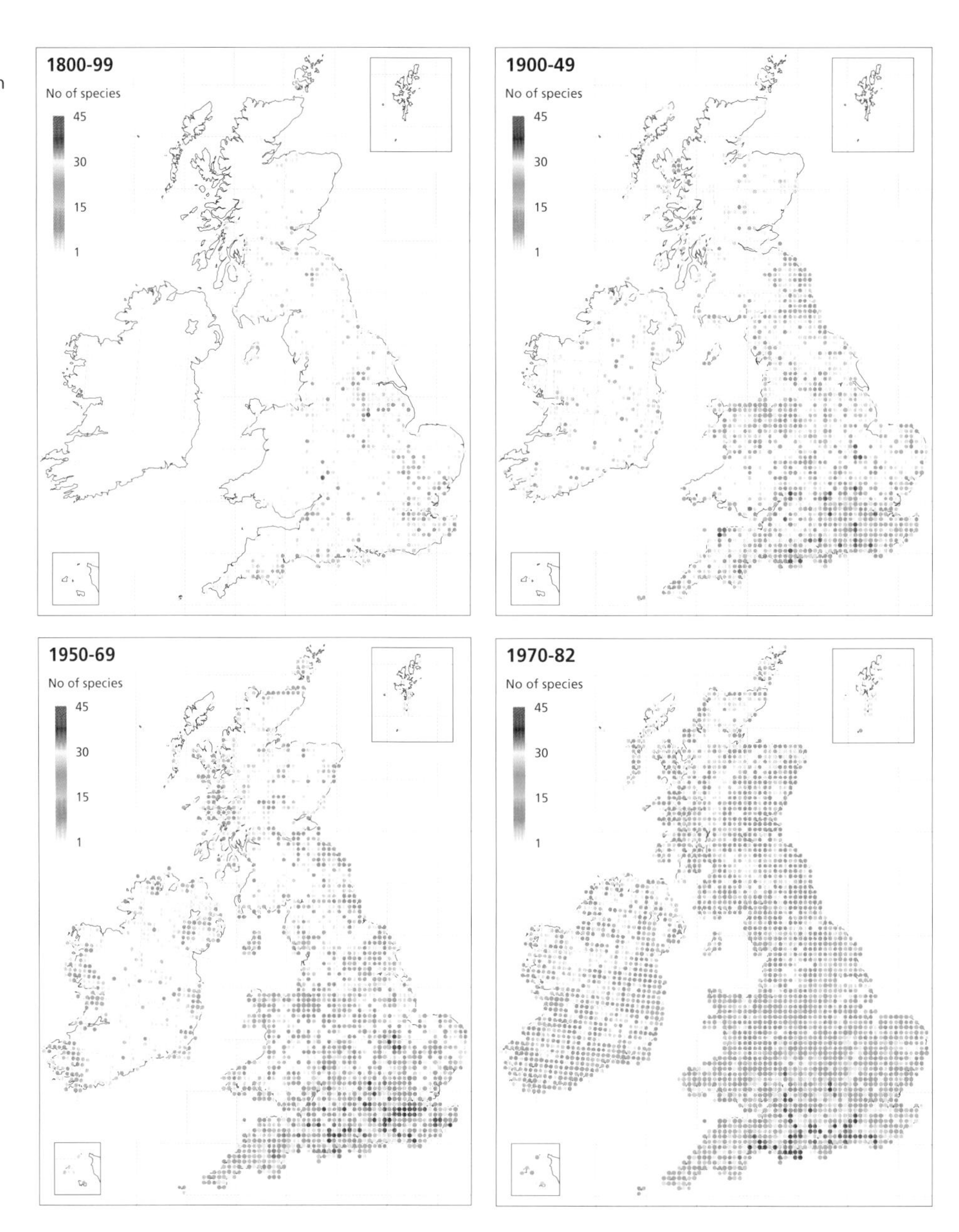

Coverage maps of records held in the central database for the periods 1800–99 (100 years), 1900–49 (50 years), 1950–69 (20 years), and 1970–82 (13 years). The symbols indicate the numbers of species recorded in each 10 km square.

Comparison with historic data sets

The BNM data set covers the period 1995–9 and provides a snapshot of species distribution. An objective of this project has been to put the current distribution of each species into a historical context, as a means of identifying long-term changes. In order to do this, butterfly records dating back to before 1800 have been incorporated in the central database and are shown on the main distribution map for each species. There is considerable variation in recording coverage within the earlier data sets, and systematic coverage was not attempted until the 1970–82 survey.

The older historical records are limited in terms of coverage and common species are under-represented, partly because they were not targeted in the 1970s collation of nineteenth century and early twentieth century records. This effect is shown by an analysis of the data for certain species with stable distributions. The number of 10 km

The percentage of recorded 10 km squares for particular species in each decade of the twentieth century in Great Britain. The most recent period is the five-year BNM survey, and therefore data from 1990–4 have been omitted, although this makes little difference to the interpretation.

	1900–9	*1910–19*	*1920–9*	*1930–9*	*1940–9*	*1950–9*	*1960–9*	*1970–9*	*1980–9*	*1995–9*
Green-veined White	0.2	2.0	7.0	24.2	9.9	39.0	61.4	80.0	69.0	92.2
Meadow Brown	1.8	3.8	17.8	30.1	25.5	42.2	59.3	78.0	64.2	86.7
Small Heath	0.1	2.3	5.4	26.3	23.1	38.7	53.2	69.2	54.7	74.6
Lulworth Skipper	1.2	0.5	0	1.1	0.3	0.8	0.5	0.5	0.5	0.4
Black Hairstreak	1.0	0	0.4	0.7	0.6	1.2	1.1	0.9	1.1	0.9

squares recorded for such species has been compared to the total number of recorded squares (i.e. the coverage) in each decade of the twentieth century. Wider countryside species such as the Green-veined White and the Meadow Brown are grossly under-represented in early decades. By contrast the proportions of recorded squares of the Lulworth Skipper and Black Hairstreak, both well-studied habitat specialist species, have remained almost constant.

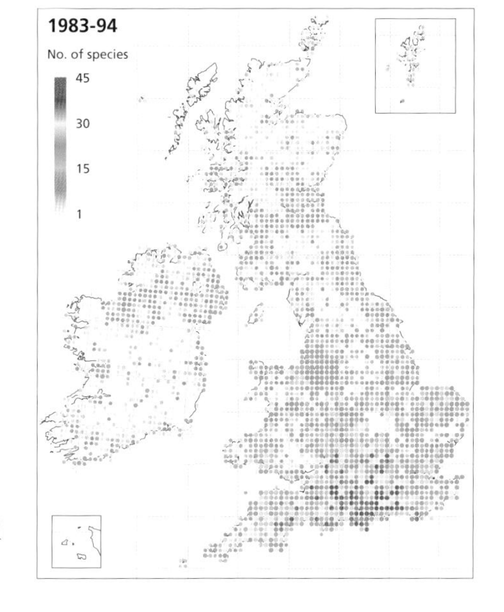

Coverage map of records held in the central database for the period 1983–94. Symbols indicate the number of species recorded in each 10 km square.

This analysis confirms that the 1970–82 period data is the most appropriate for comparison with the BNM data set. Even so, the more limited aims of the 1970–82 survey meant that the coverage at resolutions finer than 10 km squares was much poorer than that of the BNM survey (p. 41) and that the former data set contains no estimates of the numbers of butterflies seen.

Much recording was carried out between the two major surveys, mostly for local schemes. Efforts have been made to 'capture' these data and the central database contains over 400 000 records for 1983–94. Because the coverage for this period is geographically uneven, the data have not been shown on the main species maps. However, the data have been used to help to interpret some of the changes, and will be used to support conservation and ecological research in the future.

In conclusion, the BNM data set has more records and a better geographical coverage than is available from any preceding period. Together with the findings of the BMS, it allows us to identify the changes that have affected each species and to gain an overview of the current state of butterflies in Britain and Ireland. The special value of the BNM data set for the future is in the conservation of butterflies at the site level and as a comprehensive baseline against which to assess changes.

5 Species accounts

Guide to content and layout

This chapter presents detailed accounts of the resident or regular migrant species in Britain and Ireland, including the results of the BNM survey and summaries of recent research. Species that have become extinct, introduced species, and the rarer migrants and vagrants are included at the end. Exotic species, whose occurrence in Britain and Ireland is the result of occasional release or escape from captivity, are not covered. Each account follows the same layout.

Species name Vernacular names and taxonomic order follow Emmet and Heath (1990) and scientific names follow Karsholt and Razowski (1996) (see Appendix 1). Where generic names have been revised by Karsholt and Razowski, the previous genus is included in brackets.

Summary box Species are classified as Resident, Extinct, Regular migrant, Rare migrant, or Introduced. Regular migrants are species that breed in Britain and Ireland, but are unable to overwinter in sufficient numbers to be regarded as resident. The status of 'rare migrant' covers a range of migrant and vagrant species for which breeding here is a rare event.

A brief summary of the change in the range of the species since the 1970–82 survey is given.

Conservation Status

The status of the species is listed as determined by

UK Biodiversity Action Plan[1]
Butterfly Conservation[2]
European Red Data Book[3]
Bern Convention and EC Habitats and Species Directive where appropriate.

Any legal protection afforded to species by the Wildlife and Countryside Act (1981) in Great Britain and Wildlife (Northern Ireland) Order (1985) in Northern Ireland is summarized. The Republic of Ireland, Isle of Man, and the Channel Islands had no legislation that lists butterfly species at the time of writing.

European/world range

A brief outline of the range of the species is given, so that populations in Britain and Ireland can be put in perspective.

Introductory paragraph A brief description of some of the key features of the butterfly and its distribution is given. The account is not intended to be a guide to identification, as this is covered well by other books (see p. 30).

Foodplants

The known larval foodplants are listed, with vernacular and scientific names according to Stace (1997), Appendix 2. Elsewhere in the species accounts, only vernacular names of plants are used.

Habitat

The main habitats used for breeding (or other behaviour) are listed.

[1] UK Biodiversity Group (1998*a*).
[2] Warren *et al.* (1997).
[3] Swaay and Warren (1999).

Life cycle and colony structure

Together with the two preceding sections, this provides an overview of the life history, flight period, population structure, and ecology of the species.

Distribution and trends

The detailed findings of the BNM survey are given with comparisons made between the 1995–9 distributions of species, those from the 1970–82 survey, and earlier records to assess the current and long-term trends. Notable trends in abundance on the monitored sites in the Butterfly Monitoring Scheme (BMS) are also discussed.

European trend

Changes in the European distribution of the species are given, as listed in the Red Data Book of European butterflies (Swaay and Warren, 1999) and in a review by Parmesan *et al.* (1999).

Interpretation and outlook

The possible causes of the trends in distribution and abundance of species are discussed, conservation issues identified, and an assessment made of the future prospects of the species in Britain and Ireland.

Key references

The main published references are given for each species. The list does not include general texts and papers, which the authors have drawn upon for many of the species accounts. Such texts are listed below, and are recommended as further reading for many or all of the species covered in this atlas: Cowley *et al.* (1999); Dennis (1992*a*); Emmet and Heath (1989); Heath *et al.* (1984); Long (1970); Parmesan *et al.* (1999); Pollard and Yates (1993*a*); Pullin (1995); Roy and Sparks (2000); Swaay and Warren (1999); Thomas and Lewington (1991); and Tolman and Lewington (1997). In addition, local butterfly atlases (Appendix 3) have proved a source of important information for many species accounts, but in general these are not listed in the key references.

Graphics

Distribution map The main map for each species shows the 10 km square distribution, as determined by the 1995–9 survey (coloured dots), together with historical data from two periods: 1970–82 (open circles) and pre-1970 (crosses). The symbols used and the rationale behind them are discussed in Chapter 4. For each square, records from the most recent survey period take precedence over those from earlier periods.

Subsidiary maps Additional maps of butterfly distribution have been included in some species accounts. These may show a more detailed picture of the distribution, with records plotted at finer levels of resolution. In other cases, the subsidiary map shows an expansion of range or changes in range over different periods.

Other graphics A range of additional diagrams and pictures are used to provide further information for species. BMS collated index graphs are shown for many species, and information on these is given in Chapter 4. The collated index is a measure of relative abundance of the species across all monitored sites. The first year of the index (usually 1976) is given an arbitrary value of 100. As the collated index graphs presented here are on a log scale, all indices therefore start at 2. Phenograms (where they are used) are explained on p. 47.

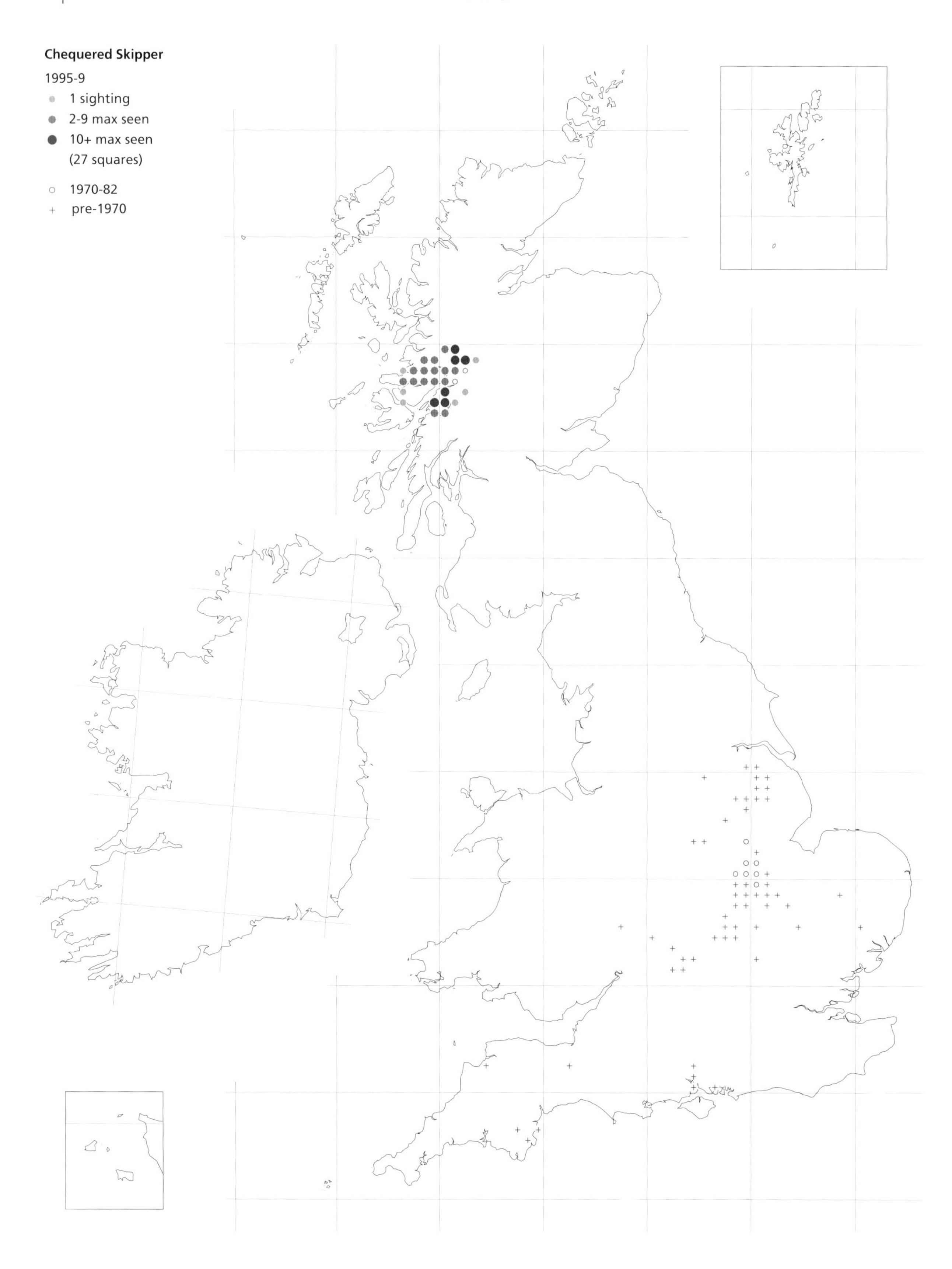
Chequered Skipper
1995-9
1 sighting
2-9 max seen
10+ max seen
(27 squares)
1970-82
pre-1970

Chequered Skipper *Carterocephalus palaemon*

Resident

Range stable in Scotland, extinct in England (1976).

Conservation status

UK BAP status: Priority Species.
Butterfly Conservation priority: high.
European status: not threatened.
Protected in Great Britain for sale only.

European/world range

Widely distributed across Europe and Asia to Japan, and also in North America as far north as Alaska. Declining in several European countries and threatened in Japan.

This small, fast-flying butterfly is now restricted to damp grassy habitats in western Scotland. Males are seen more frequently than females, perching in sheltered positions either next to wood edges or amongst light scrub or bracken. They dart out to investigate passing objects, defending their territory against other males and other butterfly species, or in the hope of locating a potential mate. Females are less conspicuous and fly low among grasses when egg-laying. The Chequered Skipper died out from England in 1976 and re-establishment trials have taken place since 1990. In Scotland, there are thought to be about ten core areas and there have been no obvious recent changes in range.

Foodplants

The main foodplant in Scotland is Purple Moor-grass (*Molinia caerulea*). In England most records were on False Brome (*Brachypodium sylvaticum*), though a range of grasses may have been used as they are in continental Europe.

Habitat

In Scotland the butterfly breeds in open grassland, dominated by tall Purple Moor-grass growing on wet but not waterlogged soils. Favoured sites are on the edges of open broad-leaved woodland where richer soils produce lusher growth of the foodplant, typically with scattered Bog-myrtle and birch scrub. Most breeding areas occur below 200 m at the base of a slope, often beside lochs or rivers. Former colonies in England occurred in woodland rides and glades, and occasionally in fens or ungrazed calcareous grassland amongst scrub. It may also have bred formerly in damp coppiced woodland as it does elsewhere in northern Europe.

Life cycle and colony structure

There is one generation per year with adults normally flying from the third week of May until the end of June, but occasionally until early July. The flight period may have been 1–2 weeks earlier in England. **Eggs** are laid singly on the foodplant and the **larvae** subsequently live within tubes formed by spinning together the edges of a leaf. The larvae make characteristic notches below the tube and feed on the grass blade above the tube, moving to a new leaf when the old one is eaten. The notches help to maintain high levels of nutrients in the leaf and may also prevent the plant's defensive chemicals from flowing into the damaged portion.

The larvae have a long development period, often feeding into November; research has shown that their survival rates are higher where the foodplant is more luxuriant and stays green well into the autumn. They over-winter as fully grown larvae, each within a larger shelter formed by spinning together several leaves. They emerge in April and, without further feeding, **pupate** concealed within grassy vegetation.

Chequered Skipper larva in its feeding tube, showing the characteristic notches made in the grass blade.

In Scotland, the butterfly does not form discrete colonies, and populations are generally scattered over wide areas of more or less continuous semi-natural habitat. However, adults often concentrate in sheltered areas where males establish territories or where nectar sources are abundant. Observations suggest that females probably range widely during the season and may even spread several kilometres away from main breeding areas, laying eggs in any suitable patches of vegetation as they go.

Distribution and trends

The butterfly is currently restricted to a small region of western Scotland centred on Fort William; ranging from Loch Linnhe and Loch Etive in the south to Loch Arkaig in the north and Glen Spean in the east, and Loch Sunart and Ardnamurchan in the west. Its distribution is thought to be limited by the combination of climatic and geological factors in this region. The climate is comparatively mild with a high rainfall and a long growing season, which encourages lush growth of the foodplant and allows full larval development.

The Chequered Skipper was formerly quite common in woods of the East Midlands of England, with scattered records as far as Dartmoor. Many of the best sites were on the heavy calcareous clays that run in a belt from Bedfordshire to Lincolnshire. The butterfly began to decline in England in the early part of the twentieth century but remained fairly abundant in its East Midlands stronghold until the 1950s. It then underwent a rapid decline that led to extinction in 1976.

The butterfly was not discovered in Scotland until 1939 and its range was poorly known until intensive surveys by the Scottish Wildlife Trust and Butterfly Conservation in the 1980s and 1990s. It is still relatively under-recorded because many sites are remote, the flight period is short, and the weather in the region can be very inclement! The most recent survey indicated that there are ten main population centres from which the butterfly disperses and occasionally establishes other smaller colonies. Present evidence suggests that the species has been more or less stable in Scotland over the last few decades.

European trend

The range of the Chequered Skipper appears to have changed little in many countries but it has declined seriously in Austria, Luxembourg, and Romania (>50% decrease in 25 years); Belgium and Lithuania (25–50% decrease); and to a lesser extent in Germany, Latvia, the Netherlands, parts of south-west Russia, and Sweden (15–25% decrease). In contrast it has expanded recently in Hungary and Italy (25–100% increase in 25 years).

Interpretation and outlook

The extinction of the Chequered Skipper in England is almost certainly related to massive changes in the management of broad-leaved woodland during the twentieth century, especially the cessation of coppicing. This system created numerous open areas within woods, which would have been suitable for the butterfly, as they are on the near continent today. Even as late as the 1940s, almost 30% of woods in the East Midlands were still actively coppiced, but this reduced to almost nothing by the 1960s. In the 1950s, the butterfly had a minor revival at some sites, many of which were cleared and replanted with conifers. It even colonized a few new sites at this time, including Woodwalton Fen National Nature Reserve. The butterfly survived in some open sunny rides for a few decades but these eventually became too small to support viable populations and its extinction was sudden.

In 1990, a programme was started by Butterfly Conservation to investigate the feasibility of reintroducing the Chequered Skipper to England. Studies have been conducted on the butterfly's ecology and requirements on the near continent, especially at damp woodland sites in the Ardennes. Potential habitats have been restored at a number of woodlands in the East Midlands following ride widening and coppice restoration. Field trials on larval survival began on a large Forestry Commission site in Lincolnshire in 1994, and some adults were released during 1997–9. The study is still regarded as experimental and despite some signs of successful breeding, it is too early to say whether it will be possible to establish a viable colony. If this experiment proves successful, surveys will be conducted to assess the feasibility of restoring the butterfly to other sites as part of the overall Species Action Plan.

Conservation effort must, however, remain focused on the Scottish colonies to ensure that they do not suffer the same fate as those in England. Although the distribution appears stable in Scotland, there are serious concerns about the changing management of the butterfly's woodland edge habitats. Increased browsing by deer is preventing the regeneration of native woodland in many areas and several recent forestry initiatives include the fencing of woods against deer. Although this encourages natural regeneration of trees, it can lead to the rapid loss of open space, including breeding areas of the Chequered Skipper. This situation is known to be adversely affecting the butterfly at a number of sites, including some nature reserves. Greater effort should be made to incorporate the needs of insects that require open spaces into such schemes, possibly by introducing some rotational clearance of woodland and maintaining open spaces in potential breeding areas.

Rotational cutting of scrub and woodland creates suitable habitat for the butterfly in Scottish woodland, and almost one-third of sites occur in wayleaves where scrub is cut on rotation beneath electricity pylons. These habitats are similar to those created by coppicing, which may once have played a role in maintaining habitat for the butterfly in western Scotland. The continuation of a regular cycle of cutting beneath wayleaves within the butterfly's range is vital, and similar management could perhaps be replicated elsewhere to create a more diverse woodland habitat. This would provide open edge habitats suitable for the Chequered Skipper as well as other insects.

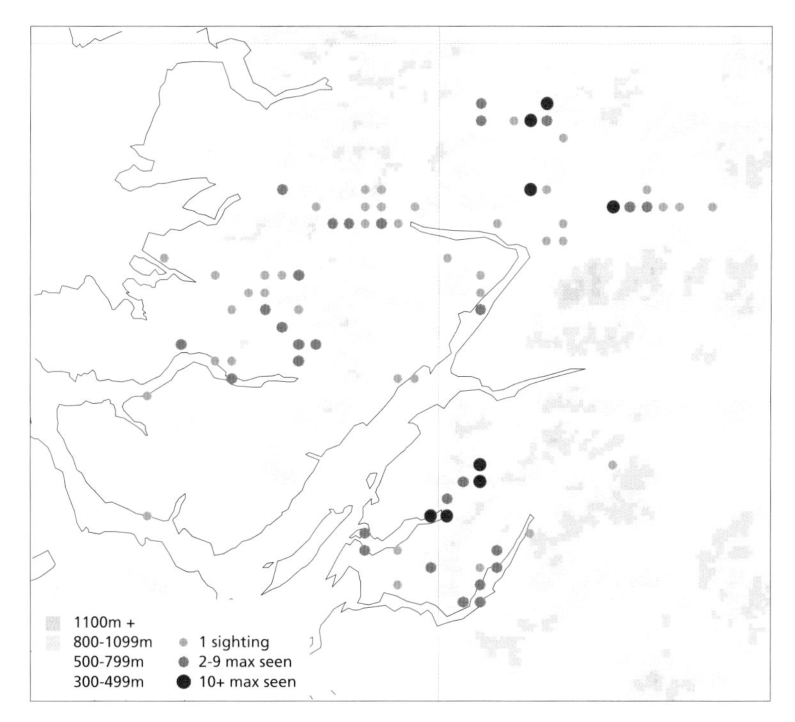

The distribution of 1995–9 records of the Chequered Skipper at 2 km square resolution in western Scotland, against a background of altitude.

Key references

Collier (1986); Ravenscroft (1994*a*,*b*,*c*); Ravenscroft and Warren (1996*a*); Warren (1990).

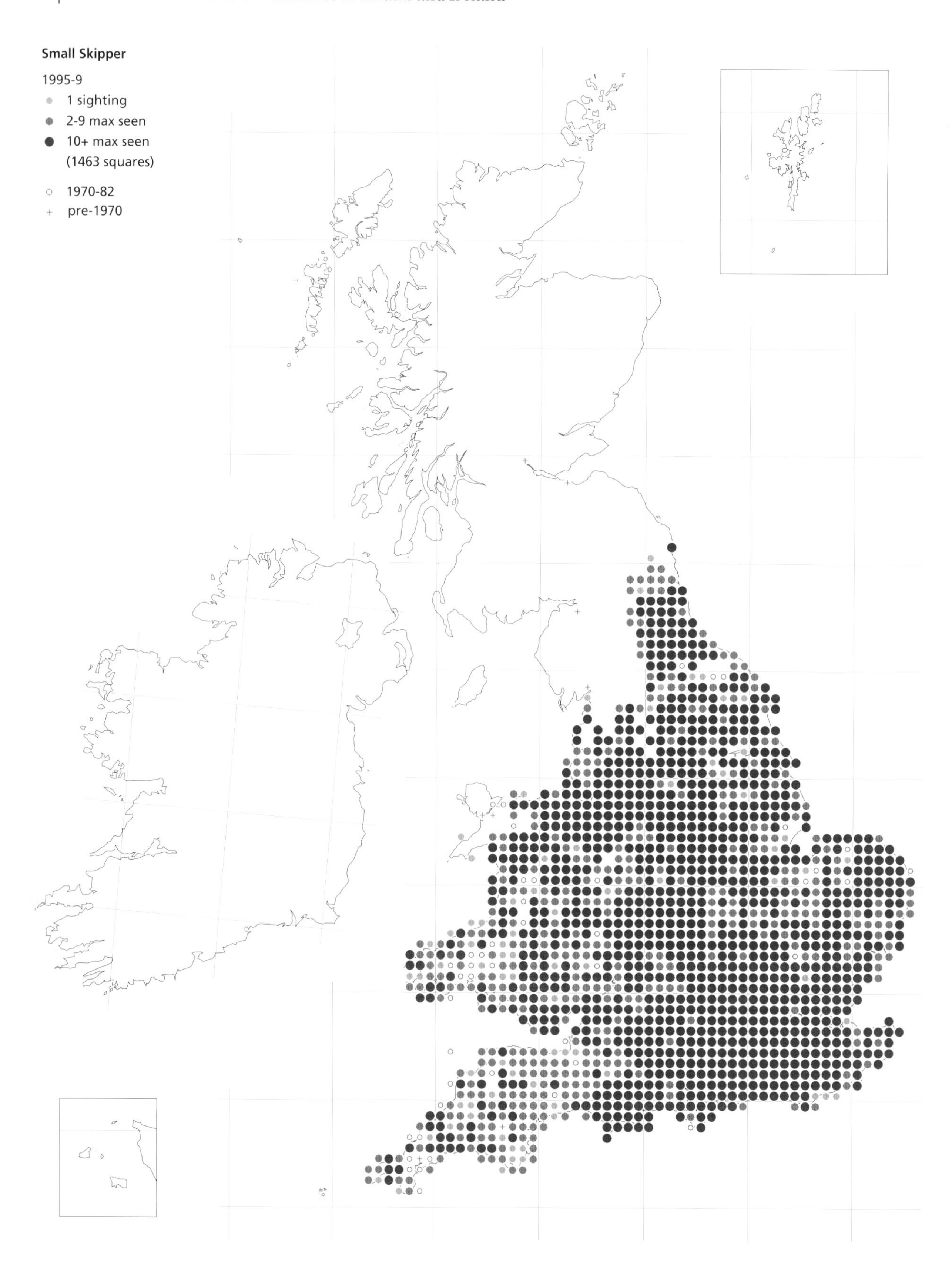
Small Skipper
1995-9
1 sighting
2-9 max seen
10+ max seen
(1463 squares)
1970-82
pre-1970

Small Skipper *Thymelicus sylvestris*

Resident
Range expanding.

Conservation status
UK BAP status: not listed.
Butterfly Conservation priority: low.
European status: not threatened.

European/world range
Much of Europe, as far north as the Baltic States and Denmark and east to the Urals. Also in North Africa and the Middle East. Its distribution in Europe is generally stable with some northward expansion.

Small Skippers are insects of high summer. Although they spend much of their time basking or resting among vegetation, they are marvellous flyers, manoeuvring expertly through tall grass stems. It is these darting flights, wings glinting golden-brown in the sunlight, that normally alert an observer to their presence. Closer examination will reveal many more individuals nectaring or basking with their wings held in the half-open posture distinctive of skipper butterflies. The butterfly is widespread in southern Britain and its range has expanded northwards in recent years.

Foodplants
The Small Skipper almost exclusively uses Yorkshire-fog (*Holcus lanatus*), although several other grasses have been recorded as foodplants, for example Timothy (*Phleum pratense*), Creeping Soft-grass (*H. mollis*), False Brome (*Brachypodium sylvaticum*), Meadow Foxtail (*Alopecurus pratensis*), and Cock's-foot (*Dactylis glomerata*).

Habitat
Small Skipper colonies are found where grasses are allowed to grow tall. Typical habitats are unimproved rough grassland, downs, verges, sunny rides, and woodland clearings. Colonies can occur on small patches of suitable habitat such as roadside verges and field margins.

Life cycle and colony structure
The species has a single generation each year and adult emergence appears to be closely synchronized across Britain, although the date varies from year to year. Typically, peak counts of adult Small Skippers occur in July, although individuals are regularly reported from mid-June through to mid-September (see p. 391).

Males appear to establish territories and will dart off a favoured perch to investigate any butterflies that approach. Females are more sedentary than males and make slow, deliberate egg-laying flights.

Small Skippers nectar on a wide range of plants (such as thistles, knapweeds, Red Clover, Common Bird's-foot-trefoil, and vetches), but individuals preferentially forage at flowers of the same species as previously visited. This strategy saves learning how to extract nectar from flowers with different structures. However, the butterflies switch to different species if nectar rewards are low or if the abundance of their previously preferred plants is reduced.

Eggs are laid in small batches in the leaf sheaths of foodplants. The female alights on a grass stem and backs downwards, inserting her abdomen into the leaf sheath and depositing several eggs inside. After hatching, the **larvae** feed on their own eggshells and then immediately enter hibernation, protected by a silken cocoon as

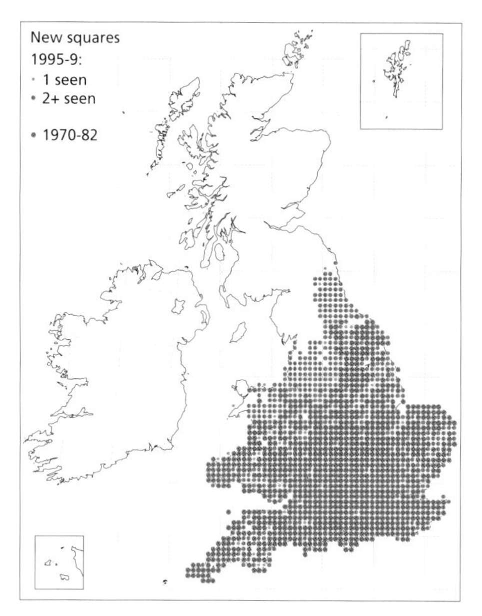

The expansion of range of the Small Skipper since 1970–82.

well as the enclosing leaf sheath. In spring the larvae become active, feeding, at least in the early stages, within protective tubes made from rolled blades of grass held together with silk. **Pupae** are formed in loosely spun tents of leaves near the bases of foodplants.

Small Skippers live in colonies and seem not to be particularly mobile. A study in southern England showed that two thirds of the population moved less than 20 m per day (the maximum distance recorded was 280 m). Elsewhere, however, Small Skippers have colonized sites isolated by farmland, woodland, or urban development and the recent expansion in northern England indicates moderate mobility. Colonies may be small, both in numbers of individuals and in the size of habitat patch occupied, but large colonies can occur where there are extensive swards of Yorkshire-fog, for example in ungrazed meadows. Experiments have demonstrated that colonies tend to be much larger than indicated by casual recording, as relatively few individuals are active, and therefore conspicuous, at any one time.

Distribution and trends

The species is widespread and common in much of England and Wales, becoming less so in the north-west. It is absent from Ireland, the Channel Islands, and the Isle of Man, and the only Scottish records are from the early nineteenth century and may be erroneous.

The Small Skipper has undergone a rapid expansion in the north-east and north-west of England since the 1970–82 survey, extending the range of the species by approximately 100 km in both regions. In the north-west the species has expanded through Cheshire, north Derbyshire, Lancashire, and West Yorkshire. In the north-east, where historically its range has always extended further north than on the western side of the Pennines, the recent expansions have seen it spread throughout Co. Durham and into Northumberland. The number of 10 km squares recorded has increased by 37% since the 1970–82 survey.

The Butterfly Monitoring Scheme (BMS) does not produce an index for the Small Skipper, as it is considered impractical for recorders to separate this species from the similar Essex Skipper where both occur at the same site. However, a graph of combined indices from BMS sites outside the current range of the Essex Skipper shows an upward trend, although the number of sites involved is small in most years. The combined BMS index for the two species across the whole of Britain also shows an increase in abundance over the monitoring period.

The confusion of Small, Essex, and Large Skippers by a few less experienced observers may introduce some recording bias. However, this probably had little effect on the distribution map for the Small Skipper. Some of the remaining gaps in the distribution of the species in lowland England and Wales are probably due to under-recording.

European trend

The species is stable in many countries, but declines have occurred in Austria (50–75% decrease in 25 years) and Luxembourg, the Netherlands, Romania, and Slovakia (15–25% decrease). Over a longer period (30–100 years) its range has extended northwards, perhaps in response to climate warming.

Interpretation and outlook

As with many British butterflies, the range of the Small Skipper is more restricted than that of its larval foodplants. It is assumed that its range is limited by climatic factors and that its current expansion is a response to climatic changes. However, the specific factors involved have yet to be determined.

The transect data available for the Small Skipper indicate that numbers increase significantly when the weather preceding and during the flight period is warm and dry. This may also account for the more northerly range of the species on the drier eastern side of England. The Small Skipper also appears to be less affected by drought than many butterfly species, presumably because its larvae enter hibernation almost immediately after hatching and are not reliant on foodplant quality until the following spring. However, the regular cutting of grasses is likely to cause high mortality of immature stages such as hibernating larvae.

The Small Skipper remains common and widespread throughout southern Britain and has made significant expansions northwards in recent decades. Its abundance at BMS sites has also increased. Although there is no cause for concern about the long-term future of this species, its overall numbers in the wider countryside have probably declined throughout much of the range due to habitat loss. Removal of hedgerows and woodlands, and their associated grassland strips and clearings, reseeding of meadows and over-grazing or frequent cutting, will undoubtedly have made much former habitat unsuitable in recent decades.

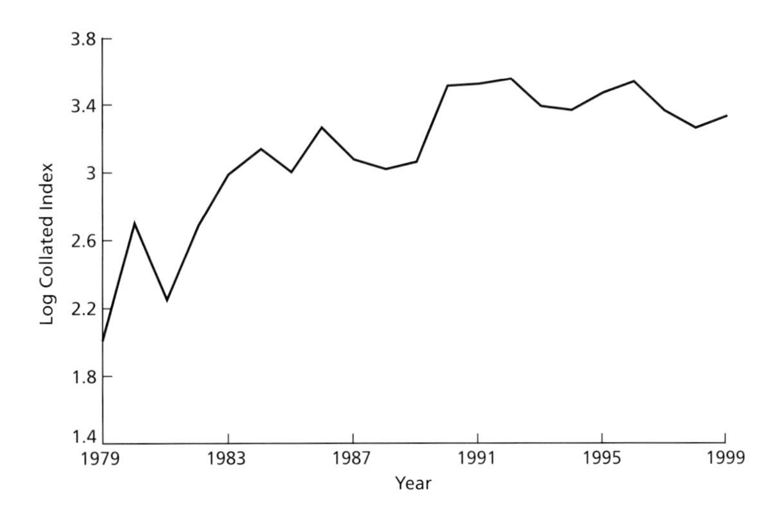

Abundance of the Small Skipper at sites outside the range of the Essex Skipper (i.e. in northern Britain, south-west England, and Wales).

Key references

H. A. Ellis (1999*a*); Goulson *et al.* (1997*a*,*b*); Hardy *et al.* (1993); Murray and Souter (1999); Warrington and Brayford (1995).

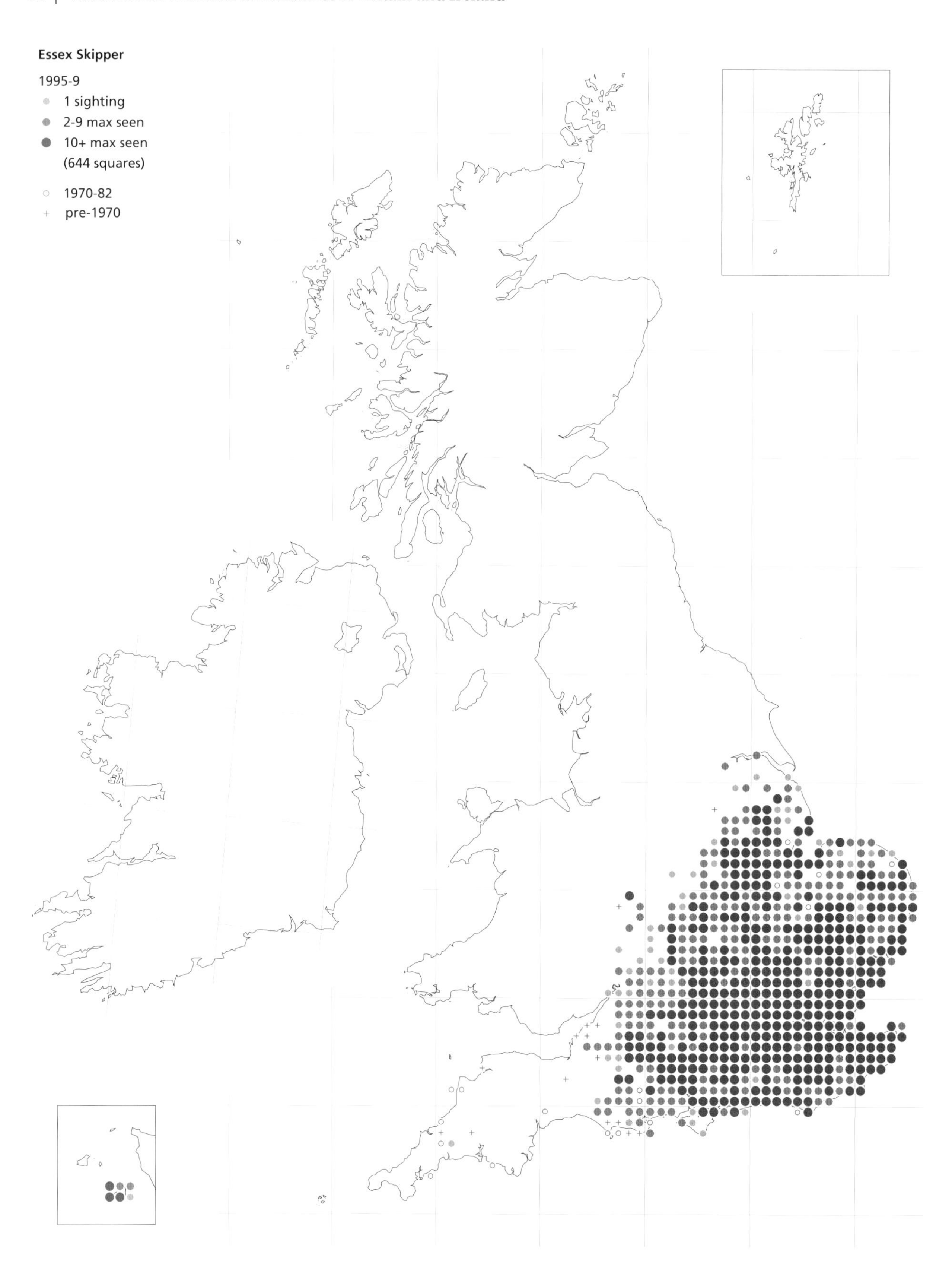
Essex Skipper
1995-9
1 sighting
2-9 max seen
10+ max seen
(644 squares)
1970-82
pre-1970

Essex Skipper *Thymelicus lineola*

Resident
Range expanding.

Conservation status
UK BAP status: not listed.
Butterfly Conservation priority: low.
European status: not threatened.

European/world range
Much of Europe, including southern Scandinavia, eastwards across much of Asia and North Africa. Distribution expanding at the northern edge of its range in Europe. Introduced to North America in 1910, since when it has spread widely and become a pest of hay crops.

Essex Skipper butterflies closely resemble and are often found in company with Small Skippers. Because of the similarities, the Essex Skipper has been overlooked both in terms of recording and ecological study, and it was the last British resident species to be described (in 1889). The simplest means of distinguishing between the two species in the field is by examining the undersides of the tips of their club-shaped antennae; they are black in the Essex Skipper and orange or brown in the Small Skipper. However misidentifications still occur. The distribution of the Essex Skipper in Britain has more than doubled in the last few decades.

Foodplants
The main species used is Cock's-foot (*Dactylis glomerata*), although the butterfly may use several other grasses including Creeping Soft-grass (*Holcus mollis*), Common Couch (*Elytrigia repens*), Timothy (*Phleum pratense*), Meadow Foxtail (*Alopecurus pratensis*), False Brome (*Brachypodium sylvaticum*), and Tor-grass (*B. pinnatum*). It rarely uses Yorkshire-fog (*Holcus lanatus*), the preferred foodplant of the Small Skipper.

Habitat
The Essex Skipper is found in tall, dry grasslands in open sunny situations, especially roadside verges, woodland rides, and acid grasslands, as well as coastal marshes.

Life cycle and colony structure
The species has a single generation each year and the flight period overlaps with that of the Small Skipper, although adult Essex Skippers typically emerge a week or so later. Peak numbers are normally recorded in late July or early August and very few adults survive through to September.

Females lay their **eggs** in small batches in the leaf sheaths of grasses, tending to select sheaths that are compact and tightly furled. This preference may explain why the Essex Skipper avoids laying on Yorkshire-fog, as its leaf sheaths are less firm than those of the other foodplants. The eggs do not hatch until the following year, whereas Small Skipper larvae hatch before entering hibernation. This difference may account for the Small Skipper's slightly earlier flight period. In spring, Essex Skipper **larvae** hatch and commence feeding. Within a few days each larva spins together the edges of a leaf blade to construct a protective tube, from which it emerges to feed. **Pupation** occurs in a silk cocoon enclosed within a tent of leaves at the base of the foodplant.

The Essex Skipper appears to be a relatively sedentary butterfly, living in discrete colonies, sometimes on small habitat patches where adult numbers are low. Nevertheless the species has exhibited significant colonization

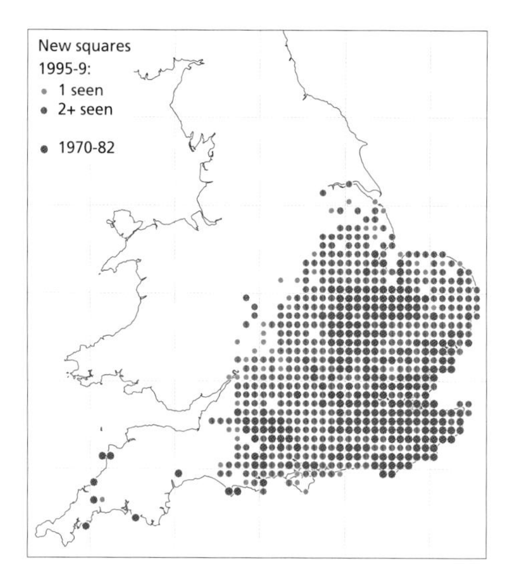

The expansion of range of the Essex Skipper since 1970–82.

abilities in recent decades. The mobility of adults differs markedly between the sexes. Females tend to be more sedentary than males and make slow flights to select sites for egg-laying. The more mobile males are thought to set up territories, which are defended from a favourite perch. In Canada males use a patrolling strategy to locate females. Essex Skipper males in Britain may use both strategies, as has been shown for the Large Skipper. Males may patrol throughout the colony area when temperatures allow and switch to perching within territories when sustained flight is impossible.

Distribution and trends

The Essex Skipper occurs only in England and Jersey. The species is well distributed south of a line drawn between the Severn and Humber estuaries, but is scarcer in the west. The recent record in Cornwall is believed to be the result of an introduction.

The historical distribution is difficult to determine because of the comparatively recent discovery of the species and problems with identification. The species probably was already well distributed in south-east England at the time of its description in the late nineteenth century, although this view has been challenged. Records from Essex made around the time of the butterfly's discovery apparently suggest an almost exclusively coastal distribution. Irrespective of its former distribution, the Essex Skipper has undergone a dramatic expansion of its range in recent years.

Comparison of 1995–9 data with those from the 1970–82 survey shows that the Essex Skipper has been recorded in nearly 400 new 10 km squares. This represents an increase in its distribution of 139% at this level of resolution (the largest percentage increase of any species recorded by the current survey). Despite the identification problems posed by this species, much of this expansion is undoubtedly real and not a result of previous under-recording.

The main recent directions of expansion have been northwards, through north Suffolk, Norfolk, and Lincolnshire, and westwards, through Leicestershire, Warwickshire, Oxfordshire, and Buckinghamshire. Many other counties have also experienced increases in the distribution of the Essex Skipper (e.g. Dorset, Hampshire, Wiltshire, and Sussex) and several have declared their first records of the species (e.g. Warwickshire in 1992, Nottinghamshire in 1995, Yorkshire in 1996, and Derbyshire in 1999).

The Butterfly Monitoring Scheme (BMS) does not produce an annual index for the Essex Skipper. The combined Small and Essex Skipper index has increased significantly since 1976, but it is impossible to ascribe this trend to one or both species. However, the transect scheme operated in the Netherlands does separate the two species and the Essex Skipper has shown increases in annual index figures in recent years, as well as a significant expansion of range. In fact the graph of Essex Skipper collated indices from the Netherlands since 1990 closely resembles the BMS plot for the combined species. This apparent synchrony suggests that a common causal factor may be operating in both countries.

Identification problems have complicated the study of this species. Nevertheless, because of careful recording in recent years, the distribution map for the Essex Skipper is believed to reflect a genuine expansion. Most of the counties into which the species has moved have good recording coverage. In addition, local survey co-ordinators and records centres have adopted a sensibly cautious approach to the acceptance of records of the Essex Skipper.

European trend

The species is stable in many countries, but has expanded at the northern edge of its range, notably in Finland and the Netherlands (>100% increase in the past 25 years) and Latvia, Norway, Sweden, and parts of Russia (25–100% increase). Elsewhere there have been some recent declines, particularly in Austria and Romania (>75% decrease in 25 years).

Interpretation and outlook

The recent expansion of the Essex Skipper has been marked, but remains unexplained. Its preferred foodplant is widespread and it breeds in a broad range of habitats. The butterfly has probably benefited from reduced grazing levels, particularly due to the decline of rabbit populations from the mid-1950s, agricultural set-aside, and reductions in roadside and trackside verge management by local authorities and railway companies.

There is increasing evidence that roadside verges are facilitating the movement of the species and many of the new colonies discovered in Dorset are on roadside verges. This theory has also been proposed in the Netherlands, although the butterfly's expansion was well underway before such management changes occurred there.

However, it seems unlikely that a lack of suitable contiguous habitat was limiting the range of the Essex Skipper in the past. Three other factors have been identified that may, singly or in combination, be driving the expansion of the species. One possibility is that the Essex Skipper is a relative newcomer to Britain, and has been expanding following a colonization or introduction event. Secondly, climatic factors may be involved. The synchronous expansion of distribution in the Netherlands and Scandinavia adds weight to such an explanation, but little work has been done to ascertain what these factors might be. A third and intriguing possibility is that human activity, specifically the long-distance road haulage of hay, may be inadvertently providing a dispersal mechanism for this rather sedentary butterfly. This theory is supported by evidence from North America, where large numbers of hibernating Essex Skipper eggs have been found in hay loads, but its impact in Britain is unknown.

Whatever the true story of this enigmatic little butterfly may be, it is clearly faring well at present and its outlook is favourable.

Key references

Murray and Souter (1999, 2000); Pivnick and McNeil (1985).

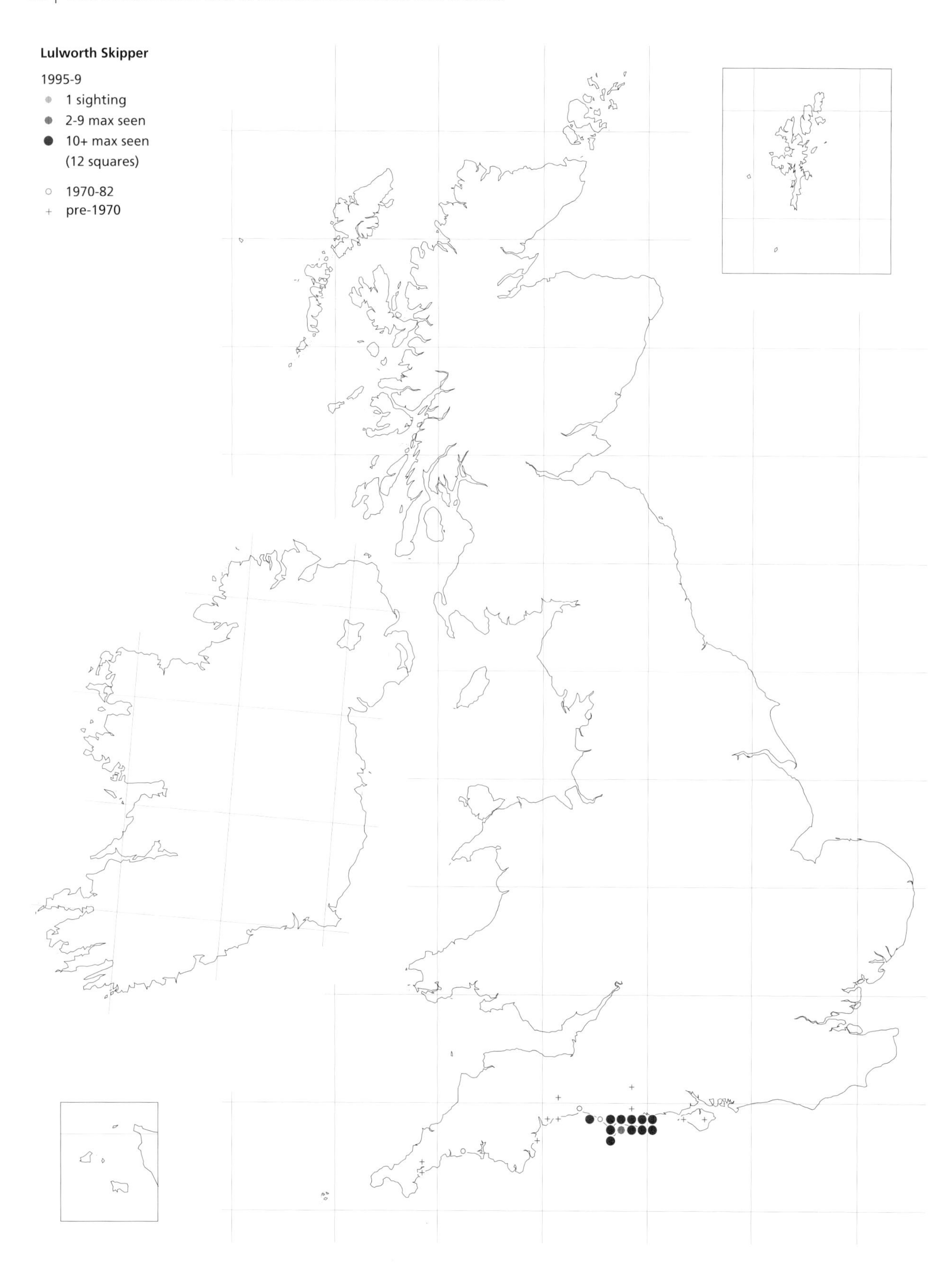
Lulworth Skipper
1995-9
1 sighting
2-9 max seen
10+ max seen
(12 squares)
1970-82
pre-1970

Lulworth Skipper *Thymelicus acteon*

Resident
Range stable.

Conservation status
UK BAP status: Species of Conservation Concern.
Butterfly Conservation priority: medium.
European status: vulnerable.
Protected in Great Britain for sale only.

European/world range
Occurs locally across south and central Europe through to Asia Minor and also in North Africa. Severe decline in the north of Europe but still locally abundant in lowland calcareous grasslands in the south.

The Lulworth Skipper is one of the smallest of our butterflies. It is restricted to the extreme south of Dorset where it can be found in large numbers along a stretch of coast centred on the village of Lulworth, where the species was first discovered in 1832. The females can be distinguished from other skippers by the pale orange 'sun-ray' markings on their forewings whereas the males have darker-brown, almost olive coloured wings. The range of the Lulworth Skipper has changed little in recent decades and it remains locally very abundant.

Foodplants
The butterfly breeds on tall patches of Tor-grass (*Brachypodium pinnatum*).

Habitat
The main habitats are unfertilized calcareous grasslands, including chalk downland, coastal grassland, and undercliffs. The butterfly also occasionally uses grasslands on calcareous clays, or even road verges where chalk or limestone ballast has been used in construction. Females prefer to lay eggs on tall foodplants (30–50 cm), and only rarely select patches 10–30 cm tall (never patches under 10 cm). Most colonies are on south-facing slopes and grasslands sheltered from onshore winds.

Life cycle and colony structure
The Lulworth Skipper has a single brood and adults fly from the end of June until late September with a definite peak in August. The **eggs** are laid in groups of up to 15 which are deposited deep in the flower sheaths of Tor-grass. The **larvae** hatch after about three weeks and immediately go into hibernation, spinning a small silken cocoon around the remains of the eggshell. They emerge during the spring and live within leaf tubes, formed by spinning together the edges of the broad leaf-blades. The larvae feed only at night, making characteristic V-shaped notches below the tube, leaving the mid-rib, and eating the leaf blade above the tube. These notches possibly maintain nutrients in the damaged leaf and make it more palatable. The larvae move and construct new shelters as they grow, but during their final instar they sometimes rest openly on the leaf blades. They **pupate** deep within the tussocks of Tor-grass, surrounded by loose cocoons of grass and silk.

Adults fly only in strong sunshine and form discrete colonies that can be very large, containing as many as 100 000 individuals. Adults rarely stray from favoured breeding areas; in 1978, of the patches of suitable habitat more than 3 km from a population in south Dorset, half had not been colonized. However, some dispersal must occur, as shown by the colonization of chalk ballast across heathland (4 km away from traditional

Lulworth Skipper larva on grass blade, showing characteristic feeding notches.

sites) and restored grassland at Durlston near Swanage.

Distribution and trends

The Lulworth Skipper is now found only in the far south of Dorset, with most colonies occurring on the coast between Weymouth and Swanage, and on the line of inland chalk hills known as the Purbeck Ridge. There are two outlying colonies on the west Dorset coast near Burton Bradstock and during the late 1990s a colony was discovered for the first time on the southern tip of the Isle of Portland. This area has been thoroughly surveyed in the past and the colony may be the result of natural colonization, or of released specimens.

Colonies formerly occurred in south Devon but apart from two single records, the Lulworth Skipper has not been seen in the county since the 1930s. There are also records for Cornwall but these may not represent native colonies. A single specimen was caught near Polperro in 1979 and is now presumed to be a migrant, an escape, or a deliberate release. There are a few other records of single individuals elsewhere but these are also thought to be releases and have not been mapped.

Detailed surveys of all known colonies were conducted in 1978 and by Butterfly Conservation in 1997. The butterfly's distribution has remained remarkably stable and although there was a slight decline in the number of colonies from 84 to 75, this was due to the loss of some very small colonies, mainly on the Purbeck Ridge. The number of medium and large colonies (over 1000 adults at peak) has shown a slight increase from 17 to 20. The stronghold of the species over the last two decades has been on the military ranges and coastal grassland around Lulworth Cove, probably the same as it has been for the last hundred or more years. Away from this coastal strip suitable habitats are smaller and more patchy, and populations tend to be small.

European trend

The Lulworth Skipper's range appears to be fairly stable in southern Europe, but it has declined seriously in many other countries. Notably: the Netherlands (now extinct); Austria, Belgium, Poland, and Romania (>50% decrease in 25 years); and the Czech Republic, Germany, Luxembourg, and Slovakia (25–50% decrease).

Interpretation and outlook

Apart from the losses in Devon, the distribution of the Lulworth Skipper appears to have been fairly stable in recent decades, limited by the butterfly's need for a warm climate and a specialized, tall grass habitat. It shows no sign of declining in Britain as it has done in continental Europe, thanks perhaps to some special factors, particularly that its coastal habitats have remained more or less intact. Most occur on slopes that are too steep to plough and many lie within large military training areas that were established in the 1940s. Restrictions on development have also reduced habitat loss in this coastal part of Dorset.

The butterfly appears to have benefited from the decline in grazing pressure and abandonment of southern hill land during the twentieth century. This has allowed Tor-grass both to spread and to grow into the taller tussocks that are essential for breeding. In fact, it seems likely that, at the time of the first thorough survey in 1979, the Lulworth Skipper was more abundant within its limited range than at any other time in its history. Although scrub had invaded and spread on many abandoned sites, most were maintained as open grassland either because of continuous coastal erosion or periodic burning (swaling) of the gorse. However, increasing sward height has been detrimental to short-turf species like the Adonis Blue.

Since 1979, grazing has been re-established or increased on many sites in order to restore the former herb-rich grassland, but this has led only to small local decreases of the Lulworth Skipper. Some populations have also benefited from recent scrub removal. Most of its core areas remain too steep or impractical to graze. Recent research suggests that the Lulworth Skipper will tolerate many types of grazing and that only very heavy grazing or regular burning is detrimental. Some grazing is also beneficial for the species because it breaks up dense mats of Tor-grass and encourages flowers that are used as adult nectar sources.

Traditionally, much of the hill grassland in south Dorset was burnt annually in early spring to produce a flush of grasses that were palatable for sheep. The annual burning and heavy sheep grazing were almost certainly detrimental to the Lulworth Skipper. In recent years swaling has become more sporadic and the butterfly tolerates the lower levels of sheep grazing as they graze selectively around large clumps of Tor-grass. However, because sheep cannot tackle coarse grassland effectively, cattle grazing is being encouraged to reduce scrub and speed the restoration of herb-rich turf. Spring and summer grazing poses the biggest risk to the Lulworth Skipper because larvae are then high up in their grass tubes.

Clearly the strategy for conserving grasslands in this important region should cater for a wide range of plants and animals, but it will have to reconcile their varied requirements. An obvious solution is to create zones with different regimes in different areas, so that the Lulworth Skipper would take preference at least in its core areas.

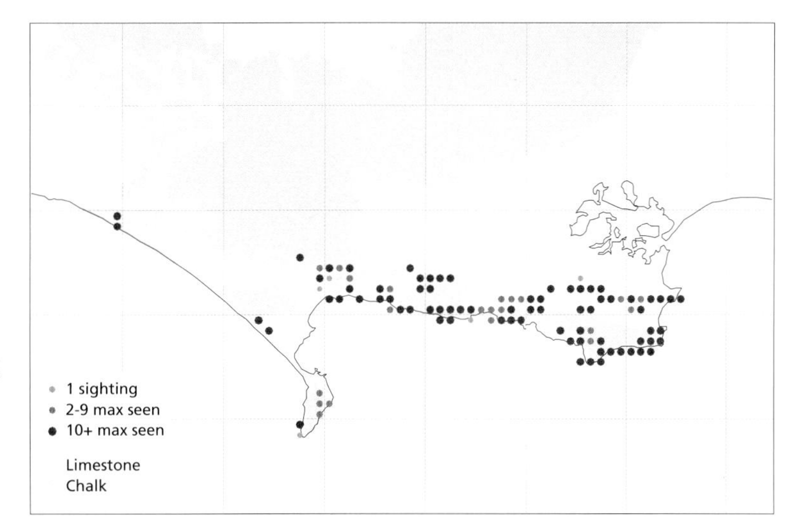

The distribution of 1995–9 records of the Lulworth Skipper at 1 km square resolution in south Dorset, against a background of chalk and limestone geology.

Key references

Bourn *et al.* (2000*b*); Bourn and Warren (1997*a*); Pearman *et al.* (1998); Thomas (1983*b*); Thomas *et al.* (1992).

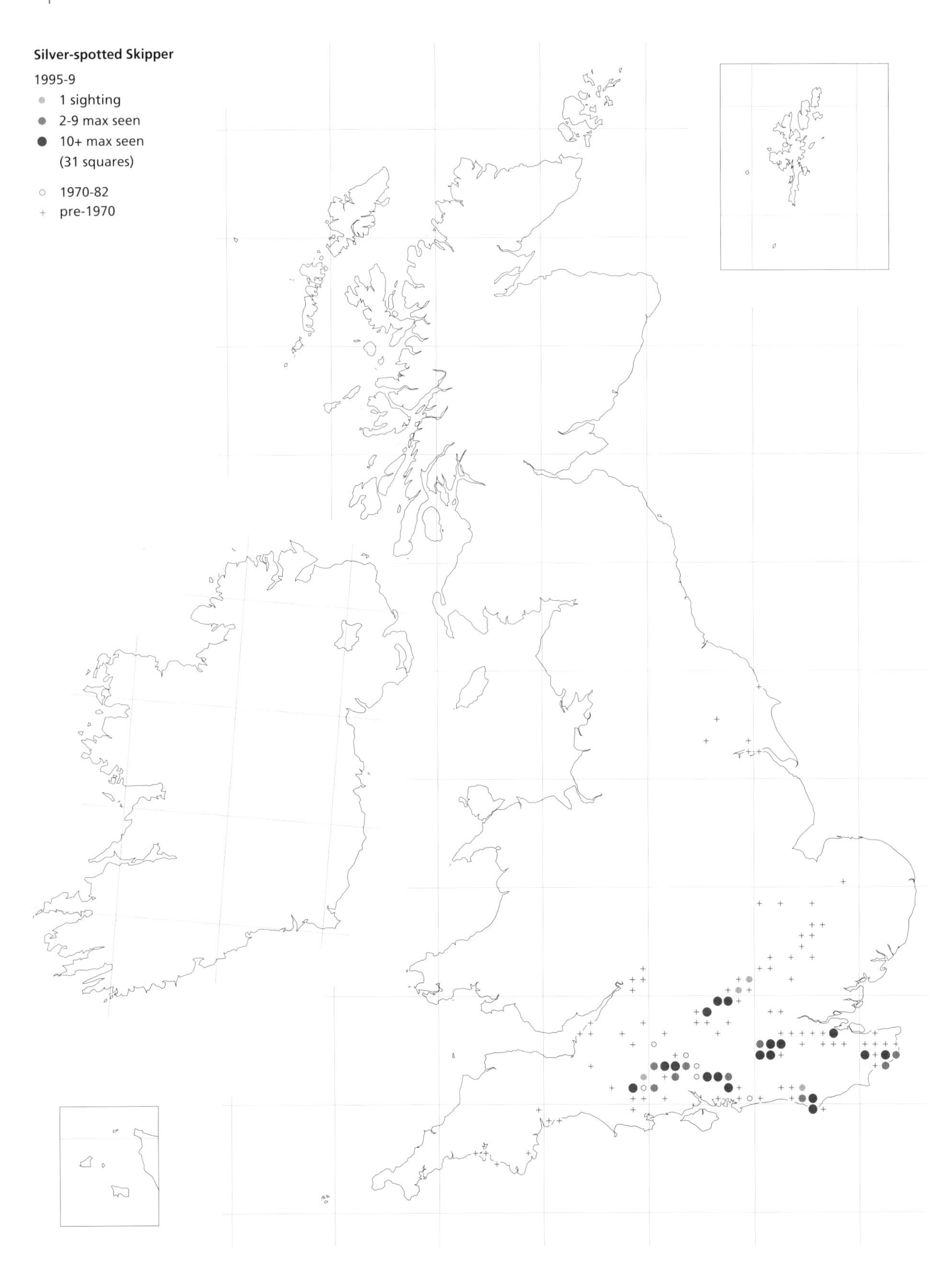

Silver-spotted Skipper
1995-9
1 sighting
2-9 max seen
10+ max seen
(31 squares)
1970-82
pre-1970

Silver-spotted Skipper *Hesperia comma*

Resident

Decline and recent partial re-expansion of range.

Conservation status

UK BAP status: Priority Species.
Butterfly Conservation priority: high.
European status: not threatened.
Protected in Great Britain for sale only.

European/world range

Across Europe from central Spain to Scandinavia and throughout temperate Asia. Also in North Africa and North America. It has declined seriously in several European countries.

This rare skipper is restricted to chalk downs in southern England where it can be seen darting low over short turf, stopping frequently to bask on bare ground or feed on flowers such as Dwarf Thistle. It can be distinguished by the numerous silver-white spots on the undersides of the hind wings, which can be seen quite easily when it rests with wings in a characteristic 'half-open' posture. The Silver-spotted Skipper has declined rapidly over the last 50 years but has re-expanded partially since 1980.

Foodplants

Sheep's-fescue (*Festuca ovina*) is the sole foodplant and the butterfly breeds only where this grows as small tufts in short or broken turf.

Habitat

The butterfly breeds in open chalk grassland which contains patches of short, sparse turf typically on thin soils. It also occurred formerly on limestone grassland in central and northern England. Warm, south facing slopes are preferred, but a few colonies occur on slopes with other aspects, including some gently north-facing ones.

Life cycle and colony structure

The butterfly is single brooded with adults flying from late July to early September. **Eggs** are laid singly on leaf blades of Sheep's-fescue or occasionally on adjacent plants, where they pass the winter. Females are highly selective and lay most of their eggs on small tufts of foodplant growing in short turf (1–4 cm tall) and often next to small patches of bare ground. Most eggs are laid next to animal tracks, old rabbit scrapes, or where there is some erosion on steep slopes or due to grazing animals. However, in hot conditions some eggs are laid on short tufts of Sheep's-fescue growing in more closed turf. The pale cream eggs are quite conspicuous and careful searches have enabled some breeding areas to be mapped accurately.

The **larvae** hatch in early spring and feed in small silken webs around the foodplant. Temperature measurements have shown that the sparse vegetation used for breeding provides warmer conditions, which

Silver-spotted Skipper eggs on a tuft of Sheep's-fescue in sparse turf.

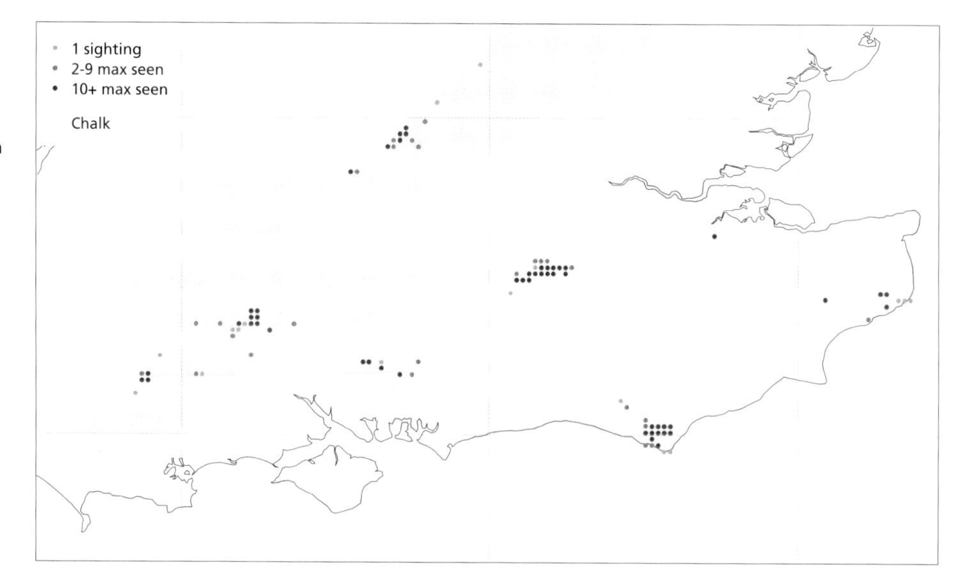

The distribution of 1995–9 records of the Silver-spotted Skipper at 2 km square resolution against a background of the chalk geology of southern England.

presumably aid larval and pupal development. They **pupate** at ground level, deep within small grass tussocks and surrounded by loose silken cocoons.

Recent detailed studies have shown that although adults tend to concentrate on suitable breeding habitat, the butterfly has reasonable powers of colonization. Since 1980, it has colonized several new sites within 1 km of existing populations, and occasionally sites up to 8.5 km away. However, large areas of habitat are more likely to be colonized than small ones and the rate of colonization tails off rapidly with distance. Small populations have also been shown to have particularly high emigration (and often immigration) rates and cannot be regarded as self-sustaining colonies. The butterfly is thus thought to form metapopulations covering relatively large areas of countryside within which it breeds in numerous discrete habitat patches. Some of these are larger and more permanently occupied core sites while others are occupied more temporarily.

Distribution and trends

The Silver-spotted Skipper is a rare butterfly, found only on chalk downland in southern England. Its range declined substantially during the twentieth century and it has become extinct in many areas including Yorkshire, the Cotswolds, the northern part of the Chilterns, and much of the North Downs. Its decline began early in the twentieth century but was particularly severe during the 1960s and 1970s. It has since re-expanded a little in the late 1980s and 1990s, mainly in south-east England.

An intensive national survey of the species, conducted in 1982, found that it had been reduced to 49 more or less distinct sites. Over half of these contained very small colonies (<200 adults at peak) and only 9 were large (>3500 at peak). A subsequent survey of the North and South Downs in 1991 found that the number of occupied habitat patches in this region of south-east England had increased by about 30%. The data for 1995–9 show that the butterfly has continued to expand and has been recorded in 48% more 10 km grid squares than in 1970–82. However, sites further than about 10 km from any refuge sites have not been colonized and the butterfly is still absent from much of its former distribution.

There are currently eight main centres of distribution: north Dorset; south Wiltshire; east and west Hampshire; Surrey; east Sussex; east Kent; and the southern part of the Chilterns. Attempts have been made to re-establish the butterfly at two nature reserves in Kent where habitat management has been improved but where natural re-colonization was thought to be unlikely.

European trend

The range of the Silver-spotted Skipper appears to be stable in many European countries but it has declined seriously in Belgium, Latvia, and Luxembourg (>50% decrease in 25 years), and also in Denmark, Germany, the Netherlands, Romania, and Sweden (25–50% in 25 years).

Interpretation and outlook

The Silver-spotted Skipper is at the north-west limit of its European range in Britain and requires some of the hottest and driest grasslands available, laying its eggs on foodplants growing in the warmest positions. This restricts it to the warmer climate of southern England and to shorter swards where it usually relies on grazing by livestock or rabbits.

Many of its habitats have been lost due to agricultural improvement but others are too steep to plough and have become unsuitable following the cessation of grazing. The impact of abandonment was ameliorated to some extent by heavy rabbit grazing which maintained patches of short turf on many sites. However, the introduction of myxomatosis in the early 1950s severely reduced rabbit populations and led to a sudden decline of the butterfly, which reached a low point in the 1970s.

The partial recovery in the butterfly's distribution during the 1980s and 1990s was due to an increase in habitat suitability on existing and former sites, caused partly by specific conservation management but also by the recovery of rabbit populations. The re-expansion has been accompanied by, and probably assisted by, a steady increase in population sizes on many extant sites over the last 20 years. Many conservation organizations have made huge efforts to restore grazing on neglected downland and there are signs that schemes such as Countryside Stewardship and Environmentally Sensitive Areas are helping the recovery of this rare skipper. However, many colonies still rely solely on rabbit grazing. This is a risky situation as rabbit numbers are notoriously unpredictable and difficult to control. Moreover, rabbit populations may be reduced in the future by new strains of myxomatosis and new diseases such as viral haemorrhagic disease.

An important part of the Species Action Plan for the Silver-spotted Skipper is to restore stock grazing and reduce the butterfly's dependency on rabbits. Ideal grazing regimes are those producing abundant short turf (1–4 cm tall) with plenty of small patches of broken turf. This can be achieved either by cattle or sheep grazing, though cattle are preferable because they are heavier and their hooves create more broken turf. The land must not be grazed too heavily with sheep in late summer, as they can remove all the flower heads needed as nectar sources by the adults and the butterfly rarely lays on heavily nibbled tufts of foodplant. The butterfly often does well under rotational systems of grazing where parts of the site are grazed heavily and then left for a few months or even a whole year.

The Silver-spotted Skipper also requires the maintenance of extensive networks of habitat patches which are capable of supporting the butterfly in the long term. There is strong evidence to show that it does not survive well on small, isolated sites, and it may be essential for networks to contain one or more large populations. An effective conservation strategy needs to consider the landscape as a whole and the relative position of existing and potential sites.

Key references

Barnett and Warren (1995*a*); Hill *et al.* (1996); Thomas *et al.* (1986); Thomas and Jones (1993); Warren *et al.* (1999); Wilson and Bourn (1998).

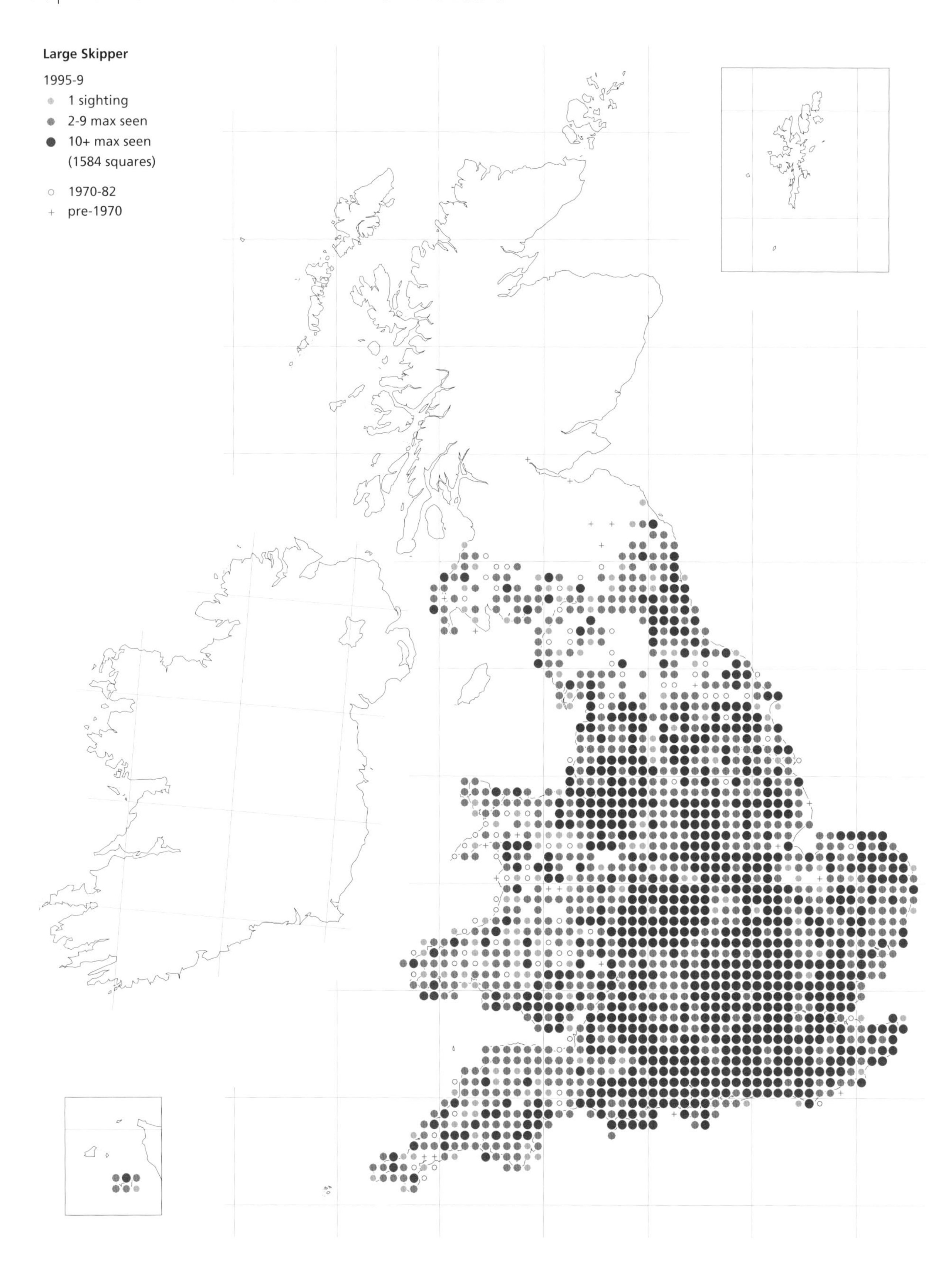
Large Skipper
1995-9
1 sighting
2-9 max seen
10+ max seen
(1584 squares)
1970-82
pre-1970

Large Skipper *Ochlodes venata*

Resident
Range expanding.

Conservation status
UK BAP status: not listed.
Butterfly Conservation priority: low.
European status: not threatened.

European/world range
Widespread through most of Europe, occurring as far north as latitude 64°N in Sweden, and extending eastwards throughout Asia to China and Japan. Stable in most European countries.

Male Large Skippers are most often found perching in a prominent, sunny position, usually on a large leaf at a boundary between taller and shorter vegetation, awaiting passing females. Females are less conspicuous, though both sexes may be seen feeding on flowers, Bramble being a favourite. The presence of a faint chequered pattern on both sides of the wings distinguishes this species from the similar Small and Essex Skippers, which fly at the same time. The Large Skipper is widespread in southern Britain and its range has extended northwards in north-east England since the 1960s.

Foodplants
Cock's-foot (*Dactylis glomerata*) and occasionally Purple Moor-grass (*Molinia caerulea*) and False Brome (*Brachypodium sylvaticum*) are used. Females have been observed laying eggs on Tor-grass (*B. pinnatum*) and Wood Small-reed (*Calamagrostis epigejos*).

Habitat
This butterfly favours grassy areas, where foodplants grow in sheltered, often damp, situations and remain tall and uncut. It is found in a wide variety of habitats where there are shrubs, tall herbs, and grasses, for example woodland rides and clearings, pastures, roadside verges, hedgerows, and wet heathland. It is also a species of urban habitats, occurring in parks, churchyards, and other places with long grasses.

Life cycle and colony structure
There is one generation a year. Adults begin to emerge in the final days of May and numbers reach a peak in early July. Some individuals may be seen as late as the end of August.

Males use both perching and patrolling strategies for mating. On sunny days they usually patrol in mid- and late morning, searching for females, whereas in the early morning and in the afternoon the males perch and await passing mates. A vantage point on an edge between vegetation of different types and heights is chosen, often close to nectar sources that will attract females. It is not unusual for the same vantage points to be used by different males in the same season and for the same perches to be used from year to year. This mate-location behaviour may change to suit circumstances; for example if females are scarce, patrolling may be abandoned in favour of perching.

Females spend much of their time basking, with short periods of nectaring and egg-laying. **Eggs** are laid singly on the undersides of leaves of foodplants growing in sunny, sheltered spots, and hatch after about two weeks. Newly-hatched **larvae** are concealed in tubes formed by joining the

Left:
Large Skipper eggs.

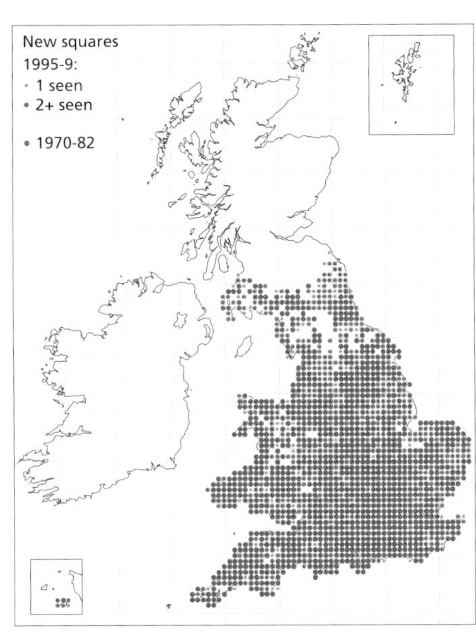

Right:
The expansion of range of the Large Skipper since 1970–82.

edges of a leaf blade together with silk, emerging from the tube only to feed. Overwintering takes place in a hibernaculum constructed in a similar way and feeding is resumed in the spring. **Pupae** are formed among leaf blades in a silk cocoon several centimetres above ground.

Large Skippers live in discrete colonies and often occur at the same sites as Small Skippers, which begin to emerge about two weeks later. Many colonies contain only a few dozen adults in a season, but where conditions are suitable, much larger populations may develop, for instance in the first few years after clearing of woodland. Little is known about their mobility, but their recent spread in north-east England indicates that they have reasonable powers of dispersal.

Distribution and trends

Though widespread in England and Wales, the Large Skipper is absent from Ireland, the Isle of Man, most of Scotland, and most offshore islands. It occurs on Jersey but not the other Channel Islands. It is absent from the highest ground in upland areas such as the Cheviots and Pennines, the Lake District, Dartmoor, and the Welsh mountains.

It extends into low-lying areas of south-west Scotland (Ayrshire and Dumfries and Galloway). Here its distribution appears to be stable and there has been no recent spread northwards. In the north-east of England it became more widespread and common in Co. Durham in the 1960s, and in the mid-1970s it began to be recorded in Northumberland, reaching the north of the county in the 1980s. It has since been seen in more locations and greater numbers.

In the current survey the species has been recorded in new 10 km squares in other northern parts of its range, mainly in Yorkshire and Lancashire, but it is not clear if this represents genuine expansion or previous under-recording. Overall, the number of recorded 10 km squares has increased by 31% since the 1970–82 survey.

Despite the recent range expansion in the north-east, colonies continue to be lost throughout its range in Britain. A study in north Wales has shown that the flight area of the Large Skipper has declined by over 70% since 1901. The rate of loss in other areas is likely to be similar. Intensive land use can result in the removal of the native grasses that the butterfly requires, while hedgerow removal and regular cutting alongside hedgerows and of verges also have an adverse effect.

European trend

The range of the Large Skipper is stable in most countries, but it has increased slightly in the Netherlands in the last 25 years.

Interpretation and outlook

It is not clear why the Large Skipper spread northwards in Northumberland in the latter part of the twentieth century. Increased availability of suitable habitat and/or more favourable climatic conditions may be responsible. Some colonization has taken place at old industrial sites, where spoil heaps, disused quarries, opencast mining sites and dismantled rail tracks have become over-grown, creating suitable habitat. However, it has also recently colonized habitats such as roadside verges, churchyards, hedgerows, ditches, and other grassy places that have been available for many years, which suggests that the expansion may be due to climatic changes. The recent availability of ruderal sites due to land use changes may provide a more suitable network of habitat, with improved opportunities for progressive colonization as the climate improves. In western Scotland, where there has been less change in land use, there has not been a similar range expansion. This suggests that the change in the north-east is not entirely due to climatic factors.

Data from the Butterfly Monitoring Scheme show a small but significant increase in the all-site index, but a mixed picture on individual sites. Although colonizations have taken place at four sites in the scheme in Norfolk, Suffolk, and Essex, increases in numbers at about a dozen sites have been balanced by a similar number of decreases. Coarse grasses may have increased at a number of monitored sites, especially in eastern England, in recent years, to the benefit of this species. This is not the case, however, in the countryside as a whole, where breeding habitats continue to be lost.

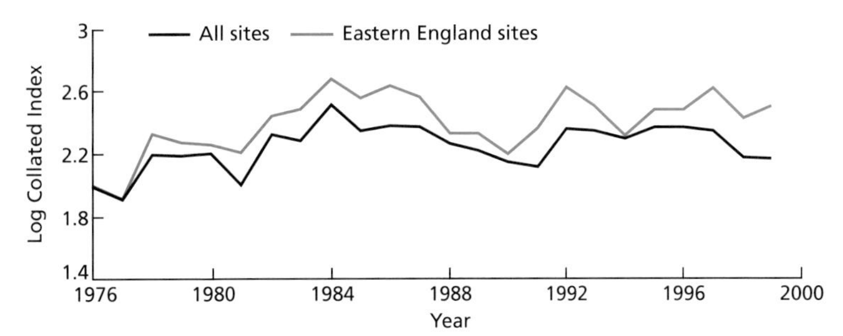

BMS collated index of abundance shows the generally higher abundance of the Large Skipper at sites in eastern England.

The Large Skipper is widespread and mobile enough in England and Wales to be able to colonize patches of land where vegetation becomes suitable for breeding. There is much scope for encouraging populations in extended field margins and conservation headlands on farmland, and in other situations, by allowing areas of native grasses to flourish. The rough grasses of set-aside land seem to provide suitable, if temporary, habitat for this butterfly, which can colonize relatively small areas. Afforestation can also create suitable breeding conditions as coarse grasses grow around newly planted trees, but this habitat is short-lived, becoming shaded out as the trees mature.

As eggs, larvae, and pupae are all found among the leaves of tall grasses, the Large Skipper is vulnerable to grass cutting at all times of year. It is thus essential that cutting of woodland rides, conservation headlands, and other situations where it breeds should be carried out on rotation, always leaving a large proportion of the vegetation uncut in any year.

Key references

Dennis and Williams (1987); Ellis (1998); Pollard *et al.* (1998*b*).

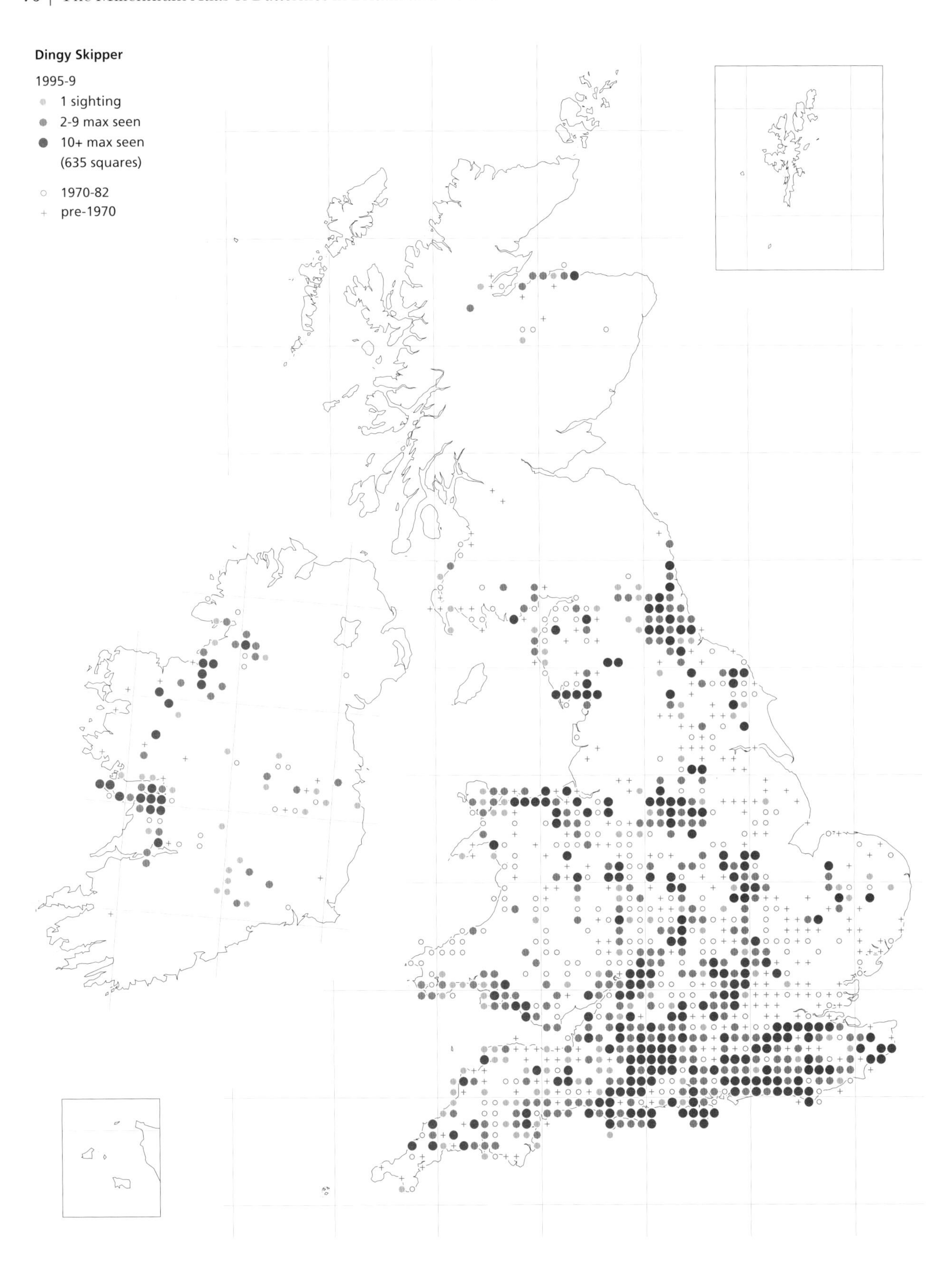
Dingy Skipper
1995-9
1 sighting
2-9 max seen
10+ max seen
(635 squares)
1970-82
pre-1970

Dingy Skipper *Erynnis tages*

Resident

Range declining.

Conservation status

UK BAP status: not listed.
Butterfly Conservation priority: low.
European status: not threatened.
Protected in Northern Ireland.

European/world range

Occurs widely in Europe to latitude 62°N and eastwards through Asia to China. Declining in several European countries.

In sunshine, the Dingy Skipper often basks on bare ground with wings spread wide. In dull weather, and at night, it perches on the tops of dead flowerheads in a moth-like fashion with wings curved in a position not seen in any other British butterfly. This small brown and grey butterfly is extremely well camouflaged. It may be confused with the Grizzled Skipper, the Mother Shipton moth, and Burnet Companion moth, which sometimes occur on the same sites at the same time. The Dingy Skipper is locally distributed throughout Britain and Ireland, but has declined seriously in recent years.

Foodplants

Common Bird's-foot-trefoil (*Lotus corniculatus*) is the usual foodplant in all habitats. Horseshoe Vetch (*Hippocrepis comosa*) is also used on calcareous soils, and Greater Bird's-foot-trefoil (*L. pedunculatus*) is used on heavier soils.

Habitat

Colonies occur in a wide range of open, sunny habitats including chalk downland, woodland rides and clearings, coastal habitats such as dunes and undercliffs, heathland, old quarries, railway lines, and waste ground. Suitable conditions occur where foodplants grow in a sparse sward, often with patches of bare ground in a sunny, sheltered situation. Taller vegetation is also required for shelter and roosting.

Life cycle and colony structure

There is one generation a year, but in hot summers there may be a partial second brood at some sites. Adults usually fly from early May until the end of June but they can begin to emerge as early as mid-April in warm springs such as in 1997. The few second brood adults emerge in August (see p. 391).

Eggs are laid singly on young leaves of the foodplants and a recent study has shown that females choose the longest shoots of large plants growing in sheltered situations. The **larvae** hide in tents formed by spinning leaves of the foodplant together and feed through the summer months. When fully grown in August, each larva spins more leaves together to form a hibernaculum in which to spend the winter. **Pupation** occurs the following spring in the hibernaculum, without further feeding.

The Dingy Skipper occurs in discrete colonies, many of which are very small, containing fewer than 50 individuals at the peak of the flight period. It is a sedentary species and unlikely to colonize new areas of habitat unless they are close to existing populations. However, observations of natural colonizations reveal that a few individuals may move several kilometres.

Dingy Skipper eggs on Bird's-foot-trefoil.

Distribution and trends

This is the most widespread skipper in Britain and Ireland, although it is more localized and patchily distributed than the Large and Small Skippers. It occurs further north in Scotland than the Chequered Skipper and is the only skipper found in Ireland. Irish colonies are frequently found in limestone areas on rough ground, notably eskers, and are widely scattered across the midlands from Co. Dublin to eastern Co. Mayo and in the Burren area of Co. Clare.

In Scotland, there are a few colonies in the north-east, mainly occurring on coastal dunes between Inverness and Banff. These are widely separated from colonies in the south-west in Dumfries and Galloway. In Wales, they are most often found in coastal habitats and rarely occur above 100 m. The central and southern counties of England, extending northwards to Derbyshire, form the stronghold of the species, with especially large colonies occurring on extensive areas of chalk downland and undercliff. It is absent from most offshore islands, the Channel Islands, and the Isle of Man.

Because many colonies are small, they may be overlooked, particularly as the butterfly is so inconspicuous. There is no doubt, however, that it has declined considerably in recent years. The current survey shows that it has probably disappeared from almost 40% of the 10 km squares in which it was recorded in 1970–82. The number of colonies has also probably decreased in many 10 km squares where the butterfly is still present. Never common in eastern England, this species became extinct in Essex around 1990 and is now very scarce in East Anglia and Lincolnshire.

In its southern stronghold of Wiltshire, Hampshire, Surrey, and Sussex, the Dingy Skipper is widespread on downland, but is scarce and declining in woodland and other habitats. In other areas, especially in Ireland, Scotland, and Wales, there appear to have been losses since the 1980s, though under-recording of the species makes the true situation difficult to interpret.

Data from Butterfly Monitoring Scheme (BMS) sites show a long-term decline in numbers since 1976, greater than for most other species. Populations are lower after a poor summer and rise after a warm one, though prolonged drought can reduce population size the following season, as happened after the 1976 drought. In 1997, numbers increased at many sites with widespread records of second brood adults in August. In that year, the spring was unusually warm and adults began to emerge in the second week of April in some counties.

European trend

The Dingy Skipper is stable through most of Europe but severe declines have occurred in several countries, especially in the Netherlands (>75% decrease in 25 years) and Belgium, Denmark, and European Turkey (50–75% decrease).

Interpretation and outlook

Many colonies have been lost during the dramatic changes in land use of recent decades. Ploughing and the replacement of native vegetation with Perennial Rye-grass and arable crops have resulted in extinctions, which have also occurred where old quarries have been filled in and areas of waste ground developed. Changes in forestry practices, especially the cessation of coppicing, have led to shadier woods without suitable open areas. Many colonies in woodland are now very small and further extinctions are likely. In southern counties of England, the development of scrub on downland has led to losses of some colonies and elsewhere intense grazing in recent years also may have caused declines.

A study in north Wales has shown that the species survives only in areas with large patches of suitable habitat that are close together, whereas small and isolated habitat patches are often not occupied. However, because surviving colonies are so small, few dispersing individuals are available to colonize any new sites that become suitable, and consequently these may remain empty for many years. The conservation of this species depends on maintaining networks of habitat patches over large areas, so that the extinctions expected to occur in small populations, may be balanced by recolonizations.

Apart from naturally disturbed coastal areas such as undercliffs and dunes, the habitats used by the Dingy Skipper require management, usually grazing of grassland or active management of rides, glades, and felled areas in woodland. These activities are often carried out only for conservation purposes. Scrub clearance on remaining fragments of chalk downland has extended the habitat on some sites, but this has done little to offset the losses elsewhere. Management that maintains and increases sparse vegetation and especially bare ground, which is readily colonized by Common Bird's-foot-trefoil, is especially beneficial for this species. The retention of taller vegetation for shelter and roosting is also important.

The need for long, ungrazed shoots of the foodplant means that constant heavy grazing is deleterious for the Dingy Skipper. Rotational grazing, which provides a fresh supply of successional habitats, but allows development of the required growth form of the foodplants, is the best regime. In north

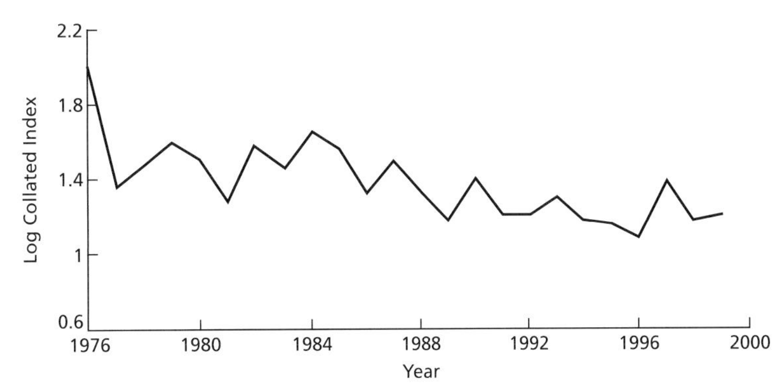

BMS collated index of abundance, showing the decline of the Dingy Skipper.

Wales, colonies have been found to be restricted to relatively ungrazed areas and it has been suggested that livestock may inadvertently eat eggs and larvae on prominent shoots. Most BMS sites are being actively managed for conservation and the declines shown by this scheme may be due partly to increased grazing since monitoring began.

Despite the declines seen during the twentieth century, much remains to be learned about the Dingy Skipper's ecology and the most beneficial management regimes. Its conservation status is currently low (except in Northern Ireland) but this will have to be revised in the light of the rapid declines identified in the current survey.

The recent research on its mobility and habitat requirements sheds some light on why the Dingy Skipper is far less widely distributed than the Common Blue, which uses the same foodplant. It is hoped that this improved knowledge will help managers prevent deterioration of the butterfly's habitats in future and that the decline can be halted.

Key references

Bourn *et al.* (2000*a*); Gutiérrez *et al.* (1999).

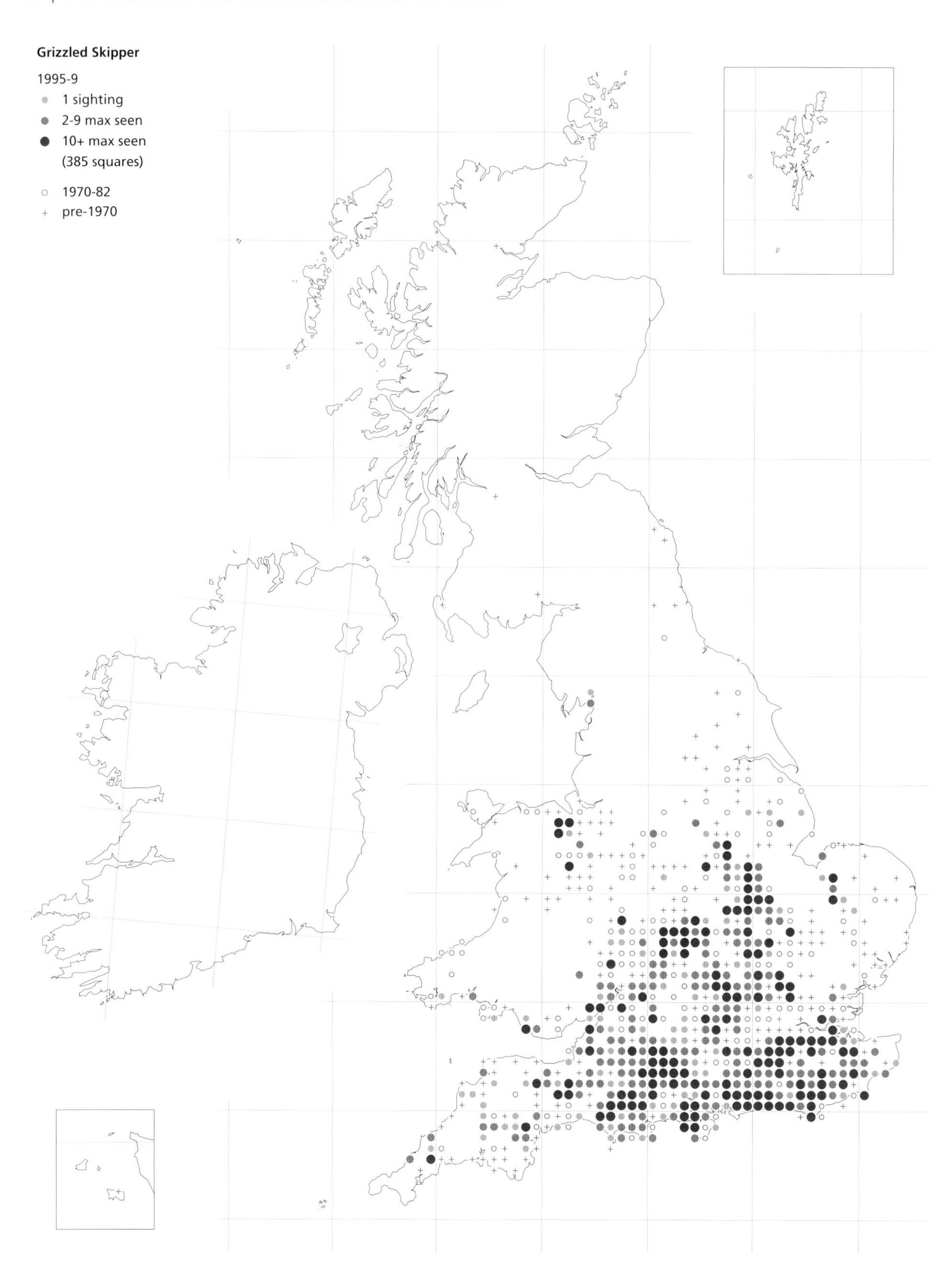
Grizzled Skipper
1995-9
1 sighting
2-9 max seen
10+ max seen
(385 squares)
1970-82
pre-1970

Grizzled Skipper *Pyrgus malvae*

Resident

Range declining.

Conservation status

UK BAP status: not listed.
Butterfly Conservation priority: medium.
European status: not threatened.

European/world range

Occurs widely in Europe, apart from the far north of Scandinavia, and eastwards across to China and Korea. It has declined recently in several European countries.

The Grizzled Skipper is a characteristic spring butterfly of southern chalk downland and other sparsely vegetated habitats. Its rapid, buzzing flight can make it difficult to follow, but it stops regularly either to perch on a prominent twig or to feed on flowers such as Common Bird's-foot-trefoil or Bugle. It can then be identified quite easily by the black and white checkerboard pattern on its wings. The butterfly occurs across southern England, commonly in small colonies, and has declined in several regions, especially away from the chalk.

Foodplants

A variety of plants from the Rosaceae family is used, mainly Agrimony (*Agrimonia eupatoria*), Creeping Cinquefoil (*Potentilla reptans*) and Wild Strawberry (*Fragaria vesca*). It may also use Barren Strawberry (*P. sterilis*), Tormentil (*P. erecta*), Salad Burnet (*Sanguisorba minor*), Bramble (*Rubus fruticosus*), Dog-rose (*Rosa canina*), and Wood Avens (*Geum urbanum*).

Habitat

Three main types are used: woodland rides, glades, and clearings; unimproved grassland, especially chalk downland but also on other calcareous soils including clays; and recently abandoned industrial sites such as disused mineral workings, spoil heaps, railway lines, and even rubbish tips. Occasionally, it breeds on heathland, damp grassland, and dunes. In all habitats it requires plentiful spring nectar plants, at least one of the main foodplants growing in short vegetation (<10 cm) usually with patches of bare ground, and patches of taller vegetation (10–50 cm) and scrub or woodland edges.

Life cycle and colony structure

The Grizzled Skipper is generally single brooded with adults flying from the end of April until mid-June. The emergence time has advanced by almost ten days in the last twenty years and in warm springs it may emerge as early as mid-March, whereas in late years it can fly until mid-July. Very occasionally, there is a small distinct second brood in August, but recent information suggests that there is a more regular partial second generation, which overlaps with the first from late June onwards.

The **eggs** are laid singly on foodplants growing in warm positions, next to either bare ground or short vegetation. They may be highly concentrated and ideal plants can contain over 20 eggs. The **larvae** build a series of tents, formed by spinning together the edges of leaves, which protect them as they grow. They leave these shelters only to make brief feeding visits to nearby leaves or to move and spin new shelters. As they grow, they become more mobile and select lush (more nutrient-rich) plants growing in taller

Grizzled Skipper larva on Wild Strawberry.

vegetation. There is also a tendency to switch to more coarse-leaved plants, notably Bramble which can become the principal foodplant on some sites. They overwinter as **pupae**, which are formed within cocoons of leaves and silk amongst low vegetation (<30 cm). Thus the butterfly thrives where there is a mosaic of short and tall vegetation.

The mobility and colony structure of the Grizzled Skipper is known from a detailed mark–recapture study over an area of fragmented downland in Sussex. This showed that populations are typically small (<100 adults at peak) but some contain as many as 1000 adults. Most adults stayed within these colonies but a few moved up to 1.5 km and are probably capable of flying much further. Discrete colonies exist where breeding areas are separated by belts of woodland, dense scrub, or improved grassland. However, colony structure is variable and populations in optimal habitats with high foodplant density tend to be more self-contained. Populations in small or sub-optimal habitats with only minor barriers to dispersal tend to have a more open structure and are more closely linked with other nearby patches. Through much of its British range, colonies are probably linked to form metapopulations covering far wider areas.

Distribution and trends

The Grizzled Skipper was formerly widely distributed in England and parts of Wales but declined substantially during the twentieth century, especially in the western and northern parts of its range. There are only a few historical records for Scotland, where it has been extinct since the nineteenth century, and it is absent from Ireland, the Isle of Man, and the Channel Islands.

The decline of the Grizzled Skipper accelerated during the twentieth century, with most losses occurring since 1950. Its status is now very well known from detailed local recording schemes and it is estimated that around 1100 colonies remain in England and only 10–15 in Wales (where it may still be under-recorded). It has declined substantially in almost every county and is now extinct in Northumberland (*c.*1918), Cheshire (1970s), Suffolk (1979), and Middlesex (1992).

The current data show that it has probably disappeared from 36% (138) of 10 km grid squares recorded in 1970–82, but over 130 new squares have been recorded, largely attributed to increased survey effort. As with other species, its decline at a population level has probably been even higher. The Butterfly Monitoring Scheme (BMS) reflects this downward trend and shows that declines have occurred on many nature reserves and protected areas.

The Grizzled Skipper's main strongholds are the southern English counties of Dorset, Somerset, Hampshire, and Wiltshire (each with over 100 colonies), followed by Buckinghamshire, Gloucestershire, the Isle of Wight, Kent, Surrey, Sussex, and Warwickshire (each with over 40 colonies). It has declined in nearly all these areas since the 1970s and the trend on monitored sites shows a general decline over the last 20 years. Moreover, the butterfly is now reduced to a handful of colonies in many counties of England and Wales, where its future is precarious.

On the positive side, there is evidence of a slight re-expansion since 1990 in several counties in central England. It also re-appeared during 1997 in north Lancashire after an absence of more than 50 years, though this is probably the result of a clandestine release. Another reintroduction has been attempted in Essex, but this failed after a few years.

European trend

Its range appears to be stable in many countries but severe declines have been recorded in the Netherlands and European Turkey (>50% decrease in 25 years), and Belgium and Croatia (25–50% decrease); and to a lesser extent in the Czech Republic, Denmark, Finland, Germany, Luxembourg, Romania, Slovakia, Sweden, and Asian Turkey (15–25% decrease).

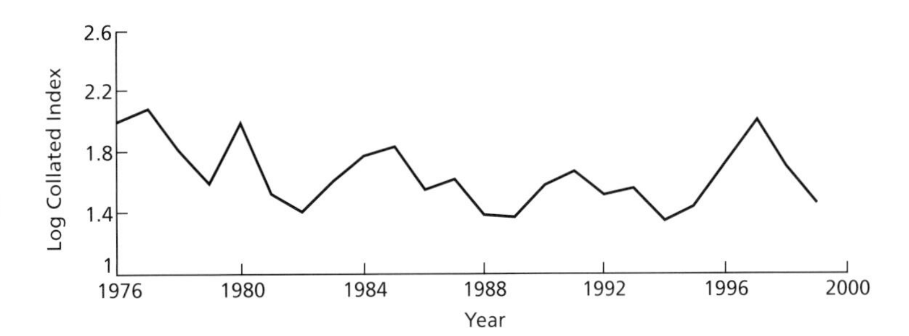

BMS collated index of abundance showing general decline in numbers of the Grizzled Skipper.

Interpretation and outlook

Although the decline of the Grizzled Skipper has not been as dramatic as that of some butterflies, it is nevertheless a matter of great concern. Measures should be taken now to conserve the species while it is still widespread in some areas.

The Grizzled Skipper needs warm, well-structured habitats that are inherently highly dynamic. It suffered not only from the wholesale loss of semi-natural grassland in lowland Britain during the twentieth century, but also from the abandonment and changing management of the fragments that remain. Like many other butterflies, it has been badly affected by the cessation of traditional woodland coppicing and the lack of regular canopy gaps in modern woodland. However, the species benefited temporarily from the clearings created by the storms of 1987 and 1990, and it may be benefiting from the recent increase in the rabbit population on many downland sites.

Populations of Grizzled Skipper have responded to general conservation measures such as the restoration of coppicing or grazing on nature reserves and other protected areas. In the immediate future, it should be helped by new Forestry Authority grants for coppice restoration and maintenance of sunny rides and glades, and agri-environment schemes that encourage farmers to graze remnants of semi-natural grassland. However, many habitats now rely on rabbit grazing to maintain suitable mosaics and numerous colonies breed on former industrial sites which may deteriorate due to successional changes or be destroyed by redevelopment.

Only a few large colonies remain and in many regions the butterfly exists in networks of small populations. Conservation needs to be implemented at the landscape level as well as on individual sites, both of which will represent a considerable challenge in coming decades.

Key references

Brereton (1997); Brereton *et al.* (1998*a*); Warren and Thomas (1992).

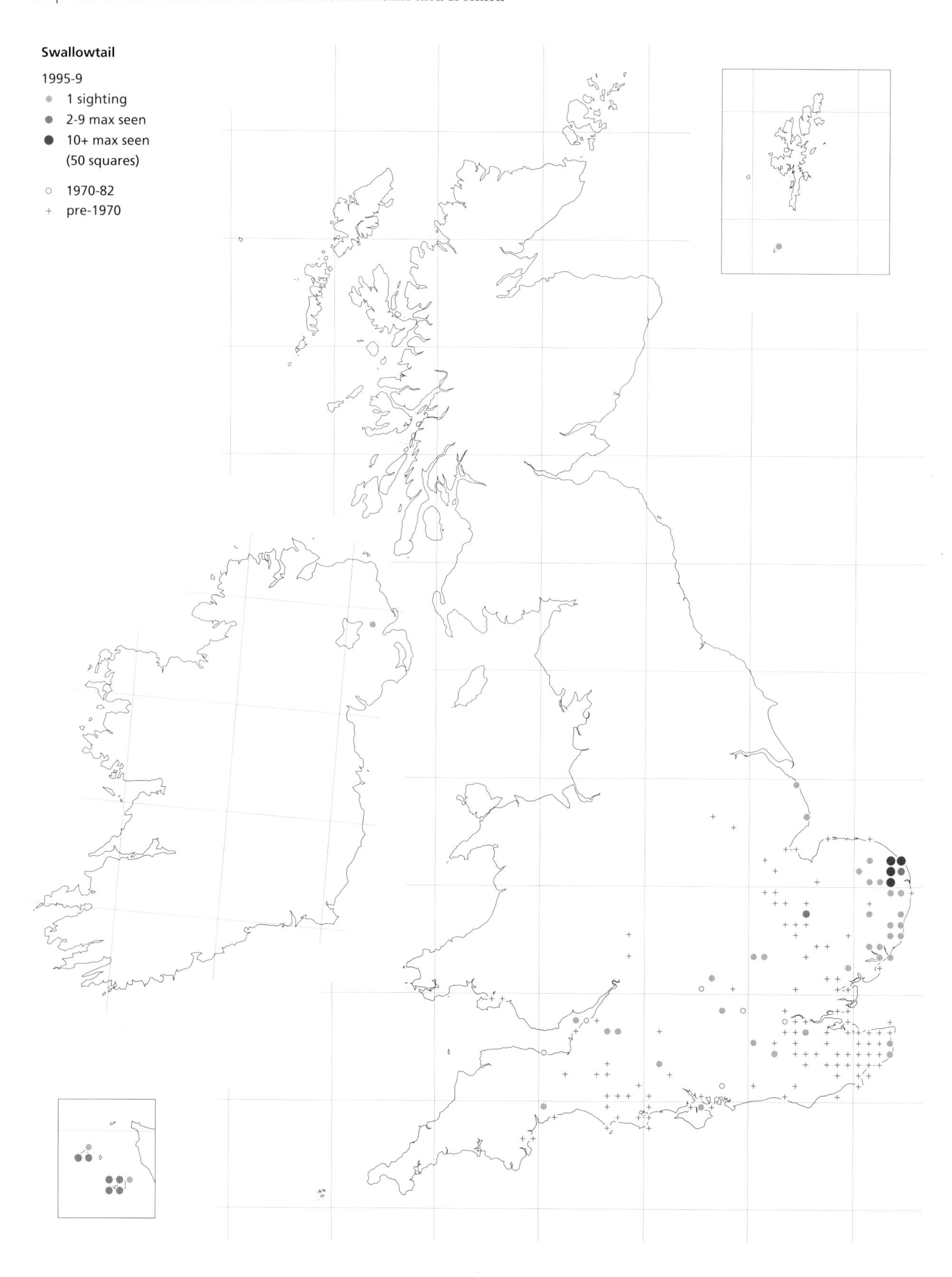
Swallowtail
1995-9
1 sighting
2-9 max seen
10+ max seen
(50 squares)
1970-82
pre-1970

Swallowtail *Papilio machaon*

Resident
Range stable.

Conservation status
UK BAP status: Species of Conservation Concern.
Butterfly Conservation priority: medium.
European status: not threatened.
Fully protected in Great Britain.

European/world range
The race *britannicus* occurs only in Britain. The continental race *gorganus* is widespread in Europe, and across Asia to Japan as well as in North America. It has declined recently in some European countries.

This is one of our rarest and most spectacular butterflies. The British race *britannicus* is a specialist of wet fenland and is currently restricted to the Norfolk Broads. Here the adults can be seen flying powerfully over open fen vegetation, stopping to feed on flowers such as thistles and Ragged-Robin. The butterfly probably declined within its range during the twentieth century but has benefited over the last few decades from conservation management aimed at increasing open fen vegetation. There are also scattered records of migrants of the continental race.

Foodplants
The native British race feeds solely on Milk-parsley (*Peucedanum palustre*). Occasional migrants of the continental race *gorganus* use a variety of umbellifers such as Wild Carrot (*Daucus carota*) and Wild Angelica (*Angelica sylvestris*).

Habitat
The British race *britannicus* breeds only in open fens and marshes that support vigorous growths of Milk-parsley. The butterfly prefers areas of mixed fen usually dominated by sedge, or sometimes reed, which are cut periodically and contain tall, prominent foodplants. The occasional migrants of *gorganus* can be found in almost any habitat but are most frequently seen on grassland near the south coast of England.

Life cycle and colony structure
The butterfly is single brooded in the Norfolk Broads with a small partial second brood in most years. Adults fly from late May to mid-July and second brood individuals in August. The **eggs** are laid on tall flowering plants of Milk-parsley that protrude above the surrounding fenland vegetation. The young **larvae** are black with a white band, resembling bird droppings, but after their second moult they become bright green with conspicuous black and orange rings. When disturbed, they erect a bright orange osmaterium from behind the head, which emits a pungent smell resembling pineapple. The larvae are quite conspicuous at this stage and often move to the top of the flower-head where they can be counted accurately by a patient recorder. They overwinter as **pupae**, low down on woody stems, or on reed or sedge.

The *britannicus* race tends to be colonial with adults concentrating in favoured breeding areas or where nectar is abundant on the edges of fenland vegetation. However, they may wander into surrounding areas and even nearby gardens. Genetic studies also suggest that there are high levels of adult mixing throughout the Norfolk Broads. The continental race *gorganus* is highly mobile and migrates to most European countries,

Fully-grown Swallowtail larva on Milk-parsley.

breeding as it goes, occasionally reaching Ireland and even as far north as Fair Isle.

Distribution and trends

The resident British race is currently found only in the Norfolk Broads with the strongest populations along the valleys of the rivers Ant, Thurne, Bure, and mid-Yare. The butterfly used to be far more widespread and probably once occurred throughout the formerly extensive East Anglian fenland south of the Wash; in the marshes along the rivers Thames and Lea; the Somerset Levels and as far north as Beverley in Yorkshire. It disappeared from these areas when the marshes were almost entirely drained and converted to intensive farmland during the eighteenth and early nineteenth centuries. A population survived at Wicken Fen nature reserve in Cambridgeshire until the 1950s but died out suddenly following the partial drying out of the habitat and reduction in the size of the breeding area. There have since been two attempts to re-establish it following improved management but these have both failed.

Migrant Swallowtails of the race *gorganus* have established temporary breeding colonies in southern England on several occasions including: during the early nineteenth century on downland in Dorset and Kent; 1857–69 near Deal (Kent); 1918–26 near Hythe (Kent); and during the mid-1940s in Kent, Dorset, and south Hampshire. However, *gorganus* is only slightly paler than the heavily marked *britannicus* and records rarely distinguish between the two races. There are records of individuals in most years outside the Norfolk Broads. Although most are probably migrants, some may be released captive bred adults or genuine native Swallowtails which have been transported accidentally around the country as pupae in thatching reed harvested from the Broads.

European trend

The Swallowtail is thought to be stable in most European countries but it has become extinct in Denmark and there have been recent declines in Latvia, Luxembourg, the Netherlands, and Romania (25–50% decrease in 25 years), and in Belgium (15–25% decrease).

Interpretation and outlook

The fenland habitat of the Swallowtail has suffered one of the biggest declines of any butterfly habitat with only about 1% remaining. The only extensive habitat left is in the Norfolk Broads and the butterfly has become extinct in other regions where fenland has been reduced to small patches. However this crucial area is at risk from a variety of factors such as a lowering water table, invasion of scrub and pollution of surface and ground water by nitrates and phosphates.

The Broads themselves were created by shallow excavations for peat in medieval times and were kept open by the regular cutting of reed and sedge for roofing, and by the harvesting of 'Marsh Hay', a mixture of plants cut annually for fodder and bedding. However, the demand for these crops diminished steadily during the twentieth century and large areas were abandoned, especially during the First World War and again during the Second World War. As a result, substantial areas of open fenland have become invaded by willows, Alder, and other trees, and have gradually reverted to carr woodland. Such woodland now occupies well over half of the 5300 ha of fen vegetation in the Norfolk Broads, leaving the remaining open habitat for the Swallowtail much reduced and fragmented.

Swallowtail populations probably reached an all-time low in the 1970s but have since recovered well due to the concerted action of conservation bodies, co-ordinated by the Broads Authority and English Nature and focusing on areas with the best potential. The expansion of the butterfly's habitat has also been helped by the revival in thatching, which has increased the demand for reed and sedge.

The viability of the farming systems in the Broads is now being improved following the designation of the entire region as an Environmentally Sensitive Area. This helps to maintain the distinctive marsh landscape through grants to landowners. The aim of a new Broads Plan is to halt the decline of fenland and restore the open fen recently lost to scrub. Central to this plan is the support for traditional regimes of reed and sedge cropping that maintain the open fen habitat. The latter is particularly suitable for the Swallowtail as the sedge is cut on a three-year rotation, allowing the Milk-parsley to become tall and lush. Under such management, the butterfly's populations have increased noticeably on the Bure Marshes National Nature Reserve and on reserves managed by the Norfolk Wildlife Trust, RSPB, the Broads Authority, and Butterfly Conservation.

Outside the Broads, most fenland habitat is probably too fragmented to support viable Swallowtail populations. Studies at Wicken Fen indicated that adults became smaller and less mobile as the site became more isolated during the twentieth century. The same process has been observed in the Broads since the 1920s. The problem is exacerbated because smaller, fragmented fen habitats tend to dry out and lose their characteristic plants, including Milk-parsley. Valiant attempts have been made by the National Trust to restore the water levels at Wicken Fen by constructing waterproof external banks, but the levels remain unpredictable and it may be many years before sufficient stable habitat is created to warrant another re-establishment attempt.

The future of the Swallowtail depends on continued conservation action in the Norfolk Broads. The signs are that this is being highly successful and the Swallowtail should continue to flourish provided current policies are maintained.

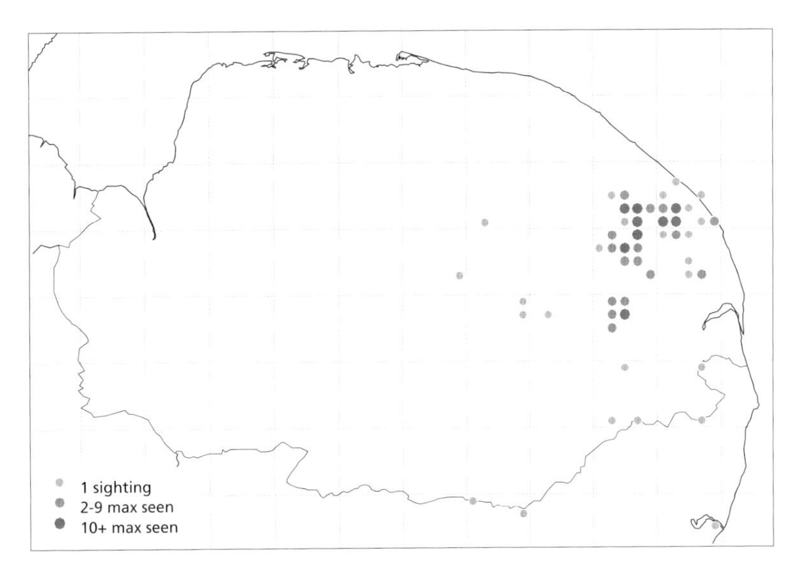

The distribution of 1995–9 records of the Swallowtail at 2 km square resolution in Norfolk.

Key references

Barnett and Warren (1995*b*); Bretherton (1951*b*); Dempster (1995); Dempster *et al.* (1976); Hoole *et al.* (1999); West (1993).

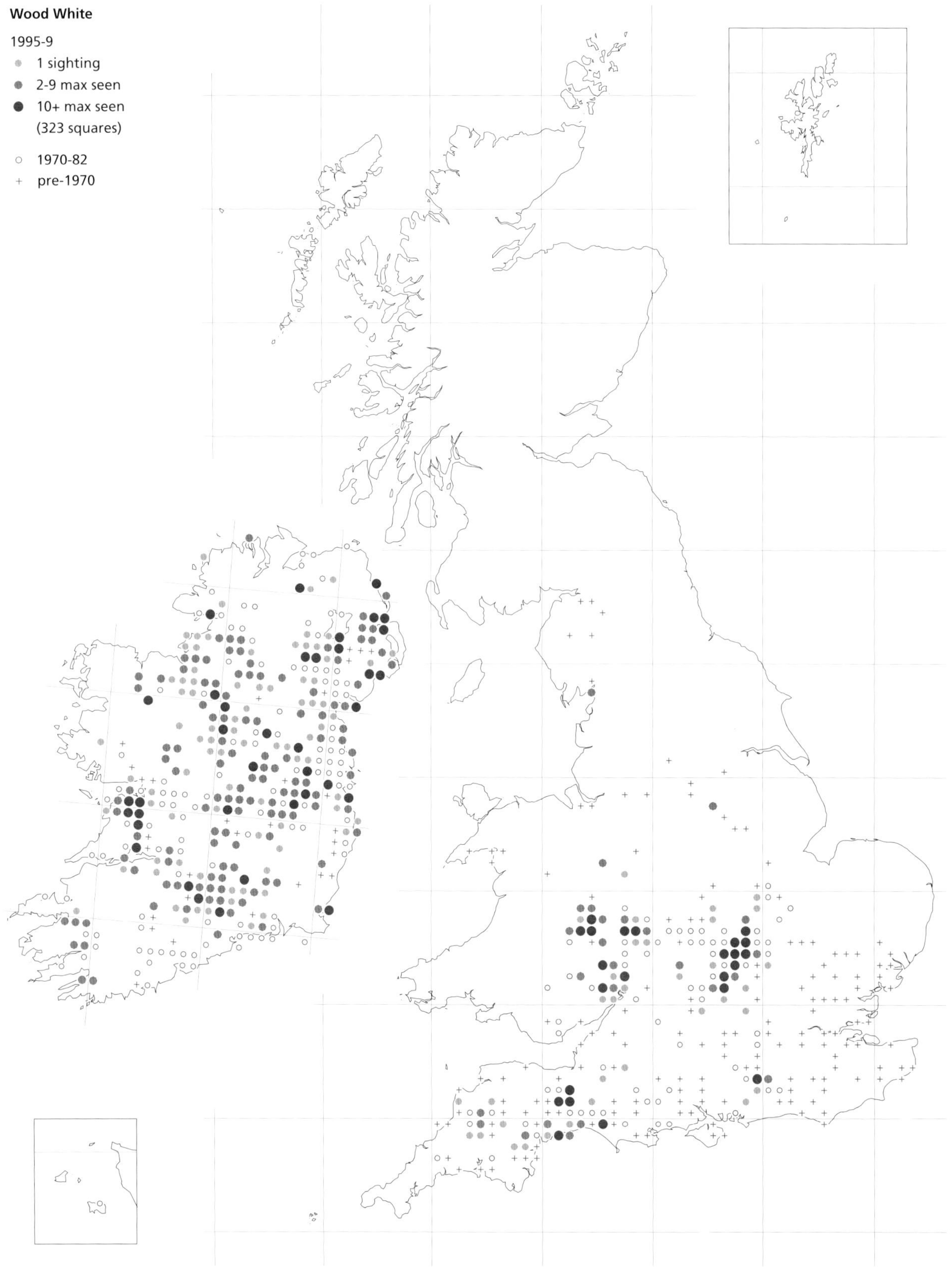

Wood White

1995-9
- 1 sighting
- 2-9 max seen
- 10+ max seen
(323 squares)

- 1970-82
- pre-1970

Wood White *Leptidea sinapis*

Resident

Range declining in Britain, stable in Ireland.

Conservation status

UK BAP status: Species of Conservation Concern.
Butterfly Conservation priority: medium.
European status: not threatened.
Protected in Great Britain for sale only.

European/world range

Widespread in Europe as far as 66°N in Scandinavia and eastwards to the Caucasus Mountains and Siberia. In some countries it overlaps with a very similar species, *Leptidea reali*, which can be distinguished only by differences in genitalia. It has declined in a few European countries.

The Wood White is a delicate, slow-flying butterfly usually encountered in sheltered situations, such as woodland rides or scrub edges. The males fly almost continuously in fine weather, patrolling to find a mate, whereas females spend much of their time feeding on flowers or resting. In the characteristic courtship display the male lands opposite the female and waves his head and antennae backwards and forwards with his proboscis extended. The butterfly has declined seriously in England and Wales. It is more widespread in Ireland where it has expanded northwards in recent decades.

Foodplants

Various legumes are used, commonly Meadow Vetchling (*Lathyrus pratensis*), Bitter-vetch (*L. linifolius*), Tufted Vetch (*Vicia cracca*), Common Bird's-foot-trefoil (*Lotus corniculatus*), and Greater Bird's-foot-trefoil (*L. pedunculatus*). (Note that some vetches are not used, notably Bush Vetch, *V. sepium*, and Common Vetch, *V. sativa*).

Habitat

The Wood White breeds in tall grassland or light scrub in partially shaded or edge habitats. In Britain, most colonies breed in woodland rides and clearings, though a few large colonies occur on coastal undercliffs. A few smaller colonies occur on disused railway lines and around rough, overgrown field edges (for example in north Devon). In Ireland, more open habitats are used, often far from woodland, including rough grassland with scrub, road verges, hedges, and disused railway lines.

Life cycle and colony structure

There is one main generation per year through most of its range, with adults flying from the end of May to the end of July. In the far south of England, and elsewhere in hot years, adults fly a few weeks earlier and there is a partial second brood in late July and August.

The **eggs** are laid singly on upper parts of foodplants, usually where they protrude above surrounding vegetation. In Britain, females prefer to lay eggs in lightly shaded and sheltered situations, either in light scrub or in the edges of sunny woodland rides. They rarely lay in very open areas and never fly into dense shade. The **larvae** are green and well camouflaged, and tend to feed on younger, more nutritious growth at the top of foodplants. They move away from the foodplant to **pupate** in surrounding tall grassland or scrub, typically suspended from grass or twigs 20–40 cm from the ground. The butterfly overwinters as a pupa.

At most woodland sites, the Wood White occurs in discrete colonies, although there

Wood White larva on Meadow Vetchling.

may be considerable movement between suitable rides and glades. In Ireland, it appears to have a looser colony structure (as it does in continental Europe), often occurring at low densities over wide areas, with concentrations only in the best breeding areas. Adults, usually males, are recorded sporadically many kilometres from known colonies and presumably represent dispersing individuals. There is little precise information on colonization range, but this must be considerable judging by its spread in Ireland over recent decades.

Distribution and trends

The Wood White has a very localized distribution in England and Wales, and declined substantially during the twentieth century. It is absent from Scotland and may now be extinct in Wales, although there are occasional sightings. In contrast, the butterfly is widespread and locally abundant in Ireland, where it has expanded northwards over the last 30 years.

The Wood White was formerly quite widely distributed in England and Wales. Its decline began earlier than many other species and by the 1900s the butterfly was extinct in several counties, including some where it had once been abundant. The Wood White continued to decline during the twentieth century and, by the late 1970s, only 87 colonies were estimated to remain. During the 1980s it declined substantially in Warwickshire and in the woodlands on the Surrey/Sussex borders (from 20 sites to fewer than 4). The current survey indicates a major recent decline in Britain where it has been lost from 62% of the 10 km squares recorded in 1970–82.

Since the 1960s, the Wood White has been introduced to a number of sites. Many of these releases failed but at least two have resulted in sizeable colonies (in north Bedfordshire and on the Buckinghamshire/Oxfordshire border). The recent records for Cumbria and Derbyshire are also suspected to be of released individuals. The butterfly's British strongholds are now in three main regions: Herefordshire and Worcestershire; Northamptonshire and Buckinghamshire; and east Devon and south Somerset.

In Ireland, the Wood White is far more widespread than in Britain but becomes sparser in the far south (for example in Co. Cork) where there is less Meadow Vetchling in road verges, and in intensively farmed areas such as Co. Carlow. It is under-recorded in some regions and is probably more widely distributed than shown. The butterfly expanded northwards during the fine summers of the 1970s and 1980s, spreading particularly along railway lines, both used and disused. These habitats are believed to have acted as linear corridors enabling the butterfly to colonize many kilometres in just a few decades. However, this period of expansion now seems to be over and there were signs of a slight contraction during the 1990s.

European trend

The Wood White is thought to be stable over much of its European range but has declined severely in Denmark (>75% decrease in 25 years) and to a lesser extent in Luxembourg (25–50%), Germany, Latvia, and European Turkey (15–25%).

Interpretation and outlook

In woodland and scrub habitats, Wood White populations are known to fluctuate in response to the growth of surrounding trees and shading of its scrub-edge breeding habitats. In former centuries, the butterfly may have been associated (in England and Wales) with the widespread practice of coppicing, which produced both shade and lightly shaded edge habitat, and allowed both the foodplants and butterfly to flourish. By the beginning of the twentieth century, coppicing was perpetuated only in larger woods with substantial ride networks and a regular cycle of clear-felling and replanting. The end of the nineteenth century was noted for its poor summer weather, which is also known to be a key factor determining the butterfly's population size.

During the 1950s and 1960s, the Wood White probably benefited from the extensive felling and replanting of ancient woodland with conifers, which opened up many shaded woods and produced abundant edge habitats. By the 1970s almost 50% of colonies occurred in rides alongside young conifer plantations. Such habitats remain very important but some have subsequently become too shaded as the surrounding plantations have grown, leading to further losses (as on the Surrey/Sussex border).

During the late twentieth century, periods of minor expansion and contraction in the Wood White's range were probably due to other more temporary changes in habitat suitability. Single individuals, usually males, have been recorded some distance from known breeding areas and occasionally new sites have been colonized. During the 1970s and 1980s, it undoubtedly benefited temporarily when the British rail network was reduced and railway lines became abandoned and overgrown. The scrub succession on railway verges provided ideal semi-shaded conditions for a period and by the late 1980s accounted for at least five colonies. In Warwickshire the butterfly spread along a number of railway cuttings and colonized several new sites, but these have since become less suitable and it has now disappeared.

The distribution of 1995–9 records of the Wood White at 5 km square resolution in southern England.

The Wood White has been conserved successfully on several nature reserves and sites managed by the Forestry Commission. Suitable management consists of the maintenance of open sunny rides and periodic cutting back of edge vegetation every 3–6 years. In commercially managed plantations it can survive well provided there is a regular cycle of clear-felling and replanting and numerous, suitably managed rides.

Although the butterfly remains widespread in Ireland, its status should continue to be monitored carefully, especially given its continuing decline in England. Changes in land use, especially to the marginal habitats such as road verges and hedges, could be detrimental to this delicate species.

Key references

Heal (1965); Warren (1984, 1985); Warren and Bourn (1998).

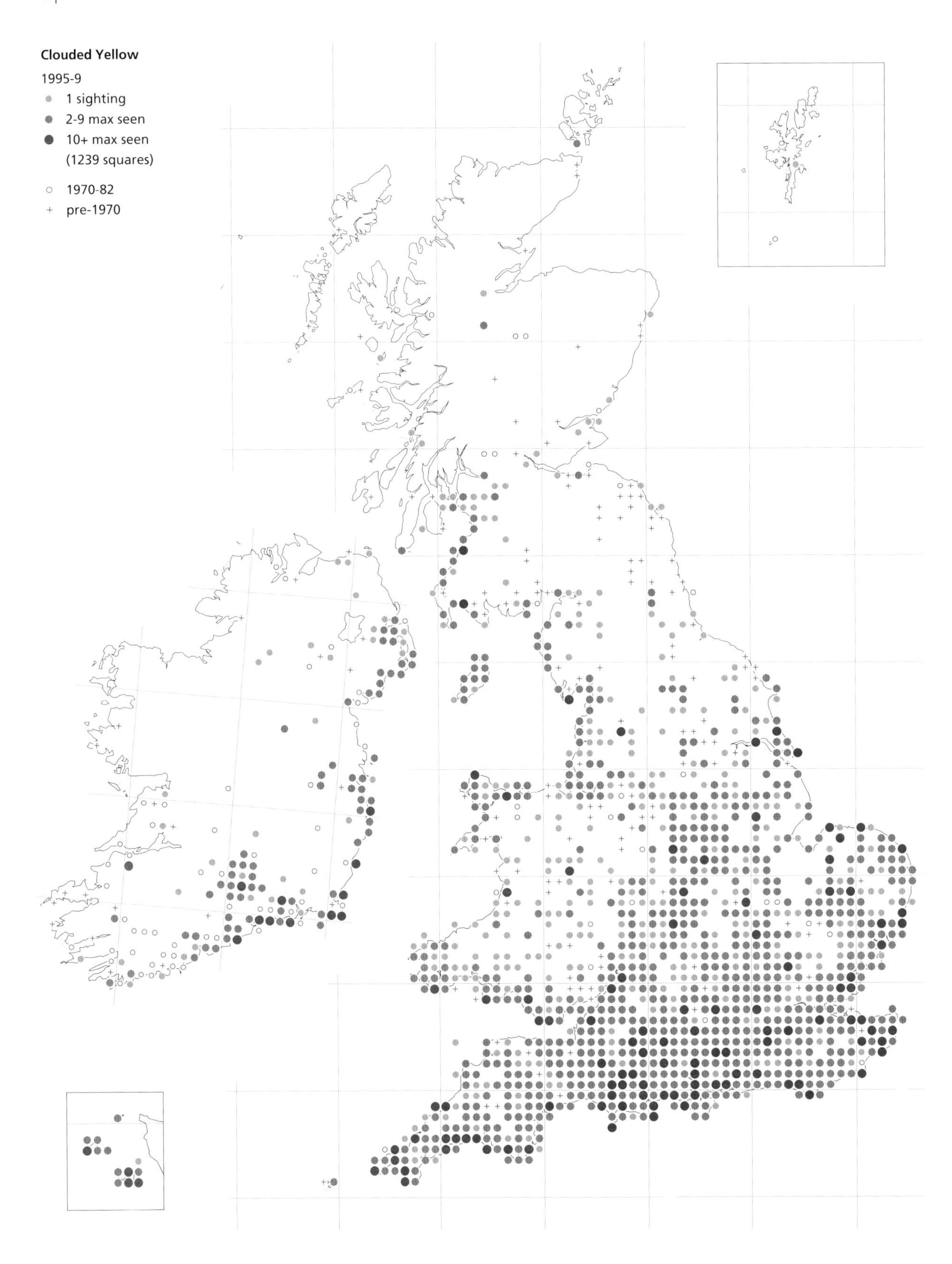
Clouded Yellow
1995-9
1 sighting
2-9 max seen
10+ max seen
(1239 squares)
1970-82
pre-1970

Clouded Yellow *Colias croceus*

Regular migrant

Conservation status

UK BAP status: not assessed.
Butterfly Conservation priority: not assessed.
European status: not assessed.

European/world range

North Africa and southern Europe and eastwards through Turkey into the Middle East. It occurs throughout much of Europe as a summer migrant, but very few individuals reach Scandinavia.

The Clouded Yellow is one of the truly migratory European butterflies and a regular visitor to Britain and Ireland. Although some of these golden-yellow butterflies are seen every year, the species is famous for occasional mass immigrations and subsequent breeding, which are fondly and long remembered as 'Clouded Yellow Years'. A small proportion of females are pale yellow (form *helice*), which can be confused with the rarer Pale and Berger's Clouded Yellows.

Foodplants

A range of leguminous plants is used, including wild and cultivated clovers (*Trifolium* spp.), Lucerne (*Medicago sativa*), and less frequently, Common Bird's-foot-trefoil (*Lotus corniculatus*).

Habitat

Clouded Yellows may be seen in any habitat, but congregate in flowery places where the larval foodplants grow. As clovers are still commonly cultivated, the Clouded Yellow is one of the few butterfly species that has no difficulty locating breeding habitat in the modern farmed countryside. In southern England there is a preference for unimproved chalk downland.

Life cycle and colony structure

The species is continuously brooded in its permanent range in North Africa and southern Europe. Every year, adult Clouded Yellows migrate northwards from these breeding grounds and attempt to colonize large areas of Europe. The first individuals normally arrive in southern parts of Britain and Ireland in May or June. Many remain near the coast, but if the immigration is large, greater numbers may disperse inland.

Females lay their **eggs** singly, normally on the upper leaf surface of foodplants. The **larvae** feed on leaves before **pupating** attached to stems by silken girdles. The early summer arrivals give rise to a generation of adults that normally reaches a peak in August or early September. In favourable years, these may go on to produce a further brood in late September, October, and even November. These home-bred generations can be extremely large, with emerging adults far outnumbering the original immigrants. In addition, abundance in late summer and autumn may be enhanced by the continued arrival of adults from continental Europe.

There is little agreement about the potential for the Clouded Yellow to overwinter in Britain and Ireland. The species certainly appears unable to survive in significant numbers, but there are convincing observations indicating that overwintering may have occurred in a few cases.

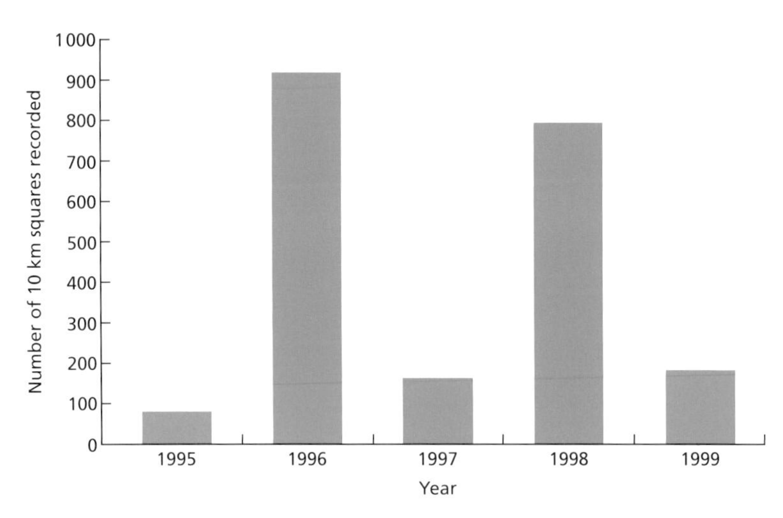

The number of 10 km squares in which the Clouded Yellow was recorded each year (1995–9).

In the winter of 1998–9, for example, Clouded Yellow larvae were watched at a coastal site near Bournemouth, Dorset. Full development was noted and the larvae disappeared, presumably into pupation, towards the end of the winter. Adults were recorded from late March and subsequently an egg was located. However, this did not hatch and no larvae were found at the site in the summer of 1999.

In common with other regular migrant species, the Clouded Yellow shows little colonial behaviour and adults range widely across the countryside. Indeed, with the exception of years of high abundance, congregations of adults at inland sites are assumed to indicate emergence from local breeding populations. Nevertheless, there is evidence of faithfulness to sheltered sites when the weather is less favourable.

Clouded Yellows are generally gregarious during migration and there are many accounts of large numbers flying northwards together. Southerly migration in the autumn has been observed both here and in other European countries, but tends to involve smaller numbers of individuals.

Distribution and trends

This migrant has reached almost every part of Britain and Ireland, including Orkney, Shetland, and the Western Isles of Scotland. However, in most years its distribution is far more restricted than that of the other regular migrants, the Red Admiral and the Painted Lady.

Clouded Yellows are recorded every year in the coastal counties of southern England, particularly Cornwall, Devon, Dorset, and on the Isle of Wight, where summer breeding populations are established regularly. The distribution in other parts of Britain, Ireland, the Isle of Man, and the Channel Islands varies considerably from year to year.

Large numbers of early immigrants and long, warm summers can result in enormous increases in Clouded Yellow abundance and distribution in particular years. Such years are unpredictable, but dramatic. Famous 'Clouded Yellow Years' were 1877, 1947, and 1983 and, more recently, 1992, 1994, and 1996. The events varied in magnitude in different areas and some were synchronous with spectacular years for other migrants (for example 1996 for the Painted Lady). The 1992 immigration was particularly notable in Scotland, where this normally rare migrant was recorded in more than 300 10 km squares (see map), and also in Ireland, Wales, and north-west England.

In the 1995–9 survey period, 1996 was the most significant year for the Clouded Yellow with records from over 900 10 km squares. In that year the butterfly was widespread in southern England, extending northwards and westwards into Northumberland and Cumbria and as far as Anglesey and Pembrokeshire in Wales. Sightings in Scotland and Ireland were fewer although under-recording may have been a factor. The preceding and subsequent years produced no more than a scattering of records by comparison (76 10 km squares in 1995 and 155 in 1997), suggesting that any successful overwintering has an insignificant effect on the population. In Ireland, 1998 proved to be the best year for the species during the current survey.

Most Clouded Yellows are distinctive in flight, as no other species in Britain and Ireland has such golden-yellow coloration. At rest with its wings closed, the species might be mistaken for the Brimstone, but this is unlikely to be a significant source of errors. More notable is the confusion caused by pale (*helice*) individuals, which may make up 10% of the female Clouded Yellow population.

European trend

The Clouded Yellow occurs only as a migrant and summer breeding species in most European countries and thus distribution trends have not been compiled.

Interpretation and outlook

The distribution of the Clouded Yellow in any one year appears to be strongly related to the number of early immigrants and the weather. Comparison between distributions in different survey periods therefore yields no information regarding the future of the species in Britain and Ireland. The larval foodplants of the Clouded Yellow remain extremely widespread in the countryside and no specific conservation measures are required.

Its continued presence as a regular migrant to Britain and Ireland depends upon the preservation of its permanent breeding grounds hundreds of miles to the south. However, it is possible that the species may be able to establish permanent populations here in the future as a result of global warming.

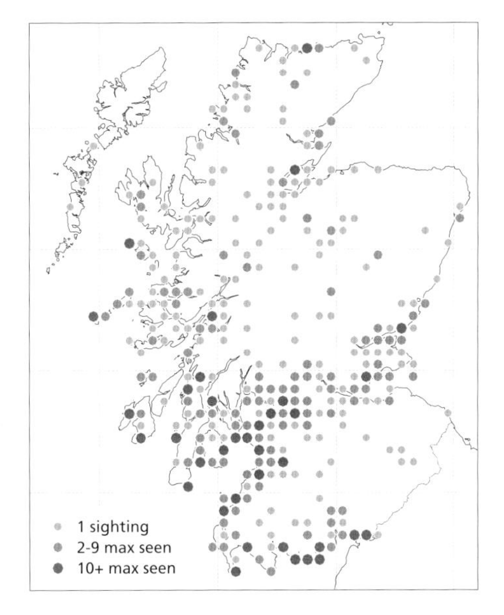

Records of the Clouded Yellow at 10 km square resolution in Scotland in 1992.

Key references

Bretherton and Chalmers-Hunt (1985); Pollard *et al.* (1984); Skelton (1999, 2000); Sutcliffe (1994).

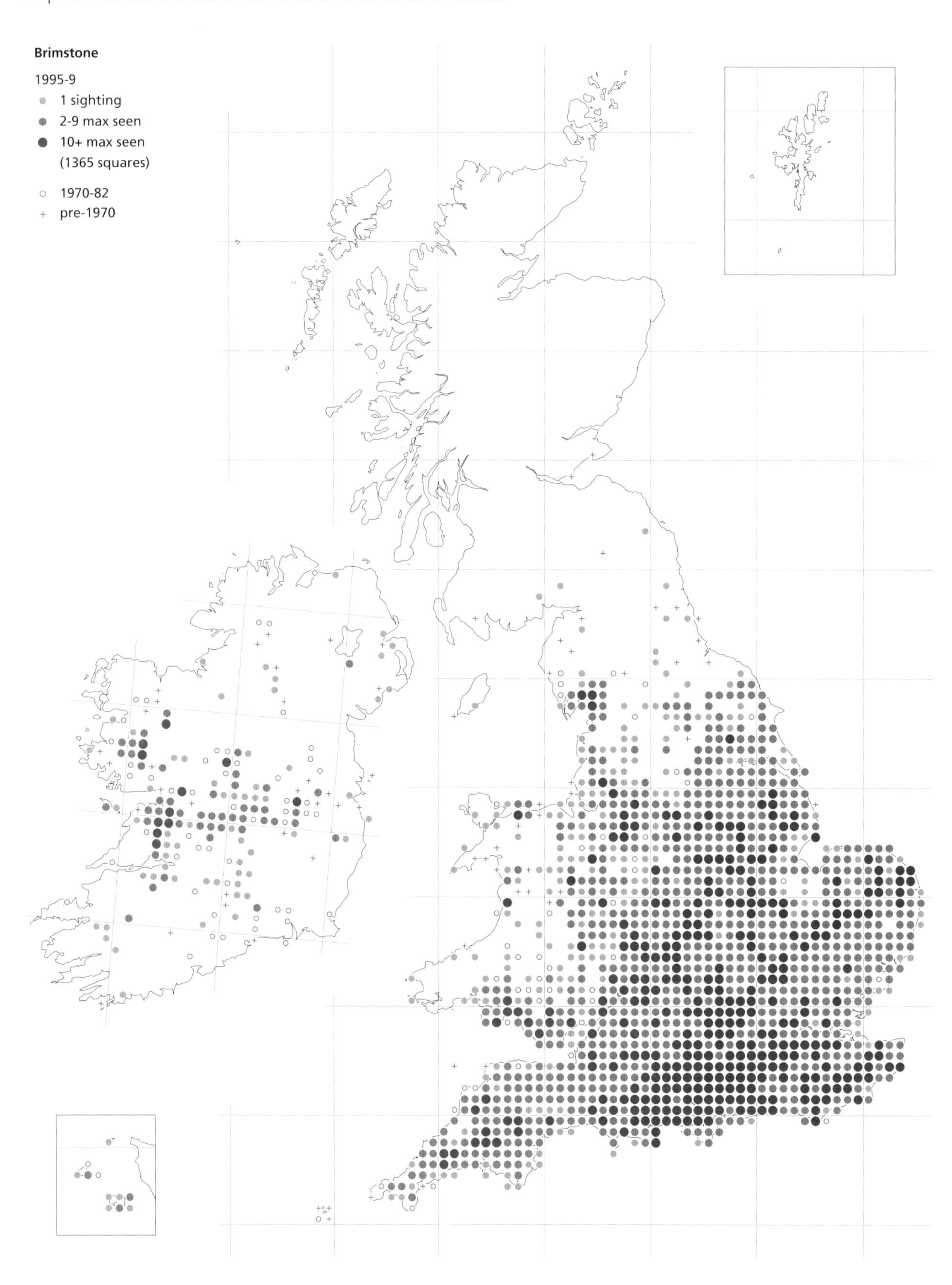
Brimstone
1995-9
1 sighting
2-9 max seen
10+ max seen
(1365 squares)
1970-82
pre-1970

Brimstone *Gonepteryx rhamni*

Resident
Range expanding.

Conservation status
UK BAP status: not listed.
Butterfly Conservation priority: low.
European status: not threatened.
Protected in Northern Ireland.

European/world distribution
Widespread through Europe as far as 64°N in Scandinavia, extending to Mongolia and North Africa. The European range is stable.

The sulphur-yellow uppersides of the wings of the male Brimstone make this species easy to identify in flight. There is a view that the word 'butterfly' originates from the yellow colour of male Brimstones. By contrast, the wings of the female are very pale green, almost white. When the butterflies roost among foliage, the angular shape and the strong veining of their wings closely resemble leaves. The Brimstone has spread in recent years, mainly in northern England.

Foodplants
The larvae feed on leaves of Buckthorn (*Rhamnus cathartica*), which occurs mainly on calcareous soils, and Alder Buckthorn (*Frangula alnus*), which is found on moist acid soils and wetlands.

Habitat
The Brimstone occurs in scrubby grassland, woodland (especially damp carr woodland), hedgerows, and open ground wherever foodplants are available in sunny positions. The butterfly ranges widely and can often be seen flying along roadside verges and tracks with hedgerows, well away from foodplants.

Life cycle and colony structure
The Brimstone has one generation each year and hibernates as an adult. It is often the first butterfly to fly during the year, appearing as early as February or March, when there are sunny days warm enough for flight. It flies throughout the spring, into June and sometimes early July. The next generation of adults emerges in late July and August and flies through to autumn and even later on warm days (see p. 392). They seek out evergreen shrubs, such as Holly and Ivy, among which to hibernate.

With an adult stage lasting up to eleven months, this is one of our longest-living butterflies and may be seen in every month of the year. In common with other species that hibernate as adults, they spend long hours nectaring in late summer, to build up fat reserves for winter. The Brimstone has an unusually long proboscis, enabling it to feed on flowers with deep nectaries such as teasels and, in gardens, the flowers of Runner Bean and buddleias.

The bottle-shaped **eggs** are laid singly on the emerging leaves of the foodplants in late

Brimstone larva in characteristic position on the mid-vein of a leaf.

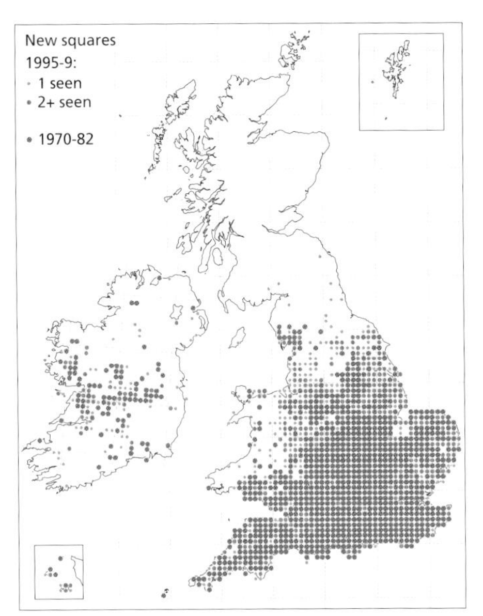

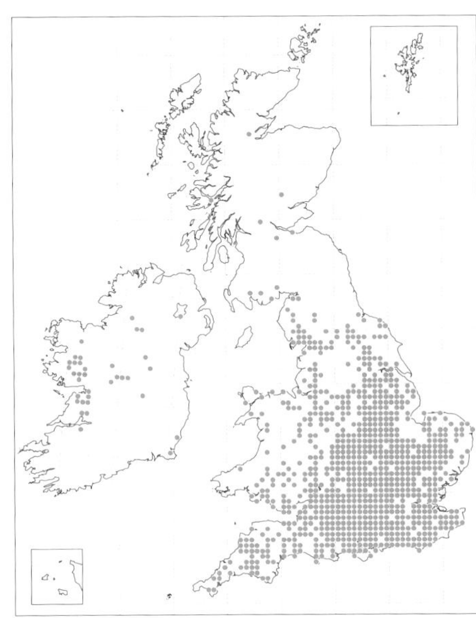

Left
The expansion of range of the Brimstone since 1970–82.

Right
The combined distributions of the Brimstone's larval foodplants, Alder Buckthorn and Buckthorn. Records, at 10 km square resolution, are from 1950 onwards.

spring. Although some literature suggests a preference for sunny, sheltered positions, research indicates a preference for isolated young trees in sunny, relatively exposed positions near ride edges. Young trees may provide more nourishing food with lower toxin levels. The young **larvae** eat holes in the leaves, progressing at later stages to consume entire leaves. The larvae leave their foodplants to **pupate** low down amongst vegetation. The pupae resemble rolled-up green leaves and are attached by single spun silk girdles to plant stems or the undersides of leaves.

The Brimstone is highly mobile and the females range widely to seek out foodplants. As a result, larvae are often found on isolated buckthorns. They do not appear to form distinct colonies, although adults may congregate at woodland and hedgerow sites where foodplants are common. Butterfly Monitoring Scheme (BMS) data show that the largest populations occur in wetlands, where Alder Buckthorn grows in carr at the water margins. The data also indicate that adults tend to migrate short distances (a few kilometres) towards woodland to hibernate and then disperse to breed in the following spring.

Distribution and trends

The Brimstone is widely distributed across southern England, within the range of the foodplants, and becomes more local further north. Since the 1970–82 survey, its range has increased by more than a third, mainly in northern England (Cheshire, Lancashire, Lincolnshire, and Yorkshire) and in the Welsh borders (particularly in Powys and Shropshire).

The Brimstone reaches the limits of its distribution in southern Cumbria and the North York Moors, although there have been occasional recent records further north, in Northumberland and southern Scotland. These records, which are outside the range of its foodplants, are likely to be of vagrants. The butterfly is uncommon in western Wales.

There are few recent records from the Channel Islands, where it seems to be less common and only one record from the Isle of Man, which may be a recent introduction or a vagrant from north-west England or Ireland.

In Ireland, it is widespread but local on limestone soils where Buckthorn grows. It has a patchy distribution in the Midland counties of Ireland and is much more common in the highly calcareous area further west, with strongholds in counties Mayo, Galway, and Clare. It may occur in more acid areas where there is Alder Buckthorn, which is an uncommon plant in Ireland. It is no longer recorded as a breeding species in Northern Ireland, although a few individuals are recorded in most years, probably vagrants from further south.

This species is easy to record, as the male is so obvious and the butterfly can be seen for most of the spring and summer months. Female Brimstones may be mistaken for Large Whites by inexperienced observers, but with a little practice they are easy to distinguish with their strong flight pattern and the absence of black markings on their wings. However, records tend to be dominated by sightings of males and the distribution map is believed to be an accurate reflection of the butterfly's range.

European trend

The Brimstone is stable in most countries with a slight increase in the Netherlands in the last 25 years.

Interpretation and outlook

There is a strong correlation between the distribution of the butterfly and that of its foodplants. The increase in the number of records in north Wales and Cheshire (and probably other areas) since the early 1980s has been linked to the planting of buckthorns. The wide ranging Brimstone has been able to colonize these plantings, some of which are beyond the natural distribution of the foodplants.

Data from the BMS show that increases in adult numbers are strongly associated with warm weather in the summer of the previous year. Such weather may help prolong activity and feeding in the autumn, so that sufficient fat is accumulated for successful hibernation. The higher average temperatures of recent years may have played a role in the expansion of the Brimstone. Further extension of its range might occur if the climate continues to warm in northern Britain, but will depend on the range of the larval foodplants also expanding northwards, whether by natural spread or planting.

Despite the recent expansion of range, the Brimstone has probably become less abundant in many areas, as a result of hedgerow destruction during the past forty years. It remains a widespread species in the south and would benefit from simple conservation measures.

Buckthorns are easy to grow in parks and gardens and can provide suitable conditions for breeding, even in developed and urban landscapes. It is, therefore, relatively easy to encourage the butterfly by planting buckthorns in sunny positions and maintaining an abundant supply of nectar in gardens and other habitats.

Key references

Dunn and Parrack (1986); Gutiérrez and Thomas (2000); McKay (1991); Pollard and Hall (1990); Shaw (1999); Sutton and Beaumont (1989).

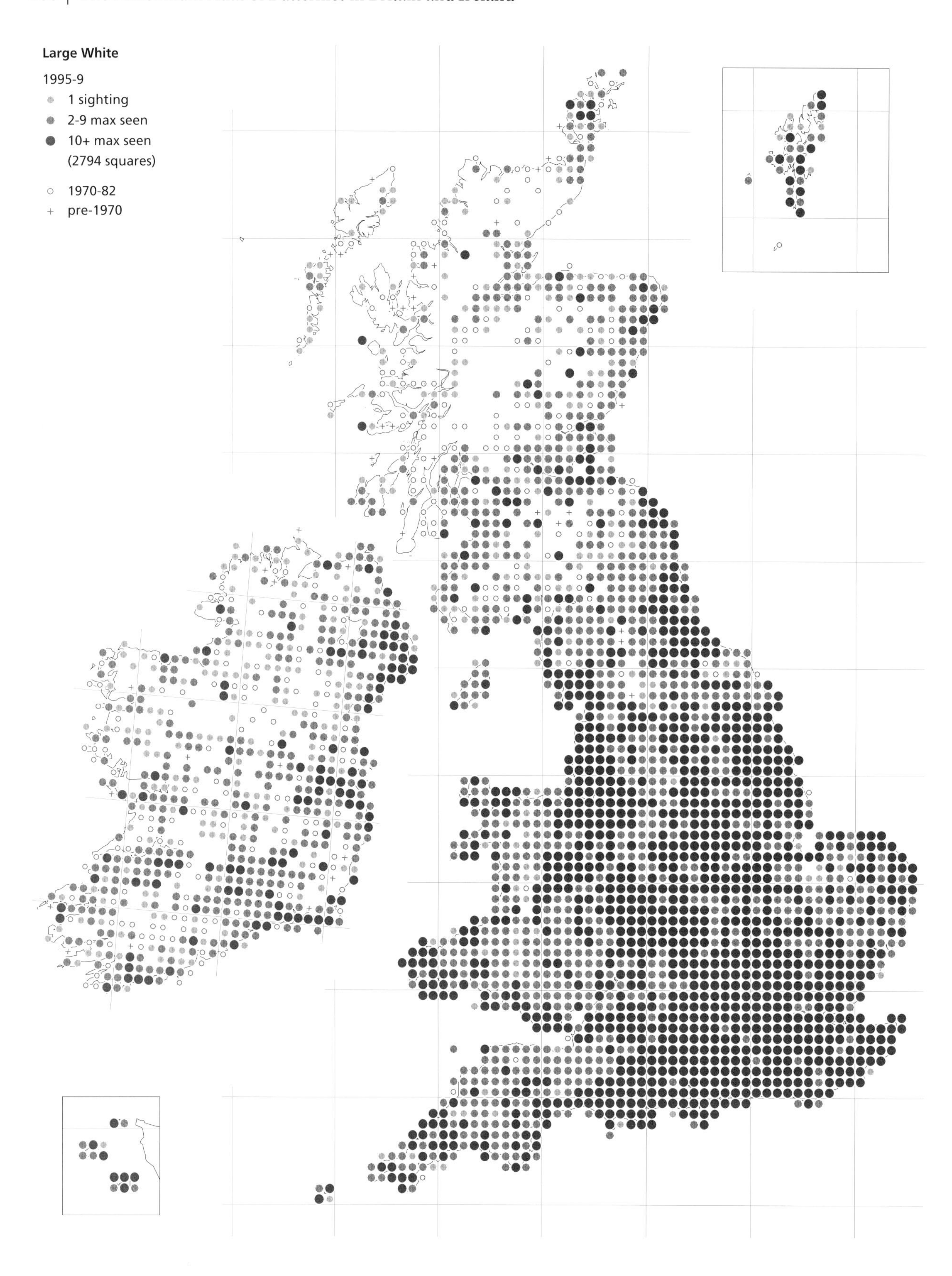
Large White
1995-9
1 sighting
2-9 max seen
10+ max seen
(2794 squares)
1970-82
pre-1970

Large White *Pieris brassicae*

Resident

Range stable.

Conservation status

UK BAP status: not listed.
Butterfly Conservation priority: low.
European status: not threatened.

European/world range

Occurs throughout Europe and North Africa and extends across Asia to the Himalayas. The distribution is stable in most European countries, but there have been declines in some countries and expansions in others.

The Large White is our largest white butterfly and is a strong flyer. It is not always welcomed in gardens and fields because of the damage its larvae inflict on brassica crops. The larvae are brightly coloured and conspicuous, a signal to warn predators of the irritant and poisonous mustard oils they have concentrated from the foodplants. Many adults seen in Britain and Ireland have flown from mainland Europe. Numbers of both residents and migrants of this common and widespread species vary considerably from year to year.

Foodplants

The larvae feed on wild or cultivated species of the Cruciferae family, with a strong preference for cultivated varieties of *Brassica oleracea* such as Cabbage and Brussels-sprout and varieties of *B. napus* such as Oil-seed Rape. Nasturtium (*Tropaeolum majus*) and Wild Mignonette (*Reseda lutea*) are also used, as is Sea-kale (*Crambe maritima*) along the coast.

Habitat

This is a strongly mobile and migrant species that may be encountered in any location, throughout Britain and Ireland, even on mountain tops. Most adults are seen close to breeding areas, in gardens, allotments, and fields where brassica crops are grown. They may congregate in large numbers in fields of Oil-seed Rape. Wild species of foodplants are thought not to be important for this species.

Life cycle and colony structure

There are usually two generations a year. Adults emerge in spring and can be seen from April onwards. The second brood emerges from July onwards and is more numerous than the first (see p. 392). There may be a third brood in the autumn after a warm summer.

The females are strongly attracted to brassica plants by the scent of mustard oils and lay **eggs** in batches of 50 or more. The resulting **larvae** are gregarious at first but gradually disperse to become solitary. They are distasteful to birds and other predators because of the mustard oils absorbed whilst feeding, but many are attacked by parasites. Offspring of the late summer broods overwinter as **pupae**, which are suspended from the underside of solid structures such as branches, wooden fences, or ledges on buildings, often many metres from the larval foodplant. At times when numbers are high, it is not unusual to find all four stages of the life cycle at once. This overlapping of broods suggests the arrival and breeding of migrants, out of step with resident individuals.

The Large White does not form colonies but flies across the countryside looking for suitable areas to nectar or lay eggs. The

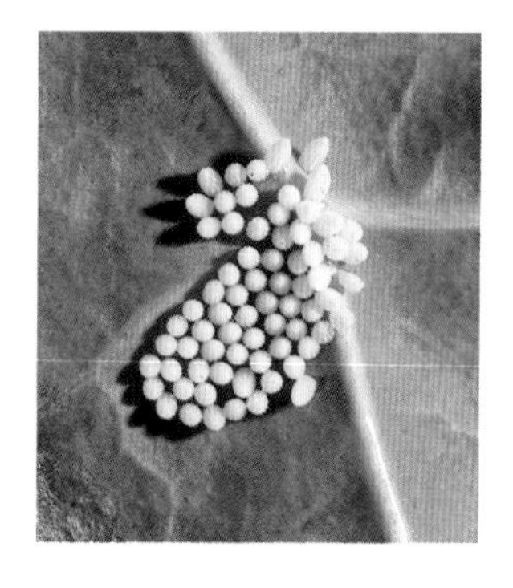

Large White eggs on Cabbage.

Left:
Gregarious larvae of the Large White.

Right:
Large White pupa.

adults are able to fly several hundred kilometres in a lifetime. They migrate here regularly from France, readily crossing the English Channel. In warm periods in late summer and autumn, adults may be seen flying south, evidence for a tendency to reverse migration.

Distribution and trends

The Large White is a strong flyer and, from the spring onwards, may be encountered in any part of Britain and Ireland, including the most remote offshore islands. A preference for cultivated foodplants means that it is most often seen close to farmland and gardens and far less often in upland areas, such as the Scottish Highlands. The lack of records in parts of Scotland reflects the scarcity both of butterflies and of recorders.

There are always more adults in the summer generation than in the spring though numbers vary greatly from year to year. In some years large numbers are lost to the small parasitic wasp, *Apanteles glomeratus*, which lays eggs in the larvae. The wasp larvae feed within the growing caterpillars and then pupate in clusters of yellow cocoons beside the dead larvae. Mortality rates of larvae through parasitism are high (>95% in some cases) and this is an important factor controlling population numbers of this species.

Until the middle of the twentieth century, populations reached very high levels, with huge numbers being seen in some years. Infection with granulosis virus, brought from continental Europe by migrating adults, may have helped to reduce numbers after the mid 1950s. Populations may also be smaller now due to increased use of pesticides on brassica crops, the intensification of farming, and loss of habitat in the near continent.

There are still records of very large numbers in some years, usually in coastal areas, and numbers in Britain are thought to be boosted by sporadic mass migrations from the continent. In 1992, for example, many observations on the coast of Suffolk and Essex (including those made from a yacht several miles offshore) gave strong support to the occurrence of a massive migration from France in mid-July in that year.

On the other hand, figures from the Butterfly Monitoring Scheme (BMS) suggest that mass immigrations have had little influence on overall numbers in recent years (including 1992 when there was a four-fold increase in the annual all-site collated index). The results show that numbers of second brood adults have been closely related to those of the first brood in each year, which would not be the case if erratic, large-scale migrations were taking place. The BMS data show that numbers at a single site show the typical pattern of building to a peak and then declining at the end of the flight period, without the sharp increases associated with the sudden arrival of large numbers of migrants.

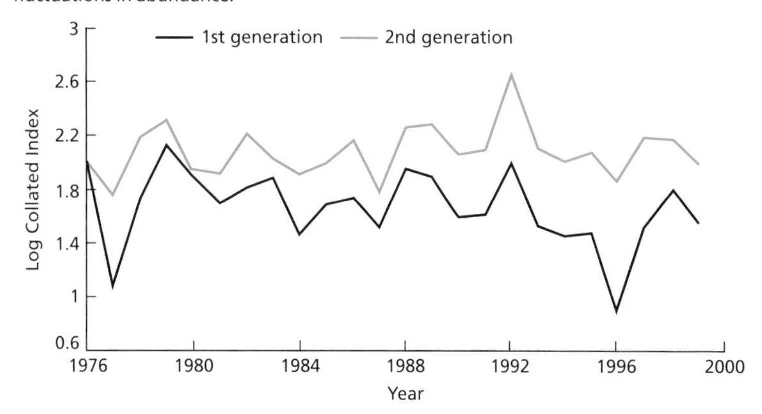

First and second broods of the Large White show similar fluctuations in abundance.

The conclusion seems to be that migration at a certain level occurs every year, occasionally and locally on a large scale, but that this now happens less than in earlier decades.

European trend

The range of the Large White is stable in most European countries but has declined in Sweden (>50% decrease in 25 years), Romania (25–50% decrease), and Austria, Finland, Luxembourg, and Malta (15–25% decrease). It has expanded recently in Latvia, the Netherlands, and north-west Russia.

Interpretation and outlook

The Large White is one of the most successful and widespread butterfly species in Britain and Ireland, and has undoubtedly benefited from the growing of brassica crops. The use of cultivated crops by the Large White also makes parasitism more prevalent, as larvae are concentrated into smaller areas and are more easily located. The link between host and parasite populations may be a key factor governing the swings in Large White abundance. The same factor probably also affects the Small White but not the Green-veined White, which has solitary larvae (and more dispersed breeding) and is less prone to parasitism.

The status of the Large White is unlikely to change in the foreseeable future, unless new pest control methods affect populations more drastically than those in current use. In gardens 'green' control methods such as inter-planting and wiping with detergents are being encouraged as alternatives to the use of insecticides. Although the Large White is an unwelcome visitor in many gardens, some gardeners grow Nasturtiums especially to encourage these large and attractive butterflies to breed.

Key references

Baker (1970, 1978); Courtney (1986); Feltwell (1982); Mendel (1995); Pollard (1994).

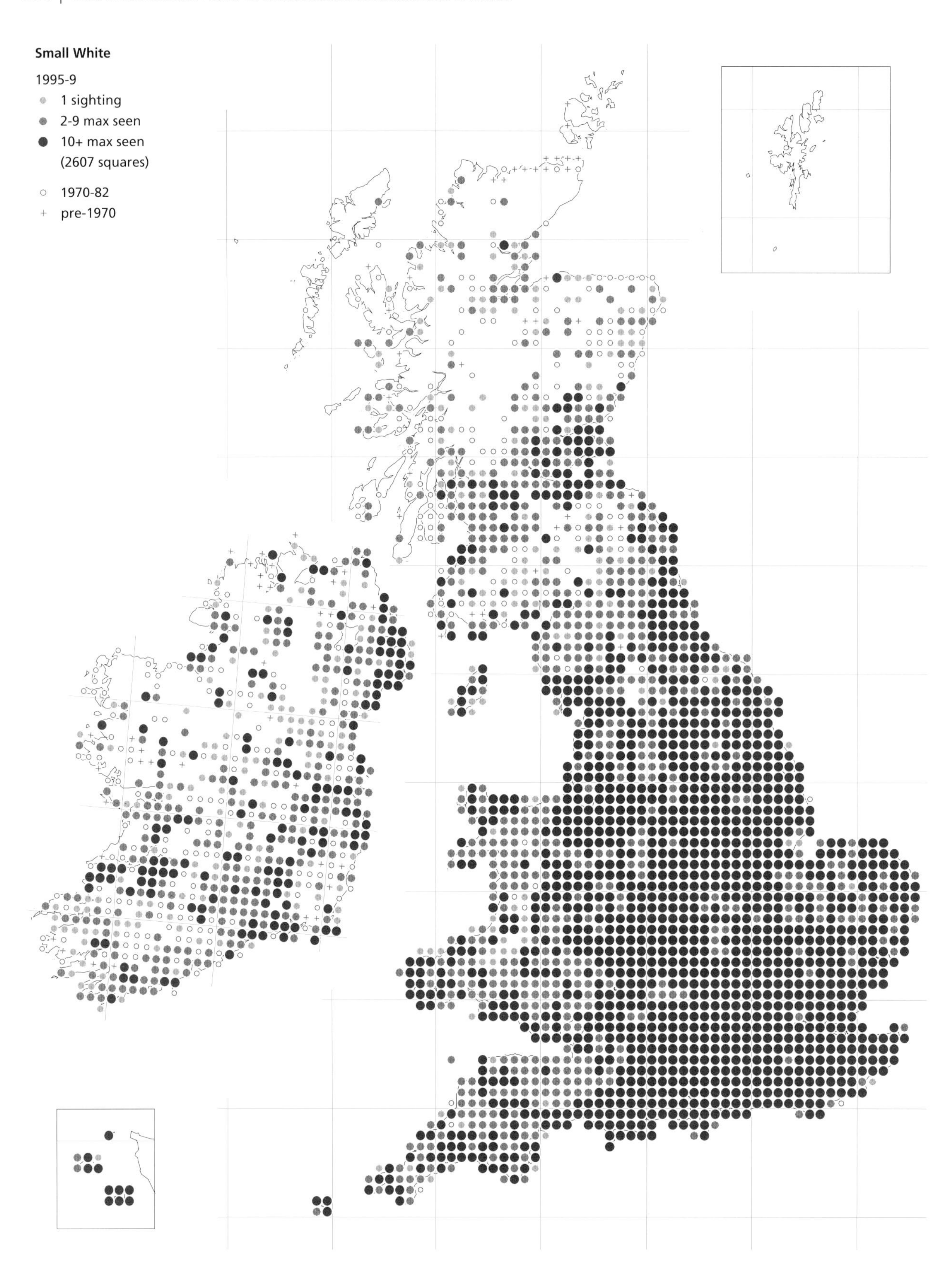
Small White
1995-9
1 sighting
2-9 max seen
10+ max seen
(2607 squares)
1970-82
pre-1970

Small White *Pieris rapae*

Resident

Range stable.

Conservation status

UK BAP status: not listed.
Butterfly Conservation priority: low.
European status: not threatened.

European/world range

Throughout Europe and north-west Africa, and east to Asia and Japan. Introduced to North America in the nineteenth century and Australia in 1939, it is now widespread in both. Its range has changed little in Europe.

The Small White is a highly mobile species and each year the resident population is boosted by individuals flying in from mainland Europe. It is a common visitor to gardens where it breeds on brassicas and Nasturtium, though it relies less on cultivated brassica crops than the Large White and breeds on a range of wild foodplants. Adult butterflies are attracted to white flowers where they feed and on which they are well camouflaged when roosting. This is a common and widespread species.

Foodplants

Cultivated brassicas are used, especially cabbages, and Nasturtium (*Tropaeoleum majus*) in gardens. Wild crucifers, including Wild Cabbage (*Brassica oleracea*), Charlock (*Sinapis arvensis*), Hedge Mustard (*Sisymbrium officinale*), Garlic Mustard (*Alliaria petiolata*), Hoary Cress (*Lepidium draba*), and Wild Mignonette (*Reseda lutea*) are used to a lesser extent.

Habitat

It occurs in almost any habitat but is most plentiful in gardens and fields where brassica crops are grown. Large numbers may congregate in fields of Oil-seed Rape. Elsewhere, it is found in smaller numbers especially in sheltered places such as hedgerows and field and wood edges where wild crucifers occur.

Life cycle and colony structure

There are two, occasionally three, broods per year. Adults may emerge during April if the weather is warm but most first brood individuals fly during May and June, almost overlapping with the second, larger, brood emerging in July. In warm summers a small third brood may be produced, with some individuals flying well into autumn (see p. 392).

The bottle-shaped **eggs** are laid only in warm weather, singly on the underside of leaves of foodplants, although there may be several eggs on each plant. The Small White has a strong preference for foodplants growing in sheltered situations, so that when large fields of brassicas are infested, larval feeding damage may be confined to the plants growing at the edge. Eggs hatch after about a week, with the **larvae** feeding on the host-plant for about two weeks. Unlike the larvae of the Large White, they are well-

Small White eggs on Nasturtium.

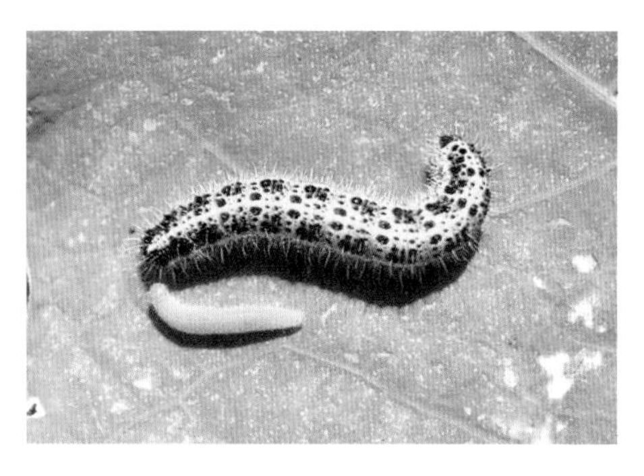

Left
The green larva of the Small White can be distinguished easily from the black and yellow larva of the Large White.

Right
Small White pupa.

camouflaged, being green and minutely hairy, similar to the stems of their foodplants, and they do not concentrate poisonous mustard oil as a defence. **Pupation** takes place in a variety of situations: the spring generation may remain on the foodplant or nearby vegetation, while overwintering pupae are generally attached to a more permanent surface such as a wall, post, or tree trunk.

The Small White does not form colonies but adults concentrate where there are abundant nectar sources or foodplants. The butterfly is highly mobile and adults fly across the countryside mingling with individuals that have flown from mainland Europe. Individuals may be seen travelling in a northerly direction early in the year and southwards in late summer in an apparent attempt at reverse migration. Though some adults may travel only a small distance in a lifetime (1 km or less), it has been estimated from the rapidity with which this species spread across Australia (3000 km in 25 generations) that an individual Small White can fly over 100 km.

Distribution and trends

This is a common and widespread butterfly that has been recorded in all areas of England, Wales, and Ireland. In Scotland, it is common in the south but scarcer than the Large White in the highlands where it is more localized and closely associated with human habitation. It is not clear whether this is due to limitations of foodplant supply or to local climatic factors. The Small White is absent from Orkney and Shetland but occurs on the Outer Hebrides (Lewis and Barra) and most islands further south, including many Irish islands, the Isle of Man, and the Channel Islands. It is encountered far less frequently than the Green-veined White on higher ground throughout Britain and Ireland.

It is under-recorded in parts of Ireland and Scotland. Some confusion between this species and the Green-veined White, especially during flight, may have contributed to under-recording in areas where the latter is more common.

Although the distribution of the Small White has remained stable, it seems that numbers were higher before the introduction of modern insecticides in the 1950s. Paradoxically, these chemicals have been shown to have a more lethal effect on the natural predators of Small White larvae such as harvestmen and beetles.

Population levels of the Small White vary greatly from year to year and numbers in the second brood are boosted by the arrival of immigrants from continental Europe. First brood individuals may be scarce in some years, particularly in parts of Wales and Scotland.

Butterfly Monitoring Scheme data confirm that the second generation is consistently larger in number than the first, though its timing varies from year to year. This increase in numbers between broods is more marked than in any other species. Counts from the scheme show a regular pattern, suggesting that immigration occurs as a steady flow of individuals rather than erratic influxes.

European trend

Its range is stable in most European countries but small declines have been recorded in Albania, Finland, Malta, and Sweden (15–25% in 25 years). It is spreading in Madeira (>100% increase in 25 years) and the

Canary Islands, the Netherlands, Romania, and north-west Russia (25–100% increase).

Interpretation and outlook

The Small White, like the Large White, is one of the most successful and widespread butterfly species in Britain and Ireland. Both species have benefited from the cultivation of brassica crops and have adapted to intensive land use by humans more successfully than other butterflies. It seems unlikely that the status of the Small White will change in the foreseeable future, unless farming practices change significantly and pest control methods are developed that affect populations more drastically than those in current use.

A preference for cultivated habitats means that conservation measures are inappropriate; indeed many gardeners who otherwise welcome butterflies often wish to control numbers of this species. However, a garden or allotment containing weeds shelters more predators than a tidy one and brassica crops may suffer less damage. Being mobile, the butterflies readily return to lay more eggs, and numbers recover more quickly after an application of pesticide spray than its more sedentary invertebrate predators and pesticide use, both on gardens and commercial crops, may have a very limited impact on this species. Although existing larvae may be killed, so are insect predators that would otherwise control Small White populations. An untidy, unsprayed garden in which there is more specific control of larvae on brassicas, may therefore suffer less from the depredations of the Small White than one that is weed-free and sprayed with pesticides.

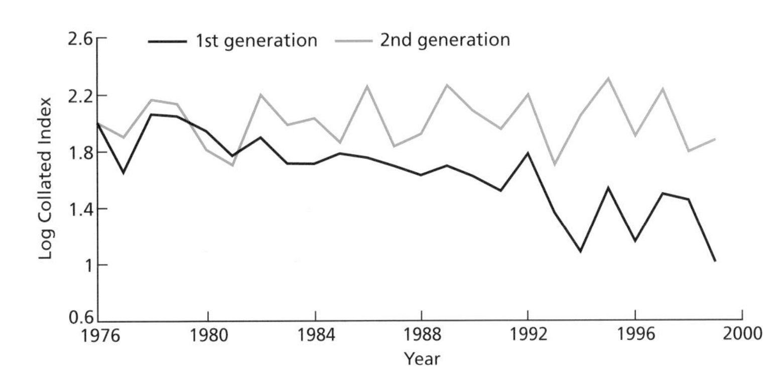

The second generation of the Small White is consistently more numerous.

Key references

Baker (1970, 1978); Courtney (1986); Dempster (1967*a*,*b*).

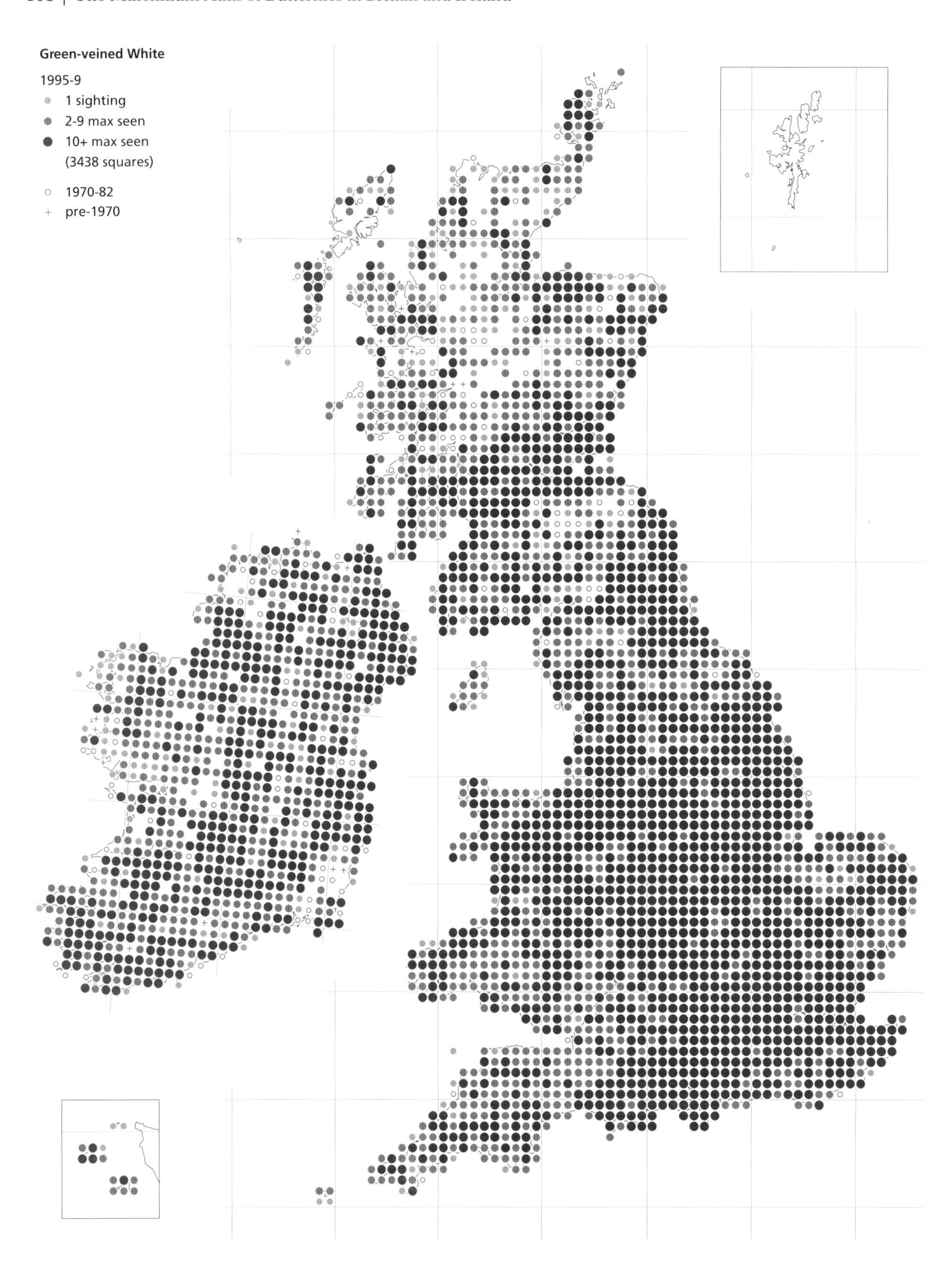
Green-veined White
1995-9
1 sighting
2-9 max seen
10+ max seen
(3438 squares)
1970-82
pre-1970

Green-veined White *Pieris napi*

Resident

Range stable.

Conservation status

UK BAP status: not listed.
Butterfly Conservation priority: low.
European status: not threatened.

European/world range

Across Europe (except for some Mediterranean islands), parts of North Africa, across Asia and in North America. Its range is stable in most European countries.

The Green-veined White can be found throughout the countryside, but prefers damp, sheltered areas. It breeds on wild crucifers and is not a pest of cabbage crops. In many areas away from human habitation, especially on higher ground and in northern latitudes, this species is more often encountered than Large or Small Whites. The dusky vein markings on the undersides of the wings are variable in colour and make it well camouflaged when it roosts among vegetation. The butterfly is common and widespread in Britain and Ireland, but it is probably less abundant than formerly due to loss of its grassland habitats.

Foodplants

A range of wild crucifers is used: Garlic Mustard (*Alliaria petiolata*), Cuckooflower (*Cardamine pratensis*), Hedge Mustard (*Sisymbrium officinale*) Water-cress (*Rorippa nasturtium-aquaticum*), Charlock (*Sinapis arvensis*), Large Bitter-cress (*C. amara*), Wild Cabbage (*Brassica oleracea*), and Wild Radish (*Raphanus raphanistrum*). Nasturtium (*Tropaeolum majus*) and cultivated crucifers are also used occasionally.

Habitat

Adults occur widely but tend to congregate in damp, lush vegetation where their foodplants are found, especially hedgerows, ditches, banks of rivers, lakes, and ponds, damp meadows and moorland, and woodland rides and edges.

Life cycle and colony structure

Throughout its range the Green-veined White has at least two generations per year. Adults begin to emerge in April and produce a second brood that is on the wing from late June or early July through to late August. In southern Britain, the first brood contains fewer adults than the second, but both broods are of similar size further north. This is shown clearly by data from the Butterfly Monitoring Scheme (BMS). In warm years the summer brood produces a third generation, which flies in September. The flight period and numbers are therefore variable and in places where good numbers are regularly seen, adults may be present continuously throughout the summer months.

Green-veined White eggs on Cuckooflower.

The first brood adults tend to have darker veins than later broods but there is also less black on the upperwings, particular in the males which may be mistaken for the Wood White. In Ireland and parts of Scotland the yellowish background colour is more evident and the veins darker.

Females usually deposit their **eggs** on small plants, laying them singly on the undersides of leaves; they hatch after 1–2 weeks. The pale, elongated eggs are easy to find in suitable habitat. The green **larvae** are well camouflaged and similar in appearance to Small White larvae, but without the yellow stripe down the side. Green-veined White larvae are sometimes found feeding on the same foodplants as Orange-tip larvae, but competition does not occur because the former feed on the leaves, while the latter eat the flower heads and developing fruits.

When fully grown, the larvae wander away from the foodplant to pupate in well-hidden positions among surrounding vegetation. The second (or third when this occurs) generation of the Green-veined White overwinters as **pupae** with adults emerging in the following spring.

In the south of Britain, the Green-veined White does not form colonies and adults fly widely across the countryside, concentrating wherever there are abundant nectar sources or foodplants. In northern areas, especially in the uplands, and in Ireland, it appears to be more colonial and sedentary. Even in the south, it seems generally to be a less mobile species than the Large White or Small White, and migration from continental Europe is not a common occurrence.

Distribution and trends

The Green-veined White is very common and widespread in most parts of Britain and Ireland, especially in the damper western areas. In the north of Scotland, it is more widely recorded than the Large and Small Whites. It was first recorded on Orkney in the mid 1930s, where it is now common. It remains absent from Shetland but is present on many other islands, including the Isle of Man and the Channel Islands. A comparison of the apparent gaps in the Green-veined White distribution in the Scottish highlands with areas of poor coverage shown in the map of recording visits (see map p. 42) suggests that the gaps are due to under-recording rather than absence.

There does not appear to have been any change in the distribution of this species in the last two decades, but it has almost certainly become less abundant in parts of its range. Many damp meadows and other patches of lush native vegetation where it once bred have been drained, ploughed, or otherwise destroyed or damaged in recent decades, reducing the amount of suitable habitat available, especially in southern and central areas.

The Green-veined White is often the only butterfly to be encountered in parts of Wales, Scotland, and Ireland, especially on moorland and other upland habitats. It has been recorded in more 10 km squares (92% of the total) than any other species during the 1995–9 survey.

As the Green-veined White is the most widespread of the Whites and is more commonly seen than the others in remote areas, it is not thought that misidentifications have affected the accuracy of the map. Nevertheless, there may be some confusion with other white butterflies in flight, especially the Small White and female Orange-tip, but when settled the characteristic green veins are easily seen.

European trend

The range of the Green-veined White is stable in most countries, but increases have been recorded in the Netherlands (>100% in 25 years) and Romania (25–100% in 25 years).

Interpretation and outlook

The Green-veined White does not show the large swings in abundance of the Large and Small Whites, perhaps because of lower levels of parasitism. Changes in abundance are probably more closely linked to climate and habitat availability. This widespread species did not suffer the large decline experienced by many butterflies during the twentieth century, probably due to its use of a range of foodplants and ready mobility.

BMS data show that it was emerging significantly earlier by the end of the 1990s, and having significantly longer flight periods than when the scheme began in 1976. Both phenomena are thought to be linked to warmer spring and summer temperatures in more recent years.

However, the BMS data also show that high summer temperatures lead to a significant reduction in Green-veined White numbers the following year. This is perhaps not surprising given the butterfly's preference for damp conditions, but it suggests that populations might decline if, as predicted, climate change brings warmer, drier summers.

The Green-veined White may also become less abundant within its range as damp grasslands continue to be drained or 'improved' for agriculture. Fortunately it breeds in a wide range of habitats and can be encouraged by simple conservation measures such as maintenance of wide rides and glades in woodland, and leaving wide, grassy verges along water courses and hedgerows.

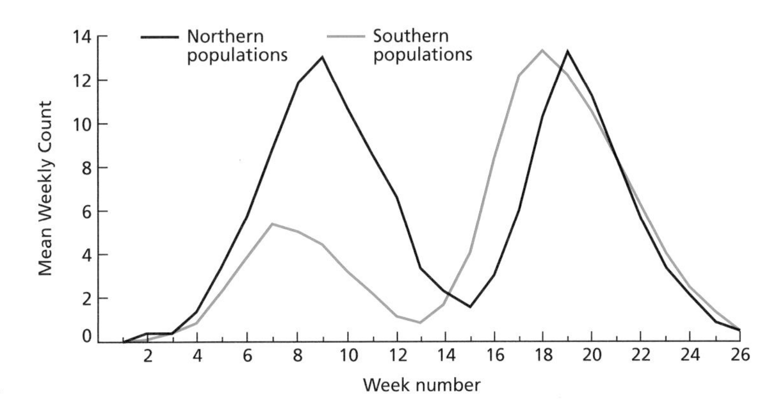

Green-veined White BMS flight period graph showing differences in brood size between populations in the north and south.

Key references

Courtney (1986); Dennis (1985*a*).

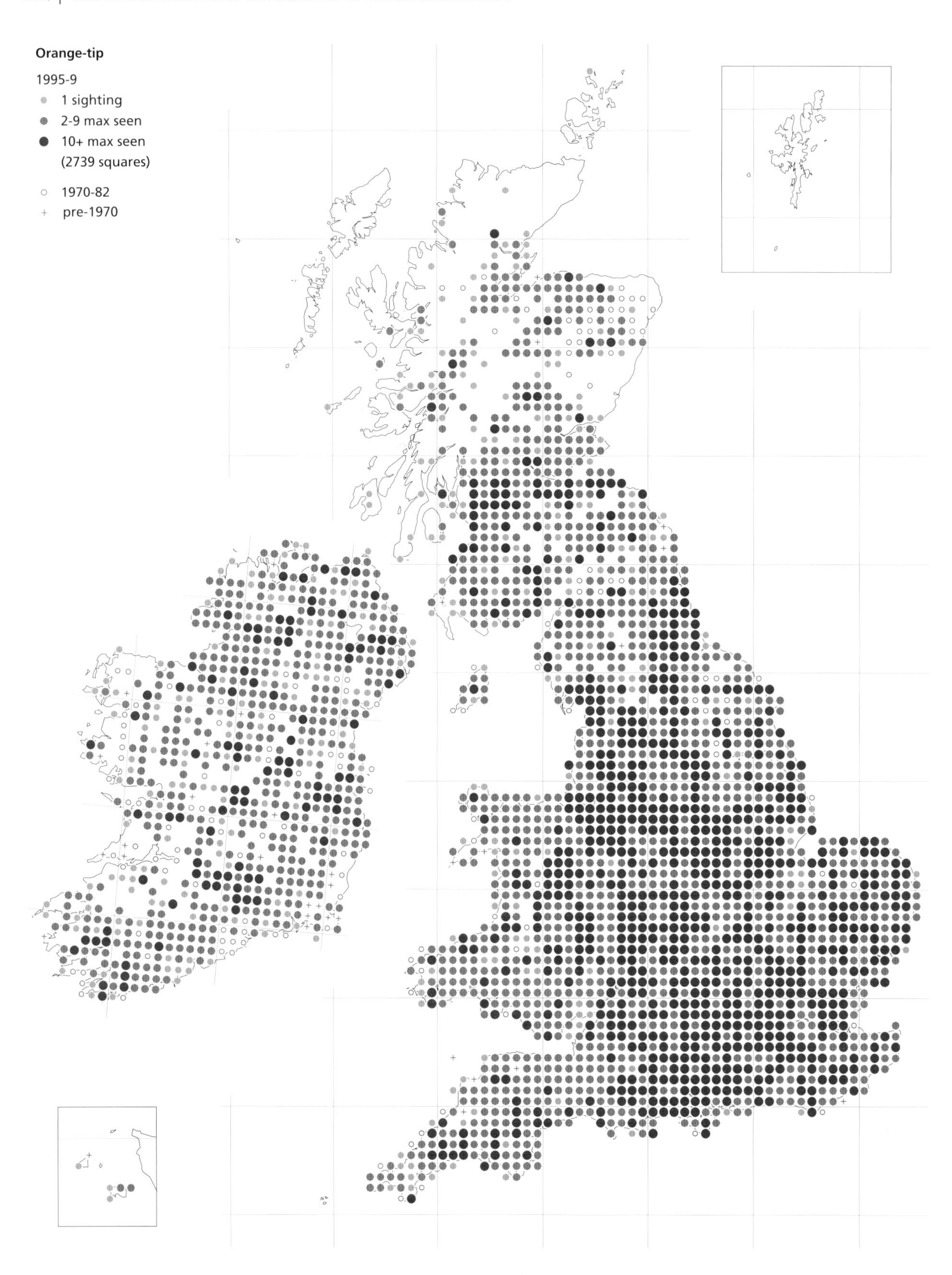
Orange-tip
1995-9
1 sighting
2-9 max seen
10+ max seen
(2739 squares)
1970-82
pre-1970

Orange-tip *Anthocharis cardamines*

Resident
Range expanding.

Conservation status
UK BAP status: not listed.
Butterfly Conservation priority: low.
European status: not threatened.

European/world range
Through most of Europe, as far north as central Scandinavia, across the Middle East and into temperate parts of Asia, as far as Japan. Its European range appears to be spreading northwards.

Orange-tips are seen commonly in early summer along hedgerows, road verges, and woodland edges. Males have vivid orange wing tips, whereas females have no orange coloration and are predominantly white on the uppersides. The mottled pattern of yellow and black scales on the underside hindwings provides excellent camouflage when they roost on flower heads such as those of Cow Parsley. The butterfly is widespread in Ireland and southern Britain and has spread north rapidly over the past 25 years, especially in Scotland.

Foodplants
Several crucifers are used, especially Cuckooflower (*Cardamine pratensis*) in damp meadows and Garlic Mustard (*Alliaria petiolata*) along road verges and ditches. Occasionally, it uses Hedge Mustard (*Sisymbrium officinale*), Winter-cress (*Barbarea vulgaris*), Turnip (*Brassica rapa*), Charlock (*Sinapis avensis*), Large Bitter-cress (*C. amara*), and Hairy Rock-cress (*Arabis hirsuta*). In addition, it lays eggs on Honesty (*Lunaria annua*) and Dame's-violet (*Hesperis matronalis*) in gardens, but larval survival is thought to be poor on these plants.

Habitat
A wide range of damp grassy habitats is used, including meadows, grassy areas in woodland, road verges and waterside habitats such as ditches and the banks of rivers and canals. Northern and western populations seem to be associated mainly with wetter habitats and Cuckooflower is the usual foodplant, perhaps because Garlic Mustard is less common.

Life cycle and colony structure
The Orange-tip has one generation each year. Adults normally fly between mid-April and mid-June, although they may be seen as early as mid-March in southern counties, if there is an early warm spell, and as late as July, if the weather is cool. In Scotland, the flight period is on average up to two weeks later than in southern England (see p. 393) and sightings were reported in August in the late 1980s, and in 1997 and 1998. If the weather is exceptionally warm during spring, as it was in 1997, the flight season in the south can be as short as a month. Data from the Butterfly Monitoring Scheme (BMS) show that the flight period of the Orange-tip now occurs significantly earlier in the year, than when the scheme began in 1976.

The **eggs** are laid at the base of flower heads on plants growing in full sun. They are initially pale in colour but turn orange within 2–3 days, making them easy to record. It is unusual for more than one egg to be laid on a single flower head and experiments have shown that egg-laying females leave a pheromone, which deters subsequent

Orange-tip egg on Cuckooflower.

females from using the same flower. The **larvae** eat the nutritious developing seeds within flower heads, while they are still tender. If larvae hatch when seed pods are more advanced, they are less likely to survive. The larvae are also known to be cannibalistic. This gives selective advantage, allowing one larva to grow to maturity if several eggs are laid on the same plant. It appears that a single Cuckooflower plant is sufficient for only one larva, although the more vigorous Garlic Mustard plants may support more than one. Usually, larvae complete their development on a single plant.

The **pupae** are formed in mid-summer and pass the winter in tall vegetation close to the larval foodplant. The long pupal cases are pointed at each end and are usually coloured pale brown or sometimes pale green-brown or green, matching their surroundings. It has been found recently, with captive-reared stock, that emergence from some pupae may be delayed by one or even two years. If this occurs in the wild, it may provide a safeguard against emergence in unsuitable conditions in a particular season.

The butterflies range widely through the countryside and can be found wherever foodplants occur in sunny positions. It has been shown that local populations are controlled by the availability of suitable foodplants. The Orange-tip does not form distinct colonies in the main parts of its distribution—the adults fly vigorously and probably range over many kilometres to find suitable foodplants. Even where foodplants are abundant, butterfly population densities remain relatively low. In the northern part of their distribution, they are believed to be less wide-ranging, forming more local populations where suitable habitats may be more restricted.

Distribution and trends

The Orange-tip is widely distributed across the southern half of Britain and is found wherever its foodplants occur. Data from the BMS show little change in abundance at monitored sites. This is consistent with its selective egg-laying behaviour, which tends to maintain a low, but relatively constant local population density.

Further north, its distribution has undergone major changes during the past 150 years. It disappeared from many parts of Scotland during the late nineteenth and early twentieth centuries, although populations persisted in Aberdeenshire and in the south. The species has been re-expanding since the 1940s, and has extended its range rapidly since the last atlas survey. This spread is particularly noticeable in Scotland where the number of 10 km squares in which it is recorded has more than doubled since 1970–82.

Overall, the Orange-tip has increased its range by over 40% in Britain and the Isle of Man, with expansions in north-west England, Borders, Dumfries and Galloway, and the central belt of Scotland accounting for most of this increase. In north-west Scotland, the butterfly has extended well into the Scottish Highlands, through the shelter of the glens, since about 1975. It has now reached Skye, Eigg, Tiree, and Islay, but remains absent as a resident species from the Outer Hebrides, Orkney, and Shetland. A single male Orange-tip was recorded in Orkney in 1999, but the origin of this individual is unknown.

Orange-tips have been recorded throughout Ireland, where the foodplant is predominantly Cuckooflower, with some use of Dame's-violet (Garlic Mustard is much less common). In Northern Ireland, Cuckooflower seems to be spreading back into some previously 'improved' pastures, possibly increasing the number of suitable breeding sites.

Although in flight the female can be confused with Small and Green-veined Whites and may be overlooked, the distinctive appearance of the male makes it easy to identify. As a result, records of this butterfly are probably dominated by sightings of males. Their easy identification prompted a scheme to distribute postcards, initially in Scotland and later in Wales and Ireland, illustrating the butterfly with the caption 'Have you seen an Orange-tip butterfly this year?' (see p. 35). The card had a pre-printed return address and space for record details. More than 1000 cards were returned with dates and locations of sightings.

The eggs and larvae are easy to find for about a month beyond the end of the flight period and therefore aid recording in poor weather conditions. Many records of immature stages were submitted during the current survey.

European trend

The range of the butterfly is stable in most European countries and it has spread in Denmark and Hungary in the past 25 years. Over a longer period (30–100 years) the European range has expanded to the north.

Interpretation and outlook

Despite the recent expansion of its range in northern Britain, the outlook for the Orange-tip is more gloomy further south. Here, it seems certain that the species has become less common in the wider countryside. Haymaking and the traditional management of flood meadows have given way to silage making, the widespread installation of drainage systems, and the application of fertilizers. As a result, Cuckooflower has declined dramatically in abundance on a local scale and suitable meadow habitats for the Orange-tip have been greatly reduced.

Increasingly, southern populations of the Orange-tip seem to rely on Garlic Mustard, a vigorous plant, which is abundant along damp roadside verges. The plant and the butterfly may have benefited from a recent reduction in verge cutting by local authorities. The butterflies are now seen frequently along many rural roadsides where this foodplant occurs.

It is unrealistic to expect a major reversal in the widespread trend towards grassland drainage and fertilizer application in agriculture, but there is increasing emphasis on the conservation of wet, unimproved, meadow habitats and on leaving unsprayed field margins, which encourage the butterfly's foodplants.

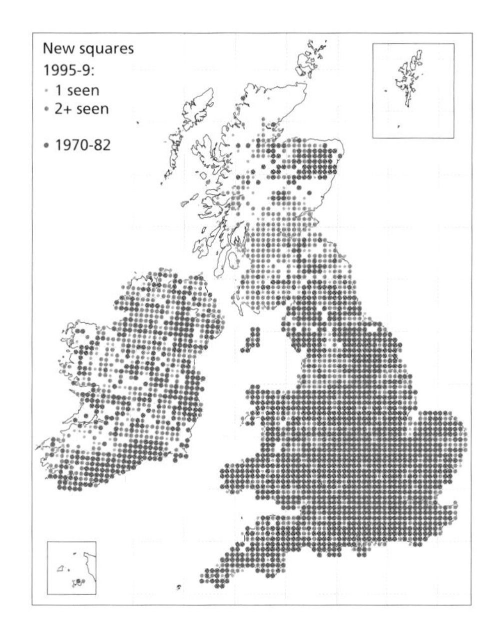

The expansion of range of the Orange-tip since 1970–82.

However, the future of this butterfly in the wider countryside may depend as much on an enlightened approach to roadside verge and hedgerow management, allowing a varied flora to develop and providing ample supplies of Garlic Mustard. It is also critical that roadside verges are left uncut during spring and early summer, when the plants are coming into flower and larvae are feeding high on the plants.

We do not know enough about the interplay of factors, such as foodplant availability and weather conditions, to understand fully the major changes in the range of the Orange-tip in the north. However, climatic factors are likely to have played a role in the decline and subsequent expansion of the species. It seems to be sensitive to the local climate in northern areas, as indicated by its occurrence in relatively sheltered, partly wooded localities that offer a mild microclimate. This species may be an indicator of long-term climatic changes in northern Britain, where abundance may be linked with annual weather patterns.

Key references

Courtney and Duggan (1983); Dempster (1992, 1997); Eales (1999*a*); Lees (1980); Stewart *et al.* (1998); Sutcliffe and Kirkland (1998); Wiklund and Åhrberg (1978).

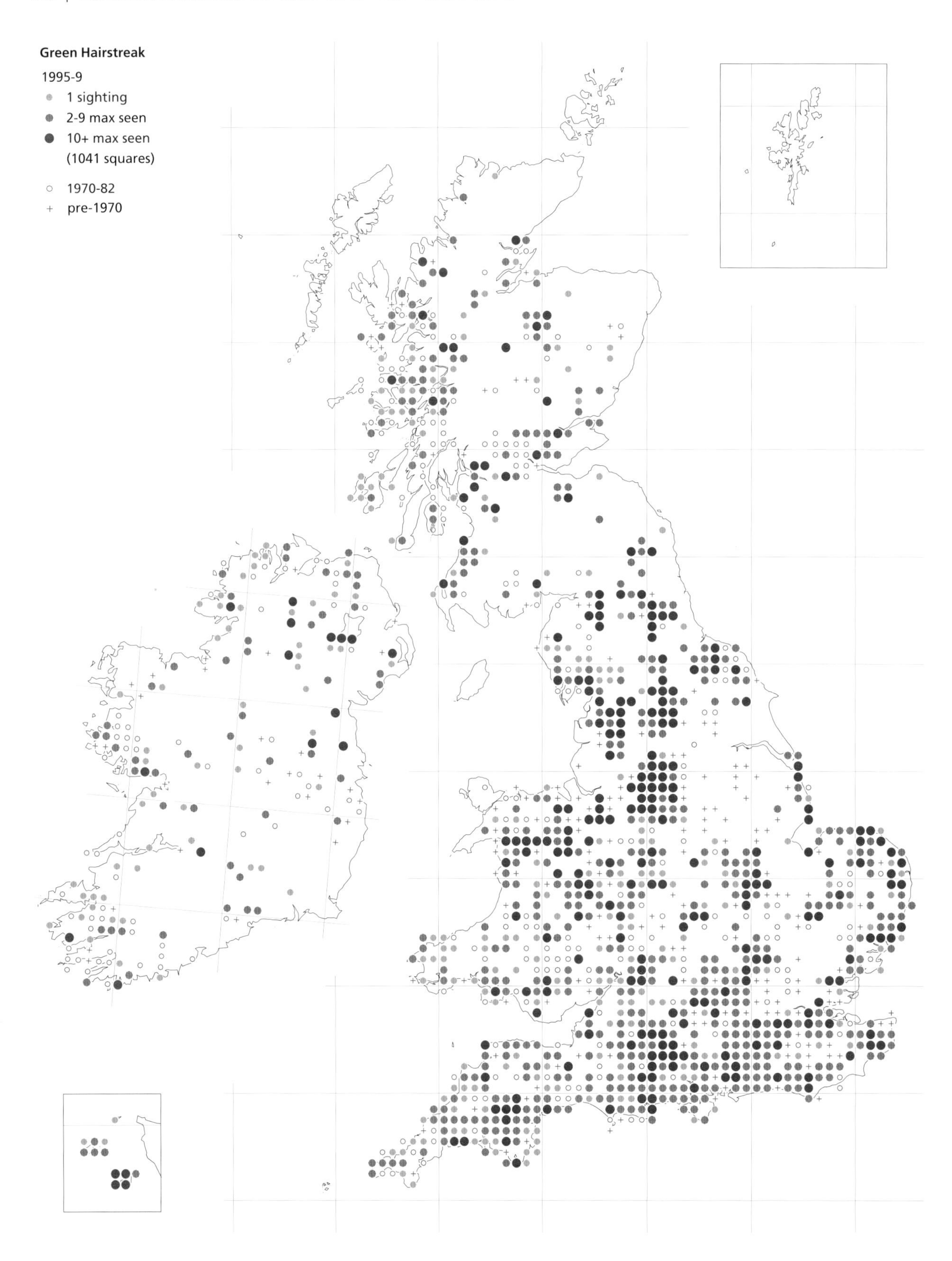
Green Hairstreak
1995-9
1 sighting
2-9 max seen
10+ max seen
(1041 squares)
1970-82
pre-1970

Green Hairstreak *Callophrys rubi*

Resident
Range stable.

Conservation status
UK BAP status: not listed.
Butterfly Conservation priority: low.
European status: not threatened.

European/world range
Throughout Europe, parts of North Africa and across Asia to Siberia. It is stable in most of Europe but has declined in several countries.

The Green Hairstreak holds its wings closed, except in flight, showing only the green underside with its faint white 'streak'. The extent of this white marking is very variable; it is frequently reduced to a few white dots and may be almost absent. Males and females look similar and are most readily told apart by their behaviour: rival males may be seen in a spiralling flight close to shrubs, while the less conspicuous females are more often encountered while laying eggs. Although this is a widespread species, it often occurs in small colonies and has undergone local losses in several regions.

Foodplants
Common Rock-rose (*Helianthemum nummularium*) and Common Bird's-foot-trefoil (*Lotus corniculatus*) are used on calcareous grassland, while Gorse (*Ulex europeaus*), Broom (*Cytisus scoparius*), and Dyer's Greenweed (*Genista tinctoria*) are used on heathland and other habitats. Bilberry (*Vaccinium myrtillus*) is used almost exclusively on moorland and throughout Scotland. Other foodplants include shrubs such as Dogwood (*Cornus sanguinea*), Buckthorn (*Rhamnus cathartica*), Cross-leaved Heath (*Erica tetralix*), and Bramble (*Rubus fruticosus*).

Habitat
Green Hairstreak colonies may be found on calcareous grassland, woodland rides and clearings, heathland, moorland, bogs, railway cuttings, old quarries, and rough, scrubby grassland. This species occurs on a wide range of soils but is strongly associated with scrub and shrubs, which are usually present at sites where it breeds.

Life cycle and colony structure
There is one generation a year. Adults begin to emerge in mid-April in southern England. The main flight period in the south is during May and early June, with a few individuals surviving into July. The flight period may be up to a month later in Scotland (see p. 393).

Males are seen more readily than females as they are very territorial and perch on shrubs, usually at the base of a slope, waiting for females to fly past. Sometimes there may be several males on a favoured bush and aerial fights take place frequently, as they defend their territories. The same perches may be used year after year by successive generations.

On chalk grassland, females fly low over the ground and they can be seen crawling over Common Rock-rose plants searching for young growth on which to lay their **eggs**. Where shrubs such as Dogwood and Buckthorn are used, eggs are laid singly in the flower buds and it is the flowers that are eaten. **Larvae** use a wide range of foodplants and the list of foodplants may lengthen as the species is studied further. Larvae are green with yellow markings when

Green Hairstreak larva on Gorse.

full grown and have the shape of a woodlouse. Their camouflage makes them hard to find, even when they are feeding on young buds and shoots. They are not known to attract ants or to be tended by them, but the **pupae**, which are formed at ground level, emit squeaks, audible to human ears, that do attract ants. One pupa was discovered in a nest of the ant *Myrmica sabuleti*, and it is thought that pupae are always buried by ants, though the range of species involved is not known. The Green Hairstreak is the only hairstreak to overwinter in the pupal stage and is consequently the earliest to emerge in spring.

Adults are rarely seen in large numbers, partly because they are inconspicuous and partly because most of their colonies are small, containing only a few dozen individuals in a season. Where there are large tracts of suitable habitat, however, larger numbers may be found in some years. Little is known about the mobility of this species, but occasional records from gardens suggest that some dispersal from breeding sites occurs in most years.

Distribution and trends

In Wales and Scotland, this is a widespread but local species, occurring at higher altitudes on moorland than many other butterflies. It also occurs on the Inner Hebrides and Arran, though not on the Outer Hebrides, Orkney, or Shetland. It is widespread in Ireland, particularly in the south and west.

Colonies in England are strongest and most widespread on calcareous grassland in the south and moorland in the north. There have been declines in recent years in eastern counties where these habitats do not occur, and former colonies were found on smaller habitat patches such as railway cuttings and old quarry sites. This has been the case in Suffolk (where colonies are now confined to heathland areas), Hertfordshire, and Essex. It is found in the Channel Islands but not on the Isle of Man.

This inconspicuous butterfly species is easily overlooked and is likely to be under-recorded even in well-worked counties in southern England; new colonies have been found here in the course of recording for this atlas. Chalk grassland colonies are well-recorded because this habitat is visited frequently by recorders during the flight period of the Green Hairstreak. Heathland colonies are more likely to be overlooked as many recorders do not visit until later in the summer when species such as the Silver-studded Blue and Grayling are flying. The Green Hairstreak is now a scarce butterfly in woodland, where most colonies are small and therefore may also be overlooked.

In more remote and less frequented habitats it is particularly under-recorded and in Ireland and western Scotland it is considered by local experts to be more widespread than the current map indicates. In Ireland this is mainly a butterfly of bog and moorland, especially where there are clumps of Gorse, while in Scotland it occurs on open moorland, particularly where there are hollows and gullies, in birch woodland, and the edges and clearings of coniferous forests.

Comparison with the 1970–82 survey shows that this species has been recorded in many different squares in the different time-periods, suggesting under-recording in both the earlier and the current survey. In Britain, it has not been re-recorded in 33% (198) of the 10 km squares where it was found in the last survey, but it has been recorded in over

500 new squares, giving an overall increase of 51%. It is suspected that these results reflect a genuine loss of colonies in some areas, as well as uneven recording. Butterfly Monitoring Scheme data do not show a clear trend in abundance at monitored sites, but because of its inconspicuous behaviour, numbers recorded on transects are generally low.

European trend

It is declining in some countries, notably in Belgium (25–50% decrease in 25 years), and also in Germany, the Netherlands, Romania, and Slovakia (15–25% decrease).

Interpretation and outlook

Although the Green Hairstreak remains a widespread butterfly, colonies have been lost in many places throughout its range due to habitat destruction. Changes in or cessation of management have also led to losses.

On chalk grassland, where Common Rock-rose and Common Bird's-foot-trefoil are used as foodplants, a short sward (less than 10 cm tall) is needed and management for this species should aim to prevent invasion of coarse grasses and scrub. Some scrub is, however, beneficial, providing shelter, perching posts for the males, and additional larval foodplants.

In other habitats there is more reliance on shrubs as foodplants and the Green Hairstreak is thus able to survive in vegetation at later stages of succession than many other butterfly species, as long as foodplants are growing in sunny situations and suitable ant colonies are present.

Many woodland sites have become too shady and the remaining colonies in woodland are perhaps the most threatened. In areas such as the Wealden woods in south-east England, only a few small colonies survive in the scarce patches of open ground provided by very young plantations and wider rides and tracks. A return to active woodland management that provides more open areas would benefit this butterfly along with many other species.

In upland areas, where sheep are kept at high density on moorland, colonies are threatened by overgrazing of the young growth of Bilberry and Gorse. Afforestation is also a threat in the uplands, while in lowland areas drainage and industrial peat stripping can also result in losses. Although the Green Hairstreak occurs in a variety of habitats it is likely that many more colonies will be lost in the future.

Little research has been carried out on the species and much remains to be learned about its life cycle, colony structure, mobility, and habitat requirements. For example, little is known about the relative importance of the different foodplants used, the temperature requirements of the early stages, and the relationship with ants. It is thought to be under-recorded in many areas and more recording is needed to gain a more complete picture of its distribution in Britain and Ireland.

Key references

Fiedler *et al.* (1993); Shaw and Bland (1994).

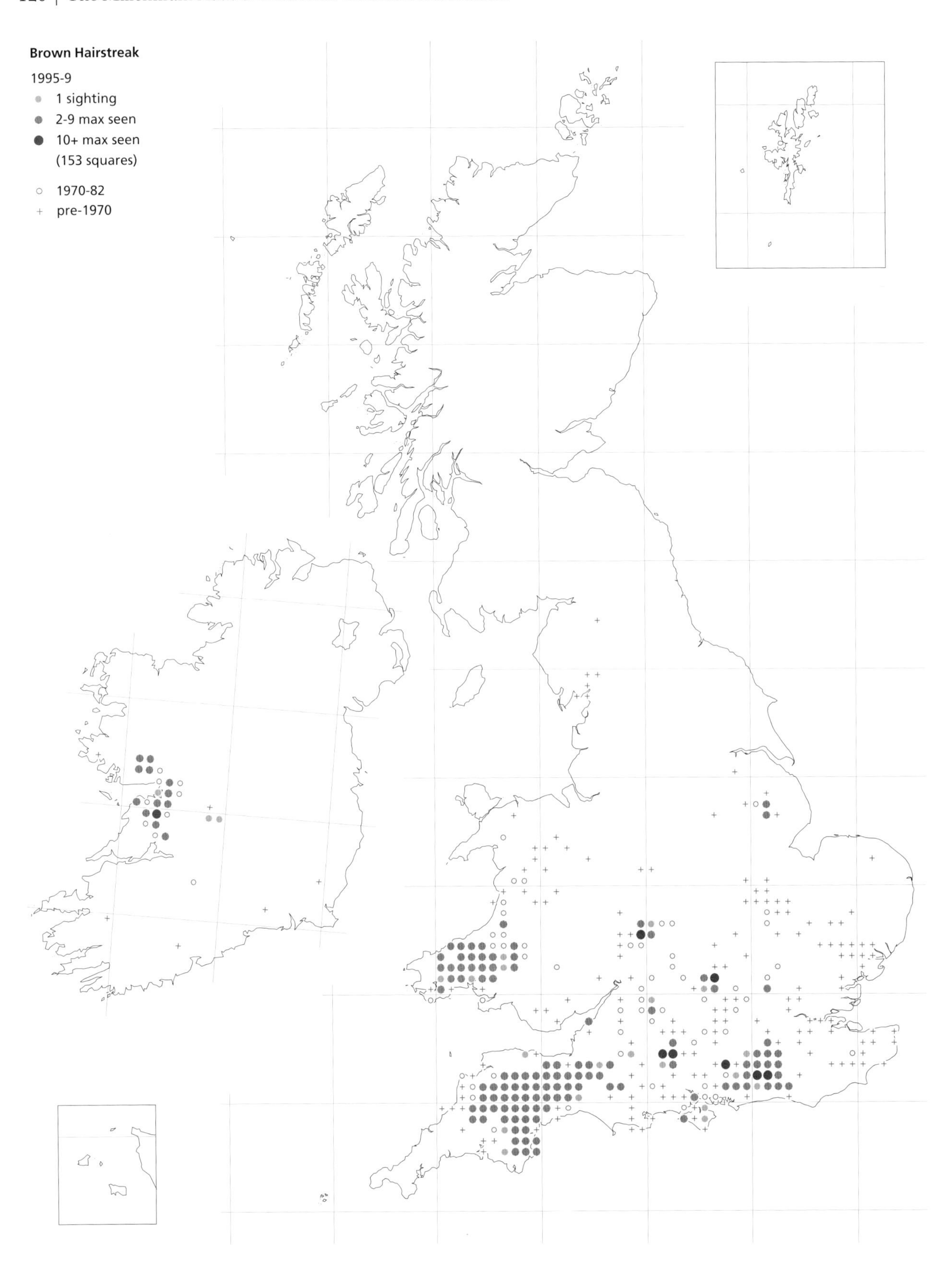
Brown Hairstreak
1995-9
1 sighting
2-9 max seen
10+ max seen
(153 squares)
1970-82
pre-1970

Brown Hairstreak *Thecla betulae*

Resident
Range declining.

Conservation status
UK BAP status: Species of Conservation Concern.
Butterfly Conservation priority: medium.
European status: not threatened.
Protected in Great Britain for sale only.

European/world range
Widely distributed across central Europe from northern Spain to southern Sweden, and east through Asia to Korea. Declining in many European countries.

The Brown Hairstreak is an elusive butterfly that spends most of its life either high in the tree canopy or hidden amongst hedgerows. It is worth looking up at prominent Ash trees along wood edges to see if small clusters of adults may be flitting around a 'master' tree where they congregate to mate and feed on aphid honeydew. Alternatively, adults sometimes feed lower down on flowers such as Hemp-agrimony, Common Fleabane, and Bramble. The females are most frequently seen as they disperse widely along hedgerows where they lay conspicuous white eggs on young Blackthorn. The butterfly is locally distributed in southern Britain and mid-west Ireland and has undergone a substantial decline due to hedgerow removal and annual flailing, which removes eggs.

Foodplants

The butterfly breeds on young growth of Blackthorn (*Prunus spinosa*) and occasionally other *Prunus* species such as Bullace (*P. domestica*).

Habitat

Hedges, scrub, and wood edges are used where Blackthorn is abundant and not too intensively managed. The butterfly typically breeds over wide areas of countryside with extensive networks of hedges and woodland, often on heavy clay soils in low-lying land. In contrast, most Irish colonies are on lighter soils over limestone bedrock.

Life cycle and colony structure

The butterfly is single brooded with adults on the wing from late July until early September, occasionally into October. In order to find mates, adults congregate early in the flight period on one or more prominent 'master' trees, usually along a wood edge. Ash appears to be a favourite, possibly because it supports abundant aphids that produce the honeydew on which the butterflies feed.

After a 6–10 day maturation period, females disperse and lay **eggs** on young Blackthorn, usually singly at the base of a spine, bud, or where one-year-old wood branches from a two-year-old stem. Egg-laying sites are mostly less than 1.5 m above ground level in sheltered areas that are exposed to the sun, generally where young growth is suckering out from the scrub edge or hedgerow. They sometimes lay eggs as high as 2–3 m and on recently cut hedges but will not lay on bushes that have been heavily browsed. The white eggs are quite conspicuous and many records have been derived from systematic egg surveys, conducted during the winter months when the Blackthorn leaves have dropped.

The eggs hatch during bud-burst in late April or early May and, after feeding initially within the unfurling buds, the **larvae** hang beneath the leaves from a silken pad. They are extremely well camouflaged and stay

Brown Hairstreak egg situated in a characteristic position on a Blackthorn twig.

motionless during the day, feeding at dusk on the tender leaf-tips. When fully grown in late June or early July they drop to the ground and **pupate** in cracks in the ground, among leaves, or in grass tussocks. Few observations have been made of wild pupae but they are attractive to ants, which bury them in loose earth cells until hatching.

Brown Hairstreak colonies tend to be quite small, with rarely more than 300 adults, and typically breed at low densities over wide areas, sometimes merging with neighbouring colonies. In contrast, a few colonies occupy small areas of scrub and appear quite isolated.

There is little information on their mobility but females appear to disperse over many kilometres whereas males are more sedentary, tending to remain around the 'master' trees. Despite the female behaviour, colonies usually breed in the same regions of countryside year after year and there has been little evidence of colonization of peripheral areas.

Distribution and trends

The Brown Hairstreak formerly occurred widely across England and Wales but declined substantially during the twentieth century, notably in central and eastern England. It died out in East Sussex in the 1870s, Suffolk and Bedfordshire in the 1940s, and Kent in the 1970s. Recent local surveys show that the butterfly is still declining in many regions, including former strongholds such as Buckinghamshire, Oxfordshire, and Hampshire where overall losses have been estimated to be 60–80%. It has also been reduced to small areas in counties such as Worcestershire and Lincolnshire.

The butterfly currently has strongholds in four main areas: the heavily wooded clays of the west Weald in West Sussex and Surrey; the sheltered low-lying valleys of north Devon and south-west Somerset; low-lying pastoral areas of south-west Wales; and in Ireland, the limestone pavements of the Burren and lowland areas to the east, including Gort and Clarinbridge.

The butterfly's distribution is fairly well recorded thanks to recent surveys, often involving systematic winter searches for eggs, but it is still under-recorded in parts of Wales and Ireland. (Note that some data from early 2000 have been included in the map, as the eggs would have been laid in 1999). Recent egg surveys have shown that the butterfly may have colonized patches of scrub on Salisbury Plain and along the North Downs, but surveys in south Wales have failed to find evidence of breeding on many former sites and it appears genuinely to have declined in this region.

In Ireland, it has recently been discovered near to its Burren stronghold on the north side of Galway Bay in Connemara where it has either spread or been overlooked previously. It has also been re-discovered around Lough Derg, east of the river Shannon.

European trend

Although data quality is poor through much of its range, the butterfly appears to be stable in many European countries. However, it has declined severely in Austria, Belgium, Luxembourg, the Netherlands, and Romania (>50% decrease in 25 years); and in Slovakia and European Turkey (25–50% decrease in 25 years).

Interpretation and outlook

The decline of the Brown Hairstreak has undoubtedly been caused by the widespread destruction of hedgerows during the twentieth century as a result of agricultural intensification, combined with the increased use of mechanical hedge cutting. Over 50% of hedgerows have been removed in Britain since 1945, but losses have not been spread evenly across the country. More hedges have been grubbed out in the predominantly arable centre and east of England, which coincides with the area of greatest decline of the butterfly.

Another significant change affecting the butterfly's survival has been the switch in hedgerow management. The tradition of periodically cutting and laying hedges has now mostly been replaced by annual flailing by tractor-mounted hedge cutters during the autumn and winter. This reduces many hedges to clipped boxes, removing most Brown Hairstreak eggs in the process. The surveys in south Wales suggest that most losses are due to regular flailing.

The conservation of the Brown Hairstreak presents a great challenge because a single colony typically breeds over wide areas of land and cannot be conserved by site based measures alone. It requires the conservation of whole landscapes and sympathetic farm management encompassing woods, hedges, and field margins. Such measures are now included within various land enhancement schemes such as Countryside Stewardship and Environmentally Sensitive Areas which cover some of the butterfly's core areas.

Butterfly Conservation and English Nature have helped the process by providing a colour leaflet explaining the butterfly's needs to landowners and countryside advisors. The advice is to avoid annual trimming and to switch either to traditional hedge laying or to trimming on rotations so that less than a third of hedges are cut in any single year. It is also essential to retain patches of scrub and broad-leaved woodland, especially wood edges that act as the focal points of most colonies. Rotational hedge and scrub cutting have already been very successful in some areas, notably Worcestershire where egg numbers have increased substantially since this management was introduced. Unfortunately, most farms remain outside land enhancement schemes and hedge loss and annual trimming continue to represent major threats to the species.

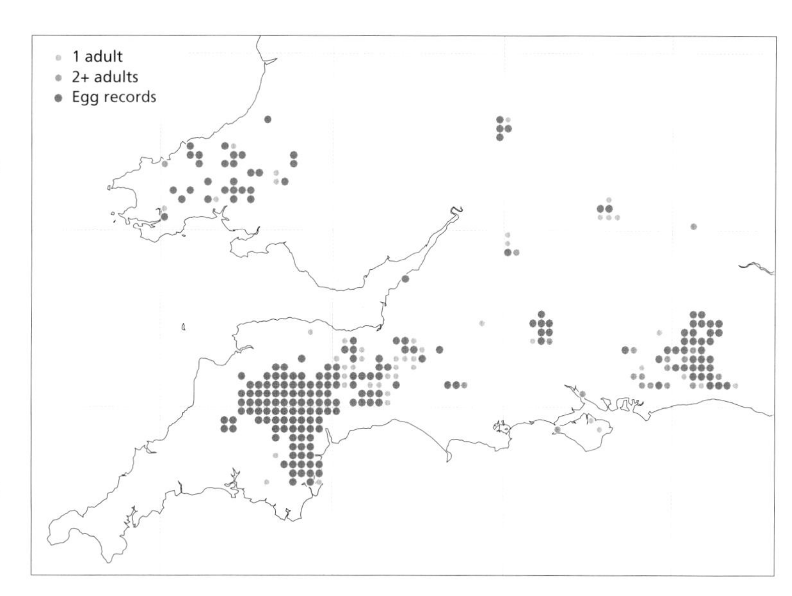

The distribution of 1995–9 records of the Brown Hairstreak at 5 km square resolution in southern Britain, showing the important contribution of egg surveys.

Key references

Barker *et al.* (1998); Bourn and Warren (1998*a*); Thomas (1974, 1991).

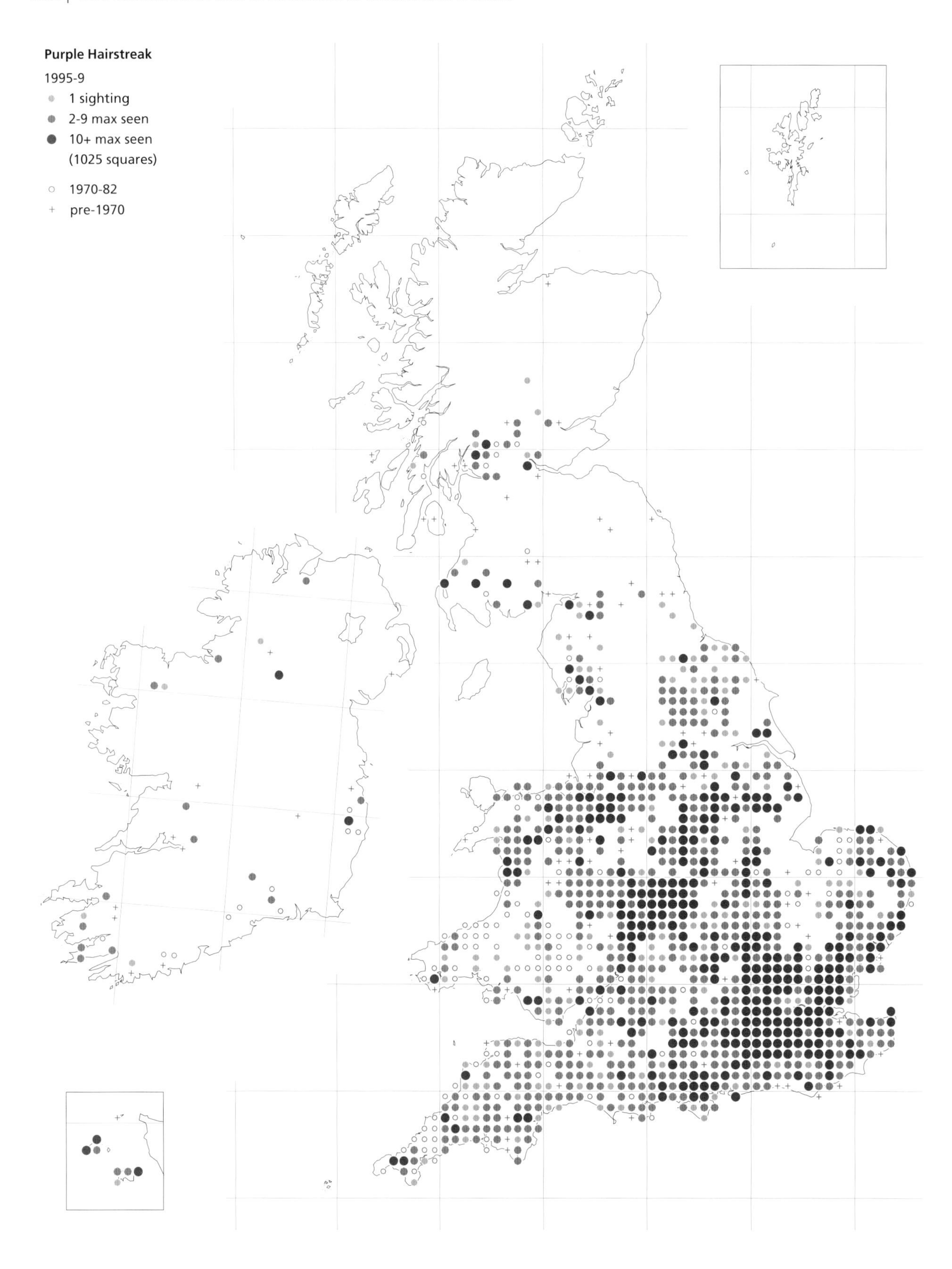
Purple Hairstreak
1995-9
1 sighting
2-9 max seen
10+ max seen
(1025 squares)
1970-82
pre-1970

Purple Hairstreak *Neozephyrus (Quercusia) quercus*

Resident
Range expanding in Britain.

Conservation status
UK BAP status: not listed.
Butterfly Conservation priority: low.
European status: not threatened.
Protected in Northern Ireland.

European/world range
The Purple Hairstreak occurs from North Africa to Southern Scandinavia and across central Europe to Asia Minor. It is stable throughout much of Europe but has declined in several countries and is spreading at the northern edge of its range.

This handsome butterfly is widely distributed throughout southern areas wherever there are oak trees; even a solitary tree may support a colony. It is frequently overlooked as adults remain largely in the canopy where the main adult food source is honeydew; they fly more commonly in the evening of a warm summer's day. They are only driven down to seek fluid and nectar during prolonged drought, as occurred in 1995–6. There has been a recent increase in records and an extension of the range of this butterfly especially in the English Midlands and south-west Scotland, even in urban areas (including London) which may be related to improvements in atmospheric quality.

Foodplants
The Purple Hairstreak is restricted to oak trees including both the native species, Sessile Oak (*Quercus petraea*) and Pedunculate Oak (*Q. robur*), and the introduced Turkey Oak (*Q. cerris*). Evergreen Oak (*Q. ilex*) also may be used.

Habitat
Colonies of Purple Hairstreak may be found in woodlands where sufficient oaks remain, in ancient park woodlands, associated with hedgerow and lane-side oaks, on heathland oak scrub, oak screens along the edge of conifer plantations, and even in towns including central London. In groups of oaks, there is usually a favourite tree and isolated oaks (which may be survivors from when the site was more heavily wooded) may support colonies that have become self-contained on the same tree for many years. In Ireland, the butterfly tends to prefer small oaks and colonies are often small. Adults will perch and feed on Ash and elms in regenerating woodland because they are fast growing and out-top oak, but oak is needed for breeding.

Life cycle and colony structure
There is one generation a year with adults flying from early July until early September, the flight period being 2–3 weeks later in northern latitudes. **Eggs** are laid, usually in late July and August, singly on the tip of an oak twig or at the base of a leaf bud. They most frequently use sunny sheltered boughs although eggs may be laid anywhere throughout the tree. The egg stage in winter can be readily identified (and counted) by examining the base of the leaf buds on accessible branches but this technique is not widely used for recording and monitoring the species.

The **larva** is fully developed within the egg in about three weeks and overwinters until early April before hatching. The timing of hatch coincides with leaf bud burst of the oak. On hatching, the larva eats part of its eggshell and then burrows into a flower bud

Purple Hairstreak larva on oak buds.

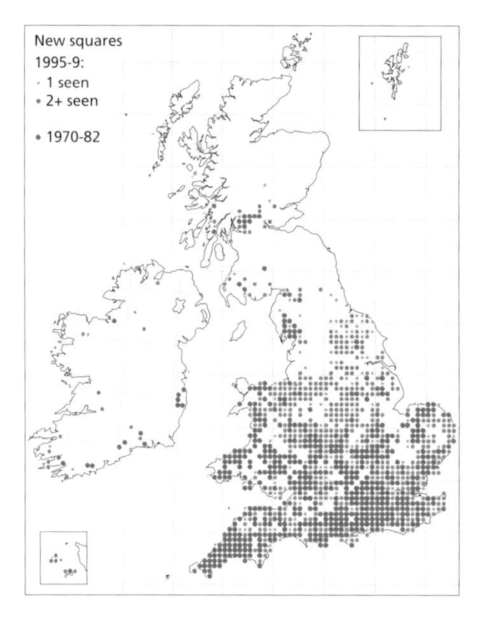

The expansion of range of the Purple Hairstreak since 1970–82.

where it feeds fully concealed. From the second instar onwards the larva lives externally in a spun web, emerging to feed at night. Pupation takes place either on the ground where the **pupa** may be tended in an ant nest (the red ant *Myrmica ruginodis* has been identified as one species that does this) or in a crevice on a larger branch of oak.

A colony may be restricted to a single oak tree. More generally the adults frequent the tops of oak or ash trees tending to congregate on a few sheltered trees in the late afternoon and early evening. The peak of activity seems to be between 18.30 and 19.30 on still, warm evenings, when tens or even hundreds may be seen darting from leaf to leaf high up in the oak trees. They are seen only rarely in large numbers at ground level, notably at the time of emergence. The Purple Hairstreak is not a highly mobile species but occasional dispersal and colonization have been observed. A mass migration was recorded following a population build up in the hot, dry summer of 1976.

Distribution and trends

The Purple Hairstreak is well distributed over southern Britain but becomes increasingly uncommon further north. In Ireland, it is a scarce butterfly being local and largely confined to hillside oaks between Wicklow and Kerry, with outposts in Sligo and Fermanagh. The species is easily overlooked, however, and is probably more widely distributed in Britain and in Ireland than is indicated on the map. It has been recorded from nine off-shore islands: Arran and Colonsay in Scotland (though there are no records in the current survey period); Anglesey in Wales; Brownsea, Hayling, and Wight in England; and three of the Channel Islands (Alderney, Guernsey, and Jersey). It is absent from the Isle of Man.

In Britain, the number of recorded 10 km squares has increased by 78% since the 1970–82 survey. There have been many new records from areas of central England, the East Midlands, and Yorkshire and from south-west Scotland. Increased recorder effort certainly contributed to this but there also appears to be a real expansion of range. The expansion has included spread into London (e.g. Highgate and Kensal Green Cemeteries, and Battersea Park) and other towns. This may be due to improvements in air quality in recent decades through reduction in industrial and domestic pollution, which are thought to have restricted aphid populations and hence honeydew supply. This theory needs to be investigated further.

These increases should be set against a background of considerable decline in the total number of colonies, as oak woods were cleared or converted to conifer plantations. Although the butterfly is periodically abundant at many of its remaining sites the species could become far more restricted to woodland in the future as ageing oaks along lanes, byways, and hedgerows disappear. Isolated colonies may not be viable in the long-term since Purple Hairstreaks rarely stray from their home trees and are unlikely to colonize new oak habitat unless it is close by.

European trend

The distribution is stable in most countries but severe declines have been reported in Austria and Romania (50–75% decrease in the past 25 years), and lesser declines in Latvia and Luxembourg (25–50% decrease), Belgium, southern Russia, and Ukraine (15–25% decrease). On the other hand, its range is expanding northwards and increases have been recorded in Finland and central parts of Russia.

Interpretation and outlook

The apparent trends in distribution of the Purple Hairstreak need to be interpreted in the light of two factors that have contributed, both in the past and in the current survey, to under-recording: the butterfly's canopy habitat and its late afternoon/evening flight behaviour. As with other species, short-term trends also depend on weather patterns at critical periods. Since the highest mortalities in the life cycle of the butterfly occur during the pupal stage, it is possible that the existence of high densities of ant nests in sunny patches on the ground is as important as the oak trees on which the butterfly breeds.

As with other Lepidoptera, the possible effects of climate change need to be considered. With each degree Celsius of warming, oak budburst is advanced by 7 to 8 days and, since egg hatch is timed so that larvae can develop before toxic tannins build up in the oak leaves, this may lead to earlier emergence.

The possible beneficial effects of improved air quality need confirmation and, for the moment, the outlook for the species appears to be dependent primarily on the abundance of oaks in the landscape. The Purple Hairstreak does not seem to be fussy about oak species, age, and aspect, and can exist on single trees and especially along strips of 'cosmetic' trees surviving in conifer plantations. Nevertheless, the future for the butterfly may not be secure, especially in Ireland where it is already rare and in a habitat under increasing pressure from agriculture and coniferization. Though widely distributed in southern Britain, continued fragmentation could lead to colonies becoming vulnerable as oaks in lanes and hedges die off. There may also be a threat from oaks dying prematurely due to disease and defoliation. In woodland, management needs to ensure that oaks remain abundant in the future.

Despite these concerns, in Britain at least, oaks as a whole are not threatened and the Purple Hairstreak remains a common and expanding, if still under-recorded, butterfly.

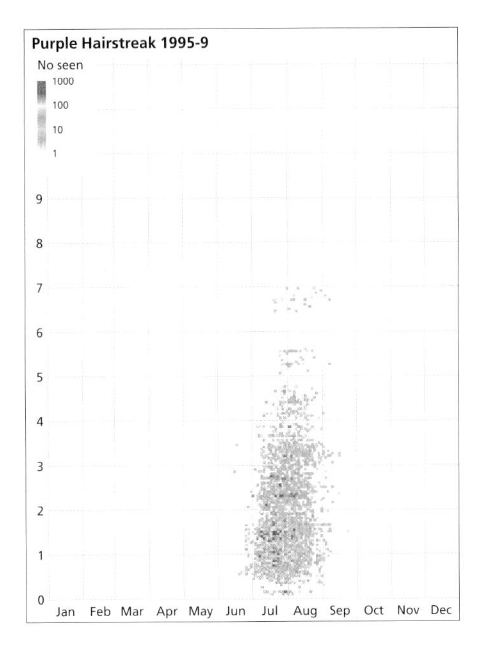

Phenogram plotting the dates of Purple Hairstreak records in 1995–9 against latitude, showing later emergence in the north.

Key References

Hillis (1975); Holloway (1980); Sparks (1997); Thomas (1975*b*).

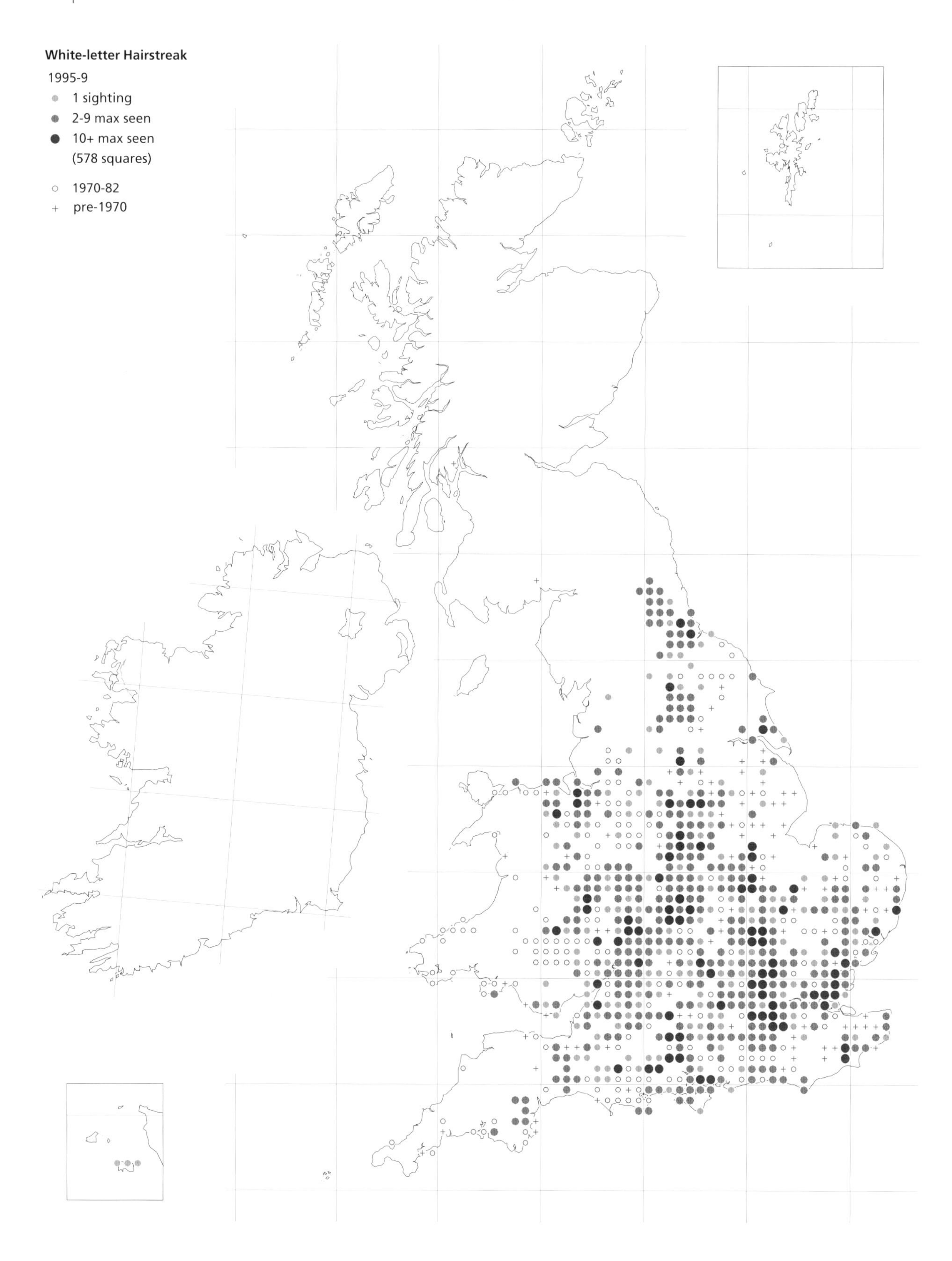
White-letter Hairstreak
1995-9
1 sighting
2-9 max seen
10+ max seen
(578 squares)
1970-82
pre-1970

White-letter Hairstreak *Satyrium w-album*

Resident
Range expanding in some areas.

Conservation status
UK BAP status: not listed.
Butterfly Conservation priority: medium.
European status: not threatened.
Protected in Great Britain for sale only.

European/world range
Across Europe as far north as southern Scandinavia, and east to Japan. Its range is stable in much of Europe, but it has declined in several countries and is spreading in Finland.

The White-letter Hairstreak is a small butterfly with an erratic, spiralling flight typical of the hairstreaks. It is distinguished by a strongly-defined white 'W' mark across the undersides. The dark uppersides are seen only in flight as the butterflies always settle with their wings closed. Adults are difficult to see because they spend so much time in the tree canopy, although they occasionally come to ground level to nectar on flowers near elm trees or scrub saplings. The species declined during the 1970s when its foodplants were reduced by Dutch Elm Disease, but it seems now to be recovering in some areas.

Foodplants
The butterfly breeds on various elm species, including Wych Elm (*Ulmus glabra*), English Elm (*U. procera*), and Small-leaved Elm (*U. minor*). Research at one site has indicated a preference for (and a higher success rate on) Wych Elm. It breeds on mature trees or abundant sucker growth near dead trees. It has also been shown to survive on the Dutch Elm Disease-resistant variety of *U. japonica*, Sapporo Autumn Gold.

Habitat
It breeds where elms occur in sheltered hedgerows, mixed scrub, and the edges of woodland rides, and also on large, isolated elms. It was previously thought to have a strong preference for large, mature elm trees, but is now found frequently on younger elm growth.

Life cycle and colony structure
The White-letter Hairstreak is single-brooded, the adults flying from early July to mid-August. After a very warm spring, they can emerge up to two weeks earlier. They spend long periods high up in trees feeding from aphid honeydew, but may be found nectaring at ground level on Creeping Thistle, ragworts, or Bramble.

Female butterflies crawl along twigs of the foodplants and probe with their abdomens to seek out suitable laying positions. Small green 'flying-saucer' shaped **eggs** are laid singly, usually around the terminal bud or where new growth joins the previous year's growth on twigs. The eggs turn dark brown within about two days and are well camouflaged as they overwinter on the twig.

The **larvae** emerge in early spring, when elm begins to come into flower, and they feed on developing flower buds. As the larvae grow, they move to feed on leaf buds and then the new leaves. The survival of larvae on non-flowering elm suckers has shown that flower buds (found on mature trees) are not an essential part of the diet as had been thought previously. Fully grown larvae are green with angled stripes, and resemble unopened leaves. The dark-brown, bristly **pupae** are normally formed under elm

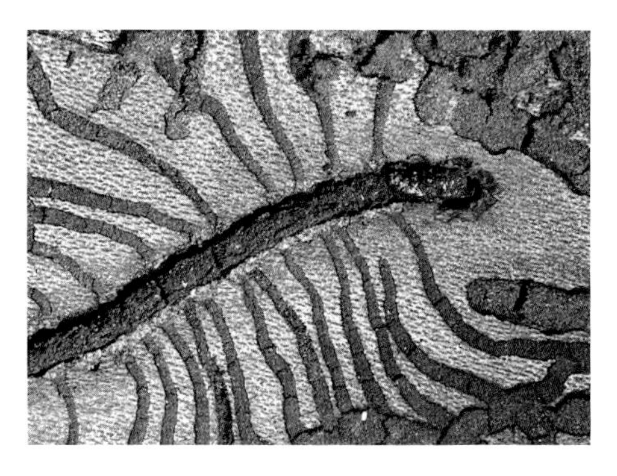

Larval feeding galleries of elm bark beetles, which carry Dutch Elm Disease.

leaves and sometimes against twigs, attached with a single silk girdle. White-letter Hairstreaks are not known to be dependent on ants, as occurs with other Lycaenid species, although Wood Ants have been seen attending larvae, perhaps to feed on the sweet liquids they can produce from a special gland.

Because of their elusive nature, information on the colony structure of White-letter Hairstreaks is sparse, but marking experiments have shown that a population along one ride edge numbered several hundred and that adults moved regularly between trees up to 300 m apart. Many colonies are restricted to a small group of trees, but dispersal appears quite common and individuals have been seen several kilometres from known breeding sites.

Distribution and status

The White-letter Hairstreak is widely, but locally, distributed across southern and central areas of England, in Yorkshire, and into north and south Wales. Populations are found mainly in lowland areas and are generally absent from land above 500 m.

The known range of the butterfly has extended further north since the early 1980s, across Co. Durham and into southern Northumberland, and a colony was discovered in north Lancashire in 1999. However, it is not clear whether these observations represent genuine expansion or improved recording.

The butterfly was recorded historically in Scotland, but there were no records during the twentieth century, despite a greater intensity of recording. There are a few recent records from the Channel Islands (Jersey), but it has never been recorded in Ireland or on the Isle of Man.

Populations were dramatically reduced and became more sparsely distributed in the 1970s. The rapid spread of a virulent new form of Dutch Elm Disease, introduced in the bark of imported elms, led to the death of over 10 million elms in Britain. Spores of the Dutch Elm Disease fungus, carried by elm bark beetles, infect tree tissues when the beetle feeds or lays eggs in the bark. New elm suckers spring from the roots of dead trees, but, when they reach 5–10 m tall, the bark becomes suitable habitat for the beetles and the tree is prone to reinfection by the disease.

Wych Elm is thought to be more resistant to the disease and has been affected less dramatically. Also, some mature elms received inoculations against the disease to conserve them, particularly in high-amenity street locations. Consequently some urban colonies, including those in London, have fared better than rural ones.

At the time of the previous national atlas, written in the early 1980s, Dutch Elm Disease was at its most damaging level. It was not known if the butterfly could survive the epidemic. In the words of the authors, 'the fate of most remaining colonies appears … to be in the balance'. A study of sites in central-southern England found that population losses increased rapidly in the 1970s (about 40% in that decade). However, recent records show that the butterfly is breeding on young elm trees (even non-flowering ones) before the disease recurs and populations seem to have recovered. The current survey shows it has been recorded in over 300 new squares since the 1970–82 survey, but some of these are probably due to the increased level of recording. It has not been re-recorded in 40% (178) of the 10 km squares recorded in the last survey, indicating that it may still not have recovered in large parts of its former range. Overall, the results show a net increase in recorded squares of 30%.

White-letter Hairstreaks spend much time high in the trees, and are difficult to identify without good binoculars and patience, unless they visit nectar plants closer to the ground. Because of this, the butterfly is almost certainly under-recorded. Although the map is probably broadly representative of its geographic range, it may be present in many of the apparently blank squares within or just beyond its known range.

Given these uncertainties, the numerical analysis of recorded squares should be treated cautiously.

European trend

The White-letter Hairstreak has declined in several European countries, especially Austria, Belgium, Luxembourg, and the Netherlands (>50% decrease in 25 years). In contrast, it has spread recently in Finland and is stable in many other countries.

Interpretation and outlook

Because of its elusive habits, the past status of the White-letter Hairstreak is difficult to determine accurately, but it was probably widespread when large elms formed a dominant feature of the lowland landscape of England and Wales. Ironically, it was probably more difficult to find then, when the trees were so tall.

Die-back of hedgerow elm caused by Dutch Elm Disease.

When the loss of elms peaked in the 1970s, it seemed that this butterfly might become rare, but it has been able to recover recently, breeding on sucker re-growth. The ability of elms to sucker in this way has helped the butterfly and it seems to have become abundant again in many areas. There may now be equilibrium between new sucker growth and the onset of the disease, and butterfly populations may be more stable. However, it remains rare in parts of southern England and Wales, and it may take longer to recover where regrowth is sparse and recolonization is therefore less probable.

The range of this butterfly has spread north in recent years, perhaps due to recent climate changes. If this trend continues, it may extend further north in England and may reach Scotland again at some time in the future. The extent of its spread near the limits of its current known range needs further investigation with carefully targeted recording effort.

The survival of the White-letter Hairstreak is linked closely to the future of elms in our countryside and the maintenance of sheltered, sunny hedgerows, woodland rides, and edges. Unfortunately, the scrubby nature of the sucker regrowth gives an 'untidy' appearance and there is a real threat to the butterfly from areas being 'tidied up', by grubbing out the remaining trees. The butterfly is also at risk from further outbreaks of disease and we cannot take its future for granted.

Key references

Brasier (1996); Davies (1985, 1986, 1992); Gibbs *et al.* (1994).

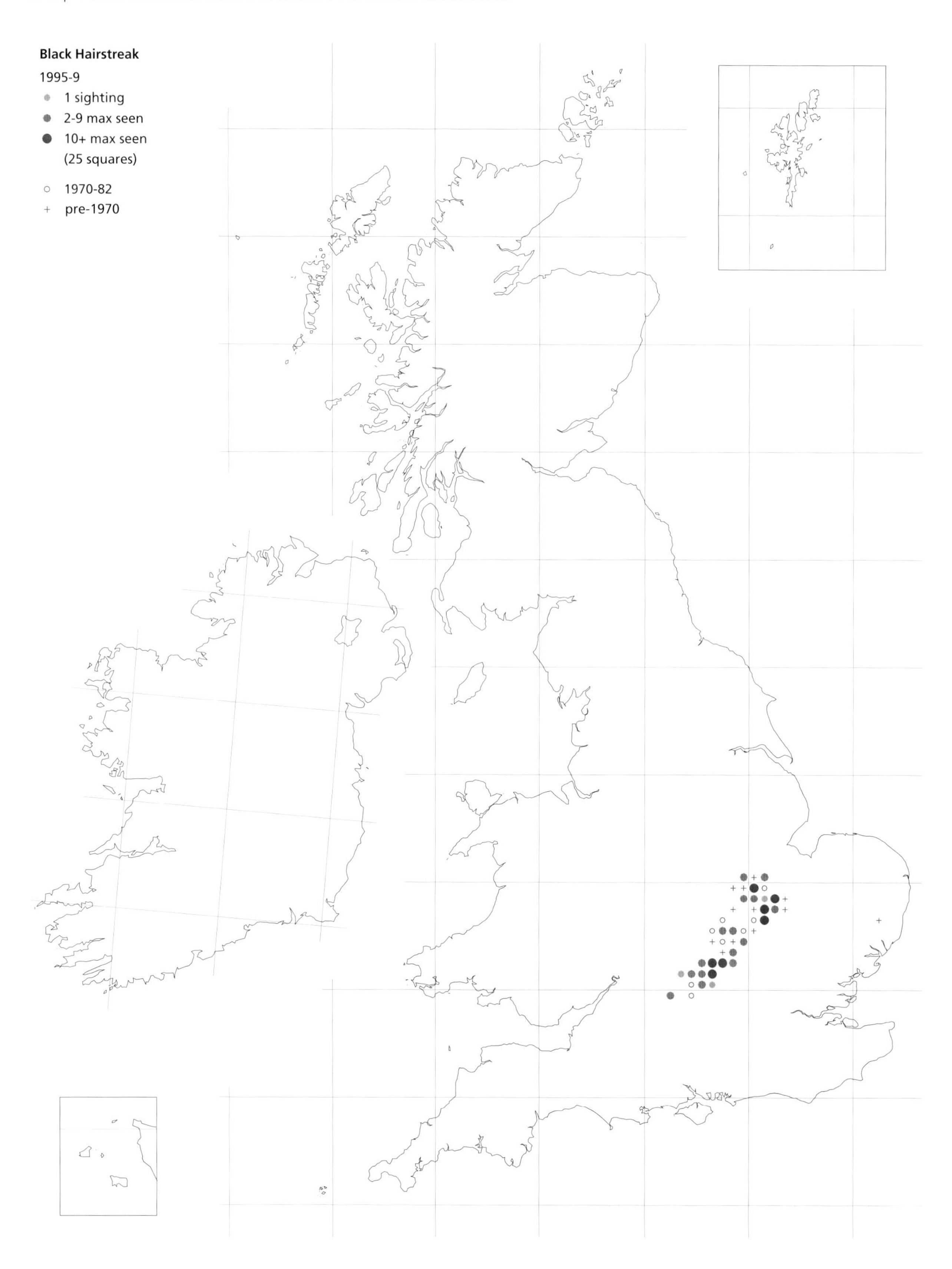
Black Hairstreak
1995-9
1 sighting
2-9 max seen
10+ max seen
(25 squares)
1970-82
pre-1970

Black Hairstreak *Satyrium pruni*

Resident
Range stable after steady decline.

Conservation status
UK BAP status: Species of Conservation Concern.
Butterfly Conservation priority: medium.
European threat status: not threatened.
Protected in Great Britain for sale only.

European/world range
Occurs widely in Europe and Asia, from the south of France to southern Finland and east to Japan. Apparently stable through much of Europe but declines have been reported in several countries.

The Black Hairstreak is one of our most elusive butterflies found only in thickets of Blackthorn in a small part of the East Midlands of England. The adults spend nearly all their time in the canopies of trees or dense scrub where they feed on honeydew secreted by aphids. At certain times they make short looping flights in and out of the tree tops, with a peak of activity between 12.00 and 14.00. The adults are easy to confuse with those of the White-letter Hairstreak and Purple Hairstreak which fly at the same time of year, so care is needed to confirm identification of the underside marking, which has a row of black spots in the outer orange margin and may have a white 'W'. The Black Hairstreak declined steadily during the twentieth century and is now reduced to around 45 sites.

Foodplants
Blackthorn (*Prunus spinosa*) is used exclusively by most colonies, but occasionally Wild Plum (*P. domestica*) and other *Prunus* species are used.

Habitat
Most colonies breed in dense mature stands of Blackthorn growing in sunny, sheltered situations, usually along wood edges or the edges of rides or glades. Smaller colonies occur in more exposed or shady situations, such as in sheltered hedgerows, small patches of scrub or canopy gaps in mature woodland.

Life cycle and colony structure
The butterfly is single brooded with adults flying in a short period from mid-June to mid-July. The adults often congregate on Field Maple or Ash, possibly because these trees contain large aphid populations that produce abundant honeydew. They are rarely seen at ground level but sometimes come down to feed on flowers of Wild Privet or Dog-rose.

Eggs are laid singly, usually on the bark of young growth on mature Blackthorn over 7–10 years old. Most eggs are laid above 1.5 m from the ground though small numbers are laid lower down on Blackthorn suckers. The eggs last through the winter and hatch in the spring, just before leaf-break, when the young **larvae** feed on the developing flower buds. Older larvae feed on the leaves, which they resemble closely. They move to **pupate** on the top of Blackthorn leaves or twigs,

Black Hairstreak pupa in typical position showing 'bird dropping' camouflage.

Ideal Blackthorn habitat along a woodland edge in Oxfordshire.

often in exposed positions, where the conspicuous black and white pupae resemble bird droppings.

The Black Hairstreak is a sedentary butterfly that often breeds in the same small, discrete part of the wood for 20 years or more. Larger woods may contain several separate breeding areas but butterflies are rarely seen outside woods, and then always within 50 m along a thick hedgerow. The Black Hairstreak has very limited powers of dispersal. One introduced colony in Surrey spread 1.5 km over more or less suitable habitat in 30 years and even slower dispersal rates are recorded in its traditional East Midlands woodlands.

Distribution and trends

In historical times, the Black Hairstreak was almost entirely restricted to a belt of woodlands on heavy clay soils between Oxford and Peterborough in eastern England. This is a comparatively well-wooded region covering the formerly extensive forests of Bernwood, Grendon Underwood, Waddon Chase, Whittlewood, Salcey, Yardley Chase, Rockingham, and Nassboro. There are a few records outside this range that are all thought to be misidentifications (usually of the very similar White-letter Hairstreak), apart from one old record in Essex. There are also several records emanating from a colony introduced in Surrey in 1952 which spread to form about five colonies by the late 1970s. However, the habitats have since been destroyed and there have been no records since 1990.

The distribution of the Black Hairstreak is well documented thanks to detailed surveys made between 1969–75 and subsequent recording for the current atlas. A comprehensive review of known colonies was also conducted by Butterfly Conservation in 1999 and 2000 as part of an Action Programme for the species. The butterfly was first discovered at Monks Wood (Cambridgeshire) in 1828 and has since been recorded at about 90 localities in the East Midlands. Because of the difficulty in recording adults, they are only reported from a small proportion of sites each year, making it difficult to determine the exact number of extant localities. The butterfly has been recorded in only about half of these since 1995.

The butterfly's decline would probably have been far worse had it not been for the activities of early entomologists such as Lord Rothschild who paid for 'large numbers' to be released between 1900 and 1917. It is possible that as many as one-third of all current colonies derive from such reintroductions, including the one at Monks Wood. This is the longest known successful reintroduction of any butterfly species, lasting from the early 1920s until the present.

There have also been several attempts to establish it outside its historical range but these have so far failed and have not been mapped. In addition to the Surrey releases described above, it has been released unsuccessfully at Knebworth (Hertfordshire), Slimbridge (Gloucestershire), and in the New Forest (Hampshire).

European trend

The Black Hairstreak's distribution appears stable in many European countries but it is declining seriously in Austria and Denmark (>50% decrease in the last 25 years), as well as in Belgium, Lithuania, Slovakia, and Romania (25–50% decrease in 25 years).

Interpretation and outlook

The restriction of the Black Hairstreak to the East Midlands of England remains a mystery, but may be due to the butterfly's requirement for mature Blackthorn stands and the historical management of woods in this region. Until the twentieth century most woods in lowland England would have been managed as coppice, cut on short cycles of 5–15 years—too short to develop any suitable habitat for the Black Hairstreak. However, the Royal forests of the East Midlands were often managed on an unusually long cycle of 20–40 years, thus allowing suitable Blackthorn stands to develop.

Most losses are attributable to clearance of Blackthorn when woods were replanted, often with conifers, or where woods have been grubbed out for agriculture. In some famous cases, such as at Monks Wood, it disappeared following the widespread clear felling that took place during the First World War.

The conservation of the Black Hairstreak is quite simple in the short term. It requires the protection of breeding areas and their immediate surrounds from major clearance. This must be a top priority especially as so few sites remain.

In the longer term, nearly all sites require some gradual management as many Blackthorn stands are even-aged and will eventually become moribund and fall over. This will lead to local extinction which, given the extremely sedentary nature of the butterfly, may well be permanent. Ideal management is thus to create new habitat by periodically clearing small patches of Blackthorn within a site and allowing these to regenerate into mature stands, or by creating new habitat nearby by allowing the foodplant to spread by sucker growth. There are also great opportunities to include Blackthorn stands in any new woodland plantings close to existing colonies.

The majority of existing sites are nature reserves or are being managed sympathetically and successfully for the species, including about half of the extant colonies, which are on land managed by the Forestry Commission. Although these measures have stabilized the number of colonies over recent decades, it is now almost entirely conservation dependent and remains vulnerable due to the small number of sites and fragmentation of suitable habitat.

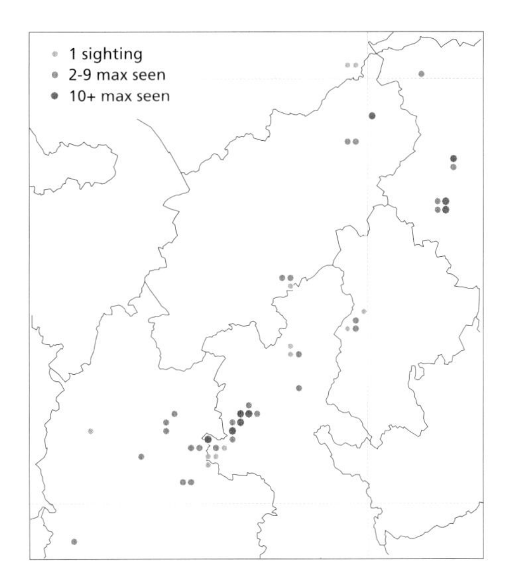

The distribution of 1995–9 records of the Black Hairstreak at 2 km square resolution in the East Midlands.

A new threat is emerging from the increase and spread of deer populations (mostly introduced species such as Muntjac and Fallow Deer), which are limiting the natural regeneration of Blackthorn and other trees in many woods. The deer are also radically changing the ground flora of many woods and causing serious problems for other wildlife species. New techniques have been developed to overcome this problem for the Black Hairstreak (such as laying the Blackthorn like a hedge rather than cutting it, or fencing regeneration plots), but a better solution is to reduce deer populations to more acceptable limits. Although difficult to achieve this would be more beneficial for the overall conservation of woodland communities.

Key references

Bourn and Warren (1998*b*); Roberts *et al.* (1999); Thomas (1974, 1975*a*); Thomas *et al.* (1992).

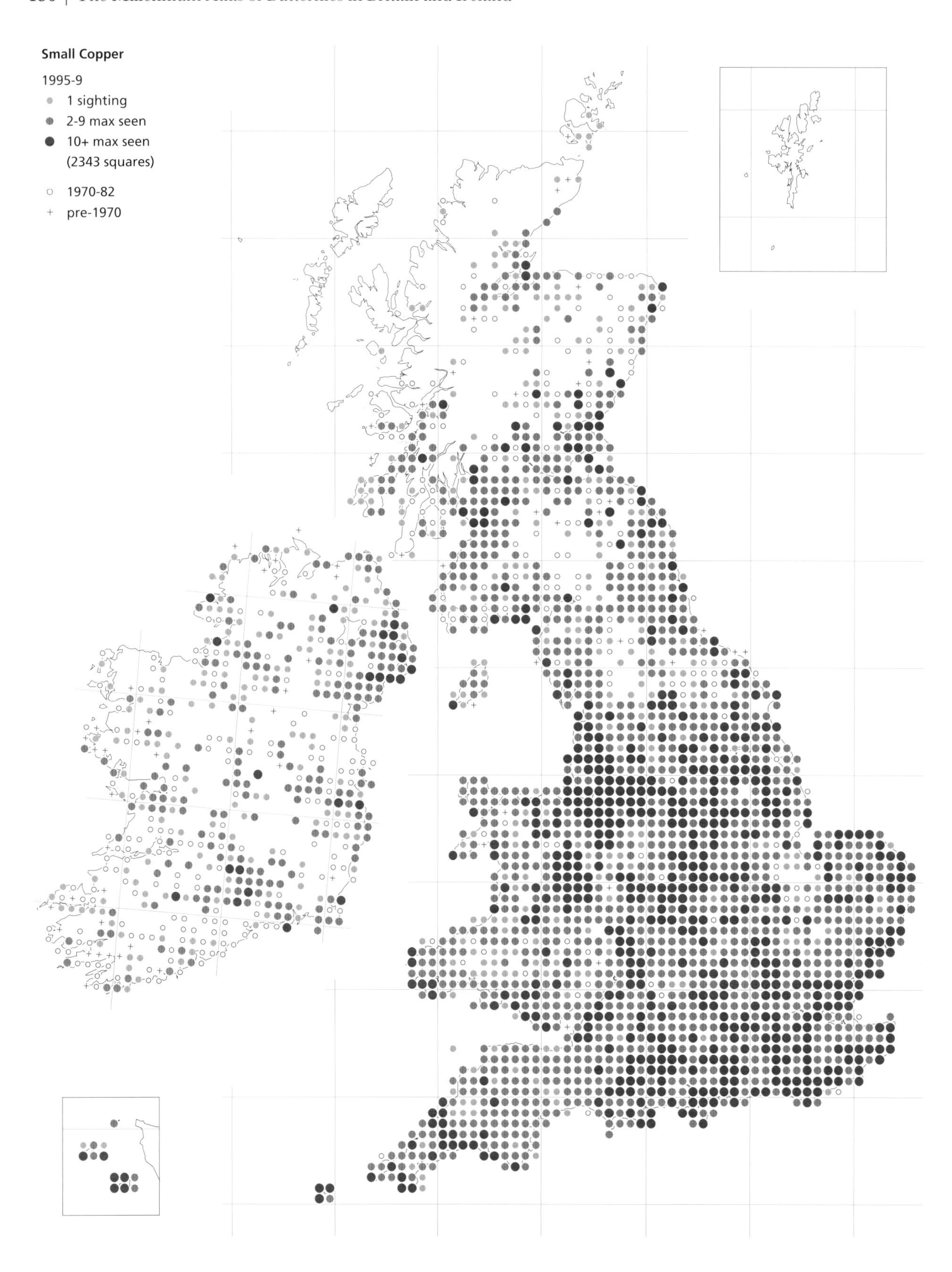
Small Copper
1995-9
1 sighting
2-9 max seen
10+ max seen
(2343 squares)
1970-82
pre-1970

Small Copper *Lycaena phlaeas*

Resident
Range stable.

Conservation status
UK BAP status: not listed.
Butterfly Conservation priority: low.
European status: not threatened.

European/world range
Widespread and common throughout most of Europe including all larger Mediterranean islands. It occurs in North Africa and across Asia to Japan, and also in North America. Stable in most European countries.

The Small Copper is usually seen in ones and twos, but in some years large numbers may be found at good sites. Males are territorial, often choosing a piece of bare ground or a stone on which to bask and await passing females. They behave aggressively towards any passing insects, returning to the same spot when the chase is over. Though it remains a common and widespread species, the Small Copper declined throughout its range during the last century.

Foodplants
Common Sorrel (*Rumex acetosa*) and Sheep's Sorrel (*R. acetosella*) are the main foodplants. Broad-leaved Dock (*R. obtusifolius*) may occasionally be used.

Habitat
It occurs in a wide variety of habitats: chalk grassland, moorland, heathland, coastal dunes and undercliffs, woodland clearings, and unimproved grassland. This species may be found also in small patches of land such as set-aside fields, roadside verges, railway embankments, allotments, churchyards, and waste ground, even in cities. Warm, dry situations are especially favoured.

Life cycle and colony structure
Over most of Britain and Ireland there are two generations per year but in the south there are commonly three in warm summers and exceptionally four following a long, warm season, provided that drought has not withered the foodplants. There are no more than two generations in northern England and Scotland. In England, adults are on the wing in May and June, with the second brood flying in August and September. Later broods continue well into October and adults may even be seen in November. In Scotland, the first brood emerges up to a month later (see p. 393).

First brood females select larger plants among a taller sward, while later broods seek out small, young sorrel plants in short sparse vegetation. The white, golf-ball shaped **eggs** are placed under the leaves and the young **larvae** make distinctive 'windowpane' feeding damage by eating grooves which leave the top surface of the leaf intact but transparent. The survival of larvae on Broad-leaved Dock has not been studied, but the contribution of these plants to

Small Copper larva on sorrel, showing feeding damage.

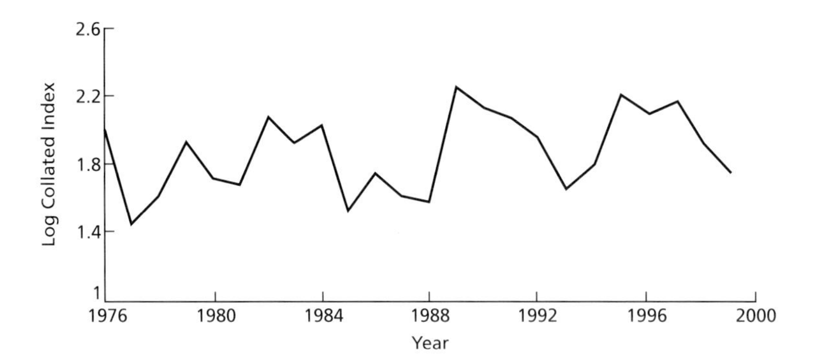

The abundance of the Small Copper fluctuates widely and is affected by the weather.

populations of the butterfly will be very small, as so few are used by ovipositing females. Larvae overwinter when very small and resume feeding in the spring, when they make larger, but still distinctive, holes in the leaves and also eat the leaf edges. It is thought that **pupae** are tended by ants among leaf litter but little is known about the pupal stage in the wild.

The Small Copper lives in discrete colonies that are usually small, often producing only a few dozen individuals in a brood. Colonies may survive in small fragments of suitable habitat such as short stretches of roadside verge. Little is known about adult mobility, but individuals may be seen some distance from breeding habitat and are commonly recorded nectaring in gardens.

Distribution and trends

The Small Copper is widespread in England, Wales, and Ireland except on the highest ground and present on the Channel Islands and the Isle of Man. In Scotland it is more restricted to warm sheltered sites and though it is widespread in the southern and eastern highlands it is seldom found in the far north and west. It is absent from the Outer Hebrides and Shetland (although an unconfirmed sighting was reported from Shetland in 1999). It was recorded in Orkney, where it is not normally resident, in 1996, following an influx of migrants of various species from mainland Scotland. Further sightings in 1997 and 1999 suggest that it may have bred on Orkney, but this has not been confirmed.

Early twentieth century accounts suggest that this species was more abundant at that time. Many colonies have undoubtedly been lost due to destruction of semi-natural grassland and the butterfly is now scarce or even absent in parts of eastern England where there is intensive arable agriculture. A detailed study in north Wales concluded that both the area occupied and the population level have declined by about 90% since the beginning of the twentieth century. Similar declines in other areas are likely, but are not apparent in distribution maps at the 10 km level.

There is some potential for colonization of new sites because Common Sorrel readily grows on fallow land and the Small Copper is reasonably mobile. It will colonize urban sites and can be seen in parks and on wasteland and commons, even in London and other cities.

Numbers of adults vary considerably from year to year. First brood numbers are usually low. Warm, dry summers favour the species and in such years numbers of later broods will be much higher while in cool, wet seasons there will be a smaller increase. The severe drought in 1976 caused a decline, which was made worse by the cool summer of 1977, and it took several years for numbers to recover. Butterfly Monitoring Scheme figures show large increases in 1989 and 1995, following poor years in 1988 and 1993–4.

Even when present at low population levels, this is a conspicuous butterfly, flying rapidly but settling frequently in a prominent position with wings open. It is distinctive in appearance and unlikely to be confused with any other species. Nevertheless, it is under-recorded in Ireland, where it is thought to be more widespread than indicated by the current map.

European trend

The Small Copper's range is stable in most European countries but has declined in Luxembourg (25–50% decrease in 25 years) and Romania (15–25% decrease).

Interpretation and outlook

More than many, this species is affected by extreme weather conditions. In cool, wet summers, populations may be lost from shadier habitats such as woodland rides, remaining only on more open habitats, such as heathland, which receive the maximum amount of warmth and sunshine. In warmer years, when populations are high, these marginal, shadier habitats are recolonized.

Sparse vegetation, with patches of bare ground for basking and male territorial perching, is particularly good for this species and bare, recently disturbed ground is also readily colonized by the larval foodplants. In southern England the strongest colonies are often found on heathland where large amounts of Sheep's Sorrel grow in ideal conditions. Current conservation management on heathlands, aimed at reversing the invasion of scrub and maintaining short vegetation, is particularly beneficial to Small Coppers.

On other grassy habitats, colonies may be found along paths, tracks, and banks where there are patches of bare ground, vegetation of varied heights, and plentiful nectar sources. Rabbit scrapes, mole hills, and other disturbances favour growth of sorrel. Management should be aimed at preventing sorrel from being shaded out in dense vegetation and this may be achieved by grazing or cutting, although the optimum timing and frequency of such management have not been studied.

The Small Copper is thought to have undergone widespread declines in Britain and Ireland during the twentieth century. Recent research shows that losses in north Wales have been high and are likely to have occurred on a similar scale elsewhere. This species is not able to survive in 'improved' grass swards, and it has become a scarce butterfly in intensively farmed parts of eastern England.

More research is needed on the ecology of this butterfly, its habitat requirements, and mobility. Although it is able to live in a variety of habitats and to colonize new sites created by human activity (such as waste ground, set-aside fields, landfill sites, and golf courses), habitat destruction continues to threaten colonies. Thus it is likely to decline further in many areas as pressures increase on marginal land, especially at the edges of towns and cities, where it often thrives.

Key references

Aspinall (1987); Dempster (1971); León-Cortés *et al.* (2000).

Large Copper
1995-9
No records
○ 1970-82
+ pre-1970

Large Copper *Lycaena dispar*

Extinct (1864).

Re-introduction attempts so far have failed (new trials underway).

Conservation status

UK BAP status: Priority Species.
Butterfly Conservation priority: high.
European status: not threatened.
Fully protected in Great Britain.
Bern Convention (Annexe IV) and EC Habitats and Species Directive (Annexe II and IV).

European/world range

Locally but widely distributed in central Europe as far north as southern Finland, and across to Asia. Divided into three races: *dispar* formerly in Britain, *batavus* in the Netherlands (both single brooded), and *rutilus* (double brooded). The latter is declining in many European countries while *batavus* is reduced to two sites.

The Large Copper is a magnificent butterfly whose shiny metallic-looking wings flash brightly as it flies over its fenland habitat. The British race of this butterfly (*dispar*) has been extinct for over 150 years and most of our knowledge of its life cycle and ecology comes from studies of the similar race (*batavus*) found in the Netherlands. There have been several reintroduction attempts to sites in both Britain and Ireland, but these have all ultimately failed. Research is now being conducted to see whether a further attempt is worthwhile in more extensive habitats available in the Norfolk Broads.

Foodplants

The sole foodplant of the single-brooded races (*dispar* and *batavus*) is Water Dock (*Rumex hydrolapathum*). Other docks are used by the *rutilus* race in continental Europe.

Habitat

The single brooded races breed in open fen vegetation where Water Dock grows at a high density in relatively open, sunny positions. The race *rutilus* breeds in a far greater range of damp habitats in Europe.

Life cycle and colony structure

The former British and current Dutch races are single brooded with adults flying from July to early August. **Eggs** are laid singly on the upperside of prominent leaves of Water Dock that are not smothered by surrounding vegetation. The young **larvae** feed on the undersurface leaving characteristic 'windows' in the upper surface which can be seen easily by the experienced observer. They overwinter from mid-September as second instar larvae amongst older leaves at the base of the foodplant. The larvae resume feeding in April on the young shoots and become fully grown in June when they **pupate** either on the foodplant or nearby vegetation.

Little is known about adult behaviour and dispersal, partly because the butterfly occurs at very low densities (1–2 sightings per hour are typical in its native Dutch habitats) and partly because adults are difficult to follow in wet fenland with intervening dykes. Evidence from the Netherlands suggests that adults range widely over extensive fenland and may be capable of colonizing habitats as far as 20 km away from existing populations. Males seem to establish territories in open patches of fenland and often perch on prominent nectar plants such as Purple-loosestrife.

Distribution and trends

The Large Copper has been recorded from many parts of Britain but was probably

Water Dock, the sole larval foodplant of the Large Copper.

already undergoing a rapid decline by the time entomology became popular in the mid-eighteenth century. The earliest known record was in 1749 near Spalding, Lincolnshire, and its range probably then included Lincolnshire, Huntingdonshire, and from Cambridgeshire across to the Norfolk Broads. There are less reliable records of more westerly populations on the Somerset levels and the Wye marshes in Monmouthshire (these have not been mapped).

It disappeared from Lincolnshire and was last recorded in Huntingdonshire at Holme Fen in 1848 and in Cambridgeshire at Bottisham Fen in 1851. The last British records are from the Norfolk Broads at Ranworth in 1860 and Woodbastwick in 1864. There are no historical records for Ireland or Scotland.

Since its extinction there have been several unsuccessful attempts to reintroduce the Large Copper using continental races. The bivoltine race *rutilus* was introduced at Wicken Fen (Cambridgeshire) in 1909, and subsequently, with little success. Attempts were made also at Greenfields, Co. Tipperary in Ireland in 1913 and 1914 where the colony survived until 1928, and at Woodbastwick in the Norfolk Broads where it survived until 1936.

The univoltine *batavus* race, which was not discovered until 1915, was introduced to Woodwalton Fen (Huntingdonshire) in 1927 where it was maintained until the early 1990s partly through protection of young larvae and partly by supplementing the populations by periodic releases of cage-reared stock.

The *batavus* race was also introduced to Wicken Fen (Cambridgeshire) in 1930 where it survived until 1942; to the Norfolk Broads in 1949 where it survived for just one year; and to Greenfields in 1926 where it survived until 1938, and again in 1943 when it survived until 1954.

European trend

The Large Copper is declining in many countries, notably Austria, Germany, Luxembourg, the Netherlands, and Romania (>50% decrease in 25 years), and Belgium, Italy, Slovenia, and Asian Turkey (25–50% decline). However, it is apparently stable in many others and has expanded its range in the Czech Republic, Estonia, Latvia, and Poland (25–100% increase in 25 years). It is not classed as threatened overall, but the Dutch race is endangered because it is restricted to two large adjacent sites.

Interpretation and outlook

The Large Copper was greatly prized by early butterfly collectors and the large numbers caught have been implicated in its demise in England. However, even if the last few individuals were caught by collectors, it would soon have become extinct due to the massive loss of core habitats in the vast East Anglian Fens. These included its last strongholds around Whittlesey Mere, once the largest natural lake in lowland England (9×5 km) which was finally drained during the eighteenth century.

With the failure of so many reintroduction attempts, the outlook for the species is not good. Apart from at Woodwalton Fen, there has been virtually no monitoring of reintroduced populations and the reasons for their failure are not well known. With the benefit of modern knowledge, it is clear that most sites were far too small, even if they contained abundant foodplant. Most remaining fenland habitats are very small and isolated, and nearly all are drying out. They often now stand several metres above the surrounding arable farmland where the peat soils have shrunk due to windblow and oxidation.

Enormous efforts have been made recently to maintain a high water table on several reserves. At Woodwalton Fen a waterproof external bank has been constructed and water is pumped into the reserve from the adjacent Great Raveley Drain. However, the levels still fluctuate greatly leading to prolonged flooding which in the past has killed the reintroduced Large Copper larvae.

Woodwalton Fen is also suffering from declining water tables and lack of traditional management such as peat cutting and rough grazing. This led to the spread of carr woodland on over half of the site, which is only now being reduced using heavy machinery. A special grazing regime has been reintroduced on many areas, using Galloway cattle to maintain a mixed fen vegetation, but it is not an easy task.

The conclusion of many years research is that the area of suitable habitat (*c.*10 ha of grazed fen) is too small to support a viable population of the Large Copper, although it remains valuable for other special fauna and flora.

In view of its extreme global rarity there are still hopes that the single-brooded race of the Large Copper might be reintroduced successfully in Britain and attention has now switched to the more extensive fenland in the Norfolk Broads. Here conservation effort, co-ordinated by the Broads Authority and English Nature, has extended the open fen habitat substantially, possibly offering greater opportunities for the Large Copper. Trials are now taking place to examine whether larval survival is high enough and habitat sufficiently extensive to warrant a reintroduction attempt. British ecologists are also working closely with colleagues in the Netherlands to ensure its survival there.

Thanks to their careful work we now know far more about the butterfly's requirements, particularly the need for extensive areas of habitat containing numerous patches where Water Dock is abundant and growing in open, sunny conditions. Ideal management regimes are also better known, using either extensive cattle grazing or rotational cutting of waterside vegetation where larval survival is best. Armed with this knowledge it may soon be possible to restore this marvellous insect to Britain, thus helping to ensure the survival of this unique race.

Key references

Barnett and Warren (1995*c*); Duffey (1968, 1977); Pullin (1997); Pullin *et al.* (1995, 1998).

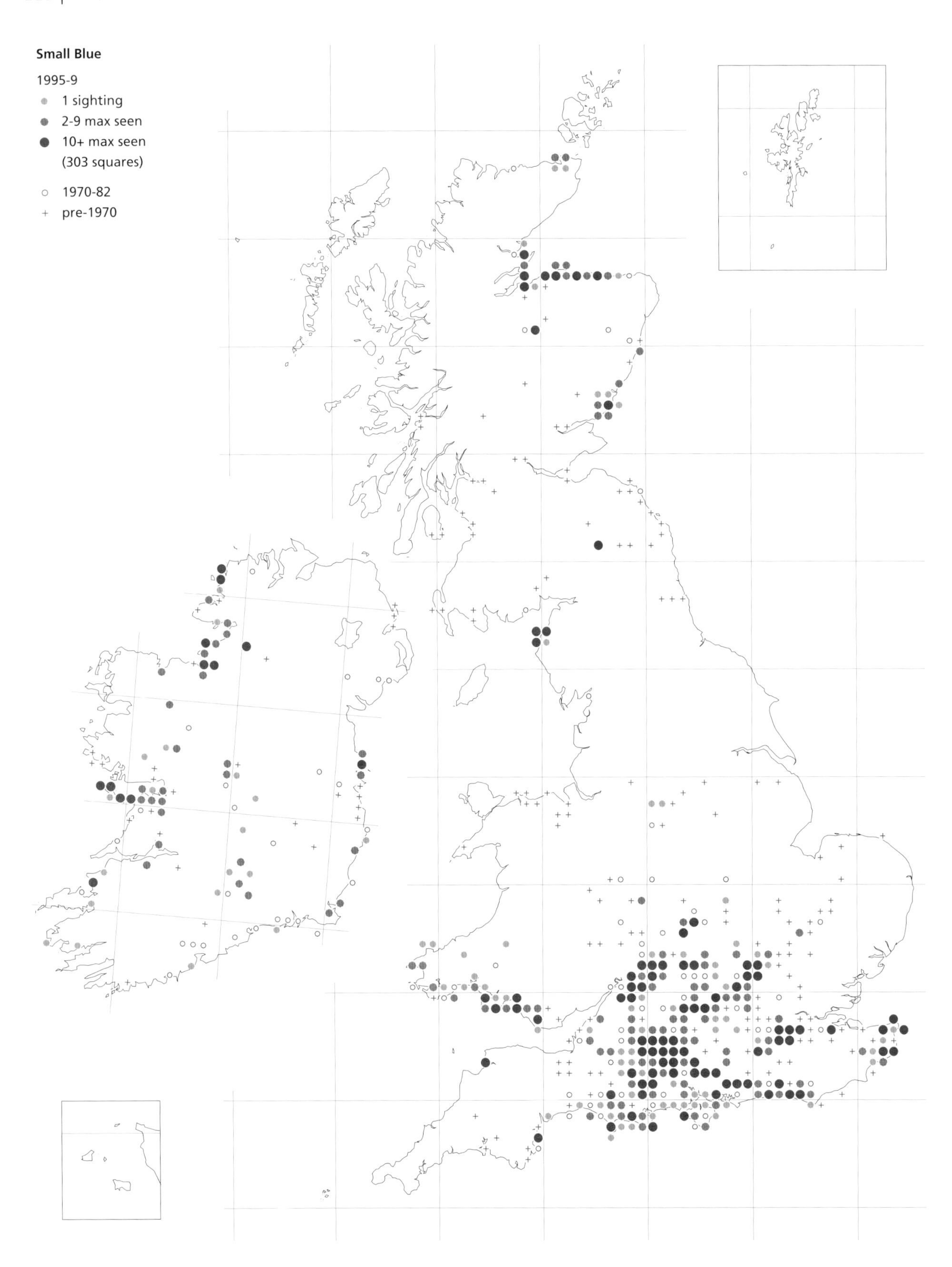
Small Blue
1995-9
1 sighting
2-9 max seen
10+ max seen
(303 squares)
1970-82
pre-1970

Small Blue *Cupido minimus*

Resident
Range declining.

Conservation status
UK BAP status: Species of Conservation Concern.
Butterfly Conservation priority: medium.
European status: not threatened.
Fully protected in Northern Ireland, protected in Great Britain for sale only.

European/world range
Throughout Europe from north Spain to Scandinavia, and across Asia and Mongolia. Declining in some countries of north-west Europe, but stable elsewhere.

Our smallest resident butterfly is easily overlooked, partly because of its size and dusky colouring, but partly because it is often confined to small patches of sheltered grassland where its sole foodplant, Kidney Vetch, is found. Males set up territories in sheltered positions, perching on tall grass or scrub. Once mated, the females disperse to lay eggs but both sexes may be found from late afternoon onwards in communal roosts, facing head down in long grass. The butterfly tends to live in small colonies and is declining in most areas.

Foodplants
The sole foodplant is Kidney Vetch (*Anthyllis vulneraria*). The larvae live only in the flower heads where they feed on developing anthers and seed.

Habitats
The butterfly breeds in a range of dry sheltered grasslands where Kidney Vetch grows, including: chalk and limestone grassland, coastal grassland and dunes, and man-made habitats such as quarries, gravel pits on eskers, road embankments, and disused railways.

Sites are usually sheltered and contain sparse or eroding vegetation where Kidney Vetch seedlings can become established and where flowering plants are abundant. The best habitats typically contain a mosaic of short and tall vegetation and patches of light scrub.

Life cycle and colony structure
Small Blues are single brooded with adults usually flying from mid-May to late June, with a partial second brood on southerly sites in late July or August. However, there is considerable variation between sites and the BNM data suggest that emergence is generally 2 weeks later in the north of Scotland compared to southern England (see p. 393).

Eggs are laid singly, tucked into the young flower heads of prominent Kidney Vetch plants. The **larvae** feed on the developing flowers, burrowing deep into the florets. The larvae are cannibalistic and eat any younger larvae they encounter. Females seldom lay on plants where an egg is present. As the larvae grow, they sit more openly on the flower heads, biting holes in the base of the flowers to eat the seed. When fully grown, they descend to the ground and pass the winter in soil crevices or under moss. They **pupate** the following spring in late April or early May, also at ground level.

Both larvae and pupae have structures that attract ants and in continental Europe they are usually tended by ants throughout their development. However, detailed observations in Britain have rarely found

Small Blue larva on flower head of Kidney Vetch, the sole larval foodplant.

ants in attendance, possibly because few native ant species forage high up in the flower-heads. There have been very few observations of the overwintering larvae and pupae but these are possibly attended by ants.

The Small Blue is widely reported to be sedentary, typically forming small, discrete colonies. Mark–recapture studies in both England and Scotland have shown that adults rarely move more than 40 m, and that males are more sedentary than females. However, some greater distances have been recorded, including a few of over 1 km between neighbouring sites and vagrants have been recorded in Wiltshire as far as 17 km from known colonies. There is also evidence that mobility may be greater and dispersal more frequent in hotter years. Moreover, populations fluctuate greatly from year to year, possibly in relation to flowering cycles in the foodplant. The butterfly has also been known to colonize newly created sites such as road cuttings where these are close to existing colonies.

Distribution and trends

This species occurs very locally throughout Britain and Ireland, typically in small populations. The main centres of distribution are the chalk and limestone grasslands of southern England, especially the Cotswolds and the extensive tracts of downland on Salisbury Plain. Elsewhere, colonies are far more scattered, often in coastal areas such as the Moray and Angus coasts in north-east Scotland, the south-west coast of Wales, the coast of northern Cumbria, the Aran Islands off western Ireland, and the coasts of Counties Donegal and Sligo in north-west Ireland.

The Small Blue declined substantially during the twentieth century and has become extinct throughout most of the northern half of England. In many regions, it is now reduced to a few small colonies which are vulnerable to extinction. There has also been a considerable decline within its main strongholds of southern England and it has disappeared from numerous sites, including several nature reserves. On the North Downs of Surrey it is reduced to about 15 colonies, most of them extremely small (<10 adults recorded at peak). Nevertheless, many colonies still occur in Wiltshire (*c.* 185 colonies), Dorset (*c.* 100), and Gloucestershire (*c.*145).

In Scotland, colonies have been lost in the last few decades but many new ones have been discovered. Overall, the BNM survey has recorded the Small Blue in 109 additional 10 km squares in Britain since the 1970–82 survey, and 40 more in Ireland. This indicates that it is probably still under-recorded in some regions due to the small size of colonies. In a few areas the situation is complicated further by deliberate reintroductions. These include twelve experimental introductions that were conducted as part of a PhD study during the 1980s in Cornwall and Sussex, leading to the establishment of several colonies in the latter county. Despite these uncertainties, it seems that the species is undergoing a steady decline.

European trend

The Small Blue's distribution appears stable in many European countries but there have been serious declines in Belgium and Finland (>50% decrease in 25 years) and in Luxembourg (25–50% decrease in 25 years).

Interpretation and outlook

The Small Blue relies on grassland habitats that have a very specific combination of shelter, to provide a warm microclimate for the adults, and early successional conditions that allow the larval foodplant, Kidney Vetch, to flourish. This limits it to very small areas and to small populations that are vulnerable to local extinction.

In the past, such habitats were far more extensive and many were lost to agricultural intensification during the twentieth century. Remaining patches have often suffered from a lack of management, which has either eliminated the larval foodplant or reduced it to critically low densities. Several colonies have been lost to built developments over recent decades, especially close to London.

However, while these losses were occurring, some new habitats were being created, for example when railway lines and quarries became disused, and when new roads with steep cuttings were built. On some road verges, the Small Blue has even been able to breed on large-flowered continental varieties of Kidney Vetch that were used in some early habitat creation schemes.

On most grassland sites, conservation measures for the Small Blue will require continuation of suitable grazing regimes to maintain high densities of flowering Kidney Vetch in a mosaic of short and tall sward. The foodplant thrives best in a sparse sward with small patches of bare ground where seeds can germinate freely. Periodic ground disturbance may be thus essential on sites that cannot be grazed.

Care has to be taken not to graze too heavily during the summer, especially with sheep, which can remove the flower heads where larvae are feeding. For this reason, the best sites are either grazed predominantly by cattle, or by sheep during the autumn or winter. Rabbits can also eat flower heads and may cause problems. On other sites, especially in coastal areas, natural processes of erosion are sufficient to maintain the foodplant together with minimal amounts of scrub removal. However, the small size and remoteness of some sites means that even such simple measures may be difficult to implement without special conservation effort.

Perhaps the biggest long-term threat to the species is from the small size and increasing isolation of its populations. Given the butterfly's need for early successional vegetation, local extinction is probably commonplace but in the past this was balanced by colonization. In many regions this balance has been altered and recolonization is far less likely.

Some losses could be replaced by deliberate reintroductions but these will not provide a long-term solution. It is far better to improve management and increase available habitats so that natural processes can maintain populations. Fortunately, breeding conditions for the Small Blue appear quite simple to create and there are many opportunities to restore connections between colonies, for example along disused railway lines, on derelict sites and quarries and on new road verges.

Key references

Bourn and Warren (2000*a*); Coulthard (1982); Morton (1985).

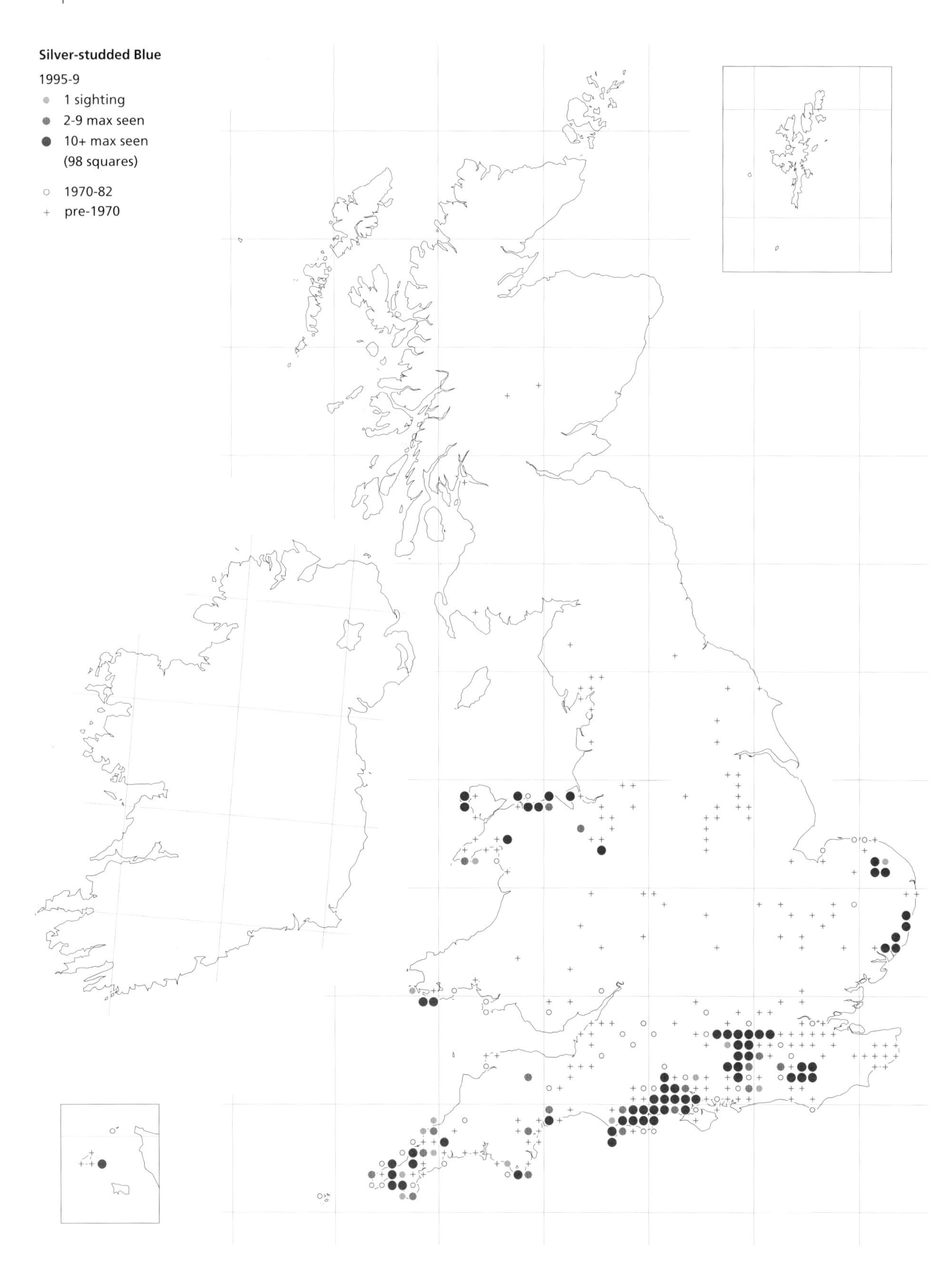
Silver-studded Blue
1995-9
1 sighting
2-9 max seen
10+ max seen
(98 squares)
1970-82
pre-1970

Silver-studded Blue *Plebeius (Plebejus) argus*

Resident

Range declining.

Conservation status

UK BAP status: Priority Species.
Butterfly Conservation priority: medium.
European status: not threatened.
Protected in Great Britain for sale only.

European/world range

Widespread across temperate Europe and Asia to Japan. Its range appears to be stable in much of Europe but declines have been recorded in some central and west European countries, and expansions in parts of south-east Europe and Russia.

This small butterfly is found mainly in heathland where the silvery-blue wings of the males provide a marvellous sight as they fly low over the heather. The females are brown and far less conspicuous but, like the male, have distinct metallic spots on the hindwing. In late afternoon the adults often congregate to roost on sheltered bushes or grass tussocks. The Silver-studded Blue has a restricted distribution but occurs in large numbers in suitable heathland and coastal habitats. It has undergone a major decline through most of its range.

Foodplants

A wide variety of ericaceous and leguminous plants are used: on heathland the most common are Heather (*Calluna vulgaris*), Bell Heather (*Erica cinerea*), Cross-leaved Heath (*E. tetralix*), gorses (*Ulex* spp.); and on calcareous sites mainly Common Bird's-foot-trefoil (*Lotus corniculatus*), Common Rock-rose (*Helianthemum nummularium*), and Horseshoe Vetch (*Hippocrepis comosa*).

Habitat

Three main habitats are used: lowland heathland (the most widely used); calcareous grasslands (in north Wales, Pembrokeshire, and the Isle of Portland in Dorset); and sand dunes (for example in Cornwall). It occasionally occurs in other habitats such as bogs. In all habitats the butterfly requires short or sparse vegetation, such as recently burnt heathland, or where there are thin, eroding soils (for example old quarries and coasts). In the south of England it is less demanding and is often associated with shorter areas of wet heath dominated by Cross-leaved Heath.

Life cycle and colony structure

The Silver-studded Blue is single brooded with adults flying from July to August on heathland and from June to mid-July on calcareous sites. Fresh individuals have also been observed on a few sites during late August in recent years, but it is not clear if this represents a genuine partial second brood.

Eggs are laid singly, close to the ground, where they pass the winter. On heathland, they are often laid on woody stems of the foodplants where there is sparse vegetation and patches of bare ground. Most eggs are laid in short vegetation below 10–15 cm, but even shorter vegetation (<7 cm) seems to be selected at the north of its range in Wales. On calcareous sites, eggs are typically laid either amongst bare rock or in very short vegetation (<3 cm).

Larvae hatch in the spring and feed on the buds, flowers, young leaves, or growing tips of the foodplants. The larvae have a close relationship with ants, whereby the ants tend and protect them in return for sugar-rich liquids produced from special glands on the

Silver-studded Blue larva attended by ants.

larvae's bodies. This relationship is highly evolved and research has shown that females choose to lay eggs where they detect suitable ant pheromones.

On heathland, the most commonly associated species are the black ants *Lasius niger* and *L. alienus*. On calcareous sites, the association is almost exclusively with *L. alienus*. The larval ecology is not fully known but the ants probably pick up the larvae soon after hatching and place them in ant chambers beneath rocks or stones. They **pupate** within or close to ant nests where they continue to produce secretions and be tended by ants until the adults emerge and expand their wings.

The population structure of the Silver-studded Blue has been extensively studied using mark–recapture techniques and by observing patterns of habitat occupancy. Adults are extremely sedentary and form colonies on discrete patches of habitat. Most adults move less than 20 m per day and only a few travel more than 50 m. However, a small proportion of adults disperse and have been known to move up to 1.5 km between colonies. Colonizations over this distance are very rare and the maximum recorded distance is 4 km. Because most of its early successional habitats are ephemeral, the butterfly often occurs as metapopulations spanning numerous nearby habitat patches.

Distribution and trends

The Silver-studded Blue was once quite widely spread across England and Wales, occurring as far north as Cumbria and possibly into Scotland (though many of the old records have indistinct place names that cannot be mapped). The butterfly declined steadily during the twentieth century and became extinct in most of central and northern England, parts of Wales and on the North Downs in Surrey and Kent.

The Silver-studded Blue is now widespread only on the heaths of southern England, although there are large populations on the limestone in north Wales and certain sand dunes in Cornwall. Outside these areas it occurs in smaller colonies on the remnant heathlands of north Wales, the Suffolk Sandlings, the Norfolk Breckland, Shropshire, and along the coast of Cornwall and south Devon. It occurs on Sark in the Channel Islands, but is absent from Scotland, Ireland, and the Isle of Man.

Apart from these natural colonies, the butterfly has been reintroduced to a number of sites, including the Dulas Valley in north Wales, where it has spread since 1942, and to six other sites in north Wales from 1978–83, two of which are still extant. Successful introductions have been made at two sites in Suffolk 1986 and in the Wirral in 1994, but a number of other attempts elsewhere have failed.

The distribution of the Silver-studded Blue is now well known, thanks to detailed surveys in all main parts of its range. The current survey suggests that it has disappeared from 44% (46) of the 10 km grid squares recorded in 1970–82, but colonies have been found in 39 new squares due to the increased level of recording. Its overall decline in range at a 10 km square level has been 71% since 1800.

European trend

The butterfly's distribution appears to be stable across much of Europe and has expanded a little in parts of south-east Europe and Russia. In contrast it has declined severely in Belgium (>50% decrease in 25 years) and Germany, Lithuania, and the Netherlands (25–50% decrease in 25 years).

Interpretation and outlook

The decline of the Silver-studded Blue appears to be due to two main factors; the destruction of heathland and the loss of early successional habitats. The overall loss of lowland heathlands in Britain is estimated to be over 60%, with large areas now converted to intensive agriculture, plantation forestry, industry, and housing. The remaining fragments have often become overgrown and unsuitable following the cessation of traditional management such as turf and furze cutting, burning, and grazing by livestock. Extensive suitable habitats have been maintained only in a few areas such as the New Forest (Hampshire). Elsewhere, accidental summer burning has caused local extinctions from which this sedentary species rarely recovers.

The Silver-studded Blue survived on many chalk downs in south-eastern England until the 1950s when the butterfly disappeared suddenly following the introduction of myxomatosis. Some of these sites are now short-grazed again following the recovery of rabbit populations and the restoration of livestock grazing, but so far the butterfly shows no signs of recolonizing.

Outside its southern strongholds, the Silver-studded Blue is highly vulnerable due to the small size and isolation of colonies. The butterfly exists as a number of races in Britain, each of which is important for the conservation of genetic diversity.

Fortunately, a concerted effort is being made to conserve Britain's remaining heathlands, which are recognized as being important in an international context. The Silver-studded Blue is often used as a flagship for this work and, throughout its range, there are teams of conservationists trying to restore heathland and reinstate traditional management. Some local populations have responded well, for example in the Suffolk Sandlings, but huge problems remain elsewhere due to continuing neglect and threats such as quarrying.

In northern heathland, the species is especially difficult to conserve because its habitats are transient and populations can be maintained only by regular, rotational management within easy colonizing distance (i.e. ideally <100 m). In warmer climates further south, the species occupies a somewhat broader niche and management is less critical, though it still has to be regular. Special effort is needed to ensure proper management where habitats are most fragmented and to attempt to extend habitat and re-establish links between sites. Experience in the Suffolk Sandlings and elsewhere shows that it is possible to reverse heathland fragmentation but it needs considerable resources and a concerted strategic approach.

Key references

Joy (1995); Ravenscroft (1990); Ravenscroft and Warren (1996*b*); Thomas (1985); C. D. Thomas *et al.* (1998, 1999).

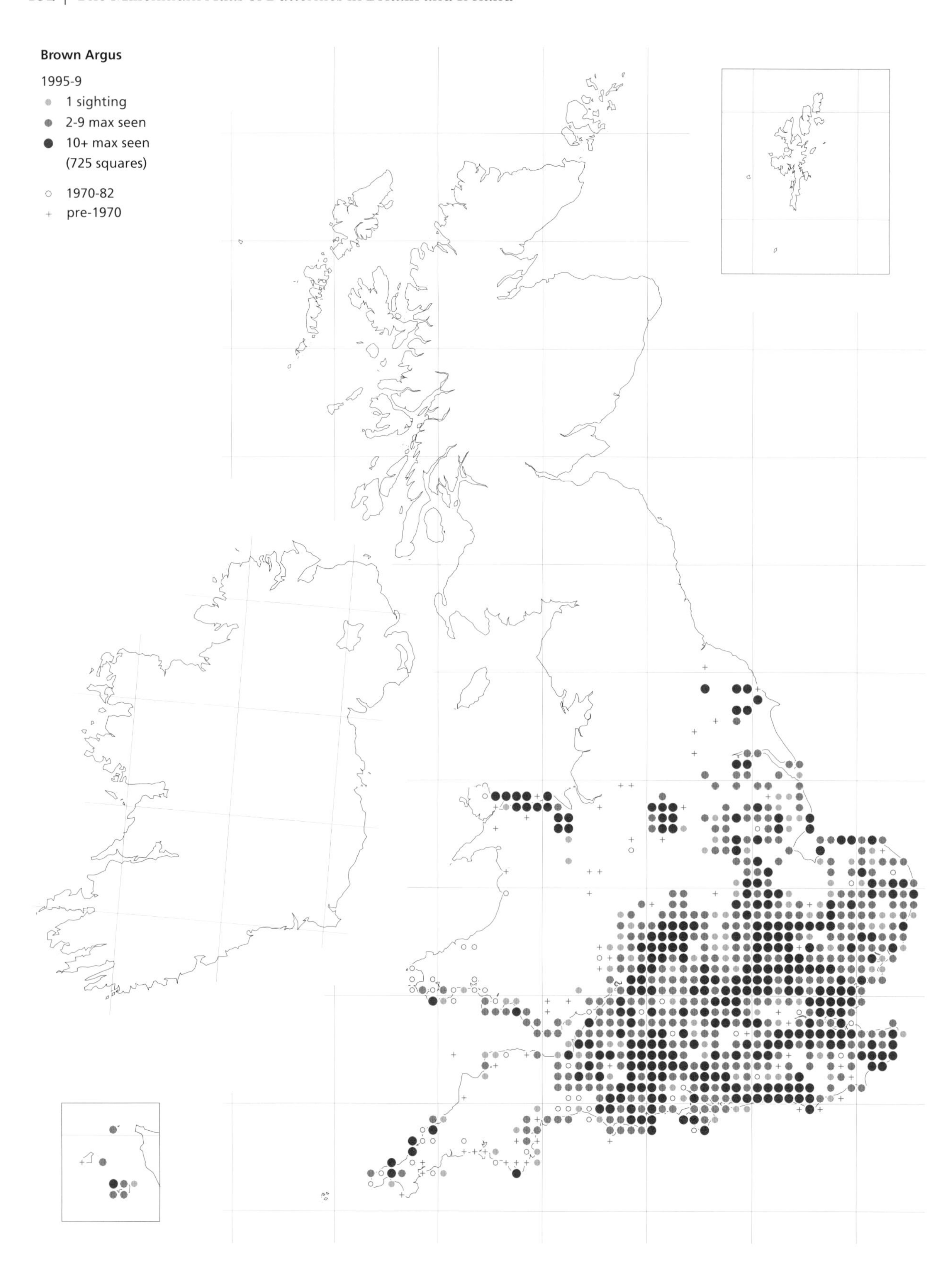
Brown Argus
1995-9
1 sighting
2-9 max seen
10+ max seen
(725 squares)
1970-82
pre-1970

Brown Argus *Aricia agestis*

Resident
Range expanding.

Conservation status
UK BAP status: not listed.
Butterfly Conservation priority: low.
European status: not threatened.

European/world range
Widespread and common across Europe as far north as Denmark and southern Sweden, across to the Middle East and Siberia, and also in North Africa. It has declined in a few western European countries.

This small butterfly is characteristic of southern chalk and limestone grassland but occurs in a variety of other open habitats as far north as north Wales and Yorkshire. It is a close relative of the Northern Brown Argus, which is restricted to Scotland and northern England. The adults have a silvery appearance as they fly low to the ground and they stop frequently either to perch or feed on flowers. They may be confused with Common Blue females, which also have brown upperwings but usually with some blue at the base. The butterfly spread rapidly in the mid-1990s but lost ground in the last three years of the twentieth century.

Foodplants
Common Rock-rose (*Helianthemum nummularium*) is used almost exclusively on calcareous grassland. In other habitats it uses annual foodplants, mainly Dove's-foot Crane's-bill (*Geranium molle*) and Common Stork's-bill (*Erodium cicutarium*). There are also recent reports of egg-laying on Cut-leaved Crane's-bill (*G. dissectum*), Meadow Crane's-bill (*G. pratense*), and Hedgerow Crane's-bill (*G. pyrenaicum*).

Habitat
The butterfly's traditional habitats are chalk and limestone grassland, but it also occurs in a range of other habitats with disturbed soils, including: coastal grassland and dunes, woodland clearings, heathland, disused railway lines, road verges, and more recently set-aside fields. Sheltered sites or slopes facing south or west are preferred.

Life cycle and colony structure
The Brown Argus is double brooded through most of its range with adults flying from early May until the end of June and again from mid-July to mid-September (see p. 394). In the north of its range in England (for example the Peak District) and parts of north Wales there are single brooded populations which fly in June and July. These may occasionally have a small second brood, as in the Peak District during 1999. Until recently, some of these were thought to be populations of the Northern Brown Argus, but new genetic analysis shows that they contain predominantly Brown Argus genes.

The **eggs** are laid singly, usually on the underside of the foodplant leaves, though some are laid on the upper surface, especially on crane's-bills and stork's-bills. Females select foodplants with larger, fleshy leaves that have higher nitrogen content. They lay eggs in fairly short swards, most being laid where the turf is 1–10 cm tall. The **larvae** feed on the inside of leaves, leaving the upper epidermis intact and creating characteristic pale patches. The larvae feed by day and are always attended by ants, which often swarm over them, consuming the sweet secretions

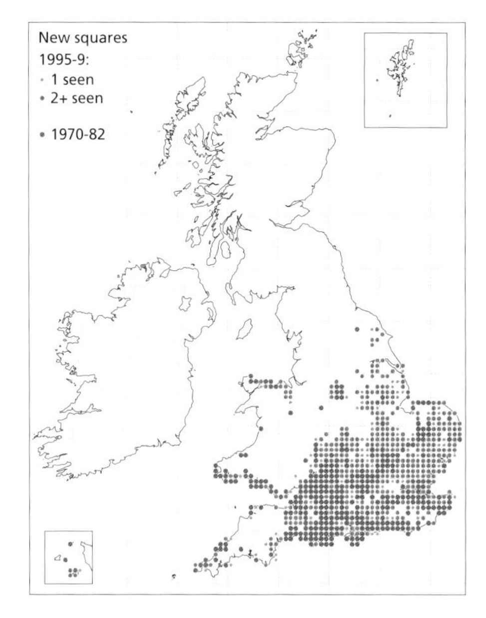

The expansion of range of the Brown Argus since 1970–82.

produced from special glands. The ants most commonly in attendance on downland are the red ant *Myrmica sabuleti* and the black ant *Lasius alienus*. In autumn, the fully-grown larvae overwinter close to the ground and **pupate** the following spring. Both hibernating larvae and pupae are probably covered in earth cells by ants, which continue to attend them until emergence.

The Brown Argus often forms discrete colonies, but it is more mobile than many other colonial species. Mark recapture experiments on downland have shown that adults regularly travel more than 100 m, and can move over 300 m of improved farmland between adjacent hillsides. There are also numerous records of vagrant individuals and of new colonies becoming established many kilometres away from traditional breeding habitats, implying considerable dispersal ability.

Distribution and trends

The Brown Argus is widespread and locally common in southern and eastern England, with scattered colonies around the coast of south-west England, north and south Wales, and the Channel Islands. It is absent from Ireland and Scotland. Recent genetic studies have shown that the single-brooded populations in the Peak District, Yorkshire Wolds, and north Wales are Brown Argus (contrary to the assumption made at the time of the last butterfly atlas). Further research is needed to confirm the identity of some populations in northern England.

The Brown Argus declined substantially during the twentieth century and an estimated 40% of populations had been lost by the late 1980s. However, this decline has since been counteracted by a rapid spread during the 1990s, especially in central and eastern England. In this well-recorded region, it was formerly restricted mainly to the chalk and limestone grassland along the Chiltern Hills and the Cotswolds, but over the last decade it has spread to additional habitats such as road verges, old gravel workings, and non-rotational set-aside fields.

The number of 10 km grid squares recorded has more than doubled since the 1970–82 survey. The spread began around 1990, but did not become very noticeable until 1994–7 when it recolonized many counties, sometimes after an absence of 50 years. It has since retracted a little, especially during the poor summers of 1997 and 1998. There is far less evidence of any expansion in south-west England and in Wales, though these areas are less thoroughly recorded.

During the 1990s the butterfly colonized 23 sites in the Butterfly Monitoring Scheme where it had not been recorded for at least 4–10 years. Analysis of trends with weather data shows that Brown Argus population increases are strongly correlated with summer temperature (June–August), and are negatively correlated with rainfall. Warm, dry summers therefore seem to benefit the species.

European trend

The Brown Argus is stable in most European countries although there have been substantial declines in Austria (>50% decrease in 25 years), Belgium and Luxembourg (25–50% decrease in 25 years).

Interpretation and outlook

Prior to 1990, the Brown Argus had declined due to the loss of its traditional calcareous grassland habitat through intensive agriculture and abandonment. In the last decade this loss has been more than compensated for by its spread onto alternative habitats where it breeds on several species of foodplant that are annuals, rather than the perennial Common Rock-rose. It seems that several factors have influenced this spread.

First, warm summers in the early 1990s may have allowed populations to build up and spread. Second, an important new habitat, non-rotational set-aside, has been created on a large scale since the 1992 reforms of the EC Common Agricultural Policy (CAP). This required farmers to set-aside 15% of their arable fields from production to qualify for arable area payments. Most set-aside fields are rotated each year and are of limited use to the Brown Argus, but a proportion was put into non-rotational set-aside where the annual foodplants such as crane's-bills can become established. The distribution of non-rotational set-aside is concentrated in the predominantly arable areas of central and eastern England, exactly where the Brown Argus has spread most. Breeding has been confirmed in many set-aside fields but the extent of use is unclear.

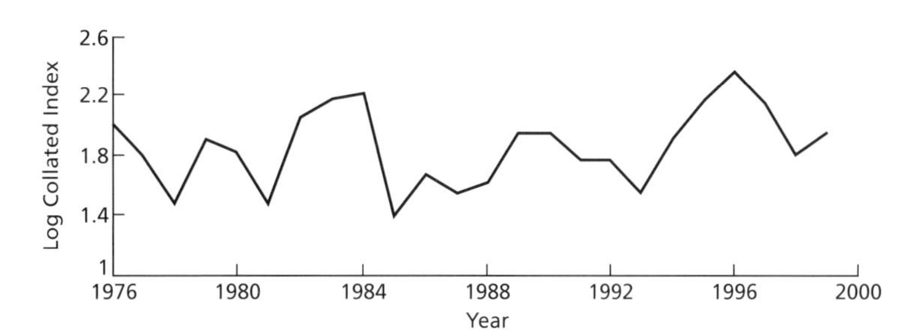

Left
Some set-aside fields have been colonized by the Brown Argus.

Right
Brown Argus abundance at BMS sites, showing high numbers during the recent period of range expansion (1994–7).

Many of the alternative habitats used by the Brown Argus are temporary and contain annual foodplants that vary considerably in abundance from year to year. The butterfly's populations on such habitats are thus prone to large fluctuations in size. Moreover set-aside will eventually be phased out, although it will continue for at least seven years as part of recent CAP reform (Agenda 2000). Over this period, the amount of set-aside is expected to be 10% of the arable area and it will be interesting to see whether the butterfly is capable of maintaining itself in this and other habitats. The recent poor numbers in 1998 and 1999 suggest that this is uncertain and the butterfly could retreat again. Alternatively, it might continue to spread if temperatures in southern England increase as a result of global climate change. If this occurs, the butterfly may overlap with and perhaps displace the Northern Brown Argus.

Whatever the fate of the Brown Argus in other habitats, the conservation of chalk and limestone grasslands is crucial. Fortunately the butterfly thrives under a variety of grazing regimes, though it favours shorter turf and deeper soils where foodplants are green and lush. Such areas are prone to scrub invasion and it is important to include them in scrub control programmes.

Key references

Bourn and Thomas (1993); Greatorex-Davies (1997); Jarvis (1959, 1966); Smyllie (1992, 1998).

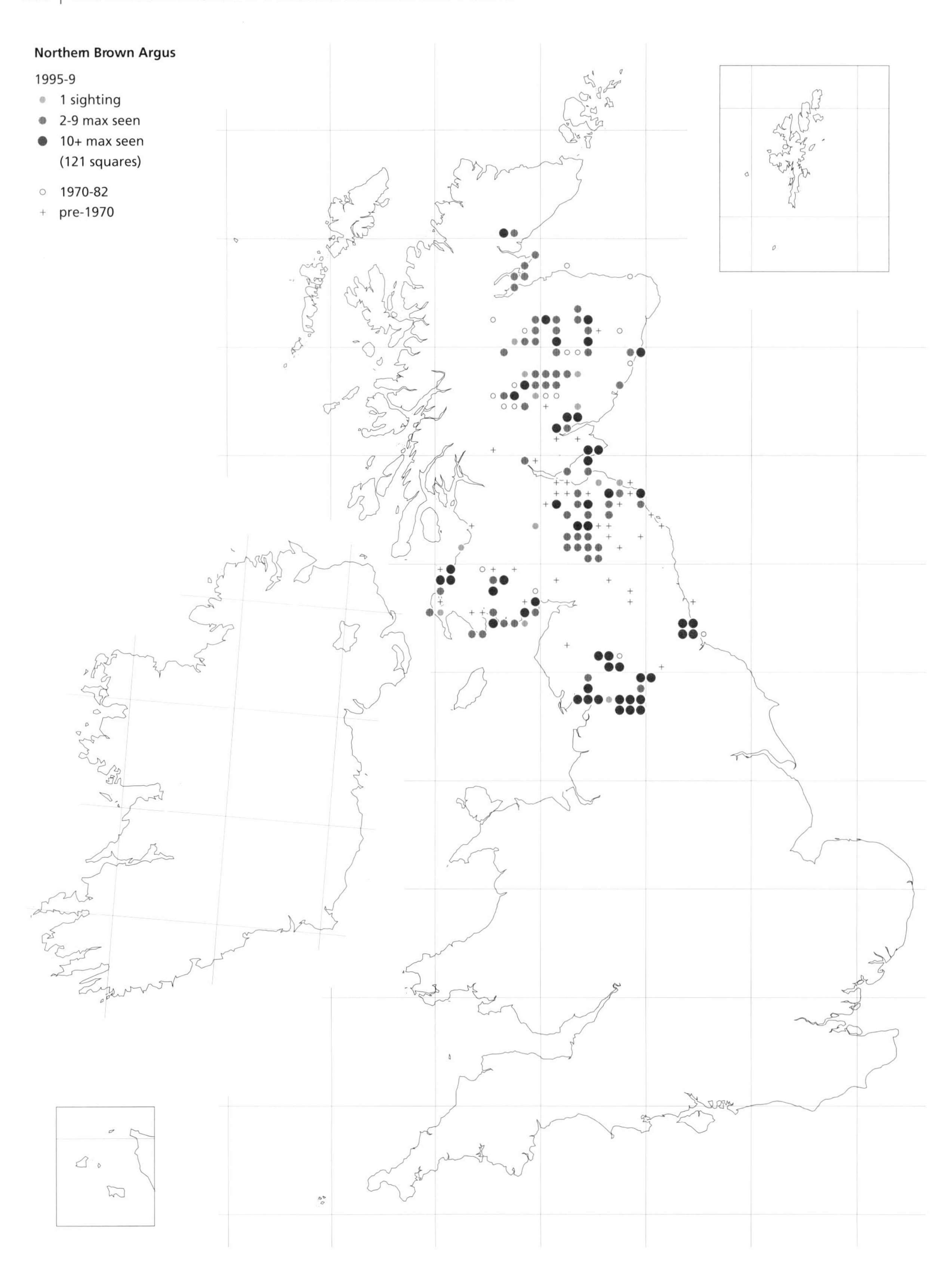
Northern Brown Argus
1995-9
1 sighting
2-9 max seen
10+ max seen
(121 squares)
1970-82
pre-1970

Northern Brown Argus *Aricia artaxerxes*

Resident

Range declining in south.

Conservation status

UK BAP status: Priority Species.
Butterfly Conservation priority: high.
European status: not threatened.
Protected in Great Britain for sale only.

European/world range

Occurs in Scandinavia and mountainous areas of central Europe as well as North Africa. Recent genetic studies suggest that the Northern Brown Argus found in Britain is not endemic, as some authors argue, but is the same species that occurs throughout Europe. However further work is needed to confirm this in all parts of its range. Declining in some countries.

This small butterfly has a silvery appearance as it flies low to the ground over sheltered flowery grasslands. In Britain and mainland Europe, the pattern of wing spots is highly variable and many local races (and sub-species) have been described. In Scotland, most individuals are of the race *artaxerxes* and have a characteristic white spot in the middle of the forewing (see photo). In northern England, this spot is generally dark brown or black. The butterfly occurs mainly as small, scattered colonies and has declined in northern England.

Foodplants

The most important foodplant is Common Rock-rose (*Helianthemum nummularium*), though there are rare records of egg-laying (probably accidental) on other plants.

Habitat

The butterfly occurs in well drained, unimproved grasslands where Common Rock-rose grows in a lightly grazed or ungrazed sward. Most sites are sheltered (often with scrub) and have thin, base-rich soils with patches of bare ground, for example coastal valleys, steep slopes, sand dunes, and quarries. In Scotland it may also occur on predominantly neutral and even acidic soils where Common Rock-rose is able to grow if there is some calcareous influence through weathering or flushing.

Life cycle and colony structure

The adults are single brooded and typically fly from early June until mid-August, with a definite peak in numbers from late June to mid-July. The flight period varies considerably between years and between regions. In northern Scotland it flies later, from July until September (see p. 394), and on some sites the flight period can be as short as 3–5 weeks.

The **eggs** are laid singly on the upperside of Common Rock-rose leaves where they are highly visible and easily counted. Females select plants that have fleshy leaves and are rich in nitrogen, typically growing in sheltered situations. They lay in swards of various heights ranging from 1–30 cm, although a mosaic of short and tall patches seems to be preferred, with swards of 6–10 cm.

The young **larvae** move to the underside of the leaf and feed on the interior, leaving the upper surface intact. They hibernate while quite small (second or early third instar) at the base of the foodplant or on the ground amongst nearby grass. The larvae start basking in early spring before recommencing feeding. They possess ant-attracting organs on the abdomen and some researchers have observed post-hibernation larvae being attended by ants. However, others have detected only occasional interest from

Northern Brown Argus eggs on the uppersides of Common Rock-rose leaves.

Northern Brown Argus larva on Common Rock-rose, with leaves showing typical larval feeding damage.

foraging ants and any association may be irregular. The larvae **pupate** in late May, often lying on the ground on a silken mat, or attached by silk threads amongst the vegetation.

The Northern Brown Argus forms discrete colonies that are generally small (<200 adults). Studies in several regions show that most colonies breed on habitat patches <1 ha and that very few sites are larger than 10 ha. Mark–recapture studies in Scotland and England show that adults tend to be highly sedentary, typically moving <20–30 m over several days but with some movement between nearby habitat patches up to 150 m. The species thus appears to have a very limited colonizing ability.

Another study showed that extinction was more frequent on isolated sites. It appears that clusters of nearby habitats are interconnected by periodic dispersal and probably support metapopulations within which periodic local extinctions and colonizations can be accommodated.

Distribution and trends

The Northern Brown Argus is found in scattered colonies across southern and eastern Scotland and in the far north of England. There is evidence of a serious decline in the Durham region with loss rates of inland colonies estimated at over 35% during the 1980s and 8% on coastal sites. Fortunately, this decline appears to have slowed during the 1990s, probably due to conservation measures at remaining sites. Numerous colonies occur on the limestone outcrops of southern Cumbria and north Lancashire (including nature reserves) where it appears to be fairly stable.

In Scotland, the butterfly has been greatly under-recorded in the past and the current survey has almost doubled the number of 10 km squares recorded. It is found in the Borders and along the coast of Dumfries and Galloway, but then has a predominantly eastern distribution to south-east Sutherland, with strong populations in Perthshire and northern Tayside. It has undoubtedly declined in the south of Scotland, especially in the Borders and around Edinburgh, but further surveys are needed to assess its true status in Scotland.

Recent genetic analysis by universities in Britain and Norway has suggested that single brooded populations in the Peak District, Yorkshire Wolds, and north Wales, once thought to be Northern Brown Argus, are predominantly Brown Argus. In contrast, the populations in Co. Durham (formerly referred to as *salmacis*) appear to be predominantly Northern Brown Argus. The main map shows the distribution as accurately as is currently known, but further analysis is required to confirm the identity of a few populations in northern England.

European trend

The species appears to be stable in most European countries but has undergone a severe decline in Poland (>75% decrease in 25 years) and also in Denmark and Latvia (25–50% decrease in 25 years).

Interpretation and outlook

The distribution of the Northern Brown Argus is limited by the availability of specific types of sheltered unimproved grassland habitats that provide enough foodplant and are not heavily grazed. Such habitats have been steadily reduced throughout its range and it is clear that many colonies are now very small and vulnerable to extinction. Conservation measures are thus needed urgently through much of its range.

The butterfly's decline in northern England and southern Scotland is attributable partly to loss and fragmentation of habitats but also to changing management on remaining patches of grassland. On most sites, the

Northern Brown Argus requires some form of light grazing either by livestock or rabbits to maintain suitable open conditions. The butterfly has become extinct at many sites following neglect and scrub invasion, whereas at others it has suffered from overgrazing. For example, it is absent from many heavily grazed sites in Scotland even though they contain abundant foodplant.

The butterfly seems to prefer an uneven sward, which is best maintained by light winter grazing at stocking rates equivalent to less than 1–2 sheep or 0.5 cattle per hectare. Summer grazing is thought to be a less satisfactory alternative that may not be suitable on all sites. However, some sites are maintained purely by rabbit grazing or by natural erosion of steep slopes and may require little additional management apart from periodic removal of scrub.

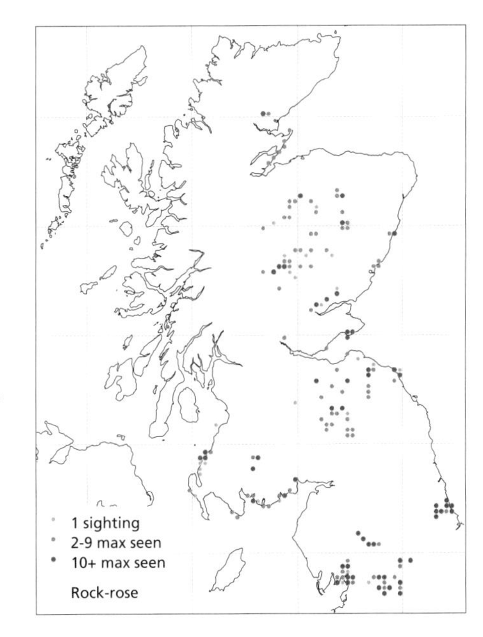

Distribution of 1995–9 records of the Northern Brown Argus at 5 km square resolution, in relation to that of Common Rock-rose (1950 onwards records) in northern Britain.

There are several recent examples of large population increases following improved management. For example, it was formerly recorded in small numbers at St Abb's Head National Nature Reserve in the Borders, but has increased dramatically since 1992 following a switch from heavy grazing by sheep to selective spring and autumn grazing. It has also increased markedly at a neglected site in Cumbria where winter grazing has been introduced. Many colonies in northern England now occur on nature reserves or on private land entered into the Countryside Stewardship scheme, which contains special management prescriptions for the butterfly.

Key references

Clunas (1986); S. Ellis (1995, 1999*a*); Ravenscroft and Warren (1996*c*); Smyllie (1992, 1998).

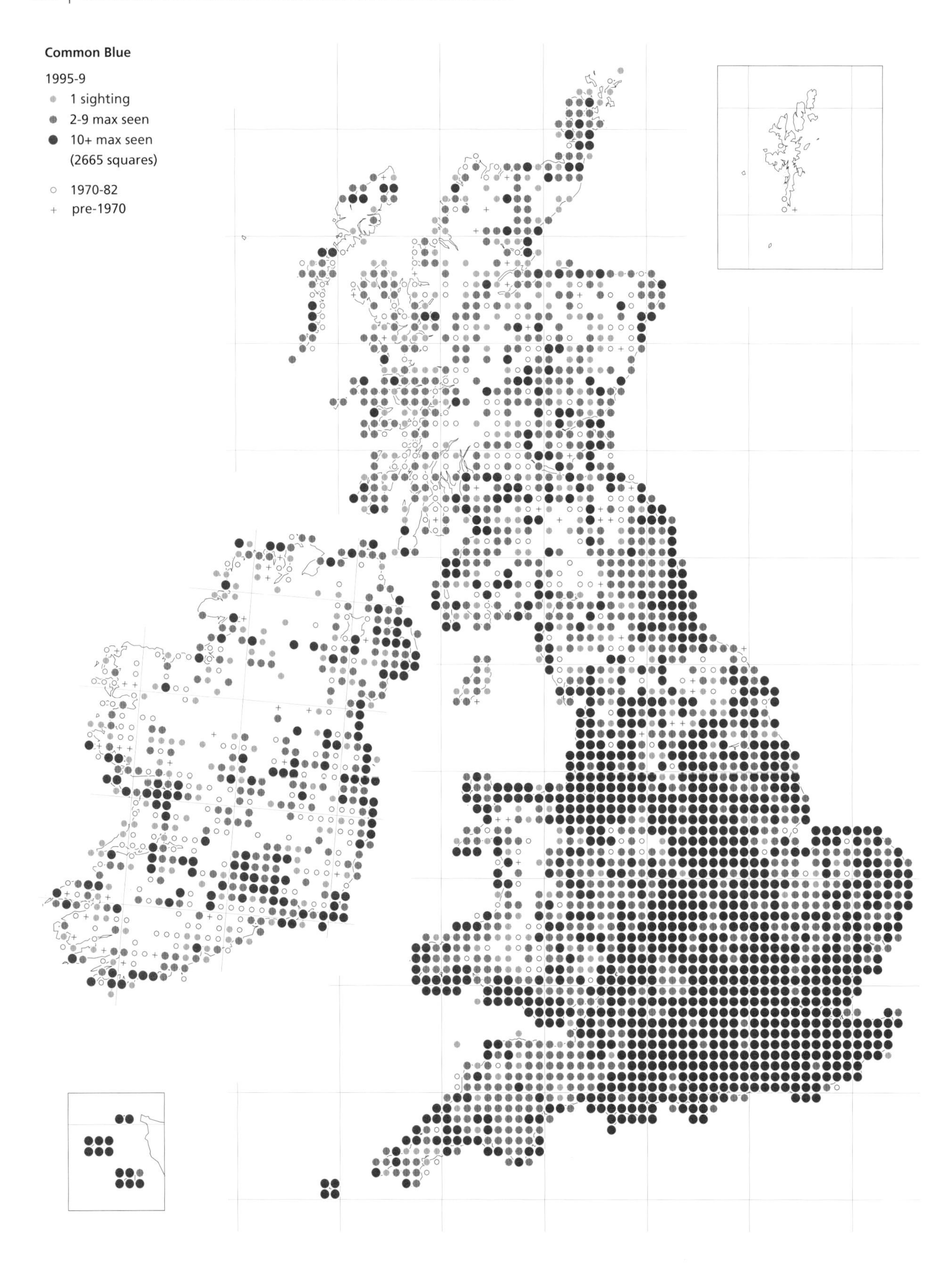
Common Blue
1995-9
1 sighting
2-9 max seen
10+ max seen
(2665 squares)
1970-82
pre-1970

Common Blue *Polyommatus icarus*

Resident

Range stable.

Conservation status

UK BAP status: not listed.
Butterfly Conservation priority: low.
European status: not threatened.

European/world range

Occurs widely throughout Europe and in North Africa and temperate Asia. It appears to be stable in most European countries, but there have been some declines and expansions.

The Common Blue is the most widespread blue butterfly in Britain and Ireland and is found in a variety of grassy habitats. The brightly coloured males are conspicuous but females are more secretive. The colour of the upperwings of females varies from almost completely brown in southern England to predominantly blue in western Ireland and Scotland, but the colour is variable within local populations with some striking examples. It remains widespread but there have been local declines within its range.

Foodplants

Common Bird's-foot-trefoil (*Lotus corniculatus*) is the main foodplant. Other plants used include: Greater Bird's-foot-trefoil (*L. pedunculatus*), Black Medick (*Medicago lupulina*), Common Restharrow (*Ononis repens*), White Clover (*Trifolium repens*), and Lesser Trefoil (*T. dubium*).

Habitat

It occurs in a range of grassy places where its foodplants grow in sunny, sheltered situations on downland, coastal dunes and undercliffs, road verges, acid grassland, and woodland clearings. It is also found in waste ground, disused pits and quarries, golf courses, and urban habitats such as cemeteries.

Life cycle and colony structure

The Common Blue is one of the few species of butterfly in which the number of generations per year varies in different parts of Britain and Ireland, though the boundary between single-brooded and double-brooded populations is not clear-cut (see p. 47). In the double-brooded populations (in most parts of England, Wales, and Ireland) adults fly in May and June and again in August and September. In the single-brooded populations (England from Yorkshire northwards, Scotland, and in parts of the north of Ireland) adults emerge between June and September, earlier in warm locations and later in cooler ones. The flight period of the single-brooded populations is longer than either the first or second broods of the double-brooded populations. The latter may have a partial third brood in warm summers, and the univoltine populations may have a partial second

Common Blue larva.

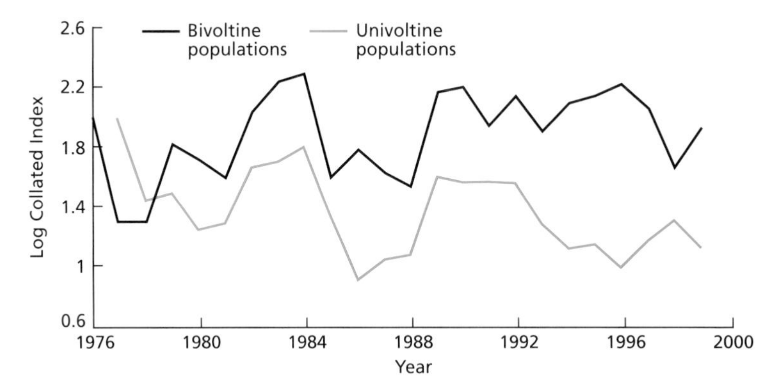

There is synchrony between single brooded (northern) and double brooded (southern) populations of the Common Blue in the BMS until 1992.

brood. This adaptability supports the view that there is an environmental, rather than a genetic, reason for the geographical differences in voltinism.

Eggs are laid singly on young shoots of the foodplants. Half-grown **larvae** overwinter on the lower stems of foodplants or at ground level amongst litter. Some larvae enter hibernation in June and do not emerge as adults until the following year. The second brood is thus partial, but as it is generally larger than the first, it is thought that most larvae complete their development in the same season. Larvae in the final instar are attractive to ants, though do not seem as dependent on them as some other blues. The **pupae** also attract ants and may be carried into their nests in the spring.

Common Blues live in discrete colonies. Comparatively little is known about their mobility, but some dispersal takes place and newly formed habitat is readily colonized in most parts of its range. Adults of the first generation may be more mobile than those of the second.

Distribution and trends

One of our most widespread butterflies, the Common Blue occurs throughout Britain and Ireland. It occurs on offshore islands, even remote islands in the Outer Hebrides, and is present on the Isle of Man and Channel Islands and also occurs on Orkney. It was recorded on Shetland during the 1970s, but has not been recorded in the current survey period. It becomes scarcer at altitudes above 300 m and is probably absent from land above 500 m. Colonies are most numerous in southern parts of England and Wales, and become more scattered in northern England and Scotland, where they are especially associated with coastal habitats, limestone areas, and river valleys. In Ireland colonies are most abundant around the coast, while gravel pits in limestone eskers are a typical habitat in the Irish midlands.

This species is thought to be under-recorded in a few areas on both current and previous distribution maps, particularly in western Scotland. In some places there is a possibility of misidentification due to the presence of closely related species. The female may be mistaken for the Brown Argus and other female blues. The male is less likely to be misidentified although confusion with the Holly Blue is possible, especially during flight.

Although its overall distribution has not changed, the number of colonies is thought to have declined in recent decades, especially in the intensively farmed parts of eastern England. The large-scale replacement of herb-rich grassland with Perennial Rye-grass or arable crops has had a significant impact on this species. However, it is able to establish small populations even in urban areas, and there are records from all London boroughs at some time in the last two decades.

European trend

Although stable in most European countries, the Common Blue has declined in the Canary Islands (25–50% decrease in the past 25 years), Finland, Luxembourg, and the Netherlands (15–25% decrease). In contrast its range has expanded in Malta and parts of western Russia.

Interpretation and outlook

Although there have been many extinctions of colonies due to habitat loss, this species remains widespread enough to find its way to areas of land that are suitable for breeding within a few kilometres of an existing population. There is scope for the Common Blue to become established on a range of man-made habitats, such as landfill sites and golf courses, even in urban areas, as well as on less intensively managed agricultural land and set-aside. A small colony may survive in as little as 0.5–1.0 ha of suitable habitat.

Common Bird's-foot-trefoil readily colonizes disturbed land and the butterfly will also breed on the continental forms of this foodplant that are widely planted on road verges and in other grassland restoration schemes. However, these colonizations do not balance the huge losses in intensively farmed areas.

The largest colonies occur on downland and on some coastal habitats where there are abundant foodplants in sheltered, warm spots, with some taller vegetation to provide roosting areas. Although numbers are generally higher in a warm season, prolonged drought may result in withering of foodplants and a sharp fall in numbers. This happened after the summer of 1976 when populations took several generations to recover.

In most situations management is required to prevent vegetation from becoming too tall and dense and to maintain a suitably varied structure. A mixture of tall (up to 15 cm) and short sward heights with patches of bare ground is required. This can be achieved by grazing or by cutting of vegetation, which encourages fresh growth of foodplants. Cattle and horse grazing are especially beneficial as they break up the sward and their hooves create small, bare depressions. However, heavy grazing which results in a uniform short sward does not suit this species.

Interesting evidence is accumulating on the link between the diet of larvae at different times and in different broods and their subsequent appeal to ants. Whether this information can be used to guide conservation strategies in the future remains to be evaluated.

Although not threatened on a national scale, the Common Blue, like many other widespread species, remains vulnerable to local extinctions through changes of land use. It is a robust species as it can breed in a variety of habitats and has a wide range of foodplants, a relatively high mobility and an adaptable phenology that allows it to respond to climate variations.

Key references

Burghardt and Fiedler (1996); Dennis (1984, 1985*b*); Dowdeswell *et al.* (1940); León-Cortés *et al.* (1999).

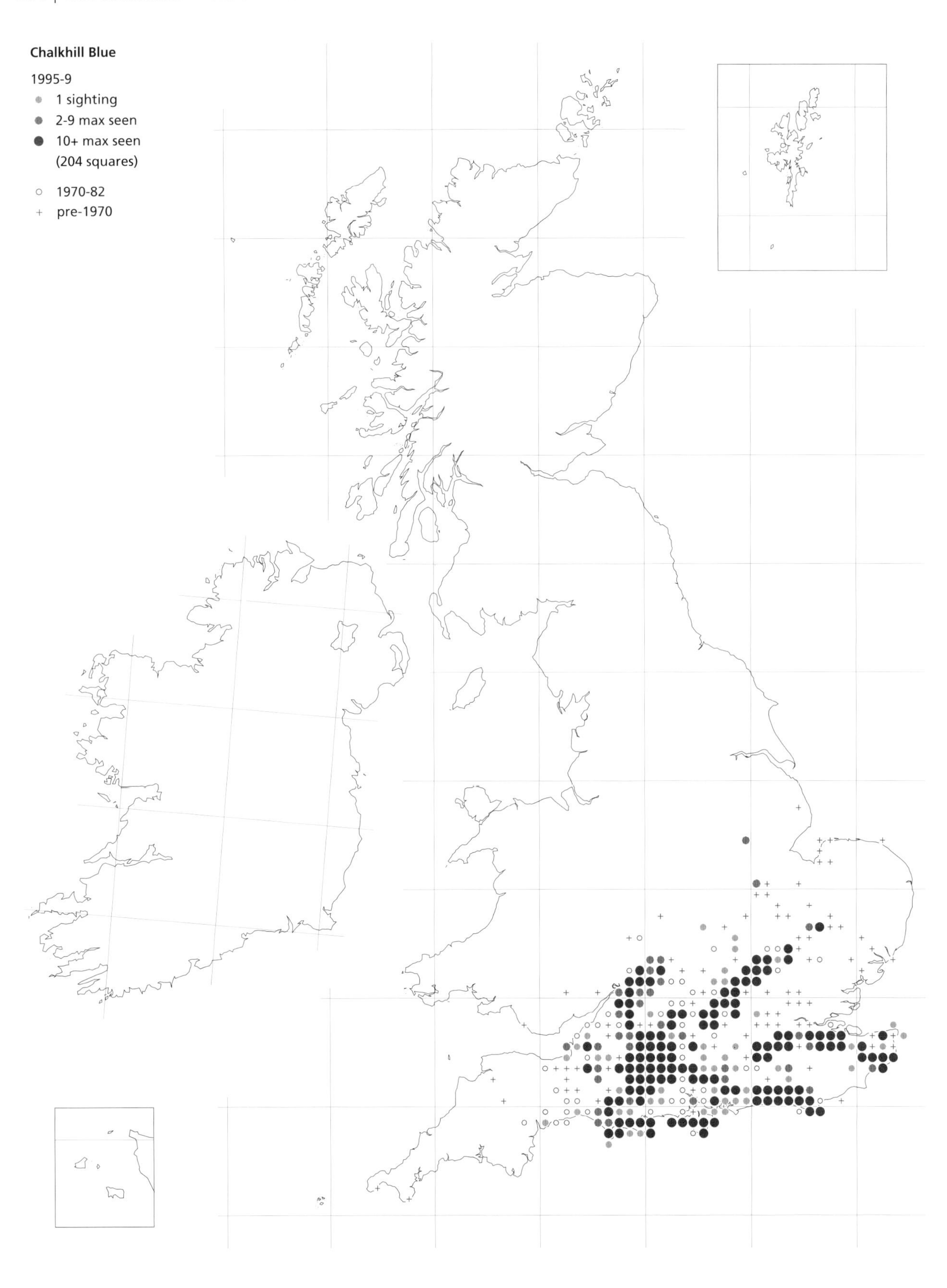
Chalkhill Blue
1995-9
1 sighting
2-9 max seen
10+ max seen
(204 squares)
1970-82
pre-1970

Chalkhill Blue *Polyommatus (Lysandra) coridon*

Resident

Range declining in some areas.

Conservation status

UK BAP status: Species of Conservation Concern.
Butterfly Conservation priority: low.
European status: not threatened.
Protected in Great Britain for sale only.

European/world range

Across most of Europe (excluding Scandinavia and areas south of 40 °N) eastwards to the Urals. Its range is stable through much of Europe but it has declined in several countries.

Foodplants

The sole foodplant is Horseshoe Vetch (*Hippocrepis comosa*).

Habitat

Both the foodplant and the butterfly are restricted to chalk and limestone grassland. The Chalkhill Blue breeds on all aspects but prefers south- and west-facing slopes and shorter turf.

Life cycle and colony structure

There is one generation a year. Adults begin to emerge in early July in warm seasons and numbers reach a peak in early August, with some individuals flying until mid-September. **Eggs** are laid singly on or near Horseshoe Vetch. As the larvae do not feed until the following year, the exact placing of eggs on the foodplant is not as critical as it is for many species, and eggs are laid on the stems of Horseshoe Vetch, or on adjacent vegetation or debris, where they pass the winter before hatching in March. Most eggs are laid where the vegetation is 2–10 cm tall. **Larvae** are tended by ants, which are attracted by sweet secretions produced by special glands. Larvae are often buried by ants during the day, and at dusk, when they resume feeding, may be accompanied by several ants. **Pupae** are also attractive to ants, which cover them in earth cells or take them into their nests.

The male Chalkhill Blue is paler and, apart from the Large Blue, larger than other blue butterflies seen in Britain and Ireland. At some sites many hundreds may be seen in August, flying just above the vegetation, searching for females. Large numbers of males may also congregate on animal dung and other sources of moisture and minerals. Females are much less conspicuous, being duller in colour, more secretive in their habits, and spending less time than the males in flight. The butterfly is confined to calcareous grassland in southern England and has declined in some areas during recent decades.

Chalkhill Blue butterflies live in discrete colonies but adult males are mobile and have been seen 10–20 km from known colonies (see 2 km square resolution distribution map). Marking experiments on three populations in Hampshire suggest that the adults disperse more readily than those of the Adonis Blue, with 1–2% of the populations exchanging between colonies 1–2 km apart.

Distribution and trends

Colonies occur mainly on the calcareous grassland of the North and South Downs, the Chilterns, the Cotswolds, the Hampshire, Wiltshire, and Dorset Downs, and the Isle of Wight, with scattered colonies in a few other counties. Horseshoe Vetch occurs much further north than the Chalkhill

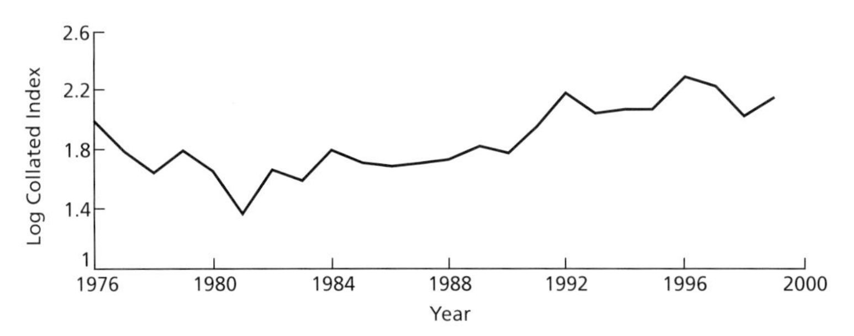

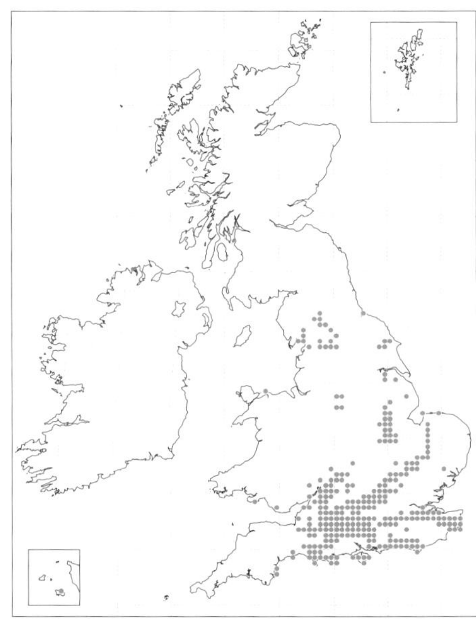

Left
Abundance of the Chalkhill Blue is increasing steadily at BMS sites despite declines in range.

Right
The distribution of the Chalkhill Blue's sole larval foodplant, Horseshoe Vetch, 1950 onwards records at 10 km square resolution.

Blue, suggesting that the northern limit of the butterfly's range is determined by climate rather than the distribution of its foodplant. However, where it does occur further north, Horseshoe Vetch becomes more fragmented in distribution and occurs in smaller patches, which may be a factor limiting the distribution of the butterfly. Other factors, such as soil type and association with ants may also limit its distribution.

Widespread losses of colonies due to the intensification of agriculture began early in the twentieth century. The species was last seen in Shropshire in 1916 and in Suffolk, where there had been several colonies in the nineteenth century, in 1923. The most northerly population on the current map, in Lincolnshire, was the result of a recent reintroduction that has since failed. Losses were greatest on flat land, for instance in East Anglia and in Wiltshire, around Salisbury Plain. Many remaining colonies occur on slopes too steep for the plough, often on old earthworks, trackways, and archaeological sites. Further losses occurred as traditional grazing of downland became uneconomic, while in the 1950s the reduction of rabbit numbers through myxomatosis meant the loss of the only remaining grazing in many places. In the following years, many slopes became too overgrown for Horseshoe Vetch to survive and further colonies were lost.

In recent years, recovery of rabbit populations and active management of downland for conservation has halted the decline in some areas. Scrub clearance and the reintroduction of grazing have allowed numbers to increase as sward with Horseshoe Vetch has been restored to shorter conditions. However, in other areas losses have continued and the current map shows that it has probably disappeared from 25% of the 10 km squares recorded in 1970–82. The Chalkhill Blue is a conspicuous species, which occurs in habitats much visited by recorders. Although the females may be confused with those of other species, the males are more readily seen and easily identified. It is unlikely therefore that many colonies have been overlooked and the species has been recorded from a number of new squares in the current survey.

European trend

While stable in many countries it has declined substantially in Belgium (25–50% decrease in 25 years) and in Austria, the Czech Republic, Germany, Lithuania, Luxembourg, and Slovakia (15–25% decrease).

Interpretation and outlook

The distribution of the Chalkhill Blue is determined by that of its foodplant, Horseshoe Vetch. Colonies often occur at the same sites as the scarcer Adonis Blue, which uses the same foodplant. The Chalkhill Blue, however, is able to breed in taller swards than the Adonis Blue can tolerate, and has therefore not undergone such a severe decline due to the cessation of grazing. In fact the very short turf (2 cm tall) preferred by the Adonis Blue seldom supports large populations of the Chalkhill Blue.

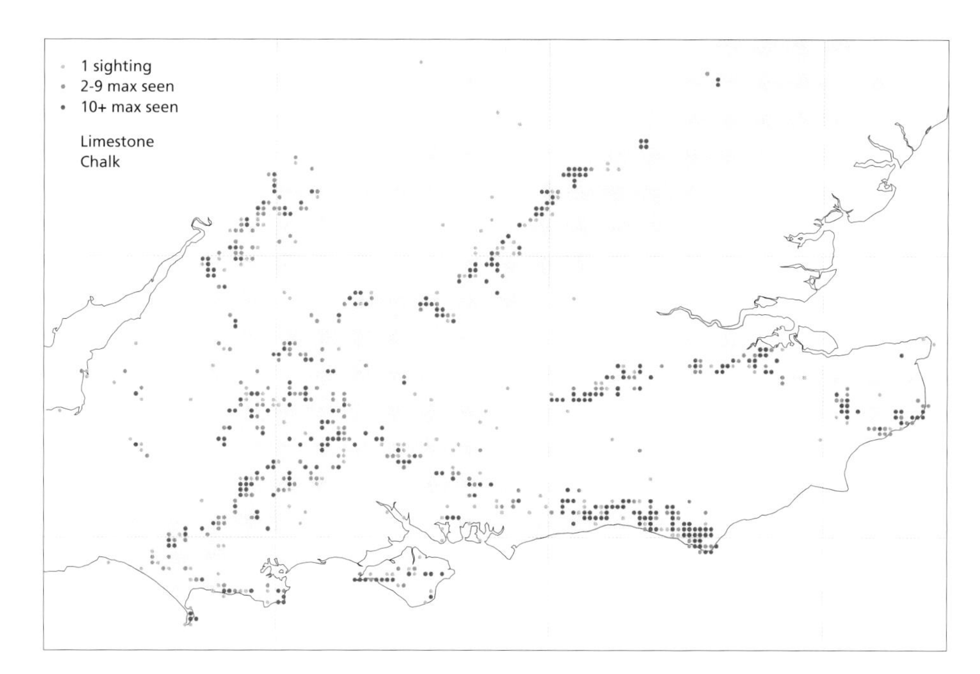

The distribution of 1995–9 records of the Chalkhill Blue at 2 km square resolution, against a background of the main chalk and limestone strata of southern England.

A sward height of up to 10 cm is tolerated by the Chalkhill Blue, but a shorter sward is preferred, especially on north-facing slopes. In contrast, Horseshoe Vetch will survive in a sward up to about 20 cm tall, but requires shorter sparse sward for the establishment of seedlings. The ideal vegetation height of 2–6 cm is best maintained by grazing, especially by animals that create an open sward with a varied height structure. This is most easily achieved by grazing with cattle or sheep during autumn or winter, when the butterfly is in the egg stage. Overwintering eggs fall to the ground when disturbed and are largely undamaged by winter grazing. Some taller grasses and scrub are beneficial for roosting adults and to provide shelter.

Horseshoe Vetch is more readily eliminated from neglected swards than many other plants and is very slow to return once it is lost. There is little chance of the butterfly recolonizing such sites, even if sympathetic management is restored. For this reason the conservation of sites supporting existing populations is particularly important, as future extinctions are unlikely to be balanced by colonizations. Fortunately, many remaining populations are on nature reserves or other land which is under sympathetic management. Schemes such as Environmentally Sensitive Areas and Countryside Stewardship are also having a positive effect by restoring grazing to neglected downland. As long as Horseshoe Vetch is still present, colonies that have dwindled in numbers respond well to appropriate management and large populations have been restored at some localities. Several sites in the Butterfly Monitoring Scheme (BMS) have shown significant increases in numbers and there has been one colonization of a site where numbers have subsequently increased steadily. Two BMS sites, however, have shown decreases. The species has done particularly well in the generally warm summers of the 1990s.

Although more is known about the ecology of the Chalkhill Blue than about many other butterflies, less is known about its foodplant, which also has exacting requirements and is a declining species in need of conservation. In some situations it sets seed readily, whereas in others no seeds are produced. At several sites it declined suddenly from being abundant to being scarce in the early 1990s, for example on the North Downs. Heavy grazing by rabbits may have been responsible for this, though at sites nearby, rabbits, which are selective feeders, appear not to eat Horseshoe Vetch. Further research into the foodplant may help in planning conservation measures for both the Chalkhill Blue and Adonis Blue.

Key references

Fearn (1973); Loertscher *et al.* (1997).

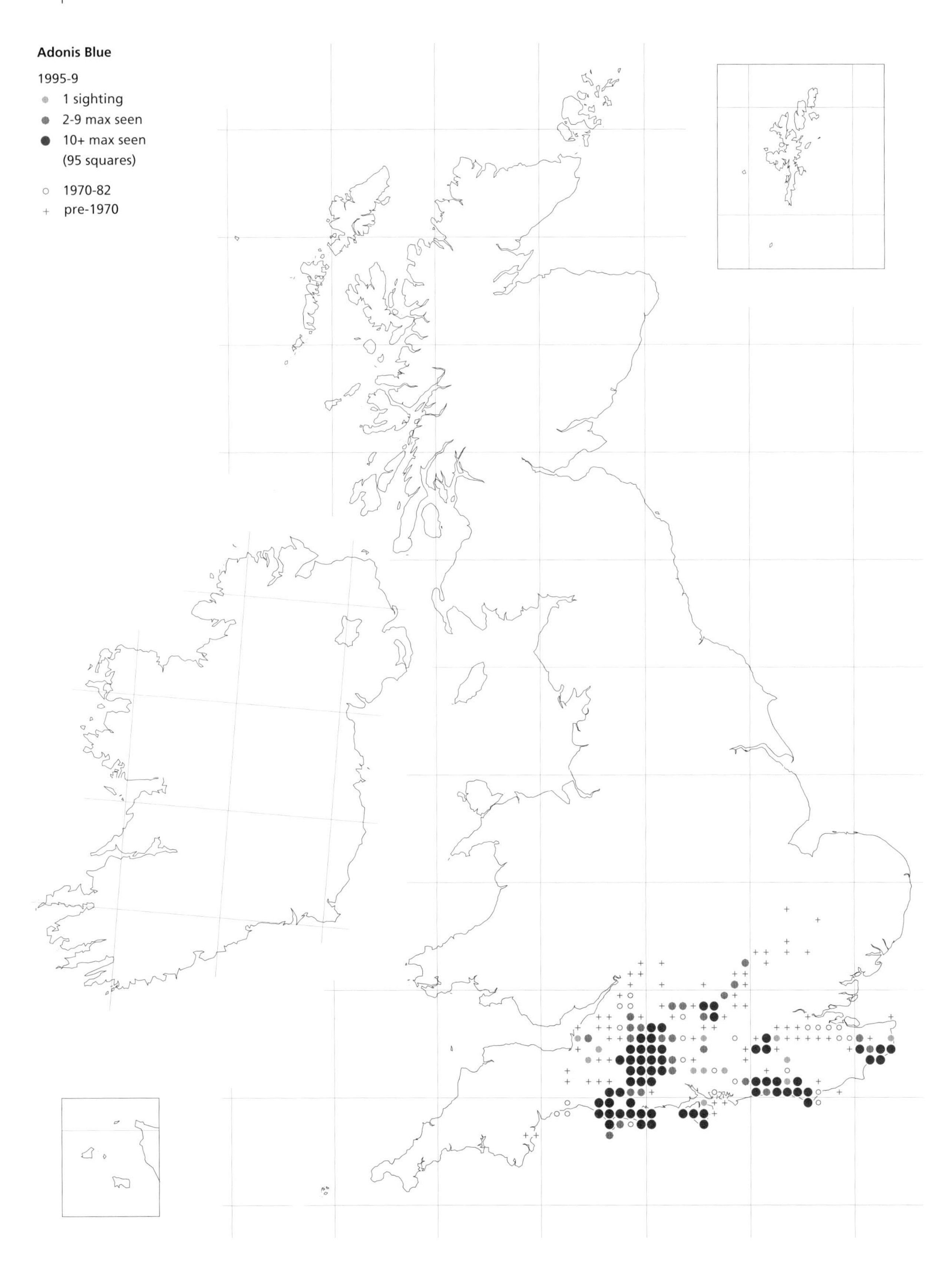
Adonis Blue
1995-9
1 sighting
2-9 max seen
10+ max seen
(95 squares)
1970-82
pre-1970

Adonis Blue *Polyommatus (Lysandra) bellargus*

Resident

Range decline and recent slight re-expansion.

Conservation status

UK BAP status: Priority Species.
Butterfly Conservation priority: medium.
European threat status: not threatened.
Protected in Great Britain for sale only.

European/world range

Across Europe from southern Spain to Lithuania (not in Scandinavia) and east across Russia. Declining in many northern European countries but more stable in south.

This beautiful butterfly is one of the most characteristic species of southern chalk downland, where it flies low over short-grazed turf. The males have brilliant sky-blue wings, while the females are brown and far less conspicuous. Both sexes have distinctive black lines that enter or cross the white fringes to the wings. Despite its restricted distribution, the butterfly can be seen in many hundreds on good sites. It has undergone a major decline through its entire range, but has recently re-expanded in some regions.

Foodplants

The sole foodplant is Horseshoe Vetch (*Hippocrepis comosa*).

Habitat

Dry, chalk or limestone grassland with abundant foodplants growing in short turf 1–4 cm tall, except in sheltered quarries where slightly taller vegetation may be used. Most colonies occur on warm, south-facing slopes where favoured breeding areas are often sheltered hollows, especially old chalk pits and quarries.

Life cycle and colony structure

There are two broods per year, with adults flying from mid-May to mid-June, and early August to mid-September. There is evidence that the average emergence period has advanced by at least twenty days during the last two decades. The **eggs** are laid singly, mainly on the underside of the terminal leaflets, on very small foodplants growing in short turf. These carefully selected conditions provide a very warm microclimate for larval development and are favoured by ants, which tend both larvae and pupae. The green **larvae** are well camouflaged and feed on the lower surface of the leaflets leaving the upper cuticle intact, causing characteristic feeding damage consisting of pale discs on the leaflets. Larger larvae feed on whole leaves, stems, and fruits.

The larvae are nearly always attended by ants, which are attracted by secretions from special 'honey' glands and pores. The larvae and pupae also emit a low rasping sound that may attract ants. Any ant species appears suitable, but the most common are the red ant *Myrmica sabuleti* and the small black ant *Lasius alienus*. The ants protect the larvae

Adonis Blue larva attended by ants.

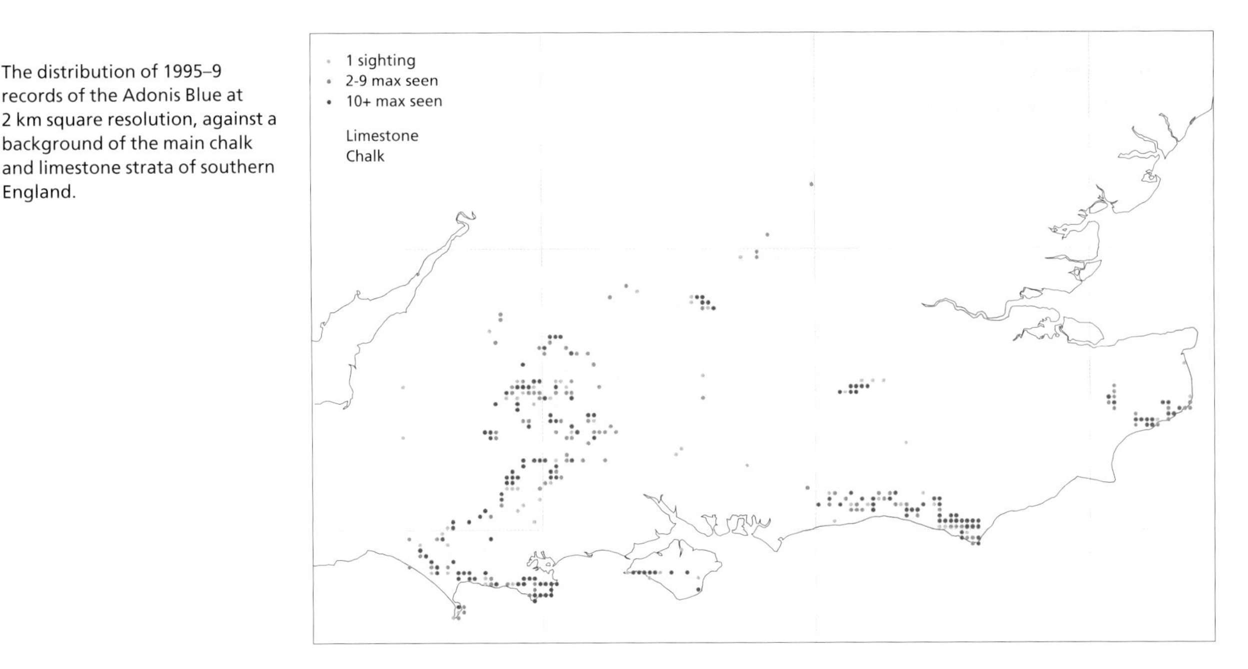

The distribution of 1995–9 records of the Adonis Blue at 2 km square resolution, against a background of the main chalk and limestone strata of southern England.

from predators and parasitoids, and even bury the larvae (in groups of up to eight) in loose earth cells at night. The Adonis Blue overwinters in the larval stage and **pupates** in the upper soil surface, often in the warm upper chambers of ant nests where pupae continue to be attended by ants until the adults emerge.

The Adonis Blue is strongly colonial and adults are concentrated where there are patches of short turf for breeding. Marking studies have shown that adults are very sedentary, often moving less than 50–100 m between captures, and that quite small barriers of scrub or hedgerow can be enough to separate colonies. However, the butterfly does have some powers of dispersal as demonstrated by its recent recolonization of sites in Dorset, sometimes over distances of 10–15 km. There is also evidence that populations on smaller and more isolated sites are more likely to become extinct, and that colony persistence may depend on being part of a larger metapopulation, covering a network of sites.

Distribution and trends

The Adonis Blue reaches the northern climatic limit of its European range in Britain, where it has always been restricted to warm, dry calcareous grassland in southern England. It has never been recorded in Scotland, Wales, Ireland, the Isle of Man, or the Channel Islands.

The butterfly has declined substantially over the last 200 years and it died out in Cambridgeshire and Essex in the early 1800s and Suffolk in the 1850s. Its decline became most severe during the 1950s and 60s, and by the time of the first detailed survey, in 1981, it had been reduced to an estimated 70–80 colonies. Over this period it became extinct in the limestone grasslands of the Cotswolds and most of the chalk downs in the Chiltern Hills, while numbers were severely depleted on the North and South Downs of Surrey, Sussex, and Kent. Even in its Dorset stronghold the loss of colonies to 1981 was estimated to be at least 70%, while nationally the figure was put at over 90%.

The early 1980s probably represented the low point in its distribution and it has since undergone a partial recovery in some areas, notably in Dorset and Wiltshire. The data for 1995–9 show that the number of recorded 10 km squares has increased by 20% since 1970–82. Thorough surveys showed that, between 1978 and 1997–9, the number of colonies in Dorset had risen from 37 to 60. Data from other county surveys during the 1990s indicate that there may now be as many as 250 colonies in Britain as a whole, the majority being found in Dorset (*c.*100

colonies), Wiltshire (*c.*90), and Sussex (*c.*40). In addition, the butterfly has recently been reintroduced in Gloucestershire and there have been casual releases in Hertfordshire and Bedfordshire, which make it difficult to ascertain its true status in these counties.

European trend

The Adonis Blue is declining in most northern European countries, especially Belgium, the Czech Republic, Luxembourg, and Poland (>50% decrease in 25 years), and Germany and Slovakia (25–50% decrease). It appears to be more stable in southern Europe and remains abundant on most unfertilized calcareous grassland.

Interpretation and outlook

The widespread decline and partial re-expansion of the Adonis Blue has many similarities with that of the Silver-spotted Skipper. Both are short turf species, whose grassland habitats have been reduced due to agricultural intensification. Even more sites have suffered from the abandonment or reduction of livestock grazing that occurred on marginal hill land throughout the twentieth century. Both butterflies were affected badly by the introduction of myxomatosis in the 1950s and the subsequent collapse of rabbit populations, which removed the only wild grazing animal capable of creating suitable short turf. The chief difference between the two species is that the Adonis Blue tended to survive better where sheep or cattle grazing continued in the western chalk of Dorset, Wiltshire, and the Isle of Wight; whereas the Silver-spotted Skipper tended to persist further east where there are very thin soils and lower rainfall (such as the North Downs).

The recent recovery of the Adonis Blue is probably largely attributable to the recovery of rabbit populations, but there is no doubt that many sites have been recolonized following the restoration of, or increase in, livestock grazing, often for conservation purposes. Good examples are the chalk grasslands of Dorset and Wiltshire which have been targeted by both the South Wessex Downs Environmentally Sensitive Area and by Countryside Stewardship. Many other sites have either become nature reserves or are under new management agreements funded by English Nature. There are indications that these measures are beginning to reverse the fortunes of the Adonis Blue. However, in some areas it may still rely on rabbit grazing and will be vulnerable if rabbit populations are reduced in the future.

Former grazing levels will have to be restored on many more sites to allow recovery of the Adonis Blue to anything like its pre-1950 levels. Even then, some sites may now be too small or isolated to support populations by themselves in the long term. The restoration of sites that have been neglected for years still presents serious practical problems, especially where there is extensive scrub invasion. The larval foodplant, Horseshoe Vetch, is extremely slow to spread, often taking decades to recover from periods of abandonment. It has disappeared altogether on some sites and its ability to regenerate from buried seeds is uncertain.

Because the Adonis Blue breeds only on the warmest slopes, it is tempting to assume that it would benefit if temperatures rise in southern England as predicted by some global warming models. However its restriction to short turf makes the foodplant very prone to drought, and data from the Butterfly Monitoring Scheme show that populations crashed after the prolonged drought of 1976. The impact of climatic change is, therefore, hard to judge until climatologists can predict the inter-relationship of rising temperatures and rainfall more precisely. A warmer, wetter climate could lead to increased numbers of Adonis Blue, but if droughts become more frequent, any benefit from rising temperature might be cancelled out.

Key references

Bourn and Warren (1998*c*); Bourn *et al.* (1999, 2000*b*); Pearman *et al.* (1998); Thomas (1983*c*, 1990*a*); Whitfield (1999).

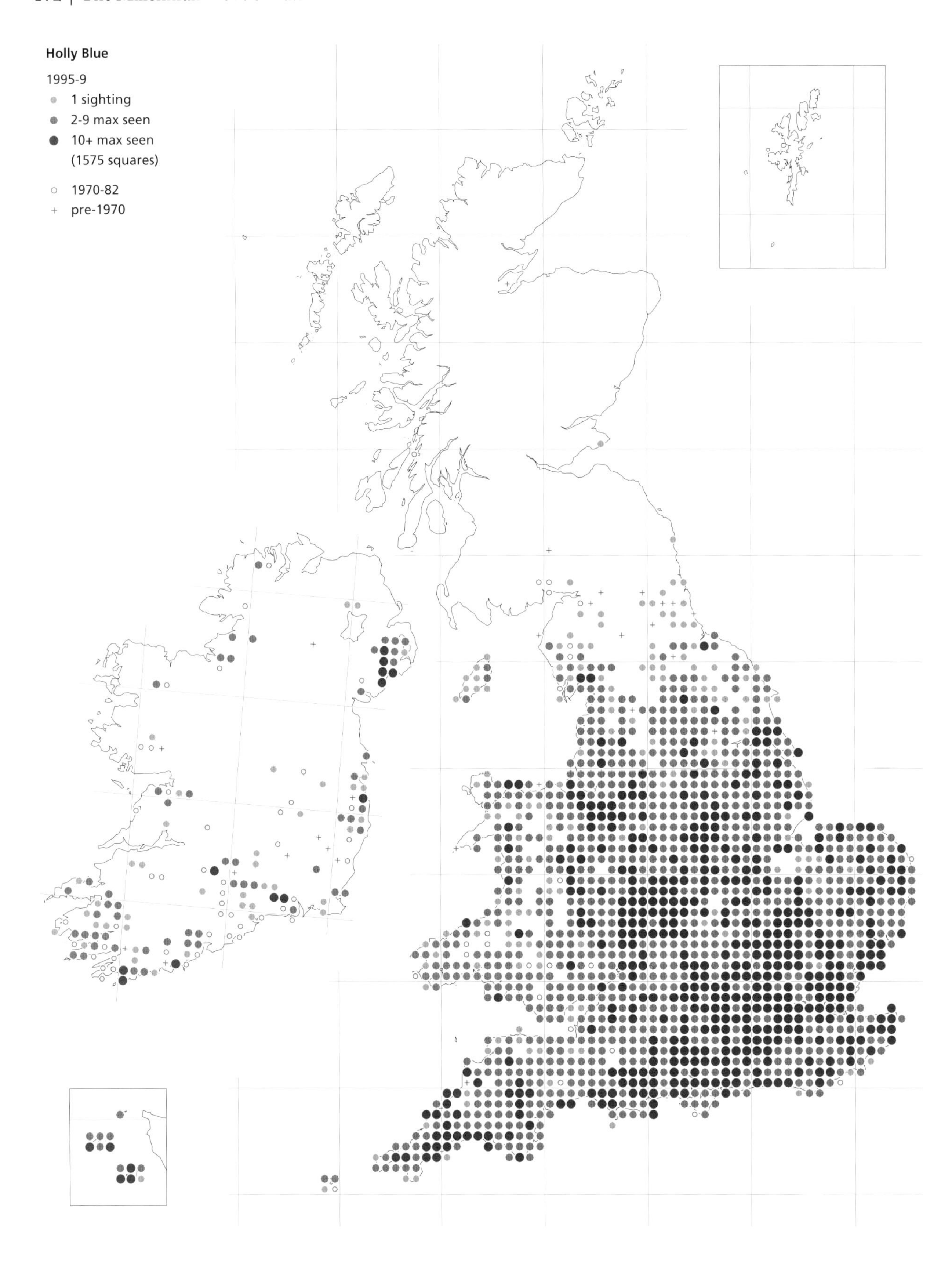
Holly Blue
1995-9
1 sighting
2-9 max seen
10+ max seen
(1575 squares)
1970-82
pre-1970

Holly Blue *Celastrina argiolus*

Resident

Range expanding in Britain.

Conservation status

UK BAP status: not listed.
Butterfly Conservation priority: low.
European status: not threatened.
Protected in Northern Ireland.

European/world range

Widespread in Europe between 40 °N and 67 °N and as far east as Japan, as well as being found in North Africa and North America. Its European range is stable, although there have been recent expansions in a few European countries.

The Holly Blue is easily identified in early spring, as it emerges well before other blue butterflies. It tends to fly high around bushes and trees, whereas other grassland blues usually stay near ground level. It is much the commonest blue found in parks and gardens where it congregates around Holly (in spring) and Ivy (in late summer). The Holly Blue is widespread, but undergoes large fluctuations in numbers from year to year. It has expanded northwards in recent years and has colonized parts of midland and northern England.

Foodplants

The larvae feed predominantly on the flower buds, berries, and terminal leaves of Holly (*Ilex aquifolium*) in the spring generation, and Ivy (*Hedera helix*) in the summer generation. The spring generation can complete larval development entirely on leaves of male Holly bushes, although female bushes are preferred. They also use a wide variety of other wild and garden plants including Spindle (*Euonymus europaeus*), dogwoods (*Cornus* spp.), snowberries (*Symphoricarpos* spp.), gorses (*Ulex* spp.), and Bramble (*Rubus fruticosus*).

Habitat

The Holly Blue occurs in a wide range of habitats, including hedgerows, field margins, woodland rides, gardens, and parks, including those in urban and suburban areas. In England, it often breeds in churchyards, many of which have Holly and Ivy. In Ireland, it is limited mainly to deciduous woods with Holly and, occasionally, country gardens.

Life cycle and colony structure

There are two generations each year over most of its British range. Adults fly between late March and late June and again in late July into September (see p. 394). There is an occasional partial third brood in October, in the south. In parts of northern England, there is only one generation each year, with adults flying between April and early July, although there may be a partial second brood in August. Older literature suggests that it was single brooded in Ireland, flying in spring, but current records show two generations in the south. **Eggs** are laid on the flower buds of its foodplants and hatch within about two weeks. The developing

Holly Blue larva on Ivy.

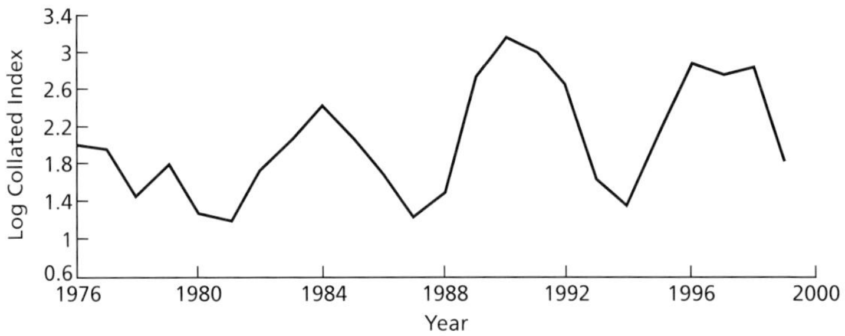

BMS index of Holly Blue abundance, showing regular cycles.

larvae are green with a black head, although the latter is normally out of sight, buried in developing flower buds, devouring the contents. Fully developed larvae leave the foodplants to form **pupae** deep in vegetation. They overwinter at the pupal stage. In common with other blues, the larvae and pupae are known to produce secretions that are attractive to ants. Larvae have been observed with red ants (*Myrmica* spp.) and black ants (*Lasius* spp.) in attendance.

A small wasp (*Listrodomus nycthemerus*) specifically parasitizes Holly Blue larvae and can cause high levels of mortality in their populations. The wasp lays an egg inside the butterfly larva, but its larva does not develop fully until the butterfly reaches the pupal stage. The wasp larva completes its development inside the butterfly pupal case, eating the entire contents and emerging as an adult.

Holly Blues range widely through the countryside and do not form distinct populations, although adults often congregate around the foodplants, especially Holly and Ivy. They are seen flying in almost any habitat and the recent expansion of range demonstrates the mobility of the species.

Distribution and trends

The Holly Blue is widespread across southern and midland Britain, and the Isle of Man, extending more patchily towards northern England and parts of Wales. Since the last atlas period (1970–82), it has been recorded in over 700 new 10 km squares in Britain alone (giving an 89% increase overall), mostly in northern England and in Wales. Part of this may be due to more intensive recording. Its range now extends significantly further north in England into Lincolnshire, the East Midlands, Lancashire, and Yorkshire and as far north as Cumbria and Northumberland. It is not currently resident in Scotland, although there were a few records near Dumfries in the 1970s and one in Fife in 1998. In good years, it can be abundant and may be common in gardens in southern England.

The abundance of the Holly Blue shows strong cyclical variations, which are sometimes accompanied by temporary changes in range. The sequence of annual maps for this species suggests that recent changes in annual abundance in southern Britain may have been progressive from the south-east towards the north-west.

In Ireland, it has a widespread but fragmented distribution. It is most abundant in oakwoods in the hills and valleys of the east and south, notably in counties Down, Wicklow, Waterford, Cork, and Kerry, but also occurs in places along the entire west coast. There is no clear evidence of a significant change of range in the current data for Ireland, although records are not sufficiently comprehensive to draw firm conclusions.

The variations in abundance of this butterfly from year to year appear clearly in the Butterfly Monitoring Scheme data, which show that in recent decades the periodicity of cycles has been around 6–7 years. There is no widely accepted explanation for these variations. One possibility, supported by a recent study, is that the cycle is due to the parasitic wasp. When the butterfly is abundant, the wasp increases, but the build-up of parasitoids causes such mortality of Holly Blue pupae that numbers of butterflies decline. A decline of the wasp follows and the reproductive success of the Holly Blue increases, so starting the cycle again.

Other possible explanations include the effects of the weather on the timing of foodplant flowering and larval feeding at critical stages. In particular, larval emergence needs to coincide with the availability of tender new flower buds. In practice, weather probably affects both butterfly development and parasitoid success, and the observed variations arise from a combination of factors.

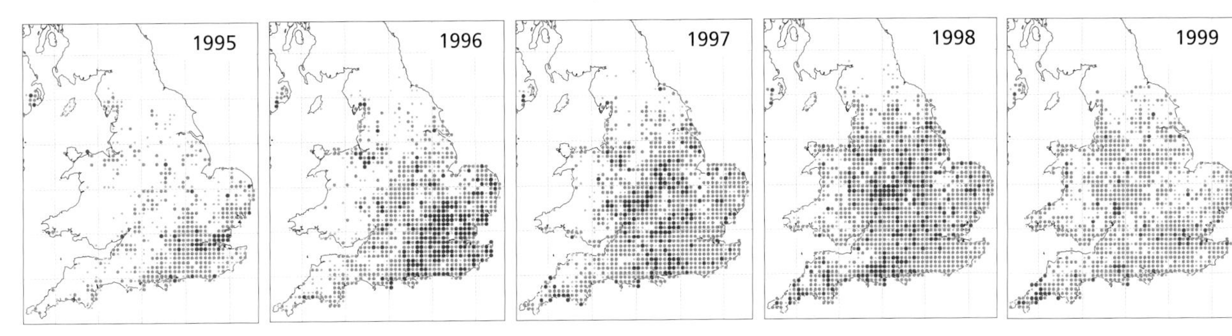

Holly Blue distribution in southern Britain in each year of the current survey (1995–9).

Despite the fluctuations, there are some places where local populations remain reasonably strong, and these may provide a base for re-expansion in the cycles of variation.

Although superficially the Holly Blue may be confused with the Common Blue, the undersides of the wings allow easy identification: those of the Holly Blue are mainly pale blue, whereas those of the Common Blue have orange markings. The Holly Blue is therefore unlikely to be misidentified. Moreover, the recording period for this atlas has coincided with a peak in its abundance, so that the true extent of its geographical range should be well represented here.

European trend

The Holly Blue is stable in most European countries and has spread recently in Germany, Malta, and Romania (>25% increase in the last 25 years).

Interpretation and outlook

Because the fluctuations in annual abundance are so marked, they tend to obscure long-term changes. Nevertheless, changes in distribution have been seen: it has extended north over the past 20 years and appears to be more widespread within its main distribution.

This species may be showing a response to climatic warming, both in terms of extension of range northwards and a possible shift towards becoming double brooded in southern Ireland. Recent work has shown that the number of Holly Blue records from 1900–66 was positively correlated with warm, dry conditions in the spring (particularly March and April) and warm weather in late summer. However, more research is needed before the influence of the weather and the parasitic wasp on the abundance and distribution of the Holly Blue can be determined.

The Holly Blue has survived the main landscape changes of the twentieth century and continues to find suitable breeding conditions in a wide range of natural and urban habitats. This is reflected in mainland Europe where this butterfly is also widely distributed. Nevertheless, in the wider countryside, the Holly Blue requires the maintenance of foodplants in sheltered hedgerow, scrub, and open woodland habitats.

Population levels may have declined over the last fifty years, with the reduction by over 50% in the length of hedgerow in lowland Britain. However, this may have been partially compensated for by the planting of Holly in parks and gardens (mainly in urban areas), and the neglect of woodlands, which has allowed Ivy to develop more vigorous flowering growth.

Key references

Pollard (1985); Pollard and Moss (1995); Pollard and Yates (1993*b*); Revels (1994); Willmott (1999).

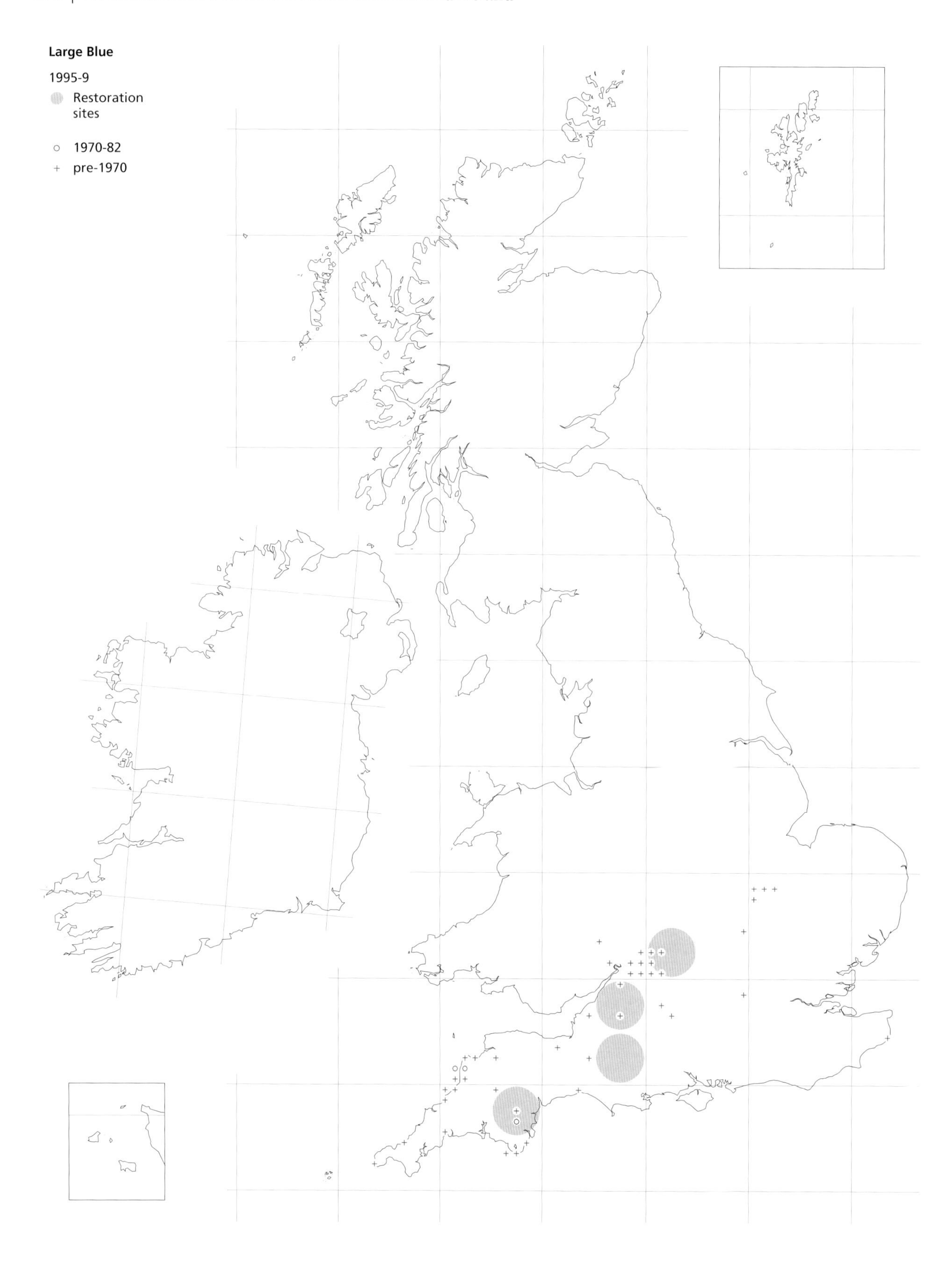
Large Blue
1995-9
Restoration sites
1970-82
pre-1970

Large Blue *Maculinea arion*

Resident

Extinct (1979) and reintroduced since 1983.

Conservation status

UK BAP status: Priority Species.
Butterfly Conservation priority: high.
European status: endangered.
Fully protected in Great Britain.
Bern Convention (Annexe IV) and EC Habitats and Species Directive (Annexe IV).

European/world range

Through Europe from central Spain to about 62 °N, and across Asia to China. Declining in most European countries and threatened throughout its global range.

This is the largest and rarest of our blue butterflies, distinguished by the unmistakable row of black spots on its upper forewing. The Large Blue is one of the most enigmatic butterflies, whose remarkable life cycle involves spending most of the year within the nests of red ants, where the larvae feed on ant grubs. It has always been rare in Britain but declined rapidly during the twentieth century and became extinct in 1979. It has since been reintroduced successfully as part of an important habitat and species conservation programme.

Foodplants

Larvae initially feed on the flower-heads of Wild Thyme (*Thymus polytrichus*) but from their fourth instar they feed on ant grubs within the nests of the *Myrmica* red ants. Survival rates are highest within nests of *Myrmica sabuleti* and much poorer in those of *M. scabrinodis*. Eggs may also be laid on Wild Marjoram (*Origanum vulgare*) but this usually flowers too late to be used on most British sites.

Habitat

The Large Blue breeds in warm and arid (well-drained) unimproved grassland, predominantly acidic coastal grassland or limestone grassland. It also used to occur on calcareous clay soils, probably on dry outcrops or mounds.

Life cycle and colony structure

The Large Blue is single brooded with adults flying from mid-June until late July. **Eggs** are laid on the tight young flower buds of Wild Thyme. The **larvae** subsequently burrow into the flower head to feed on the flowers and developing seeds. Females lay eggs on plants growing in a range of vegetation heights, but survival is best in short turf where the host ant *M. sabuleti* is most abundant. When the larvae are still only around 4 mm long (early fourth instar) they drop to the ground and wait to be found by foraging red ants, attracting them with sweet secretions from a special 'honey' gland. After a short while, the larvae hunch up mimicking ant grubs, whereupon they are picked up, taken below ground and placed within the brood chamber. The larvae then feed on ant grubs to achieve most of their final body weight, hibernating deep within the ant's nest. The larvae **pupate** in early May within the nest and the newly emerging adults have to crawl up above ground in order to expand their wings.

The Large Blue typically forms discrete colonies on small patches of suitable habitat (typically 2–5 ha) from which the adults stray only rarely. There is also evidence that the butterfly became more sedentary in Britain as its habitats became reduced in size. However, the re-established adults clearly have some powers of dispersal and one of the newly

Large Blue larva being adopted by a red ant.

founded colonies has spread 2–3 km, covering numerous small patches of suitable habitat.

Distribution and trends

The Large Blue once occurred in scattered colonies across southern England, but its strongholds were always on the north coast of Cornwall and in the Cotswolds in Gloucestershire. The butterfly began declining in the nineteenth century and by the 1850s was causing concern amongst entomologists. It became extinct in Northamptonshire around 1860 and its decline accelerated during the twentieth century. It died out in south Devon in 1906, Somerset in the late 1950s, the Cotswolds in 1960–4, and the Atlantic coast of Cornwall and Devon in 1973. Despite concerted conservation attempts, the Large Blue became extinct in Britain in 1979.

Following detailed research and improved habitat management, the Large Blue has been re-established on several former sites using stock from Sweden. The first trial re-establishment was conducted in 1983 at the last former site on the edge of Dartmoor (Devon) and was followed by a major release in 1986. It has since been reintroduced or has spread to seven other sites in the Cotswolds (Gloucestershire), the Polden Hills and the Mendip Hills (Somerset).

European trend

The Large Blue has declined markedly in most European countries and has become extinct in the Netherlands. Other severe declines have been recorded in Belgium, the Czech Republic, Denmark, Finland, Germany, Latvia, Luxembourg, Poland, and Romania (>50% decrease in 25 years); Croatia, Slovakia, Sweden, European Turkey, and Ukraine (25–50% decrease).

Interpretation and outlook

The demise of the Large Blue provided one of the most salutary lessons in modern conservation. Early efforts to save the species began in the 1930s and were based on simple habitat protection, usually in the form of erecting fences and the exclusion of collectors. The reasons for continued losses on protected sites were a complete mystery until the pioneering research of Jeremy Thomas who discovered the butterfly's reliance on one species of ant. Knowledge of the requirements of this ant finally explained the Large Blue's decline and eventual disappearance even from protected areas.

In brief, the butterfly needs continuous heavy grazing in order to maintain high densities of the correct ant, which is quickly displaced by other unsuitable species of red ant if grazing is relaxed, sometimes even for one season. It is a matter of lasting regret that this vital research was started just too late to save the British Large Blue from extinction.

Thanks to this research, it is now clear that the butterfly declined not only because of the agricultural 'improvement' of rough grazing land, but also because of a relaxation of grazing pressure when marginal pastures were abandoned during the twentieth century. By the 1940s most Large Blue sites relied solely on rabbit grazing and the butterfly was then badly hit by the severe effect of myxomatosis on rabbit numbers in the 1950s. This reduced it to one site on the edge of Dartmoor where cattle and pony grazing had, quite fortuitously, been continued. However adult numbers were already low and the population was reduced by the 1976 drought from which it never recovered.

Research is also helping to explain why the butterfly has never been recorded from other types of traditionally short grazed habitats containing abundant Wild Thyme, notably chalk downland. The host red ant, *M. sabuleti*, is very susceptible to drought and in Britain seems only able to maintain high densities on sites with slightly deeper, moister soils, notably limestone and coastal grasslands.

Armed with good ecological knowledge, it has been possible to restore suitable habitats on several former localities that are now in the ownership of conservation bodies, including the Somerset Wildlife Trust and the National Trust. The work is conducted by a consortium, led by Butterfly Conservation and the Centre for Ecology and Hydrology, with support from English Nature and ICI. The aim is to restore the butterfly to at least 10 sites by 2005 and to encourage population survival by extending the habitat available on each site.

Although the early results of the conservation programme for the Large Blue are very promising, severe problems remain. The butterfly is extraordinarily sensitive to changing grazing pressure and even a single year with no grazing can result in a huge drop in the population size. There are also now great difficulties in locating breeds of grazing animals suitable for nutrient-poor grassland habitats. Most Large Blue sites rely on using traditional breeds of cattle, ponies, or sheep that often have to be brought in specially.

Another problem is that most potential sites are still small and isolated, and priority is being given to restoring clusters of sites to ensure long-term population viability. Such networks will not only be able to accommodate occasional problems in management but will contain a greater diversity of habitat and topography, thereby increasing the butterfly's chances of surviving droughts. Because the Large Blue relies on a warmth-loving ant species, restored populations may benefit if the British climate warms as predicted during this century. On the other hand they could be badly hit if there are more droughts.

Former Large Blue site being restored in north Cornwall.

Although the management of many reintroduction sites has been geared towards the Large Blue, this has greatly benefited whole communities of plants and animals associated with traditionally grazed rough pastures. They include threatened butterflies such as the High Brown and Pearl-bordered Fritillary, as well as other rare insects and plants such as the Pale Dog-violet. The conservation of the Large Blue is therefore an integral part of restoring unique habitats for wildlife as a whole.

Key references

Barnett and Warren (1995*i*); J. A. Thomas (1980, 1990*b*, 1991, 1995, 1999*a*,*b*); J. A. Thomas *et al.* (1998).

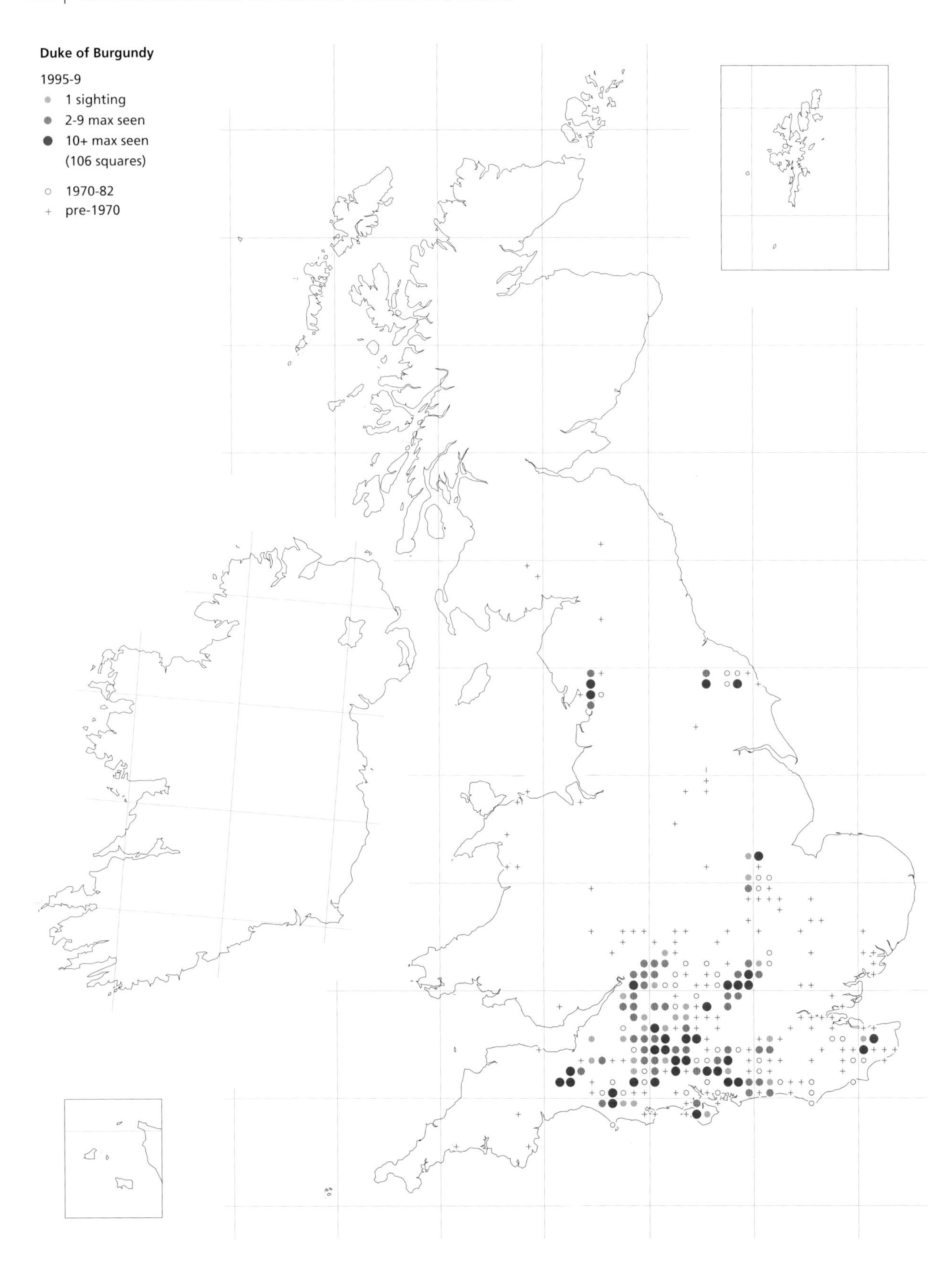
Duke of Burgundy
1995-9
1 sighting
2-9 max seen
10+ max seen
(106 squares)
1970-82
pre-1970

Duke of Burgundy *Hamearis lucina*

Resident

Range declining.

Conservation status

UK BAP status: Species of Conservation Concern.
Butterfly Conservation priority: medium.
European status: near threatened.
Protected in Great Britain for sale only.

European/world range

Widespread in Europe, from northern Spain to southern Sweden, and east as far as the Urals. Serious declines reported in at least ten European countries.

This small butterfly frequents scrubby grassland and sunny woodland clearings, typically in very low numbers. The adults rarely visit flowers and most sightings are of the territorial males as they perch on a prominent leaf at the edge of scrub. The females are elusive and spend much of their time resting or flying low to the ground looking for suitable egg-laying sites. The Duke of Burgundy is found in scattered colonies across southern England, with more isolated colonies in the southern Lake District and North Yorkshire. It has declined substantially in recent decades, especially in woodlands where it is reduced to fewer than 20 sites.

Foodplants

The main foodplants are Cowslip (*Primula veris*) and Primrose (*P. vulgaris*). It occasionally uses the so-called 'False Oxlip', the hybrid of these two *Primula* species.

Habitat

Two principal habitats are used: chalk or limestone grassland, with either extensive areas of scrub or topographical shelter; or clearings on ancient woodland sites, either regenerating coppice, young plantations, sizeable glades or wide rides. In both habitats it requires foodplants growing among tussocky vegetation and on downland it prefers north- or west-facing slopes, possibly because the humid conditions encourage lusher growths of the foodplant.

Life cycle and colony structure

The butterfly is single brooded with adults flying from early May until the middle of June, reaching a peak in most years during the third week of May. The flight period is typically 1–2 weeks earlier on southern chalk downland and later in woodland where numbers often peak in early June.

Eggs are laid on the underside edge of foodplant leaves, usually in small batches of up to 8. Females tend to select plants that are prominent amongst the surrounding vegetation, though smaller plants can be used where they grow abundantly on calcareous downland. The **larvae** feed at night and produce distinctive feeding damage on the leaves comprising a peppering of small holes, leaving the midrib and main veins intact. They strongly favour green foodplants and wander away from exposed plants before the leaves begin to yellow and wither. The larvae can wander several metres from plant to plant, eating small *Primula* seedlings on the way. They **pupate** from mid-July to early August, usually well concealed, 2–5 cm above ground, among dense tussocks of fine grasses where they pass the winter.

The Duke of Burgundy tends to form very small colonies, limited by the availability of

Duke of Burgundy larva and larval feeding damage on Primrose.

suitable breeding areas. Mobility appears to be very different between the sexes. Males are territorial and rarely move far from sheltered positions, often next to scrub that catches the early morning sun. In contrast, females are far less conspicuous but appear to be more mobile and move more freely over the habitat. In a mark–recapture study in southern England, over half of recaptured females had moved more than 250 m. Some females may disperse several kilometres and sites have been colonized up to 5 km from known populations. However, another study in north-west England suggests that they may be less mobile at the northern edge of their range.

Distribution and trends

The Duke of Burgundy still occurs widely in central-southern England, with outlying colonies in north Lancashire and Yorkshire. There are a few scattered records for north Wales and southern Scotland, but it disappeared from these countries before 1940. There are no records for Ireland, the Isle of Man or the Channel Islands. Its current strongholds are on limestone grassland in the Cotswolds, and the chalk downs of Wiltshire, Hampshire, and Sussex.

The butterfly declined steadily throughout the twentieth century with substantial losses since 1950. It is now extinct in most of the Midlands and East Anglia, and is declining in most other regions. The most serious losses have occurred in woodland, which was probably once the most important habitat, but now supports fewer than 20 colonies. In a review of extant sites, the butterfly's decline in woodland was estimated to have been 98% between 1950–90. Most remaining colonies now exist in grassland habitats, but here too the butterfly has declined substantially in recent decades. The data from the 1995–9 survey indicate that it has probably disappeared from 39% (44) of the 10 km squares recorded in 1970–82.

The distribution of the Duke of Burgundy is relatively well known thanks to special surveys and substantial recorder effort in most regions. These suggest that there are now fewer than 200 colonies remaining, mostly in southern England, with around 15 in North Yorkshire and 10 in north Lancashire and south Cumbria. Moreover, most colonies are believed to be very small and there may be only three sites with genuinely large populations (>1000 adults at peak). On the plus side, colonies have been discovered in 37 new 10 km squares in the current survey period, and a few new sites have been colonized in Hampshire.

European trend

The Duke of Burgundy has become extinct in Lithuania and has declined severely in Austria, Belgium, Latvia, and Luxembourg (>50% decrease in 25 years), Denmark, Poland, and Sweden (25–50% decrease), and to a lesser extent in the Czech Republic, Moldova, and Slovakia (15–25% decrease). Its range is thought to have changed little in other countries.

Interpretation and outlook

The decline of the Duke of Burgundy has been attributed to various factors and its fortunes have waxed and waned differently in each major habitat. During the nineteenth century, it was predominantly a woodland butterfly, thriving in the regular clearings created through the traditional practice of coppicing. At that time most chalk and limestone grassland would have been too heavily grazed to support populations. However, this situation reversed during the twentieth century: coppicing declined drastically and although the butterfly thrived during the 1950s and 1960s in some young conifer plantations, it has since disappeared from most woods as these have grown too shady.

In contrast, downlands became more suitable after the last World War as they were grazed less heavily or were abandoned. The butterfly may also have benefited from two additional factors on downland: the virtual eradication of rabbits by myxomatosis during the 1950s; and the switch from sheep to cattle grazing, which produced a more varied, tussocky sward.

In the last few decades, the Duke of Burgundy has continued to be affected by the conversion of chalk and limestone grassland to intensive agriculture and the loss of foodplants as neglected sites have become dominated by coarse grasses and scrub. The latter is currently a threat to the butterfly in the Cotswolds and on some southern chalk downs. However, the most serious new threat to emerge during the 1980s and 1990s was the recovery of rabbit populations whose selective grazing can reduce sward height and eliminate suitable prominent foodplants. Careful control of rabbit numbers will thus be essential to ensure the butterfly's survival on many sites.

The management of grassland for the Duke of Burgundy presents a dilemma to conservationists as the maintenance of the foodplant requires active grazing (or other management) whereas the butterfly thrives best in taller, unmanaged swards. Suitable management therefore involves rotations of either grazing or periodic scrub cutting, which produce a mosaic of conditions where both the butterfly and its foodplant can thrive.

Typical Duke of Burgundy habitat in Hampshire.

The butterfly can also thrive under late summer or winter cattle grazing regimes. Unfortunately, cattle grazing suffered a serious setback with the BSE crisis of the 1990s, but special Government grants are now available to support extensive grazing, notably in Environmentally Sensitive Areas in the Cotswolds and the South Wessex Downs. In woodlands, the butterfly may benefit from the modest revival of coppicing that has occurred since the 1980s, and also the open space that is being encouraged under new grants from the Forestry Commission.

Despite such measures, it seems likely that the Duke of Burgundy will continue to decline in the next few decades because so many sites are small and isolated. Although the butterfly has an amazing ability to persist in small populations, these are prone to extinction. The conservation of the Duke of Burgundy thus requires continued effort, focused initially on existing grassland sites but hopefully extending to the restoration of its habitats in woodland.

Key references

Bourn and Warren (1998*d*); Oates (1986, 2000); Sparks *et al.* (1994).

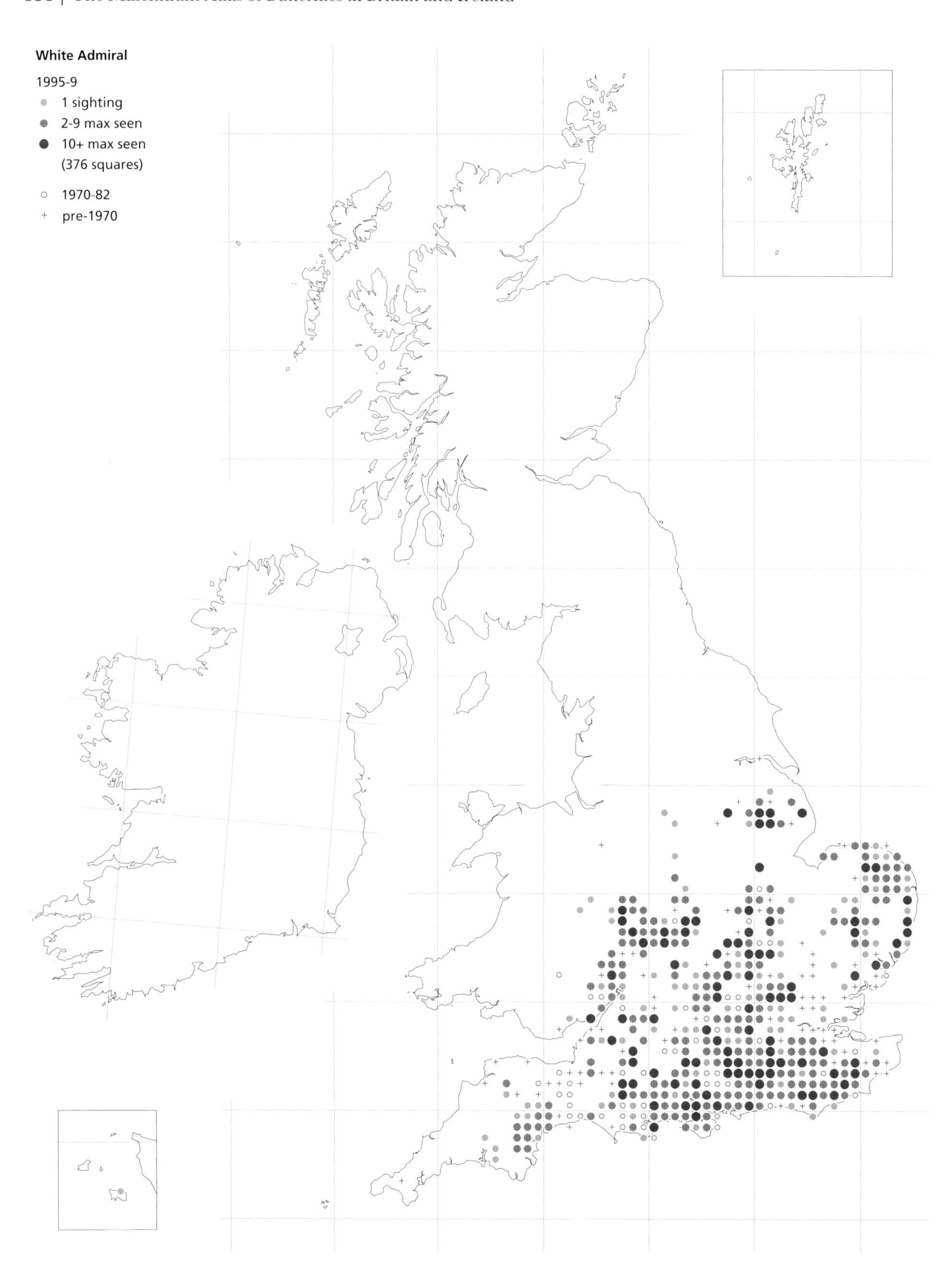
White Admiral
1995-9
1 sighting
2-9 max seen
10+ max seen
(376 squares)
1970-82
pre-1970

White Admiral *Limenitis (Ladoga) camilla*

Resident
Range expanding.

Conservation status
UK BAP status: not listed.
Butterfly Conservation priority: low.
European status: not threatened.

European/world range
Across central Europe from northern Spain to Turkey and as far north as Latvia, but absent from Scandinavia. Extends throughout Asia as far east as Japan. European range appears to be expanding northwards, but there have been declines in several countries.

The White Admiral is a spectacular woodland butterfly, with white-banded black wings and a distinctive delicate flight: short periods of wing beats, followed by long glides. Adults are often found nectaring on Bramble flowers in rides and clearings. It is a fairly shade-tolerant butterfly, flying in dappled sunlight to lay eggs on Honeysuckle. The White Admiral occurs widely in southern Britain and has spread rapidly since the 1920s, after an earlier contraction. It continued to spread in the 1980s and 90s.

Foodplants
The sole foodplant is Honeysuckle (*Lonicera periclymenum*), usually in shady positions.

Habitat
The butterfly uses shady woodland and ride edges, often associated with neglected or mature woodland where there are sunny glades with large patches of Bramble to provide nectar for the adults. It is found in both deciduous and mixed deciduous/coniferous woodland.

Life cycle and colony structure
White Admirals are single-brooded with adults flying from late June to mid-August (see p. 395).

Eggs are laid singly on the leaves of straggly growths of Honeysuckle, usually in shade at 1–2 m above ground level. A recent study, at two sites close to the northern limit of its range, has shown that it can also use low-growing plants in more sunny positions at ride edges.

The newly emerged **larvae** feed from the tips of the leaves, retreating to the uneaten central vein, which seems to provide camouflage. In late summer, the partially developed larvae use silk to construct leaf-shelters (hibernacula) secured to plant stems in which they spend the winter months. This strategy may reduce attack by predators. The larvae emerge again in spring, feeding on the new growth of Honeysuckle to complete their development. The **pupae**, formed hanging from stems, are coloured green and deep red, matching the coloration of the foodplant. They have bright gold patches which give them a wet appearance.

When disturbed the White Admiral larva takes up this distinctive posture.

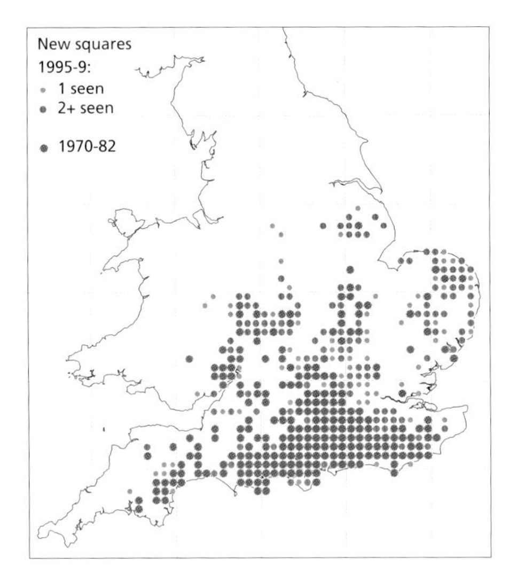

The expansion of range of the White Admiral in southern Britain since 1970–82.

White Admirals form discrete colonies within suitable blocks of woodland habitat and typically occur at low densities, with rarely more than two or three seen at a time. Within the main part of their range, many colonies have become isolated.

The mobility of adults has not been studied in detail, but the spread of the butterfly during the twentieth century indicates that it can colonize over distances of many kilometres. Between the 1920s and 1930s, the distribution extended by distances up to 100 km, implying an average spread of up to 10 km per year. This mobility is also indicated in the current survey by sightings some kilometres from known colonies.

Distribution and status

White Admiral colonies are found in England, south of a line between the Humber estuary and the Malvern Hills, and extending into Monmouthshire and Brecknockshire in south Wales. The south Midland populations are found mainly in woodland on damp, clay soils, and there are few colonies in woods on calcareous soils on southern chalk hills, in Gloucestershire and in Wiltshire. The distribution becomes more patchy in the West Country. They are found in south Devon, but are absent from Cornwall. There was a single record from Jersey in 1969 and it was recorded there again in 1999. It has never been recorded in Scotland, Ireland, or the Isle of Man.

The range of the White Admiral has undergone several substantial changes over the last 150 years, but it is now more widespread than in any previous known period. In the nineteenth century, it was found throughout most of its current range, but by the early 1900s had become confined to a few counties in central-southern England. It then spread rapidly northwards again in the 1930s and 1940s, once more reaching as far north as Lincolnshire and Worcestershire.

It has continued to spread since the 1980s, particularly in Norfolk, Suffolk, Lincolnshire, and the West Midlands. The current survey shows that the number of recorded 10 km squares has increased by 56% since 1970–82. Although it may be overlooked because of low population densities, the White Admiral is easy to identify and the map is thought to be an accurate representation of its distribution.

European trend

The White Admiral has declined in several countries, including Austria (>50% decrease in 25 years), Belgium, Germany, Latvia, Luxembourg, and the Netherlands (25–50% decrease). However, its range is expanding in European Turkey and Ukraine (25–100% increase in 25 years). Over a longer period (30–100 years) its European range has expanded northwards.

Interpretation and outlook

The spread of the White Admiral during the twentieth century has been linked to two main factors: periods of favourably warm springs and summers, and changes in woodland management leading to increasingly shady conditions where the butterfly can breed.

A detailed population study has shown that a critical stage in the life cycle is during the late larval and pupal stages in June, when higher temperatures lead to improved survival. This is possibly because development is accelerated and there is a reduced chance of predation at these vulnerable stages. Thus, the spread in the 1930s and 1940s may be linked with a series of hotter Junes, which allowed populations to build rapidly and spread more widely.

Another analysis has shown that the number of records submitted annually between 1900–66 was also closely related to June temperature. However, data from the Butterfly Monitoring Scheme show no correlation between abundance and summer weather, and no overall trend in recent decades. This suggests that its recent spread may be due more to increased availability of suitable habitat than favourable weather.

The level of habitat availability for this species increased steadily during the twentieth century, as traditional coppicing dwindled and deciduous woodland either became neglected or managed as high forest. The White Admiral thrives particularly well in neglected coppices where the foodplant, Honeysuckle, trails down in shady conditions. The adults still need open glades and clearings where they can find abundant nectar, especially from Bramble. The species can also breed in conifer woods where Honeysuckle grows in shady ride edges. Thus, like the Speckled Wood, the White Admiral benefited from changing management during the twentieth century, whereas other, more light-demanding species (such as the fritillaries) disappeared.

The outlook for the White Admiral appears good at least in the short term, and it should continue to benefit from a government forestry strategy that increasingly favours deciduous woodland.

White Admiral abundance at BMS sites showing no overall increase despite range expansion.

There are potential threats from increasing deer populations, especially of Muntjac, which have been found to eat Honeysuckle leaves that trail close to the ground (up to 80–120 cm). Muntjac populations have increased rapidly in recent years and spread across much of central England. Fortunately, it is a small deer and many eggs are laid out of reach, so the impact may not be serious unless they begin to reduce the density of the foodplant. A study in Monks Wood National Nature Reserve (Cambridgeshire) showed that Muntjac browsing was occurring widely, but there was no evidence of a decline in White Admiral numbers.

Other threats to the butterfly are from clear-felling and from regular cutting back of Honeysuckle, traditionally regarded by some foresters as a weed, as part of forest management.

Key references

Joy *et al.* (1999); Mendel and Piotrowski (1986); Pollard (1979); Pollard and Cooke (1994); Pollard and Moss (1995).

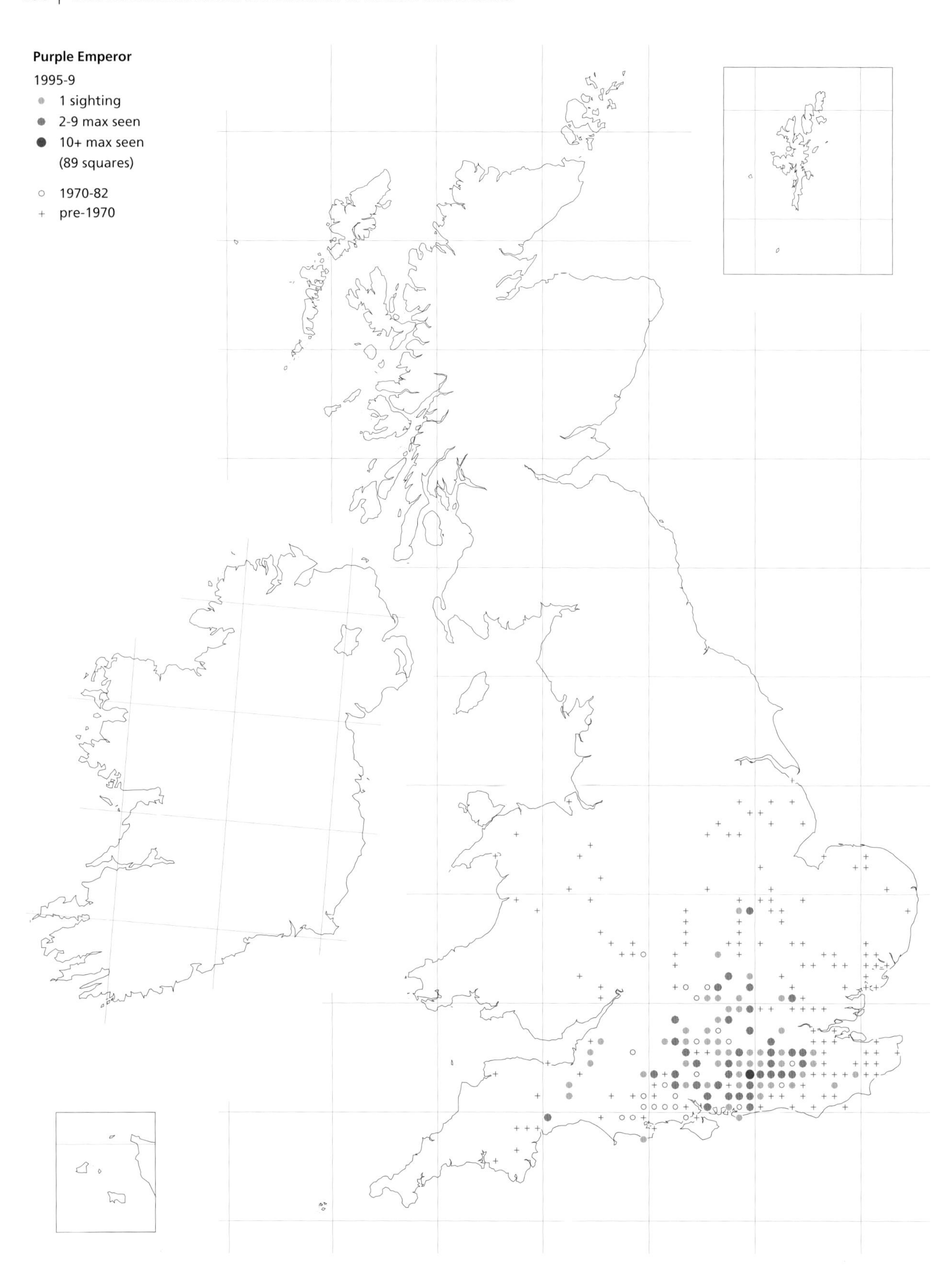
Purple Emperor
1995-9
1 sighting
2-9 max seen
10+ max seen
(89 squares)
1970-82
pre-1970

Purple Emperor *Apatura iris*

Resident

Decline and slight re-expansion of range.

Conservation status

UK BAP status: Species of Conservation Concern.
Conservation priority: medium.
European status: not threatened.
Protected in Great Britain for sale only.

European/world range

Throughout central Europe from north Spain to central Russia (but absent from most of Spain and Italy) and eastwards to China and Korea. It is declining in several western and central European countries but spreading at the north of its range in Scandinavia and Russia.

This magnificent butterfly flies high in the tree-tops of well-wooded landscapes in central-southern England where it feeds on aphid honeydew and tree sap. The adults are extremely elusive and occur at low densities over large areas. The males occasionally descend to the ground, usually in mid-morning, where they probe for salts either from road surfaces or from animal dung. The Purple Emperor declined steadily during the twentieth century and is now restricted to some of the larger woods in southern England. There has been a recent slight re-expansion in some areas.

Foodplants

Goat Willow (*Salix caprea*) is the most widely used foodplant although it also breeds on Grey Willow (*S. cinerea*) and more rarely, Crack-willow (*S. fragilis*). Eggs are laid on a wide range of tree sizes, ranging from medium-sized shrubs to tall canopy trees.

Habitat

The butterfly requires large blocks of broad-leaved woodland or clusters of smaller woods and/or dense scrub where the foodplants are abundant.

Life cycle and colony structure

The Purple Emperor is single brooded with adults flying from late June until mid-August. They are seen most frequently on the ground in the first half of the season. Males set up territories on prominent 'master' trees from midday onwards and can occasionally be seen perching on outer, south-facing branches, or soaring gracefully between trees. Some of these trees are used in successive years but others are used more temporarily. Contrary to popular belief, such trees are not always oaks and alternatives are nearly always found if a 'master' tree is felled (though loss of such trees is still undesirable).

The bright green **eggs** are laid singly, during the middle of the day, on the upper surface of willow leaves, always on the shady side of the tree or within its crown. The **larvae** rest along the midrib, facing the stalk and sitting on a silken pad with the front of the body raised from the leaf. Young larvae feed on the resting leaf but larger larvae leave this intact and feed on nearby leaves. They hibernate as small larvae, typically pressed against a willow bud or in the crotch of a forked branch. They resume feeding when the buds burst in April and **pupate** in June, suspended from the underside of a leaf.

There is little detailed information on the structure of Purple Emperor populations. This is partly because the adults are difficult to study using traditional marking methods and because even their presence or absence in a particular wood is hard to determine. They appear to be highly mobile and have

Purple Emperor larva on a willow leaf.

been seen dispersing along mature hedgerows as well as across open fields between nearby patches of woodland. Populations probably extend over quite large areas of land, spanning either single large woods or complexes of smaller woods. Adult density is also difficult to assess but is probably low, and it is rare to see more than one or two at a time. This is reflected by the large number of single records on the map. In its strongholds, the butterfly seems to occur in almost every wood where there are suitable willows. Elsewhere it is usually restricted to large woods. More research is needed to determine the size of populations and how far the adults range.

Distribution and trends

Once found throughout much of England, as far north as the Humber and in a few parts of central Wales, the butterfly's elusive nature means that it has probably always been under-recorded. The Purple Emperor declined throughout the early twentieth century and disappeared from many central and eastern areas by the 1940s. Major losses occurred in counties such as Lincolnshire and Northamptonshire where it was reported to be abundant in the early 1900s, as well as Norfolk, Suffolk, and Essex, where it survived until the 1950s. Its range has continued to contract and it has retreated to core areas of central-southern England. In Wales, only a few historic records are known and it died out during the 1930s. It has never been recorded in Scotland, Ireland, the Isle of Man, or the Channel Islands.

The butterfly is now well established only in the heavily wooded Weald of Surrey and Sussex, and in several large woods in Hampshire, south Wiltshire, and on the border of Buckinghamshire and Oxfordshire. Outside these strongholds, its distribution is very patchy, but it is possible that small colonies may still be overlooked.

The Purple Emperor has undergone a modest re-expansion during the last two decades and has reappeared (or has been rediscovered) in several counties after long absences: including east Devon, Kent, Nottinghamshire, and Northamptonshire. In the latter it is reported to have expanded in the last five years. The overall number of 10 km squares recorded in the current survey is a third greater than in 1970–82, although some of this may be due to increased recording effort as well as reflecting a genuine expansion.

European trend

The butterfly has declined in several central European countries including Austria, Croatia, and the Netherlands (>50% decrease in 25 years), Belgium, Luxembourg, and Slovakia (25–50% decrease), but has spread at the northern edge of its range in Denmark and Sweden (25–100% increase in 25 years) and in Finland and parts of north-west Russia (>100% increase).

Interpretation and outlook

The Purple Emperor has always been associated with well-wooded districts, often on clay soils or in valleys where damp conditions promote abundant growth of suitable willows. The butterfly's decline coincided with a period of profound change in woodland composition and management. Although the species may have benefited from the abandonment of coppicing during the twentieth century, and the gradual change to high forest systems of management, this was more than offset by the overall loss of ancient broad-leaved woodland. Since the 1940s, the area of such woodland in England has been reduced by 30–50%, and losses have been higher in the intensive arable region in central and eastern England. This process has left remaining fragments of woodland smaller and more isolated at the edge of its range where perhaps it was most vulnerable to change.

The butterfly's range has re-expanded recently, although it is too early to say whether this is a temporary resurgence or part of a sustained recovery. This may mirror the situation in northern Europe where the Purple Emperor has expanded northwards recently, possibly linked with climate change. The last few decades have been warmer than average, which may have suited this species, but its potential for recovery may now be limited due to the fragmentation of its woodland habitats. Further monitoring will be needed to resolve these issues.

The conservation of the Purple Emperor depends on the protection and sympathetic management of ancient woodland, especially the retention of numerous willows. It is also important to ensure continuity because these trees are relatively short-lived. They can regenerate freely in damp conditions but, if natural regrowth is sparse, they can be readily propagated by pushing cuttings into the ground during the autumn. Unfortunately, low growth is threatened by deer browsing and young trees may need to be protected.

The overriding factor determining the butterfly's survival may be its need for large areas of habitat which require conservation measures at the landscape scale as well as within individual woods. The outlook for the butterfly has improved since the introduction of the Broadleaves Policy by the Forestry Commission in the 1980s, which introduced stronger safeguards for ancient woodland and encouraged the planting of broad-leaved trees. This has slowed the rate of woodland loss and the cover of broad-leaved woodland is now gradually increasing. However, there are still threats from unsympathetic forest practices that result in more uniform woods and the removal of willows from ride edges.

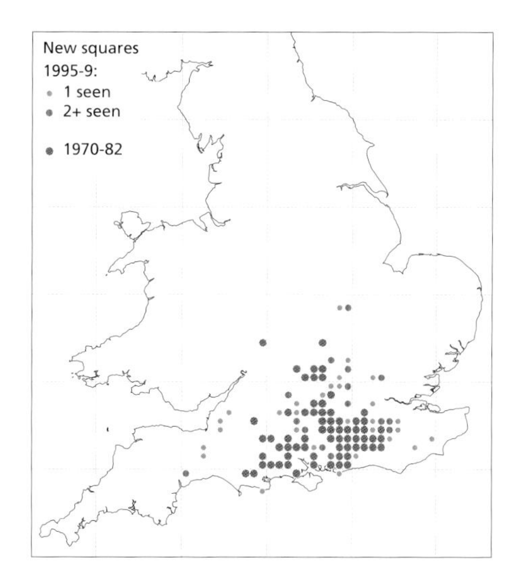

The expansion of range of the Purple Emperor in southern England since 1970–82.

Fortunately the butterfly is now being actively conserved in several key areas and has survived well in some commercially managed ancient woodland sites where willows have been retained. Forest Enterprise is also now incorporating the butterfly's requirements in its management of relevant woods, but effective conservation still requires a major shift in standard forestry practice.

Key references

Bourn and Warren (2000*b*); Heslop *et al.* (1964); Willmott (1990, 1994).

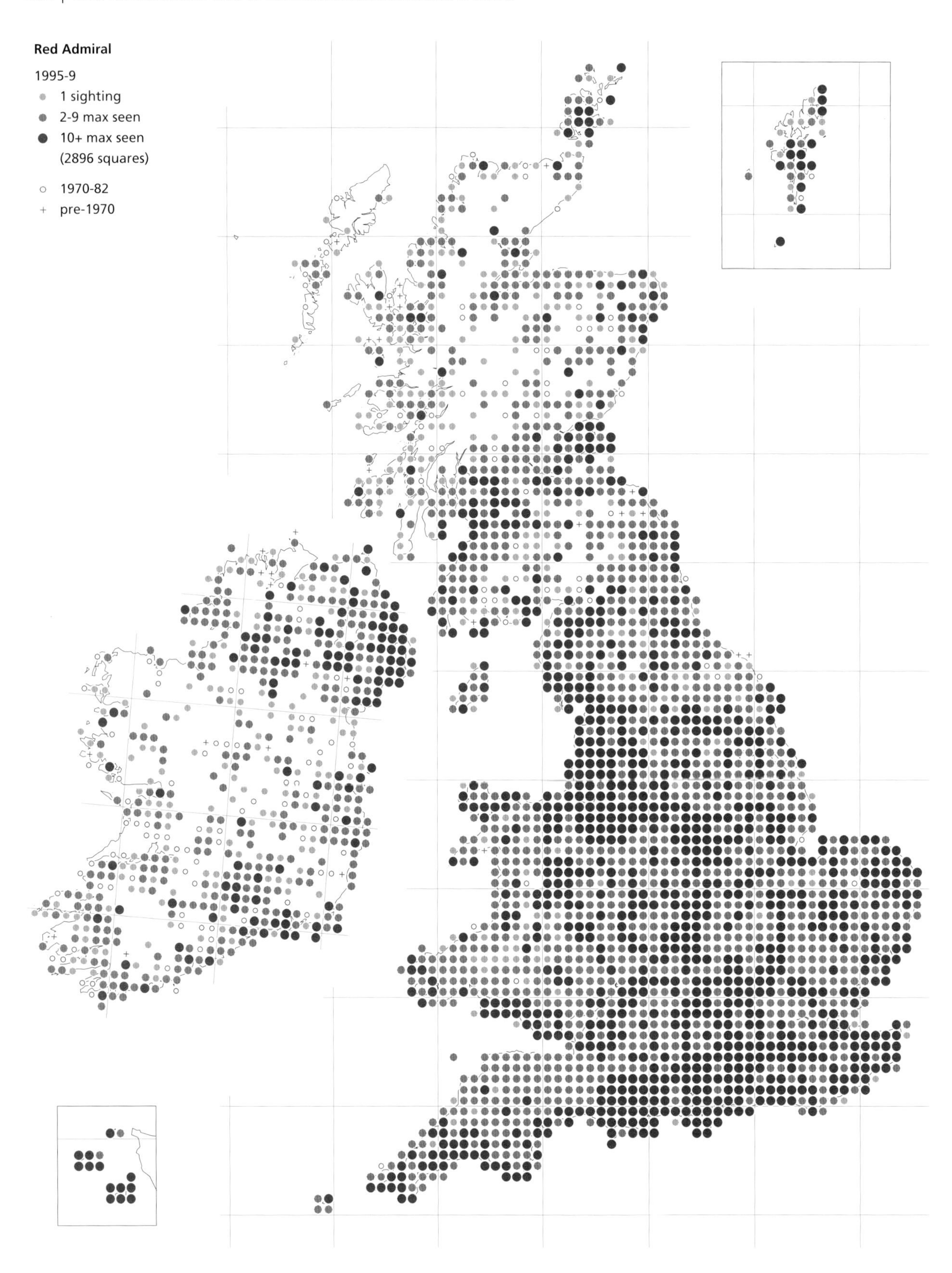
Red Admiral
1995-9
1 sighting
2-9 max seen
10+ max seen
(2896 squares)
1970-82
pre-1970

Red Admiral *Vanessa atalanta*

Regular migrant

Conservation status

UK BAP status: not assessed.
Butterfly Conservation priority: not assessed.
European status: not assessed.

European/world range

From North America across central and southern Europe and Asia to Iran, and Africa north of the Sahara. In western Europe it is a widespread and sometimes common resident from the Mediterranean north at least to central Germany where it overwinters in most years.

This familiar and distinctive insect may be found anywhere in Britain and Ireland and in all habitat types. Starting each spring and continuing through the summer there are northward migrations, which are variable in extent and timing, from North Africa and continental Europe. The immigrant females lay eggs and consequently there is an emergence of fresh butterflies, from about July onwards. They continue flying into October or November and are typically seen nectaring on garden buddleias or flowering Ivy and on rotting fruit. There is an indication that numbers have increased in recent years and that overwintering has occurred in the far south of England.

Foodplants

In Britain and Ireland the most important and widely available larval foodplant is Common Nettle (*Urtica dioica*). However Small Nettle (*U. urens*) and the related species, Pellitory-of-the-wall (*Parietaria judaica*) and Hop (*Humulus lupulus*) may also be used.

Habitat

This strong-flying migratory species may be seen throughout Britain and Ireland and in almost any habitat, from sea-shore to town centres and the tops of mountains. In spring, each newly arrived male defends its chosen territory vigorously. These territories are situated initially close to the south coast, then further inland and typically on bushy hillsides, in corners of sheltered gardens, or in sunny clearings in woodland or parkland, and may be held for a week or more if conditions are suitable for flight. Females are usually seen near nettle beds except when nectaring. Later in the season, any flower-rich habitat is likely to attract the butterfly, including gardens where buddleias, stonecrops, and Michaelmas-daisies are all popular with Red Admirals. They also favour orchards where fruit is rotting on the ground.

Life cycle and colony structure

This is more complex than for most other species. Being a migrant there are no colonies, though large numbers may collect together to feed (mainly in autumn). There is no well-defined overwintering strategy. In Britain and Ireland the variable waves of adult migratory patterns, post-winter and summer emergences and autumn reverse migration, make the interpretation of records complex.

In the early part of the year (January to March) in Britain and Ireland, the occasional sightings of adults may be overwintering individuals, especially in the relatively warm areas close to the south coast. Towards the end of this period, the first migrants may start to fly in, with larger numbers arriving in late May and June. Their offspring contribute to the even larger numbers seen

Red Admiral larval tent formed from a nettle leaf.

later in the summer and autumn (see p. 395) but migrations may also continue through to autumn. In some years of abundance they are accompanied by a large migration of Painted Lady butterflies but this is not the rule and a time difference between the peaks of the two migrants suggests that the two species have different geographical origins.

Eggs are laid singly on the upperside of young foodplant leaves (usually isolated, small plants are selected in open positions) and hatch in about a week during summer weather. The **larvae** pass through five instars and live and feed in increasingly large neat nettle-leaf tents which are of a distinctive shape and easy to find. Sometimes, they pupate also within a nettle tent and the **pupae** are also relatively easily found in the wild by man, and by parasites, particularly black ichneumon wasps.

The butterfly usually roosts on tree trunks where its cryptic underside makes it difficult to spot except in profile; when the adult butterfly overwinters it is generally seen in a similar habitat. Recent field studies of overwintering behaviour in Britain has shown that all four stages may be found in southern England and eggs and larvae are able to pass through a complete winter to pupate and then emerge as adults the following spring. Development of the early stages takes much longer during the winter than the summer. For example, eggs were laid quite widely in southern England during October 1997 and they took at least a month to hatch. During the following winter, larvae were found from December through to April feeding and growing slowly. Pupae lasted for about a month. Most resulting adults emerge in April or May and are smaller than those emerging in summer. Slow development at these stages makes them vulnerable to predation and disease and losses are high.

It has been suggested that mating of Red Admirals usually occurs immediately after overwintering and migrants from North Africa and continental Europe normally arrive mated. This may explain why mating is rarely observed in Britain and Ireland and why adults that emerge from eggs laid here (via larvae and pupae) attempt to fly south for the winter before mating at the beginning of the following season.

Distribution and trends

Each year, the earliest migrants to Britain and Ireland originate in North Africa, southern Europe, and the warmer Mediterranean islands, where there is a large post-winter emergence. The May to June migrations come from Spain and Portugal whilst later influxes probably come from France and central Europe. Sometimes there are western-moving migrations across the east coast of England from Scandinavia or eastern Europe. The Red Admiral may be found anywhere in these islands, though numbers vary from year to year, mainly determined by the weather, not only in Britain and Ireland but also in North Africa and the whole of continental Europe.

With such complex origins and the effects of changing weather conditions, it is not surprising that annual numbers vary so much. Most adults are seen in the period May to October but this species is one of the few that may be seen as an adult in any month of the year (as in southern England in 1997). Peak numbers generally occur in September, continuing in diminishing numbers into October. For this reason the Butterfly Monitoring Scheme (BMS), which runs only to the end of September, does not fully reflect annual trends in abundance for this species.

From mid-August onwards, a change in behaviour is observed in Britain and Ireland with movement increasingly, and by September almost entirely, one-way to the south or south-west, possibly triggered by shortening daylength. In Ireland, on the coast of Co. Dublin, Red Admirals have been observed flying purposefully south at intervals of one to three minutes at heights of up to 10 m, never alighting. Large numbers may accumulate along the south coast of England in autumn and in calm conditions, or when the wind is favourable, many attempt to cross the Channel. Some succeed and have been observed from the north French coast coming off the sea and then flying in a southerly direction.

During the past 20 years, the Red Admiral has increased in abundance at the monitored sites in the BMS. Abundance in spring is not however correlated with that of the previous autumn, evidence that overwintering butterflies are not important contributors to this increase. Increased immigration is the probable cause, possibly coupled with the effects of global warming and lengthening of the adult flight period.

European trend

The Red Admiral occurs only as a migrant and summer breeding species in some European countries and thus distribution trends have not been compiled.

Interpretation and outlook

With such a widespread foodplant, no special habitat requirements, and a large pool of breeding adults on the continent of Europe, there are no particular threats to this species at present and it has only

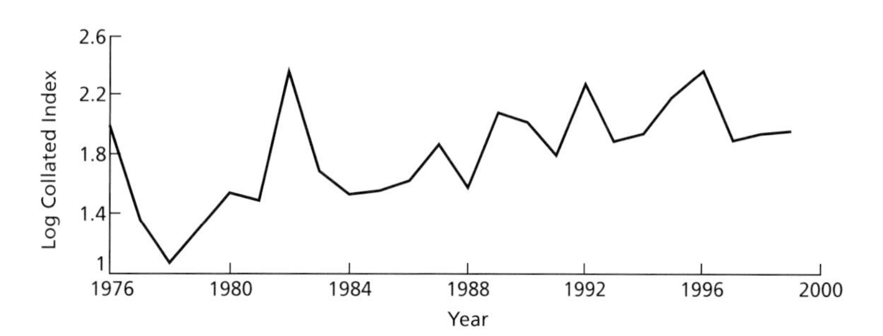

Red Admiral numbers have increased since the BMS began.

minimal need for active conservation. However nettles are not popular plants and it is an increasingly common sight throughout western Europe to see beds of nettles sprayed or cut down by property owners, farmers, or local authorities. If this trend continues the effect on the Red Admiral (and other nettle-feeding larvae) can only be damaging.

At the local and individual levels, adults may be attracted to feed in parks or gardens by suitable planting of nectar plants such as buddleias and stonecrops and to breed by allowing nettles to grow. These need to be managed by cutting back at appropriate times to provide a supply of new growth in a relatively sunny aspect.

The effects of climate change and global warming are difficult to interpret and predict for the Red Admiral. A 20-year analysis of data from the BMS has shown that the first annual record of the Red Admiral on the monitored sites was an average of nearly 40 days earlier in the mid-1990s compared with the mid-1970s. Similarly the flight period has lengthened by an average of 38 days during the same period but whether these observations reflect early emergence of overwintered adults or earlier immigration is yet to be determined.

Key references

Baker (1972*a*); Pollard and Greatorex-Davies (1998); Roy and Sparks (2000); Tucker (1991, 1997); Williams (1958).

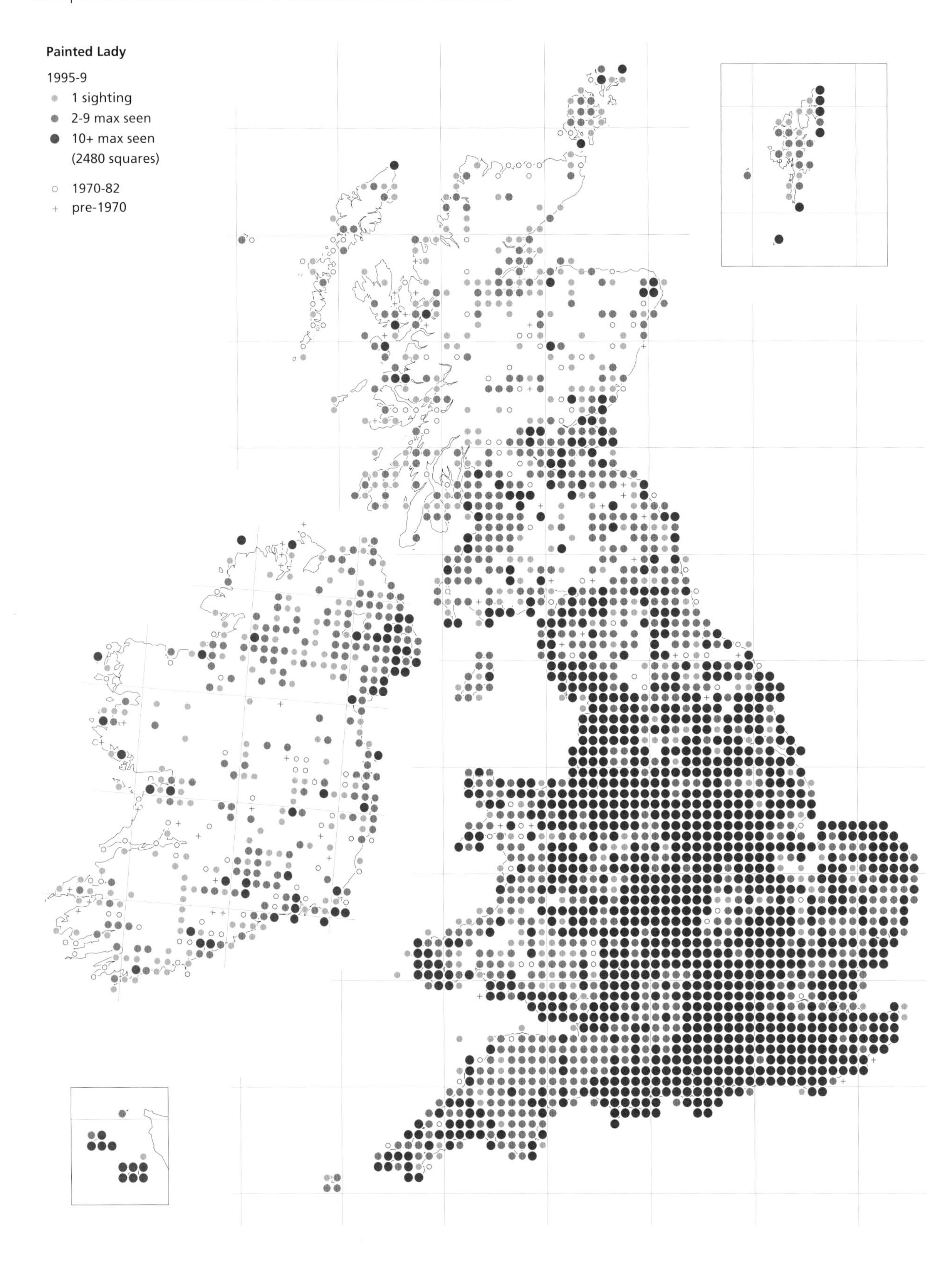
Painted Lady
1995-9
1 sighting
2-9 max seen
10+ max seen
(2480 squares)
1970-82
pre-1970

Painted Lady *Vanessa (Cynthia) cardui*

Regular migrant

Conservation status

UK BAP status: not assessed.
Butterfly Conservation priority: not assessed.
European status: not assessed.

European/world range

Worldwide, with the exception of South America. The Australian form has been classified as a separate species by some authors. Occurs in most of Europe only as a migrant and summer breeding species.

The Painted Lady is a long-distance migrant, which causes the most spectacular butterfly migrations observed in Britain and Ireland. Each year, it spreads northwards from the desert fringes of North Africa, the Middle East, and central Asia, recolonizing mainland Europe and reaching Britain and Ireland. In some years it is an abundant butterfly, frequenting gardens and other flowery places in late summer.

Foodplants

A wide range of foodplants may be used, with thistles (*Cirsium* spp. and *Carduus* spp.) being preferred in Britain and Ireland. Mallows (*Malva* spp.), Common Nettle (*Urtica dioica*), Viper's-bugloss (*Echium vulgare*), and various cultivated plants also have been recorded as larval foodplants here.

Habitat

Because it is a wide-ranging migrant, the Painted Lady may be seen in any habitat. Adults tend to congregate in open areas with plenty of thistles, which serve both as larval foodplants and nectar sources for adults.

Life cycle and colony structure

The Painted Lady is continuously brooded in its permanent southern range, and has no hibernation stage. Adults may be seen in Britain and Ireland in any month, but the peak of immigration usually occurs in June. Newly arrived individuals will reproduce and, given favourable weather, several broods may be possible during the summer (see p. 395).

Eggs are laid singly on the upper surfaces of suitable plants, and the **larvae** feed from silk tents on the undersides of the leaves. As they grow, the larvae make new and larger tents of leaves held together with silk. **Pupation** also occurs in such tents. The whole life cycle may take as little as a month to complete, hence the potential for multiple broods and rapid increases in numbers of adults.

Populations in Britain and Ireland tend to peak in late summer, when further immigrants join the progeny of earlier arrivals. For decades there has been uncertainty about the ability of Painted Ladies to overwinter successfully in Britain and Ireland. However, in the winter of 1997–8 a marked adult overwintered successfully at Hayle in Cornwall. It is not known whether the butterfly was capable of breeding after surviving the winter. Nevertheless overwintering may be a more common event than was previously thought.

The Painted Lady does not form discrete breeding colonies in Britain and Ireland, although it can occur in enormous numbers when migrating or at suitable nectar sources.

Painted Lady larva on thistle.

Distribution and trends

This intercontinental migrant may reach any part of Britain and Ireland, including remote islands (it was the only butterfly recorded from St Kilda in the BNM survey). Most areas receive some immigration and subsequently support breeding every year, but in the far north of Scotland, records tend to be restricted only to years favourable for the butterfly. There are breeding records from both Orkney and Shetland.

It appears that some Painted Ladies, particularly those seen in May and June, arrive directly from North Africa, while the peak in numbers in late summer derives from local breeding and influxes from populations in nearby European countries. These continental populations will themselves be the progeny of earlier migrants from further south.

The distribution and abundance of the species varies greatly from year to year. The best year for the Painted Lady in the current survey was 1996, when it was recorded in over 2100 10 km squares in Britain and Ireland. This compares to an average of around 1000 in each of the other four years of the project. Prior to 1996, other recent good years for this migrant were 1945–8, 1966, 1969, 1980, 1985, and 1988.

The collated index values from the Butterfly Monitoring Scheme show no clear pattern and there is no significant relationship between the numbers recorded in successive years. The enormous numbers of Painted Ladies recorded in 1996 were followed by average-level counts in 1997 and 1998. This suggests that the British-bred contribution to long-term numbers is low and that successful overwintering is insignificant.

Painted Lady butterflies are unlikely to be confused with any other resident species and the current distribution map is believed to be reliable. A vagrant, the American Painted Lady (*Cynthia virginiensis*) has a similar appearance, but it is extremely rare. There have been reports, often unconfirmed, of this species from the south-west of Britain in recent years.

European trend

The species occurs only as a migrant and summer breeding species in most European countries and thus distribution trends have not been compiled.

Interpretation and outlook

The immigration patterns of the Painted Lady are determined mainly by climatic conditions. For example, in 1980 there were two large immigrations. The first, in June, was associated with an anticyclone and it appears likely that butterflies were carried directly from North Africa or Spain, reaching first landfall in the Western Isles of Scotland, Ireland, and west Wales. Later in 1980 the immigration pattern was completely different, apparently bringing individuals from central Europe or further afield to east coast sites on an easterly wind.

The immigration of 1996 was one of the largest ever recorded and made headline news in the British national press. The number of butterflies involved was very difficult to estimate, but simple calculations suggest that many millions of Painted Ladies arrived during this influx. It provided the impetus for attempts to describe and understand the movements at a European scale, by combining evidence from national or regional transect monitoring schemes in several countries (Belgium, Finland, the Netherlands, Spain, and UK). This study found that annual abundance was closely correlated between countries, suggesting that the number of migrants from Africa determines annual abundance in Europe as a whole. The breeding success of the source populations may be related to high rainfall, and good growth and nutritional quality of foodplants, but this idea has yet to be tested.

In 1996 the speed of the migration movement provided evidence that some individuals flew directly from Africa to northern Europe, at speeds of about 150 km per day. The timing of the second, late summer, peak in numbers in northern Europe strongly suggested that this was primarily due to the progeny of the early immigrants, rather than further migrations from the south.

An unexplained aspect of the butterfly's ecology is the lack of evidence for southerly migration at the end of the season (unlike the Red Admiral, which clearly exhibits this pattern). Such behaviour would be expected to evolve by natural selection favouring individuals that return south and, consequently, survive the winter to breed in the following year. Given the high populations in northern Europe in some years, any southward mass migration should have been detected. However, these movements are very rarely recorded and the monitoring data in Spain give no evidence of such a pattern. Closer examination may eventually reveal such behaviour, as was the case for the Monarch butterfly in America, or it may remain an evolutionary paradox.

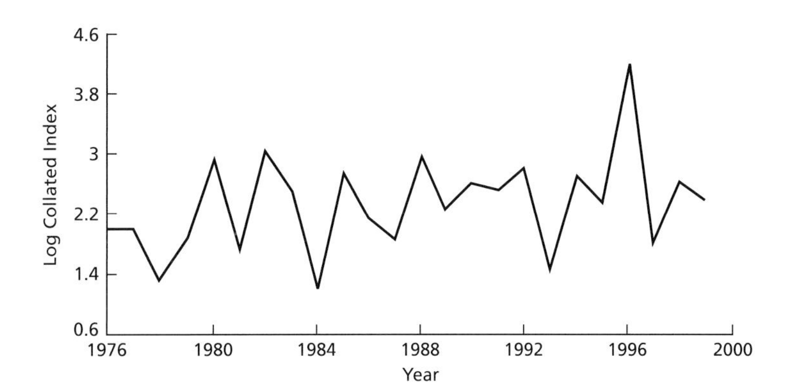

The abundance of the Painted Lady at BMS sites fluctuates according to levels of migration.

Neither distribution nor abundance measures are of much relevance in the case of the Painted Lady. There is no shortage of potential habitat in Britain and Ireland, but its occurrence here will remain dependent on the survival of suitable breeding areas outside Europe.

Key references

Bretherton and Chalmers-Hunt (1981); Pollard *et al.* (1998*a*); Wacher (1998).

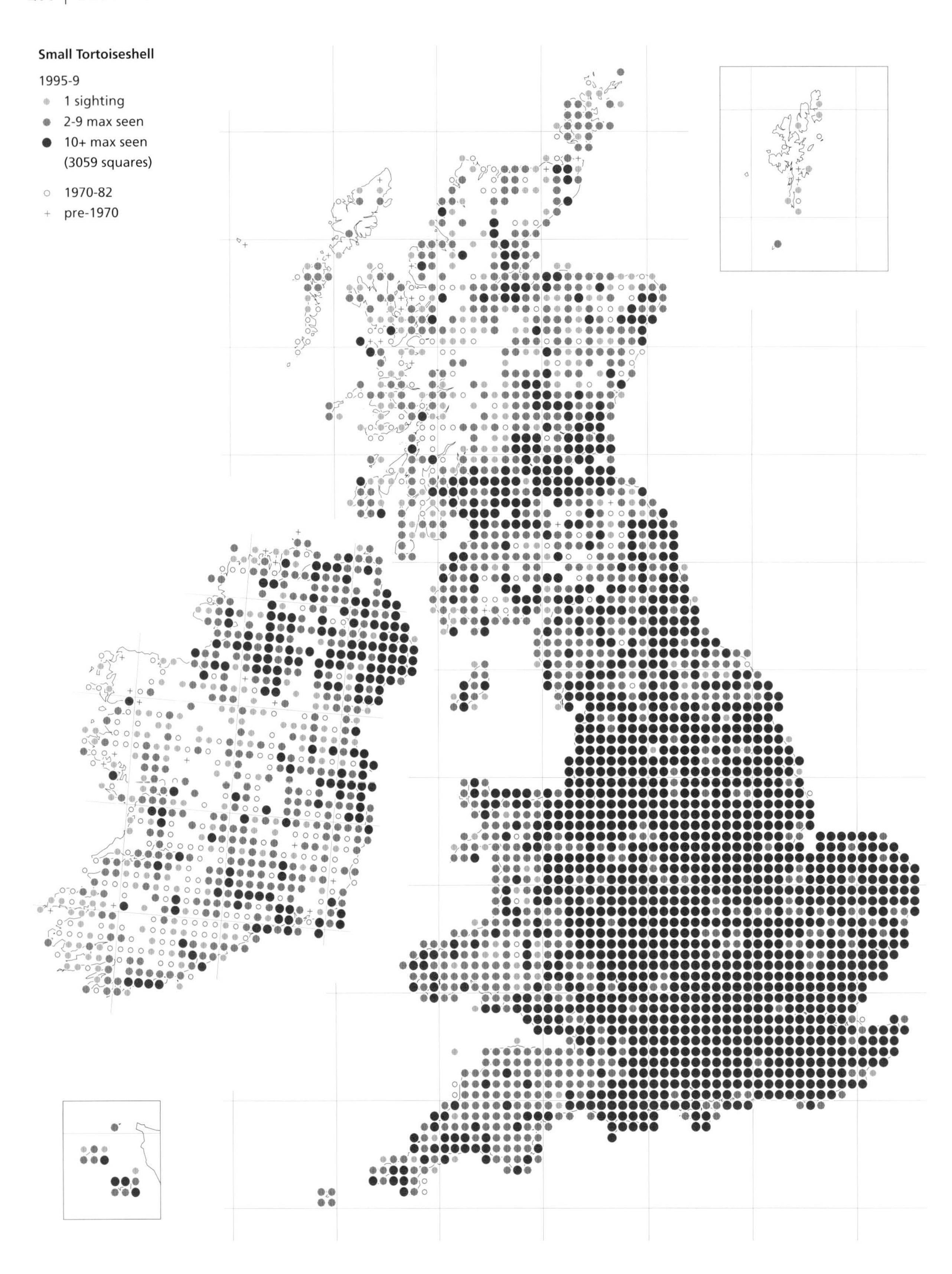
Small Tortoiseshell
1995-9
1 sighting
2-9 max seen
10+ max seen
(3059 squares)
1970-82
pre-1970

Small Tortoiseshell *Aglais urticae*

Resident
Range stable.

Conservation status
UK BAP status: not listed.
Butterfly Conservation priority: low.
European status: not threatened.

European/world range
Widespread from the Atlantic coast of Europe through to the Pacific coast of Asia. The distribution in Europe is stable.

The Small Tortoiseshell is among the most well known butterflies in Britain and Ireland. The striking and attractive patterning, and its appearance at almost any time of the year in urban areas have made it a familiar species. It is one of the first butterflies to be seen in spring and in the autumn it often visits garden flowers in large numbers. The Small Tortoiseshell is one of our most widespread species and has shown little overall change in range.

Foodplants
Common Nettle (*Urtica dioica*) and Small Nettle (*U. urens*) are used.

Habitat
The adult butterflies can be seen in any habitat, from mountain summits above 1000 m to city centres. The foodplants prosper in nutrient-enriched soils and breeding habitats are often associated with human activity, even areas of intensive agriculture. Breeding has been recorded at altitudes of over 300 m.

Life cycle and colony structure
The Small Tortoiseshell has a complex flight period that results from a variable number of generations per year in response to weather conditions and latitude (see p. 395). Hibernating adults typically emerge in March and April, although high temperatures may rouse them prematurely. In much of Britain and Ireland there are two generations per year (late June or July and late August onwards), with only Scottish populations regularly having just one. In northern England, however, cold, wet spring weather may cause a delayed emergence of the first brood of adults, which will then simply feed up and enter hibernation.

Temperature determines the speed of development and hence the number of possible generations. However, daylength has been shown to be the cue that determines whether larvae develop into adults that will breed immediately, or ones that will hibernate and reproduce the following season. In laboratory experiments, larvae exposed to long periods of artificial daylight tended to develop into sexually mature adults, but larvae taken from Scottish populations always produced adults destined for hibernation, irrespective of the amount of daylight they received. This reinforces field

Gregarious larvae of the Small Tortoiseshell on Common Nettle.

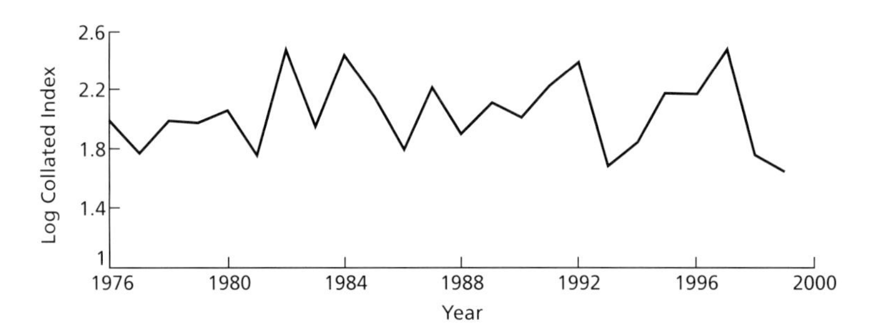

There are large annual fluctuations in the abundance of the Small Tortoiseshell at BMS sites.

observations and suggests that in Scotland, populations have adapted to produce only a single generation each year. Further south, Small Tortoiseshells have the ability to increase the number of broods in favourable years. The experiments suggest that a third brood is theoretically possible in southern England in very good years, a finding that is also backed by field observations.

Eggs are laid in large batches on the underside of young leaves of foodplants growing in full sunlight, often on the edge of nettle beds. **Larvae** are gregarious, sheltering within silk webs spun around the tops of nettle plants. When fully grown, the larvae disperse to finish feeding and **pupate** suspended from vegetation or other suitable objects.

Mature adults display daily cycles of activity that influence their likelihood of being recorded. Males, in particular, are more mobile in the morning, when they disperse, bask, and feed. They often set up territories in the early afternoon and spend the rest of the day defending these from rival males, and courting any females that pass by. Adults destined for hibernation show no reproductive behaviour, but spend their time feeding in order to build up fat reserves. Caves, hollow trees, and buildings such as churches, garden sheds, and attics are typical locations for overwintering.

Small Tortoiseshells range widely over the countryside and do not form discrete colonies. Adults may travel several kilometres per day and, in studies in Germany, marked individuals have been recaptured 150 km away from their point of release. The species is migratory with northward movements in spring and early summer and southwards in the autumn. Immigration from continental Europe is a frequent occurrence.

Distribution and trends

The butterfly is widespread and common throughout Britain and Ireland. Its wide-ranging behaviour means that individuals may be recorded from any 10 km square, even if there is no breeding population in the vicinity. The breeding range extends as far north as Orkney, but there are no known breeding records from Shetland, even though adults are recorded there.

Population levels of the Small Tortoiseshell fluctuate from year to year, although no change in distribution has been recorded and data from the Butterfly Monitoring Scheme (BMS) indicate that there is no long-term trend in overall abundance. There is, however, geographical variation within these fluctuations. For example, BMS data up to 1992 show a significant increase in abundance in eastern Britain. The species also became relatively scarce in the north of its range during the mid-1980s, although there was no equivalent decline further south.

The recorded distribution of the butterfly is likely to be accurate but may overestimate the permanent breeding range. The species is well known and unlikely to be misidentified, although some members of the public may call it the Red Admiral. Gaps in distribution almost certainly represent under-recording rather than absence, and coincide with areas where few or no visits were made in the 1995–9 survey.

European trend

The distribution of the Small Tortoiseshell is stable in most countries with expansions in the Netherlands, northern Russia, Slovenia, and Sweden in the last 25 years. Its distribution has contracted only in Slovakia during the same period.

Interpretation and outlook

The range of the Small Tortoiseshell extends further north in Britain than that of any closely-related resident species. In keeping with this, laboratory research has shown that Small Tortoiseshell larvae develop more rapidly at a given temperature than those of the Peacock or Comma (i.e. they require less heat for complete development). The adults are also better able to survive cold periods during hibernation than Peacocks. However, other findings suggest that the Small Tortoiseshell is also better adapted to warm climates than the other species. Small Tortoiseshell larvae grow more rapidly and survive better in very warm conditions, and the minimum temperature needed for larval growth is higher than that of the other species.

This apparent adaption to both high and low temperatures has been explained by the discovery that, in the wild, the gregarious larvae can raise their body temperatures substantially above that of their surroundings. This means that in relatively cool climates, they can achieve more rapid development, reducing the risk of predation or parasitism and raising the potential for an extra generation during the year.

The variability in population size from year to year is probably caused by several factors. BMS data show that the breeding success of the summer generation is reduced by drought and improved by high rainfall in May and June. This is presumably due to changes in the quality and availability of larval foodplants. The survival of overwintering adults is closely linked to feeding conditions in the autumn, when the butterflies must build up sufficient fat reserves. Warm, dry weather provides good conditions, but drought might reduce survival rates, if nectar sources are severely affected. It is also clear that predation and parasitism can take a heavy toll at each stage of the life cycle.

The ability of the species to switch from a single to a double brood in some northern parts of Britain when the weather conditions are favourable, is one of the chief mechanisms by which it quickly regains high population levels following declines. Immigration of individuals from continental Europe also may add to population levels in certain years.

No persuasive single explanation has been put forward for the upward trend in abundance at BMS sites in eastern Britain. The effect remains even when the influence of the 1976 drought is removed. One possibility is that the Small Tortoiseshell is capitalizing on an increase in the abundance of its foodplants. There is some evidence that nettles are becoming more common in Britain and nutrient enrichment of the environment may be the cause. Eastern England, with its intensive agriculture and high input of fertilizers, might be expected to be one of the first areas to show such an effect. Specialized nettle feeders, such as Small Tortoiseshell larvae, might benefit from the improved nutritional quality, as well as an increase in the abundance of food-plants. Alternatively, the trend might simply reflect the disproportionate attraction of monitored sites to wide ranging butterflies in a landscape with little semi-natural habitat, as has been suggested for the Peacock.

Flexibility of life history in response to weather conditions, a nomadic lifestyle, and widespread larval foodplants have made the Small Tortoiseshell one of the most successful butterfly species in Britain and Ireland. Despite great fluctuations in abundance, it remains ubiquitous and its future seems assured.

Key references

Baker (1972*b*); Bryant *et al.* (1997, 1998, 2000); Pollard *et al.* (1997); Pullin (1988); Pullin and Bale (1989).

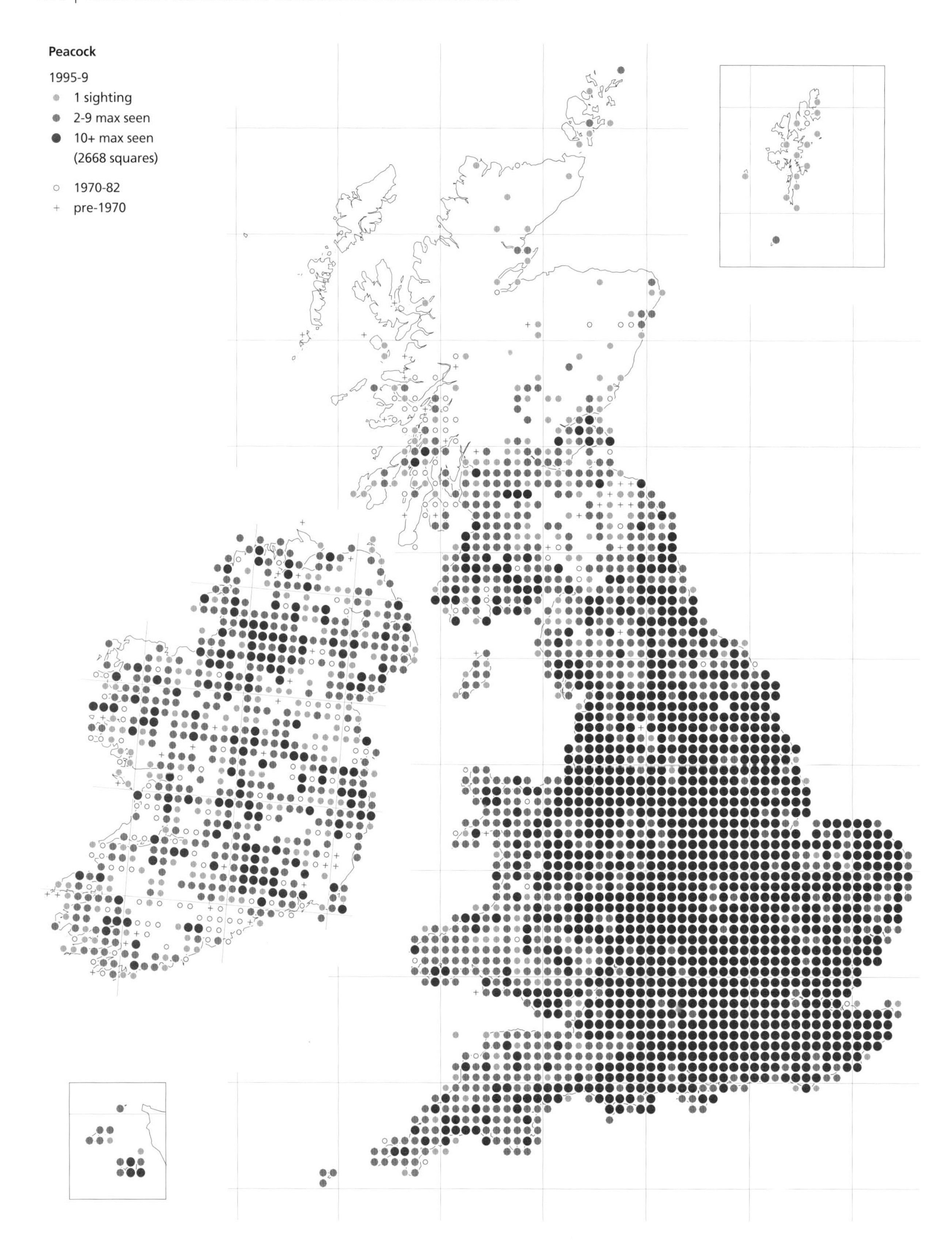
Peacock
1995-9
1 sighting
2-9 max seen
10+ max seen
(2668 squares)
1970-82
pre-1970

Peacock *Inachis io*

Resident
Range expanding.

Conservation status
UK BAP status: not listed.
Butterfly Conservation priority: low.
European status: not threatened.

European/world range
Most of temperate Europe and Asia to Japan. The northern limit in Europe occurs in southern Scandinavia and the butterfly is absent from parts of the extreme south (e.g. southern Greece, Portugal, Spain, and Turkey). Expanding at northern edge of European range.

The Peacock's spectacular pattern of eyespots, evolved to startle or confuse predators, make it one of the most easily recognized and best known species. It is from these wing markings that the butterfly gained its common name. Although a familiar visitor to garden buddleias in late summer, the Peacock's strong flight and nomadic instincts lead it to range widely through the countryside, often finding its preferred habitats in the shelter of woodland clearings, rides, and edges. The species is widespread and has continued to expand its range in northern parts of Britain and Ireland.

Foodplants
Common Nettle (*Urtica dioica*), although eggs and larvae are occasionally reported on Small Nettle (*U. urens*) and Hop (*Humulus lupulus*).

Habitat
Peacock butterflies may be seen almost anywhere, searching for suitable breeding or nectaring sites. These are often open, sunny places in woodland where the preferred nectar plants are found, e.g. willows in spring and teasels, thistles, and Hemp-agrimony in late summer. Large nettle patches are normally chosen for egg laying and these too are often located in sunny positions in the shelter of woodland or hedgerows.

Life cycle and colony structure
Generally considered to have one generation each year, Peacock adults emerge from late July onwards (later in the north). Adults hibernate and are regularly seen as early as February, although most emerge from March onwards with numbers peaking in late April. Data from the Butterfly Monitoring Scheme (BMS) show that Peacocks have been emerging from hibernation significantly earlier in recent years than when the scheme began in 1976. Field evidence suggests that a small second brood is possible in the south in favourable years. This is supported by experiments that have demonstrated the role of daylength in determining the development of larvae into adults that either will breed immediately or will hibernate. In the laboratory, larvae from southern England have been induced to develop into reproductive adults when exposed to long periods of artificial daylight (equivalent to the longest days of summer). Therefore, a partial second generation may occur when favourable weather allows the rapid development of the first brood during the long days of mid-summer.

Males feed and disperse in the mornings, then establish territories on the ground in the early afternoon. These territories are invariably in sunny spots in the corners of woodland or hedges and are vigorously defended against other males. When a female is encountered, the male will abandon his territory and give chase.

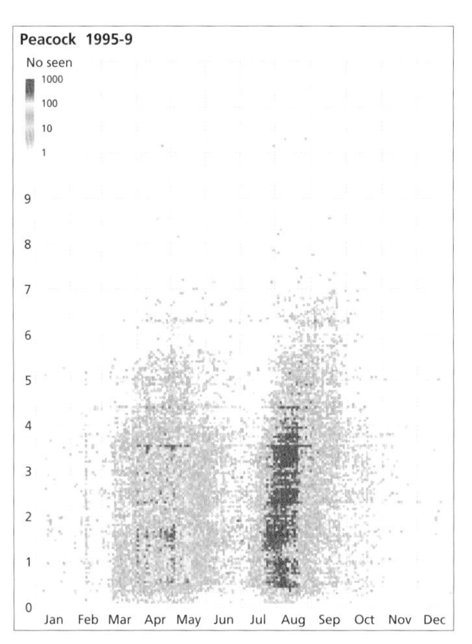

Left
Phenogram of dates of 1995–9 Peacock records plotted against latitude, showing two main flight periods and a scatter of records throughout the rest of the year.

Right
Peacock larvae on Common Nettle.

Eggs are laid in large, irregular clusters on the undersides of young nettle leaves. In contrast to the laying sites of the Small Tortoiseshell, large vigorous plants in the middle of nettle beds are often selected, although these must be in full sunlight at midday, when egg laying usually occurs. The **larvae** build communal webs over the growing tips of the nettles on which they feed, migrating to fresh plants as each is stripped bare. The feeding damage, swathes of web, and the black, spiny larvae are very conspicuous in June. Eventually the larvae disperse to **pupate** suspended from vegetation.

The emerging adults congregate in large numbers at good nectar sources, feeding voraciously to accumulate fat reserves for hibernation. The butterflies also spend time prospecting for suitable hibernation sites, often in hollow trees, crevices in bark or stone, and in unheated buildings. Once a site has been found, the individual will remain in the vicinity, roosting in the site and feeding nearby during the day, until it enters hibernation. Most appear to begin hibernation in early September, although individuals may be seen in October.

The Peacock does not live in discrete colonies and individuals range widely across the countryside in search of suitable habitats. Observations have revealed a tendency to fly to the north-west in the early part of the season and to the south-east later in the year, and some migration to and from continental Europe is likely. Nevertheless, fewer individuals are probably involved in these movements than is the case for the Small Tortoiseshell or the true migrants, the Red Admiral and Painted Lady.

Distribution and trends

The Peacock is widespread and common in the south of Britain and Ireland, but less frequent to the north and at higher altitudes. In Scotland, it is resident in the west and on islands of the west coast, as far north as Argyll. It is probably also resident in the south-east of Scotland. Scattered modern and historical records show that the butterfly has reached many northern parts of Scotland, including Orkney and Shetland, but these individuals are probably vagrants.

Little change in distribution occurred in the southern part of its range during the twentieth century. However, in the north, the Peacock expanded its range from an apparent low point in the latter part of the nineteenth century. This expansion began in the 1930s and has continued to the present day in some places, notably Northern Ireland. Here the butterfly was recorded in only one-quarter of 10 km squares in 1970–82, but in over three-quarters of squares in the period 1995–9. The butterfly is more widely recorded now across the Pennines and has become much more common in the central belt of Scotland in recent years. The current survey shows that it has been recorded in 34% more 10 km squares in Britain than in 1970–82.

As with many butterflies that are currently expanding their range, data from the BMS show a significant overall increase in the abundance of the Peacock since 1976. This trend is strong despite dramatic population crashes following wet and dull summers in 1985 and 1986, which caused extinction at several northern BMS sites and scarcity in many areas, for example around Dublin. Further falls in the collated index, such as in 1993 and 1994, did not prevent a record value being achieved in 1996. As with the Small Tortoiseshell, the largest increases in abundance have been in eastern England.

The large size and unique markings of the adult butterfly make misidentification extremely unlikely. The distribution presented is therefore believed to be accurate. The breeding distribution is undoubtedly smaller than that suggested by records of the highly mobile adults, particularly in northern parts of Britain and Ireland.

European trend

The Peacock's distribution is stable in most countries but there is evidence of expansion at the northern edge of its range, particularly in Finland (>100% increase in 25 years) and also in Latvia and Lithuania (25–100% increase). There has been a small decrease in Sweden (15–25% in 25 years).

Interpretation and outlook

The changes in the distribution of the Peacock appear to be driven by climate. The timing of its contraction and expansion relate well to periods of minor climatic deterioration and amelioration. Furthermore, although less dramatic, the Peacock's expansion in the northern part of Britain and Ireland has occurred in broad synchrony with that of other species, such as the Speckled Wood and the Comma. The Peacock, like the Small Tortoiseshell, has gregarious larvae that are able to raise their body temperatures above that of their surroundings. This lesser reliance on ambient temperature may explain the relative stability of the Peacock's distribution in comparison with that of the Comma.

New squares 1995-9:
• 1 seen
• 2+ seen
• 1970-82

The expansion of range of the Peacock since 1970–82.

The butterfly's distribution is not constrained by the range of its larval foodplant, which is common and widespread across Britain and Ireland. In eastern England, agricultural intensification may have improved conditions for nettles and this may explain the significant increases in abundance at BMS sites in this region. However, these trends may not reflect higher abundance in the countryside as a whole. Monitored sites tend to be areas of semi-natural habitat and Peacocks rely on such areas for critical periods of their life cycles, notably courtship, mating, and pre-hibernation feeding. Suitable areas are rarer in eastern England, so the densities of Peacocks congregating at them might be expected to be greater, even if the actual number in the wider countryside is the same as elsewhere.

During the twentieth century, the Peacock expanded to occupy all of its former range by the early 1980s. The BNM survey indicates that this expansion has continued in parts of Northern Ireland and Scotland and the species may now occupy more territory than at any time in its recorded past.

Key references

Baker (1972*b*); Bryant *et al.* (2000); Pullin (1986).

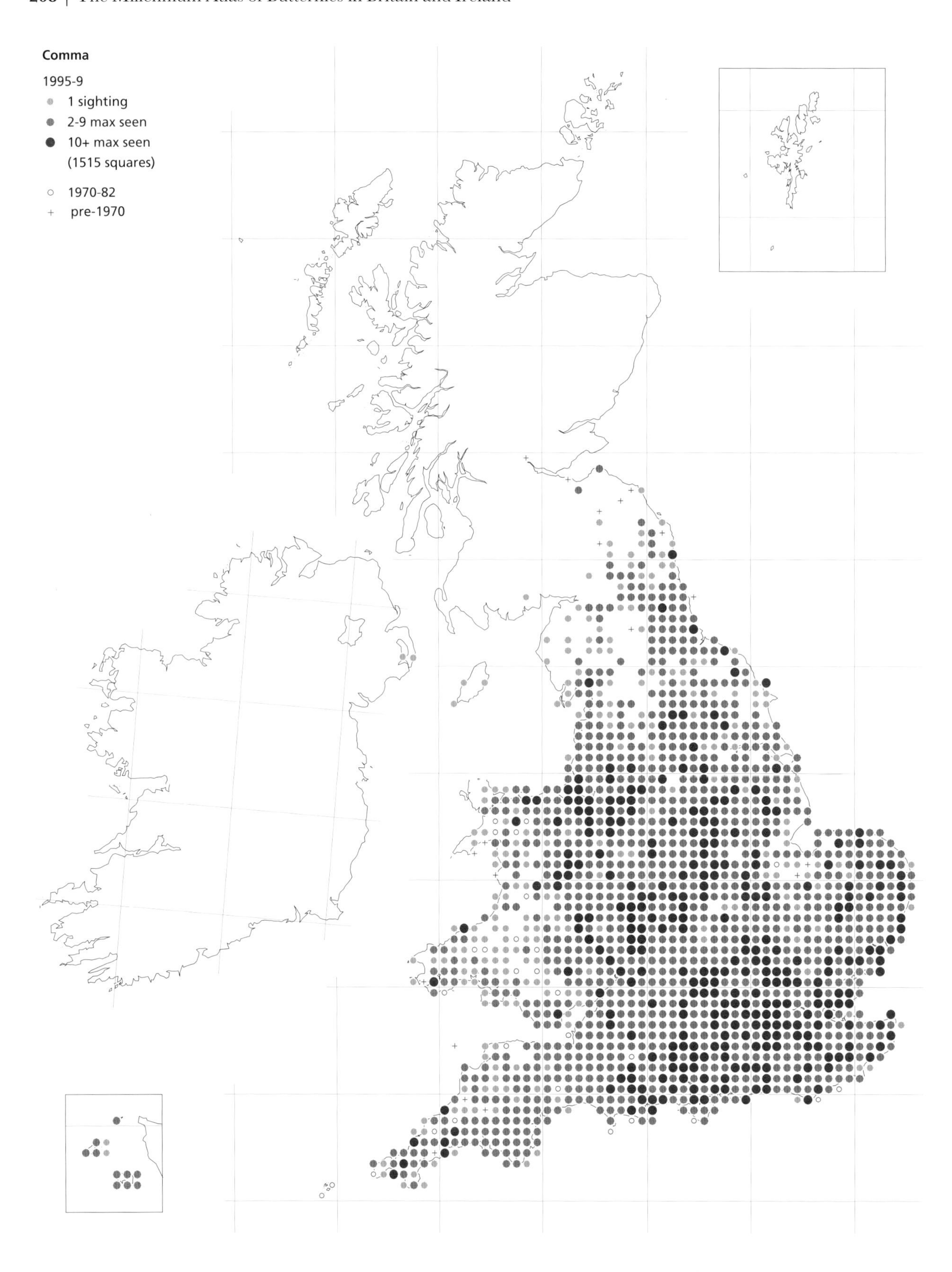
Comma
1995-9
1 sighting
2-9 max seen
10+ max seen
(1515 squares)
1970-82
pre-1970

Comma *Polygonia c-album*

Resident
Range expanding.

Conservation status
UK BAP status: not listed.
Butterfly Conservation priority: low.
European status: not threatened.

European/world range
Most of Europe, including Scandinavia, across Asia to Japan. It is also present in North Africa. European range appears to be shifting northwards.

The Comma is a fascinating butterfly. The scalloped edges and cryptic colouring of the wings conceal hibernating adults amongst dead leaves, while the larvae, flecked with brown and white markings, bear close resemblance to bird droppings. The species has a flexible life cycle, which allows it to capitalize on favourable weather conditions. However, the most remarkable feature of the Comma has been its severe decline in the twentieth century and subsequent comeback. It is now widespread in southern Britain and its range is expanding northwards.

Foodplants
The most widely used foodplant is Common Nettle (*Urtica dioica*). Other species used include Hop (*Humulus lupulus*), elms (*Ulmus* spp.), currants (*Ribes* spp.), and willows (*Salix* spp.).

Habitat
Open woodland and wood edges are the main habitats for both breeding and hibernation. Pre-hibernation individuals range more widely in search of nectar and rotting fruit, and are seen regularly in gardens and many other habitats.

Life cycle and colony structure
The life cycle of the Comma is complex. It normally has two broods per year (flying in July with a partial brood in late August and September, see p. 396). Adults hibernate on tree trunks and branches and most become active again in late March or April. Butterfly Monitoring Scheme data show a significant trend towards earlier emergence in recent years.

Eggs are laid singly on the uppersides of foodplant leaves. Females select prominent nettles growing in the shelter of woodland or hedgerows, although these need not be in full sunshine. The **larvae** spin fine webs and initially feed on the undersides of leaves, moving into the open as they grow and develop their camouflage. **Pupae** are suspended in thick vegetation, often from a foodplant.

Some of the adults in the first brood are not sexually mature. They show no reproductive behaviour but concentrate on feeding and eventually enter hibernation. The remainder mate quickly, resulting in a second generation of adults in late summer. These second brood individuals then build up food reserves and hibernate. The adults that emerge in the following spring are therefore a mixture of first and second brood individuals.

The first generation individuals that breed immediately not only have a different life history compared with those that hibernate, but also differ in appearance. Their upperwings are paler and the wing edges are

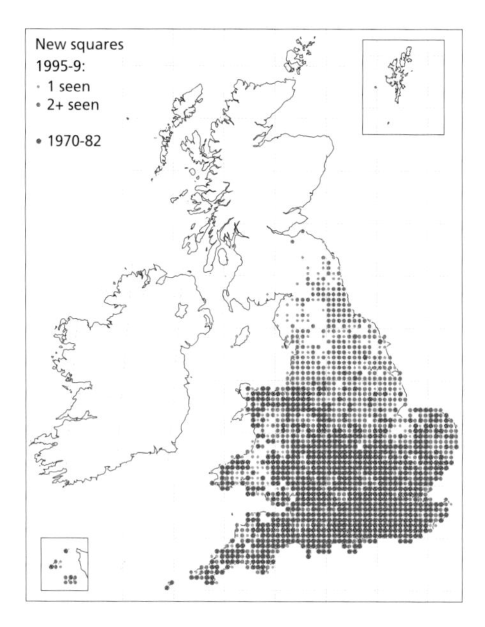

The expansion of range of the Comma since 1970–82.

less deeply scalloped. This form, termed *hutchinsoni*, makes up a variable proportion of the midsummer brood. Development into normal (dark) or *hutchinsoni* individuals is not genetically predetermined but governed by daylength during the larval and pupal stages. Larvae that develop while daylength is increasing tend to become *hutchinsoni* adults, while those experiencing shortening days generally emerge as hibernators. The result is that in favourable weather conditions e.g. an early spring, the first brood will contain more *hutchinsoni* individuals and numbers in the second brood should be larger. However, the control of the Comma's development is complex (temperature and foodplant also influence the outcome) and is not fully understood.

The Comma does not form discrete colonies and individuals are quite mobile, particularly late in the season. However, they are not as mobile as many closely related species and immigration from continental Europe is thought to be a rare event.

Distribution and trends

The Comma is widespread and common in much of England and Wales, becoming less so in the west and north, and at altitude. It is extinct in Scotland and does not occur as a resident species in Ireland or the Isle of Man. However, during the current survey vagrants have been recorded in all of these countries, prompting hopes of colonization.

The distribution of the Comma has undergone enormous change over the past 200 years. In the early nineteenth century it covered much of England and Wales and scattered localities in Scotland, as far north as Fife. Periods of decline reduced the Comma to the status of a rarity across most of Britain by the early part of the twentieth century. At its lowest point, in the 1910s, the species was limited to the counties along the Welsh border, plus a few isolated populations in south-east England. The Comma then underwent a dramatic re-expansion, first eastwards across England and then to the north. This expansion was halted or reversed during the 1950s and 1960s, but resumed in the mid-1970s. In the Channel Islands the butterfly was rarely reported prior to 1970, but is now more frequently sighted.

The pace of recent change has been remarkable, with recolonization of almost a third of Britain (in terms of latitude) since 1982. Comparison of current data with the 1970–82 data shows that almost 700 new 10 km squares have been recorded and an overall expansion of over 79% has taken place. This expansion has occurred in many parts of Britain, most notably in northern England, but also in East Anglia, mid and south-west Wales, and the West Country. In 1984 there were few records north of a line from the Wash to the Mersey, but by 1999 the Scottish border had become the demarcation of the established range.

The few recent records from Scotland are believed to be vagrants, but they nevertheless represent the first Scottish records for 130 years. In addition to the growing number of Scottish sightings there have been records from the Isle of Man in three of the five years of the current survey and, remarkably, two records from Portaferry, Co. Down. These sightings, less than 40 miles from the Isle of Man, are also believed to be of vagrants.

Butterfly Monitoring Scheme (BMS) data show a significant increase in abundance at sites located in the east and north of the butterfly's range. In addition to these increases, at least 10 BMS sites have been colonized since monitoring began. These are also mostly in eastern and northern England.

Although the Comma might be mistaken for a fritillary or even a Small Tortoiseshell in flight, its ragged wings are easily recognized and the distribution map is an accurate representation of the current range of the butterfly.

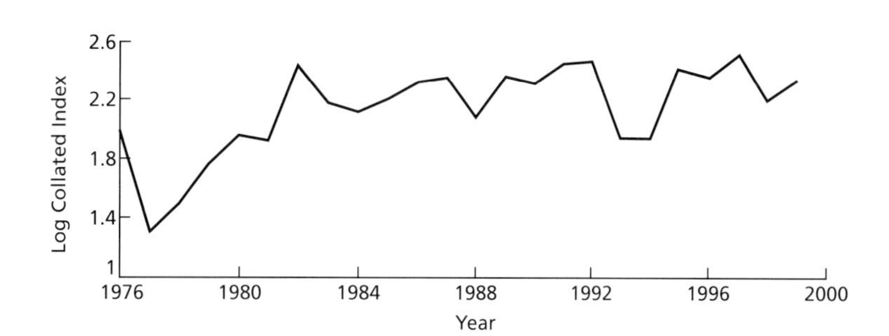

There was a large drop in numbers of the Comma at BMS sites following the 1976 drought, and a subsequent increase.

European trend

The Comma is stable in most countries, but there is evidence of a shift in range northwards over the past 30–100 years. In the past 25 years, expansions have occurred in Denmark, Moldova, the Netherlands, north-eastern Russia, and Slovakia. In contrast the species has become extinct at more than half of its historically recorded sites in North Africa and suffered recent declines in the Czech Republic and the European part of Turkey (15–25% decrease in 25 years).

Interpretation and outlook

The reason for the Comma's fall and rise is far from clear. Most nineteenth century authors regarded Hop as the main foodplant of the Comma. The species' nineteenth century decline coincided with the demise of the hop-growing industry in many areas, along with the introduction of new management techniques and pesticides at the hop-gardens that remained. However, the decline of the butterfly in southern England pre-dated these changes.

It has been suggested that in the early nineteenth century most Commas belonged to a race predisposed to using Hop as its larval foodplant. Their decline led to recolonization by other races, which had previously been geographically restricted. In recent years, Common Nettle has been regarded as the main foodplant, but in Northumberland and Durham, one of the major zones of recent expansion, most larval records have been from Wych Elm. Foodplant choice can affect larval development time, survival to adulthood, and the reproductive success of both sexes, but evidence for the existence of distinct races is still lacking.

Several climatic factors may also be relevant. Winter humidity and temperature have both been proposed as causal factors in the contraction and subsequent expansion of the Comma. In particular, the changes in distribution appear to reflect long-term winter temperature trends and successive cold winters coincide with the periods of decline. Low spring temperatures may also have a negative effect on numbers. Unlike the gregarious larvae of the Small Tortoiseshell and the Peacock, solitary Comma larvae are not able to raise their own body temperature above that of their surroundings. This dependence upon external temperature might explain the extreme fluctuation of the Comma's distribution in relation to the stability of the other two species.

The current expansion of the Comma is no less dramatic for the absence of a clear explanation. The species is almost certainly experiencing favourable climatic conditions and benefiting from the ability to vary the number of generations per year. The fact that it has two broods in its current zone of expansion in the north of England (whilst only having a single generation at its northern range margin in Scandinavia) suggests that the butterfly may have the potential to continue moving northwards. It appears to be on the verge of recolonizing Scotland and its immediate future seems favourable.

Key references

Bryant *et al.* (2000); H. A. Ellis (1999*b*); Nylin (1992); Nylin *et al.* (1996); Pratt (1986, 1987); Wedell *et al.* (1997).

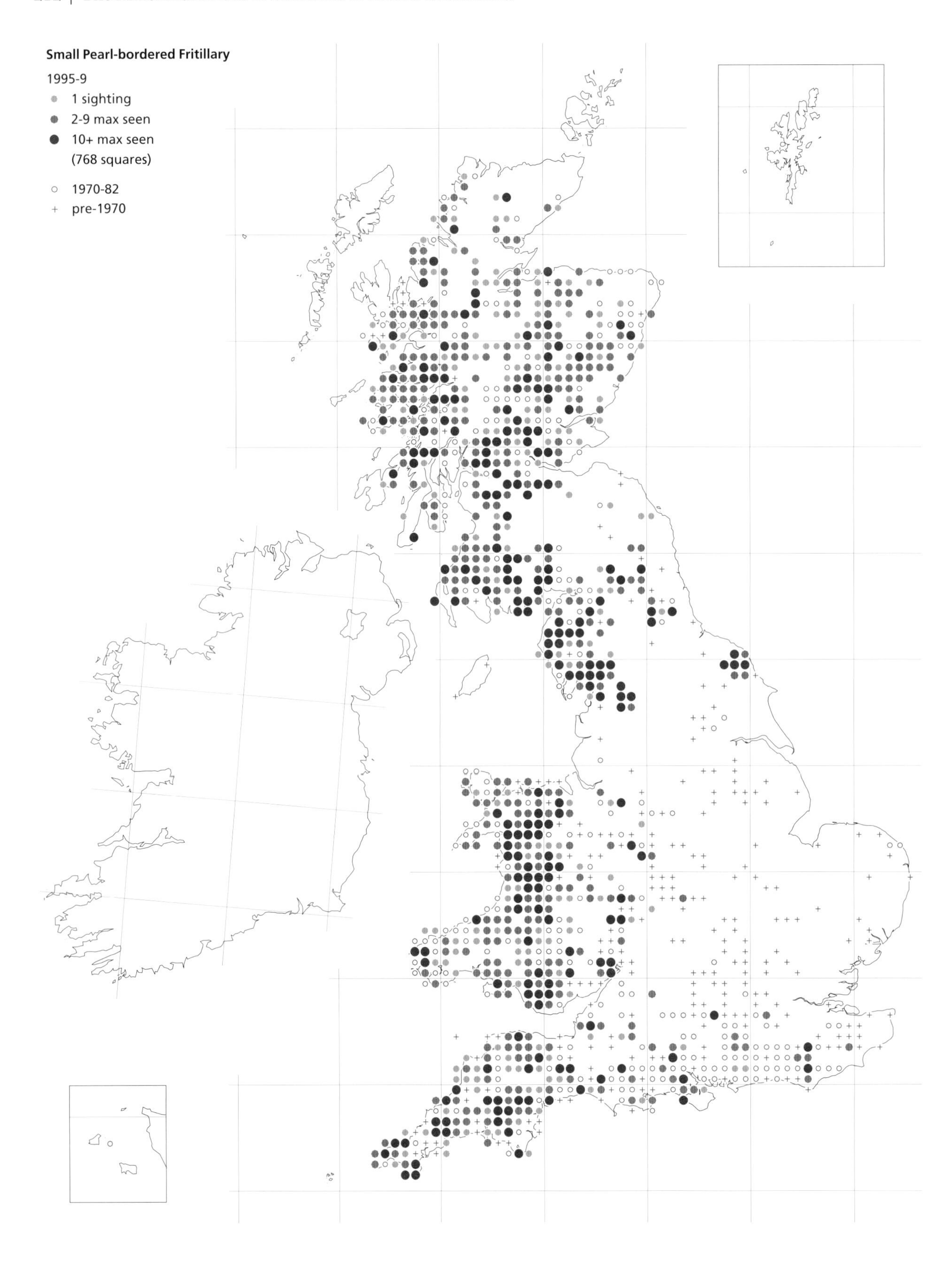
Small Pearl-bordered Fritillary
1995-9
1 sighting
2-9 max seen
10+ max seen
(768 squares)
1970-82
pre-1970

Small Pearl-bordered Fritillary *Boloria selene*

Resident

Range declining in England.

Conservation status

UK BAP status: Species of Conservation Concern.
Butterfly Conservation priority: medium.
European status: not threatened.

European/world range

Widespread across central and northern Europe and through Asia to Korea. Also occurs in North America. Range appears stable through much of Europe but declines have been reported in at least nine countries.

Foodplants

The most widely used foodplants are Common Dog-violet (*Viola riviniana*) and Marsh Violet (*V. palustris*). It may occasionally feed on other violet species.

Habitats

There are four main habitats: woodland glades and clearings (mainly in southern Britain); damp grassland and moorland (in western and northern Britain); grassland with bracken and/or patches of scrub; and open wood-pasture and wood edges in Scotland, usually where there is some grazing by deer and/or sheep. Other habitats used include dune slacks and coastal cliffs. In all habitats it breeds in damp, grassy vegetation where there is abundant foodplant growing in a fairly lush sward.

Life cycle and colony structure

The Small Pearl-bordered Fritillary is single brooded with adults flying on average from late May until the end of June, but the precise timing varies with latitude and habitat (see p. 396). In south-west England it can emerge as early as late April, whereas in Scotland it flies almost one month later from late June until the end of July. There is a partial second brood in parts of southern England, and occasionally in south Wales, with adults appearing in August.

This fritillary is similar in size and habits to the Pearl-bordered Fritillary but is more widespread and occurs in damper, grassy habitats as well as woodland clearings and moorland. The adults fly close to the ground, stopping frequently to take nectar from flowers such as Bramble and thistles. It can be identified from the more numerous whitish pearls on the underside hind wings, the outer ones bordered by black chevrons, and from the larger black central dot. The butterfly remains widespread and locally abundant in Scotland and Wales, but has undergone a severe decline in England.

The **eggs** are laid singly, either on plants or dead vegetation near to violets and sometimes on the foodplant itself, or by dropping them while crawling amongst low vegetation. The **larvae** feed until their fourth instar when they hibernate, probably amongst the leaf litter. They emerge to feed again during spring but, unlike Pearl-bordered Fritillary larvae, they rarely bask and spend most of their time concealed amongst the vegetation, only coming out for short bouts of feeding. They **pupate** close to the ground, hidden deep within vegetation.

The colony structure and mobility of the Small Pearl-bordered Fritillary have been studied only on one woodland site in Dorset where adults appeared to be highly sedentary, remaining for most of their lives in the same small clearing. They did not move

Small Pearl-bordered Fritillary larva on Common Dog-violet.

at all into new clearings made only a few hundred metres away, although they stayed there having been moved artificially. No studies have been carried out in more extensive and open habitats such as those in Scotland and western Britain, but circumstantial evidence suggests that this species ranges far more widely in these habitats. In such extensive habitats they are often found scattered over wide areas and may exist as larger metapopulations within which adults concentrate around favoured breeding areas.

Distribution and trends

The Small Pearl-bordered Fritillary is still a widespread butterfly in Scotland and Wales, but has declined dramatically in England. It is now extinct in many central and eastern counties, and is on the verge of extinction throughout central-southern England. Its decline began in the late nineteenth century but accelerated during the twentieth century, becoming very severe after the 1950s. A study of over 300 important butterfly sites in central-southern England showed that losses have risen steadily since 1940 and reached 41% per decade by the 1980s. The butterfly's decline has been less severe in the south-west and north-west of England and it is still widespread in parts of Devon (e.g. Dartmoor) and Cumbria.

In contrast, there has been only a minor reduction in the range of the butterfly in Scotland and Wales. These countries are under-recorded and it is almost certainly more widespread than shown. However there have probably been substantial declines within its range in both countries due to habitat loss, though these are difficult to quantify due to the low level of past recording. It is absent from many Scottish islands, the Isle of Man, the Channel Islands, and Ireland.

The current map shows that the Small Pearl-bordered Fritillary has disappeared from 39% (248) of the 10 km squares recorded in 1970–82, with most of the losses occurring in England. However, 373 new squares have been recorded since the last survey period, mostly in Scotland.

European trend

The butterfly appears to be stable through much of Europe but it has declined severely in the Netherlands (> 75% decrease in 25 years), Belgium and Luxembourg (50–75% decrease); and to a lesser extent in central European countries such as the Czech Republic, Germany, Hungary, Latvia, Slovakia, and Slovenia (15–25% decrease).

Interpretation and outlook

The varying fortunes of the Small Pearl-bordered Fritillary appear to be related to its habitat use in different parts of Britain, and the rate at which habitats have changed over recent decades. In Scotland and Wales, the butterfly breeds in a variety of habitats ranging from moorland and damp grassland to open wood-pasture. Although these habitats have undoubtedly been reduced in extent, they are still widespread and their management has changed comparatively little during recent decades.

In contrast, such habitats have nearly all been destroyed by intensive agriculture in lowland England and the butterfly has survived only in woodland clearings. During the early twentieth century, these were created in abundance by traditional coppicing but this practice has now all but ceased. Most woods are now either neglected or have been converted to high forest systems, which contain fewer clearings. Like several other woodland butterflies, it prospered during the 1950s and 1960s, when many woods were cleared and replanted (often with conifers), but has since gone into a steep decline as these became too shady. Its survival is now in serious doubt across much of southern England and it will be saved only by a concerted effort to restore suitable woodland management.

The use of a wider range of habitats also explains why the butterfly has declined far less severely in Britain as a whole than the Pearl-bordered Fritillary. Although the two species often occur on the same sites, they select different areas in which to breed, the latter selecting the drier and hotter areas with violets, the former choosing damper areas. However, both need open sunny habitats and both benefit from similar types of management.

In bracken habitats, continued grazing is essential to maintain a mosaic of bracken and grass with abundant violets, whereas in woodland both butterflies need a regular supply of new clearings preferably interconnected by wide rides. The Small Pearl-bordered Fritillary should thus benefit from some of the measures being undertaken to conserve the more highly threatened Pearl-bordered Fritillary, especially in southern England where the need is greatest.

In Scotland and Wales, the chief danger is complacency. Suitable damp habitats are still widespread, but they are disappearing steadily with changes in land use. For example, moorland is being lost to afforestation in Scotland and large areas of hill land in Wales are being improved or overgrazed. Losses of unfertilized damp grassland have also been substantial, indicating that the butterfly may already be declining locally, although this has yet to be reflected in its distribution when plotted at 10 km square resolution. Because such regions tend to be under-recorded, there are no finer scale historical records to examine the possible impact.

Both countries have an extensive infrastructure of nature reserves and protected areas, but similar measures have not prevented the decline of the Small Pearl-bordered Fritillary in England. It would be preferable to examine the threats now and take precautionary action soon so that this beautiful butterfly does not decline throughout its range.

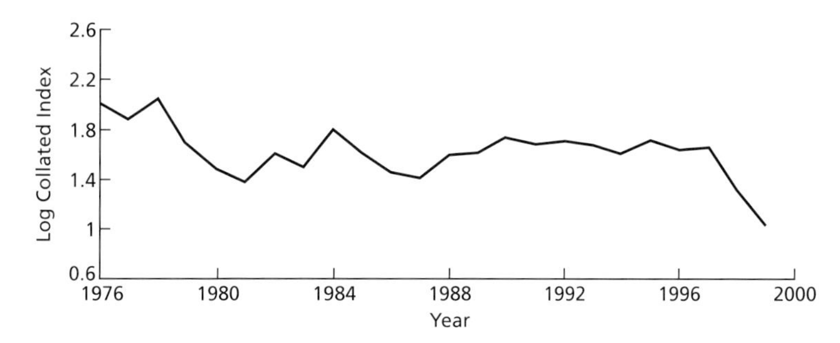

Decline of the Small Pearl-bordered Fritillary at BMS sites.

Key references

Barnett and Warren (1995*d*); Joy (1998*a*); Thomas and Snazell (1989); Thomas (1991); Warren (1992*b*, 1993*a*).

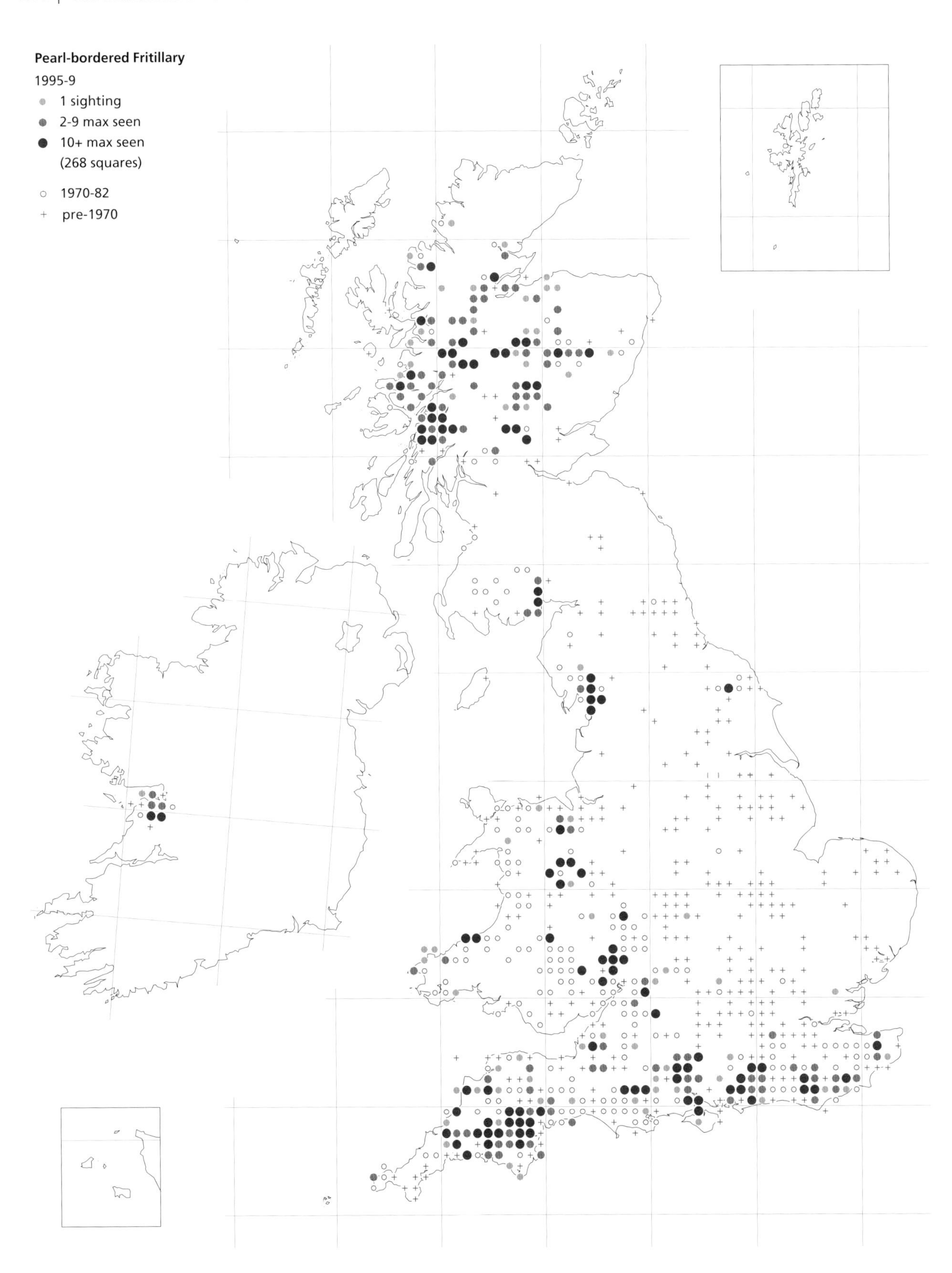
Pearl-bordered Fritillary
1995-9
1 sighting
2-9 max seen
10+ max seen
(268 squares)
1970-82
pre-1970

Pearl-bordered Fritillary *Boloria euphrosyne*

Resident

Range declining in England and Wales.

Conservation status

UK BAP status: Priority Species.
Butterfly Conservation priority: high.
European status: not threatened.
Protected in Great Britain for sale only.

European/world range

Widespread across Europe from northern Spain to Scandinavia, and eastwards to Russia and Asia. Apparently stable through much of Europe but declines reported in at least 12 countries.

This is one of the earliest fritillaries to emerge and can be found as early as April in woodland clearings or rough hillsides with bracken. It flies close to the ground, stopping regularly to feed on spring flowers such as Bugle. It can be distinguished from the Small Pearl-bordered Fritillary by the two large silver 'pearls' and row of seven outer 'pearls' on the underside hind wing, and also the red (as opposed to black) chevrons around the outer pearls and the small central spot on the hind wing. The butterfly was once very widespread but has declined rapidly in recent decades, and is now highly threatened in England and Wales.

Foodplants

The most widely used foodplant is Common Dog-violet (*Viola riviniana*) although it can use other violets such as Heath Dog-violet (*V. canina*) and in the north, Marsh Violet (*V. palustris*).

Habitat

Three main habitats are used: woodland clearings, usually in recently coppiced or clear-felled woodland; well-drained habitats with mosaics of grass, dense bracken, and light scrub; and open deciduous wood-pasture in Scotland, typically on south-facing edges of birch or oak woodland where there are patches of dense bracken and grazing by deer and/or sheep. In all habitats it requires abundant foodplants growing in short, sparse vegetation, where there is abundant leaf litter.

Life cycle and colony structure

The butterfly is normally single brooded with adults flying for about six weeks between late April and late July, with the precise timing varying according to latitude and habitat (see p. 396). There is also evidence that average emergence dates have advanced by over ten days in the last twenty years. In exceptionally warm springs, it can emerge in early April in southern bracken habitats and it flies several weeks later in the far north. There is a partial second brood in the south during August.

The **eggs** are laid singly, usually on dead bracken or leaf litter near to violets, though a few are laid on the foodplant itself. The **larvae** develop until the early fourth instar when they hibernate amongst dead leaves or bracken. They emerge in early spring when they spend much of their time basking on dead litter, interspersed with short bouts of feeding. The selection of warm, dry habitats and the basking behaviour of larvae enable them to develop rapidly even in cool spring weather. The larger larvae can move tens of metres in search of suitable basking sites and foodplants. They **pupate** in the leaf litter and emerge after a few weeks.

Pearl-bordered Fritillary larva basking on a dead leaf.

The Pearl-bordered Fritillary usually forms discrete colonies around suitable breeding areas, often comprising many hundreds of adults. Mark–recapture studies in both Scotland and England show that adults move freely within their colonies regularly covering 100 m or more. A significant proportion also disperses and can move at least 4.5 km between adjacent colonies. Nearby colonies are thus linked and the butterfly almost certainly forms metapopulations covering networks of discrete breeding areas. Where breeding habitats are small and widely scattered, such as in parts of Scotland, it may have a looser population structure, occurring at low densities over large areas.

Distribution and trends

The Pearl-bordered Fritillary was once widespread and abundant across most of England, Wales, and Scotland but has undergone a dramatic decline. The decline has been most severe in Wales, where it has died out in all but a few areas, and in England, where it has disappeared from most central and eastern counties. In Scotland it remains widespread in the west and Highland regions, whereas in Ireland it has always been confined to the thin limestone soils and scattered scrub of the Burren in Co. Clare.

The decline of the Pearl-bordered Fritillary began in northern England during the early part of the twentieth century but the loss rate has increased substantially since the 1950s. The butterfly's distribution in Britain is well known thanks to targeted recording for the BNM survey, and a survey conducted by Butterfly Conservation in 1997–8. The results show that the butterfly has probably disappeared from 60% (198) of the squares recorded in 1970–82, but has been recorded in over 130 new squares, most of them in Scotland.

The surveys have identified around 350 discrete breeding colonies in Britain of which around 50% are in woodland clearings, 30% in bracken/grass habitats, and 20% in open woodland (in Scotland). The results from the Butterfly Monitoring Scheme reflect this downward trend and show that losses have occurred on many reserves and protected areas.

The Pearl-bordered Fritillary appears to be more stable in Scotland, although it has declined in lowland areas in Dumfries and Galloway, and around Glasgow. Elsewhere, changes in distribution are difficult to determine, due to low levels of past recording and occasional misidentifications of the Small Pearl-bordered Fritillary. The species also remains under-recorded in some remote areas.

European trend

The butterfly appears to be stable in many countries but has undergone serious declines in the Netherlands (extinct); Belgium and Denmark (>50% decrease in 25 years); Germany, Lithuania, and Luxembourg (25–50% decrease). Declines have also been reported in the Czech Republic, Hungary, Latvia, Poland, Slovakia, and Sweden (15–25% decrease).

Interpretation and outlook

The Pearl-bordered Fritillary is one of the most rapidly declining butterflies in Britain and Ireland. As for several other fritillaries, a significant factor in England has been the change in woodland management during the twentieth century, particularly the cessation of coppicing and consequent reduction of sunny clearings. The replanting of woods, often with non-native conifers, provided new habitats for the butterfly during the 1950s and 1960s. However, these plantations have subsequently grown dense and shady, leading to a surge of local extinctions in the last 20–30 years.

The bracken habitats used by the Pearl-bordered Fritillary have also deteriorated due to changing management and large areas have been abandoned allowing bracken to become dominant, and scrub to invade. In other regions, such as the Welsh hills, the reverse phenomenon has occurred; bracken areas have been eliminated, and the grassland fertilized and grazed intensively with sheep.

Recent research has shown that the butterfly breeds in a mosaic of bracken and grass, where suitable conditions are maintained by grazing and trampling by livestock. Cattle and ponies are especially good at trampling dense bracken litter and the butterfly often breeds along animal paths. A crucial component of the conservation strategy is to maintain traditional grazing in key areas such as Dartmoor, and to restore grazing on recently abandoned sites as encouraged by the Countryside Stewardship and Environmentally Sensitive Areas schemes. However, it is difficult to find suitable stock and financial incentives are still not high enough for farmers to reinstate grazing of marginal hill land at an appropriate level.

The butterfly's requirements in woodland are comparatively well known and populations have responded well on several nature reserves where coppicing has been re-instated and also in some forests where rides have been widened. However, coppicing will almost inevitably remain a rare and uneconomic type of woodland management in many areas and the challenge is to provide a regular sequence of open clearings in modern high forest systems. The Forestry Commission is now planning to do this at a number of important sites.

Despite some encouraging developments, the decline of the Pearl-bordered Fritillary is likely to continue unless conservation action can be accelerated. In Scotland, there is a growing threat from recent efforts to extend native woodland and encourage natural regeneration by fencing. Many large colonies are currently at risk because the edge habitats where they breed are deteriorating rapidly now that deer and other grazing animals have been excluded. Although the extension of native woodland is undoubtedly important to conserve other wildlife, these measures will inevitably lead to the loss of woodland clearings and edge habitats. Similar factors caused the decline of many woodland butterflies in England and the practice needs to be modified urgently.

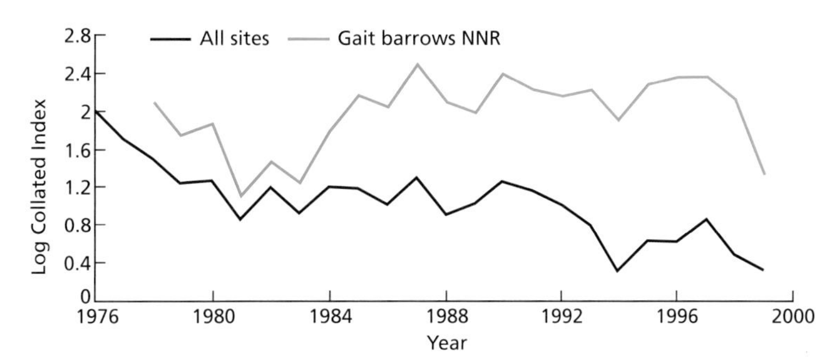

Pearl-bordered Fritillary abundance on BMS sites showing overall decline and the positive effects of conservation management at Gait Barrows National Nature Reserve, Lancashire.

Key references

Barnett and Warren (1995*e*); Brereton *et al.* (1999); Brereton and Warren (1999); Joy (1998*a*).

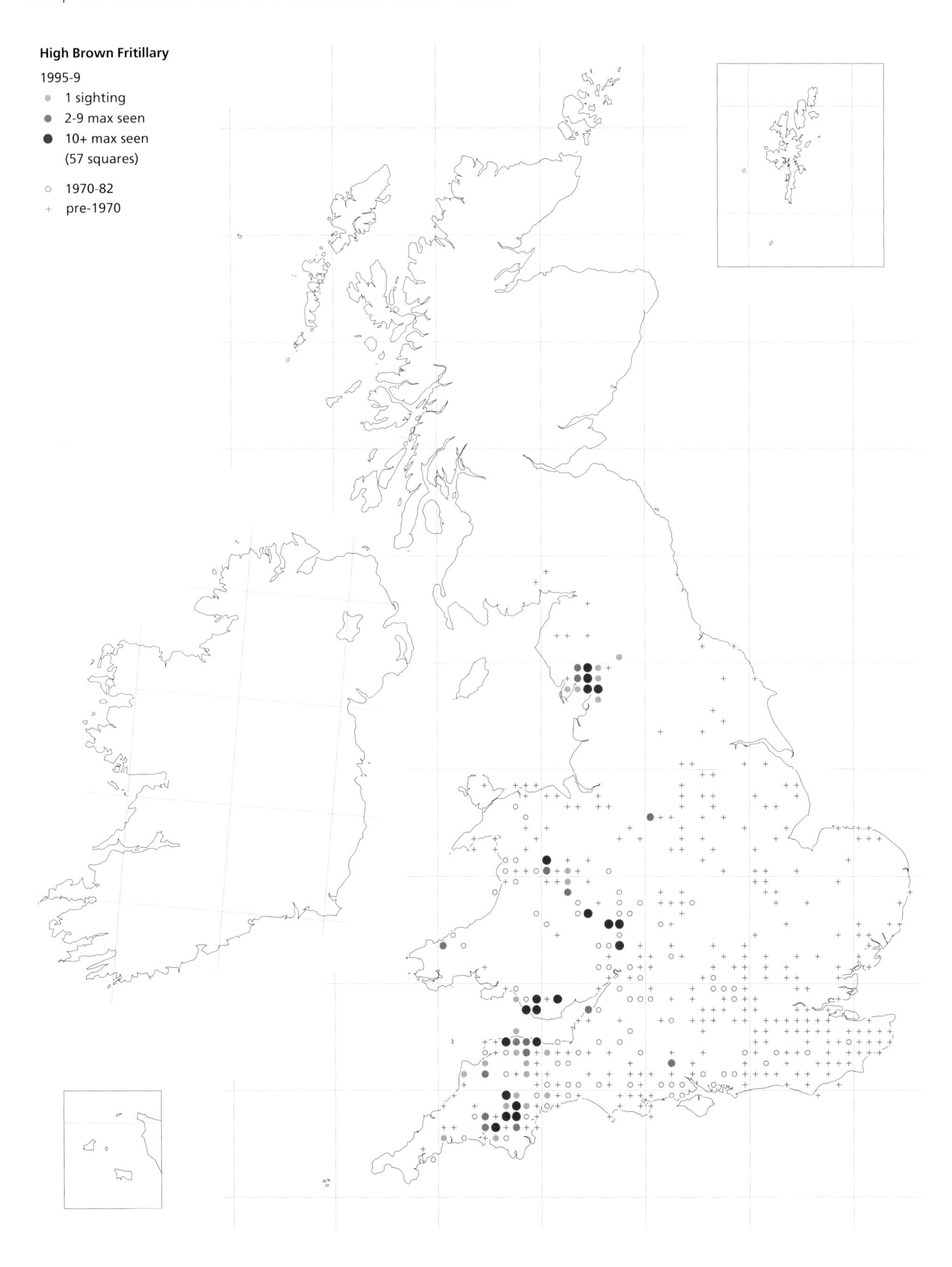
High Brown Fritillary
1995-9
1 sighting
2-9 max seen
10+ max seen
(57 squares)
1970-82
pre-1970

High Brown Fritillary *Argynnis adippe*

Resident

Range declining.

Conservation status

UK BAP status: Priority Species.
Conservation priority: high.
European status: not threatened.
Fully protected in Great Britain.

European/world range

Occurs widely through Europe and across temperate Asia to Japan. Although locally abundant in Europe, it has declined in at least eight countries.

This large, powerful butterfly is usually seen flying swiftly over the tops of bracken or low vegetation in woodland clearings. In flight, the males are almost impossible to separate from those of the Dark Green Fritillary, which often share the same habitats. However, both species frequently visit flowers such as thistles and Bramble where it is possible to see their distinctive underside wing markings. The High Brown Fritillary was once widespread in England and Wales but since the 1950s has undergone a dramatic decline. It is now reduced to around 50 sites where conservationists are working to save it from extinction.

Foodplants

Common Dog-violet (*Viola riviniana*) is used in all habitats, but Hairy Violet (*V. hirta*) is also used in limestone areas. It may occasionally use Heath Dog-violet (*V. canina*) and Pale Dog-violet (*V. lactea*).

Habitat

Two main habitats are used: bracken-dominated habitats or grass/bracken mosaics, and limestone rock outcrops, usually where scrub or woodland has recently been cleared or coppiced. Formerly the butterfly occurred in woodland clearings, probably where bracken was also present.

Life cycle and colony structure

There is one generation per year, with adults flying mainly from mid-June until early August in most localities and slightly later in north-west England, from late June to late August or even early September.

The High Brown Fritillary overwinters as **eggs** which are laid singly on leaf litter (often dead bracken), or amongst moss growing on limestone outcrops. The **larvae** hatch in early spring and spend long periods basking on dead bracken or in short, sparse vegetation. The temperatures in these microhabitats can be 15–20 °C higher than in surrounding grassy vegetation, allowing the larvae to develop quickly in otherwise cool spring weather. The larvae are cryptically coloured and have feathered brown spines that make them resemble dead bracken fronds. They **pupate** close to the ground under dead bracken or leaves.

The High Brown Fritillary forms discrete colonies that rarely contain more than a few hundred adults. However, the adults are highly mobile and are often seen feeding on flowers 1–2 km away from main breeding areas. They seem to travel freely between such areas and the flight area of many colonies is over 50–100 ha. Marking studies have also shown that they can move several kilometres between colonies and in favourable years the butterfly may establish

High Brown Fritillary larva basking on bracken litter.

temporary colonies in smaller or marginally suitable habitats.

Distribution and trends

The High Brown Fritillary was widespread and locally common across much of England and Wales until the 1950s, but has since undergone one of the most sudden and widespread declines of all British butterflies. It has never been recorded from Ireland, the Isle of Man, or the Channel Islands, and the historic records from Scotland are thought to be doubtful, possibly misidentifications of the Dark Green Fritillary.

Intensive surveys conducted through the 1980s and 1990s found that it is now reduced to around 50 definite localities. There are sporadic records of a few individuals from another 30 sites, but it is unclear whether these represent very small colonies or migrants from other sites. The butterfly's overall loss of range at the 10 km grid square level has been over 85%, and it has disappeared from 77% of the squares recorded in 1970–82. The decline at a population level is likely to have been even more dramatic.

The butterfly's main strongholds are now on the bracken-covered fringes of Dartmoor and Exmoor, and the carboniferous limestone hills around Morecambe Bay in north-west England. The latter sites contain a mixture of breeding habitats ranging from dense bracken on deeper more acidic soils to sparse grassy vegetation on thin soils, typically where scrub has been cleared recently or woodland has been coppiced. Scattered colonies also survive on other bracken-covered slopes in the Malvern Hills and a few commons in Herefordshire.

About five sites are known in Wales, three of them discovered only during detailed surveys in the 1990s. A few other colonies exist in remote areas, but thorough surveys have failed to discover the species in most formerly occupied regions.

European trend

The High Brown Fritillary is quite common and its range is apparently stable in many European countries. Nevertheless it has undergone severe declines in Austria and Belgium(>50% decrease in 25 years); Denmark, Germany, and Luxembourg (25–50% decrease); and lesser declines in the Czech Republic, Latvia, and Slovakia (15–25% decrease).

Interpretation and outlook

Until the 1950s, the High Brown Fritillary was regarded primarily as a butterfly of woodland clearings, and early books typically refer to it as being common in most large woods. However, there are few records of egg-laying to confirm the precise areas it once used for breeding. It was probably associated with traditional coppicing, which was the dominant form of woodland management over most of lowland Britain until the early half of the twentieth century. In many regions, colonies formerly occurred in wood pasture, and on acidic soils and commons where bracken was abundant. These habitats have probably always been important, though not recognized as such.

The habitat specificity of the High Brown Fritillary is thought to be caused primarily by the temperature requirements of the larvae. Throughout the spring, they bask on patches of dense bracken to get warm, only moving onto nearby violets for short periods to feed. On hot days they do not need to bask but hide in the deeper bracken litter. The most favoured breeding habitat is thus in dense, broken-down bracken litter with abundant violets, and interspersed with grassy patches containing Bramble or thistles that provide nectar for the adults. Similarly warm ground conditions characterize the limestone habitats, where most egg-laying takes place on or near small rock outcrops or very sparse vegetation.

The High Brown Fritillary seems to require fairly large flight areas and colonies are rarely found in habitat patches of less than 5 ha, although the actual breeding areas within these patches may be far smaller. This requirement for relatively extensive habitats, combined with the highly specific larval temperature requirements, may explain why the butterfly has declined more rapidly than other violet-feeding fritillaries associated with coppicing.

Nearly all bracken-dominated habitats that still support substantial populations are part of extensive livestock grazing systems, involving cattle, sheep, or ponies (or more usually a mixture of two or more). These animals break up the dead bracken litter and create an intimate mosaic of grass and bracken where violets can flourish. If grazing ceases, the bracken litter quickly builds up and eventually eliminates violets as well as the butterfly.

The cessation of traditional grazing has been a significant cause of loss in the west of England and Wales since the 1950s. In Wales, many bracken habitats are still grazed by sheep, but it is thought that they do not usually break up the litter enough to create good breeding habitat. Grazing by cattle and/or ponies is preferable. Some ungrazed sites are being managed by rotational cutting of bracken either with a tractor-drawn flail or 'bracken breaker', with promising early results.

The High Brown Fritillary is being actively conserved on several nature reserves and now that its requirements are reasonably well known there have been some notable successes, especially in the Morecambe Bay area. The requirements of the butterfly are now also recognized within several important Environmentally Sensitive Areas (e.g. those covering Dartmoor, Exmoor, and the Lake District) where there are payments for maintaining extensive grazing regimes and where bracken management is strictly controlled.

Butterfly Conservation co-ordinates a national conservation programme for this highly endangered species, in conjunction with other conservation organizations. The immediate aim is to halt its decline and restore habitats within its present range to allow natural spread. The longer term aim is to restore habitats elsewhere in its former range, but this is made difficult by their fragmentation and often lengthy periods of neglect. There are signs that the butterfly's rapid decline has slowed in recent years, thanks in part to this substantial conservation effort but it remains a highly vulnerable species which will require continued attention for the foreseeable future.

Key references

Barnett and Warren (1995*f*); Joy (1998*a*); Warren (1992*b*, 1994*a*, 1995*a*); Warren and Oates (1995).

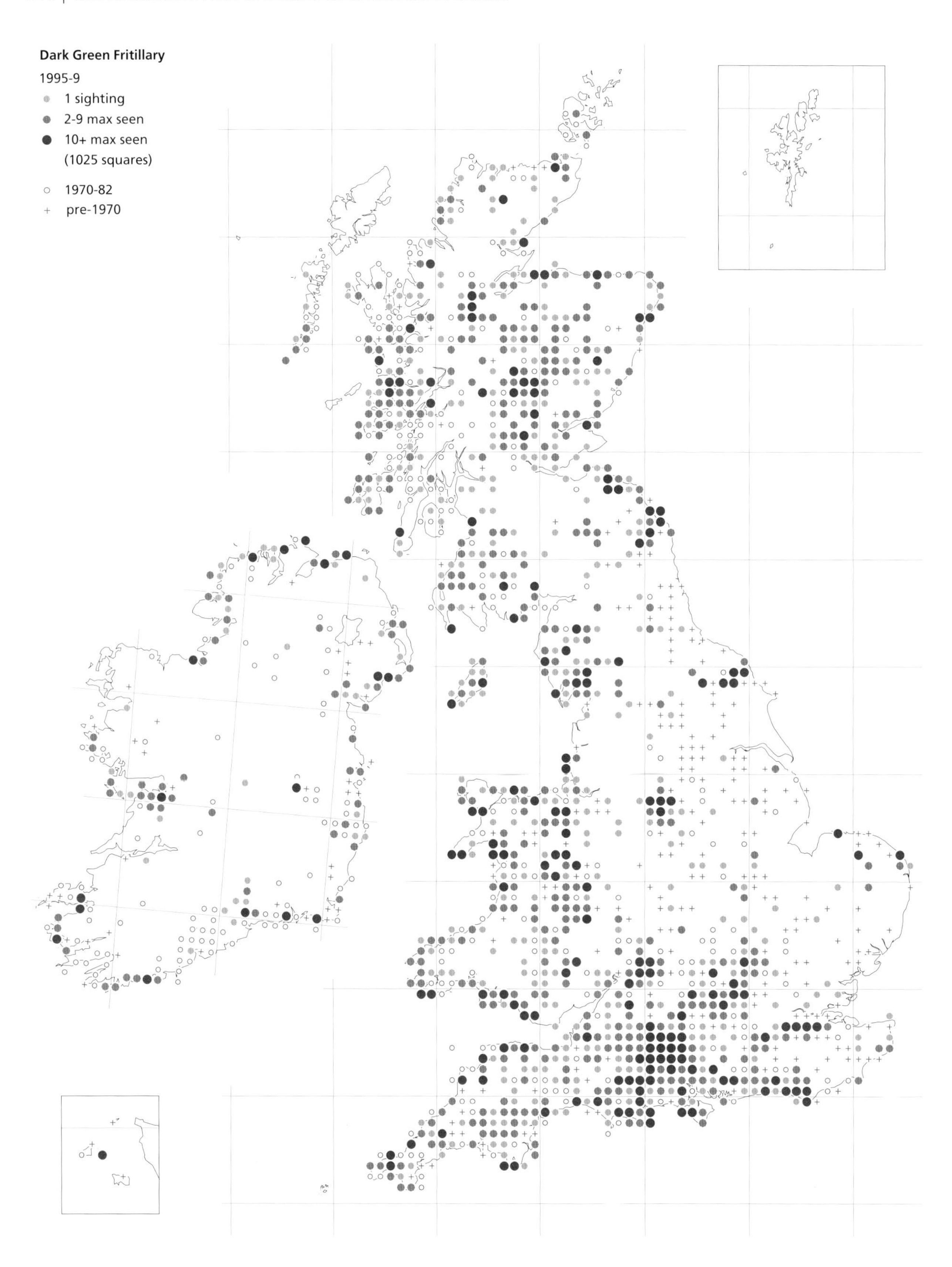
Dark Green Fritillary
1995-9
1 sighting
2-9 max seen
10+ max seen
(1025 squares)
1970-82
pre-1970

Dark Green Fritillary *Argynnis aglaja*

Resident

Range declining in some areas.

Conservation status

UK BAP status: not listed.
Butterfly Conservation priority: low.
European status: not threatened.

European/world range

Throughout Europe as far north as the Arctic Circle and eastwards across Asia to China and Japan. Range appears stable through much of Europe, but declines reported in at least eleven countries.

This large and powerful butterfly is one of our most widespread fritillaries and can be seen flying rapidly in a range of open sunny habitats. The males look similar to the High Brown Fritillary, which is far rarer but sometimes flies with them on bracken-covered hillsides. The two can be distinguished from the underwing markings, visible when they are feeding on flowers such as thistles. Although the Dark Green Fritillary is still locally abundant in some regions, it has declined in many others, notably central and eastern England.

Foodplants

Common Dog-violet (*Viola riviniana*) is used in many habitats but Hairy Violet (*V. hirta*) is used on calcareous grasslands, and Marsh Violet (*V. palustris*) on moorland and wetter habitats in the north and west. Other violets may be used occasionally.

Habitat

The butterfly occurs in a range of flower-rich grasslands, often with patches of scrub, including: coastal grassland, dunes and scrub; chalk and limestone grassland; moorland and wet flushes; acid grassland with bracken; and occasionally woodland rides and clearings.

Life cycle and colony structure

The Dark Green Fritillary is single brooded, with adults flying from early June until late August. In warmer and more southerly locations the peak is usually from mid-June to mid-July, but it can be several weeks later at cooler, more northerly sites (see p. 396).

The **eggs** are laid singly either on the foodplant or more usually on a nearby plant, dead leaves, or dead bracken. Immediately after hatching, the **larvae** enter hibernation amongst the dead grass or leaf litter. They begin feeding on the first warm days of spring and, in cool conditions, the larvae bask on top of the vegetation. In warmer weather, they remain concealed and feed in short bouts on young growths of violet.

The Dark Green Fritillary breeds in cooler vegetation than other violet-feeding fritillaries (except possibly the Small Pearl-bordered Fritillary) and selects fairly tall grassy vegetation (8–15 cm), often next to patches of scrub where there are abundant large foodplants. The larvae **pupate** in the leaf litter or within grass tussocks.

The adults are highly mobile and the butterflies tend to occur at low densities over large areas within which there are small pockets of suitable breeding habitat. On the best sites, with greater concentrations of breeding habitat, adults can become very numerous and colonies tend to occupy more discrete areas. There has only been one detailed mark–recapture study, on bracken slopes on Dartmoor, where there were three

Dark Green Fritillary larva.

distinct populations of 300–500 adults separated by 1–2 km. The butterflies moved freely over hundreds of metres within each breeding area, but most seemed to stay within their colony. A few adults were found to have moved between colonies and the butterfly has occasionally been found up to 5 km from known breeding areas. This indicates that the species has reasonable powers of colonization and that nearby colonies are probably linked, though this has yet to be quantified.

Distribution and trends

The Dark Green Fritillary is the most widespread of the fritillaries and occurs widely in Britain and Ireland, although it is rarely found in great numbers. It has a distinctly coastal distribution through much of its range, especially in Ireland, but is found through the interior of much of Wales, northern Scotland, and southern England. It also occurs on the Isle of Man and a number of Scottish islands including Arran, Bute, Islay, Jura, the Outer Hebrides (North and South Uist, and Barra), and Orkney. It is under-recorded in Scotland, Wales, and on the south coast of Ireland, and is probably more widespread there than is shown.

Its current strongholds include the chalk downs of southern England, especially Salisbury Plain; the Cotswold limestones; the bracken slopes of Dartmoor and Exmoor; the Peak District; the coast of Wales; the commons and uplands of north-east Wales; the west coast of Scotland; and the Scottish Highlands. In Ireland, important areas are coastal sand dunes, offshore islands (such as Cape Clear and the Aran Islands), and inland bogs, eskers, and limestone pavements.

Over the past hundred years, the butterfly has declined substantially in England, especially in the north and east where it is now highly restricted. This decline began in the early part of the twentieth century but has accelerated since the 1950s. In several parts of its range (e.g. north-east England and on southern downland) there was evidence of a rapid decline during the 1960s and 1970s, but it has since recovered and may now be spreading a little. The recent survey shows that it may have disappeared from 36% (234) of the 10 km squares recorded in Britain in 1970–82, but over 500 additional squares have been recorded, mainly in Scotland and Wales. Results from the Butterfly Monitoring Scheme show no overall trend over the last 20 years, and there has been considerable variation from site to site.

Dark Green Fritillaries are very similar in size and appearance to High Brown Fritillaries, especially the males. They are best identified by the markings on the underside of the hind wing, the former having a greenish suffusion near the base and lacking the row of red-ringed spots present on the High Brown. The Dark Green also has rounder, less pointed forewings and the males lack any noticeable scent-brands. The females have heavier black markings around the wing margins. The butterfly is variable in appearance and different races have been described, for example *scotica* described from Scotland, which has darker markings and brighter silver spots.

European trend

The Dark Green Fritillary appears to be stable in several countries but it has declined severely in Austria, Belgium, and the Netherlands (>50% decrease in 25 years), Denmark (25–50% decrease), and to a lesser extent the Czech Republic, Germany, Latvia, Moldova, Slovakia, Sweden, and European Turkey (15–25% decrease).

Interpretation and outlook

The Dark Green Fritillary has not undergone the dramatic declines of many other fritillaries, probably because it occurs in a wider range of habitats and has less demanding ecological requirements. It is able to breed in wetter and cooler climates than many other fritillaries and is the only one to have reached many of the offshore islands.

Moreover, although many populations have been lost due to the widespread loss of flower-rich habitats, some have benefited for a while where hill land has become neglected. The butterfly thrives in taller, transitional vegetation following reduction in grazing, but successional changes gradually cause such habitats to deteriorate. For example, many coastal sites are now covered with dense scrub while other abandoned sites have gradually lost their floristic richness. This may explain the delayed decline of the Dark Green Fritillary during the 1960s and 1970s. Similarly, the butterfly may have benefited temporarily from the loss of rabbits due to myxomatosis in the 1950s but began to decline as rabbit populations recovered in the 1980s and 1990s, especially on downland.

In certain parts of the country (mainly central and eastern England), flower-rich grasslands have become extremely scarce and the Dark Green Fritillary has become increasingly reliant on woodland clearings. It thrived in the large scale clearings that occurred on many ancient woodland sites during the post-war plantings, but many colonies died out when these grew up in the 1970s and 1980s. A few colonies remain in woodland sites where there are still regular, large scale clearances and numerous wide, sunny rides.

Striking a balance of management for the Dark Green Fritillary is not easy but it seems to thrive under low intensity grazing regimes, both with cattle and sheep (and, more rarely, ponies). Its essential requirements are for a lightly grazed (or recently neglected) sward that produces a patchy turf and preferably contains areas of scrub. Rotational scrub clearance has boosted numbers on several nature reserves, as has the restoration of light grazing or relaxation of heavy grazing.

The chief future threat is the continuing loss of flowery grassland and moorland, which will continue to fragment and isolate breeding habitats. Although the species appears to be quite mobile, it often occurs at low densities and will eventually succumb to fragmentation. Other threats include the continued recovery of rabbit populations, which may reduce habitat suitability on chalk downland. In Ireland the chief threats are from intensive sheep grazing and afforestation.

Key references

Joy (1998*a*); Thomas (1991); Warren (1992*b*, 1994*a*).

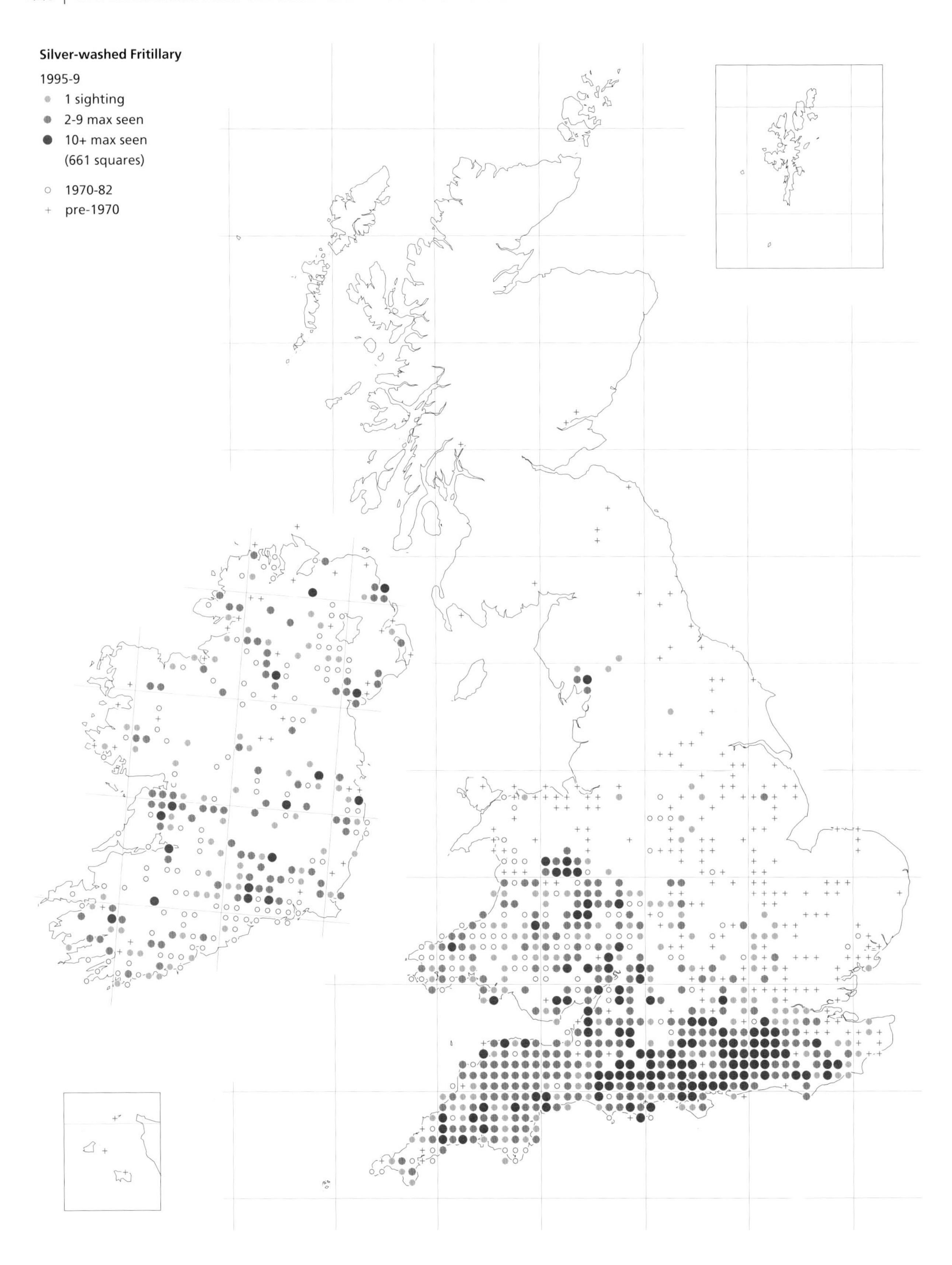
Silver-washed Fritillary
1995-9
1 sighting
2-9 max seen
10+ max seen
(661 squares)
1970-82
pre-1970

Silver-washed Fritillary *Argynnis paphia*

Resident

Decline and slight re-expansion of range.

Conservation status

UK BAP status: Species of Conservation Concern.
Butterfly Conservation priority: low.
European status: not threatened.

European/world range

Across Europe, to 62 °N in Sweden and throughout the Mediterranean to North Africa, and eastwards across to China and Japan. Declining in several European countries but spreading northwards in Sweden and Finland.

The swooping flight of this large and graceful butterfly is one of the most beautiful sights to be found in woodland during high summer. It is named after the silver streaks on the underside which can be viewed as it stops to feed on flowers such as Bramble. Although the butterfly is seen mostly in sunny glades and rides, it actually breeds in the shadier parts of adjacent woodland. In southern England, a small proportion of females have wings that are bronze-green, known as the form *valezina*. The Silver-washed Fritillary declined during the twentieth century, especially in England and Wales, but has spread noticeably during recent decades.

Foodplants

The main foodplant is Common Dog-violet (*Viola riviniana*) growing in shady or semi-shady positions on the woodland floor.

Habitat

The butterfly breeds in broad-leaved woodland, especially oak woodland or woods with sunny rides and glades. It occasionally uses mixed broad-leaved and conifer plantations and, in parts of south-west England and Ireland, also breeds in wooded hedgerows and sheltered lanes near to woods.

Life cycle and colony structure

The Silver-washed Fritillary is single brooded with adults flying from mid-June until late August. When egg-laying, the females fly into semi-shady woodland and flutter slowly around the bases of trees, alighting to explore violet growth on the woodland floor. The **eggs** are laid singly in the crevices of tree bark or amongst moss, usually on the north or west side of the trunk. Most eggs are laid 1–2 m above the ground but may be laid as high as 6 m up a tree, or on other substrates such as dead gorse. The **larvae** hatch after a few weeks but immediately enter hibernation. In the spring, they drop or crawl to the ground where they move over sparse vegetation seeking out violets on which to feed. During the day, they bask in patches of sunlight amongst the leaf litter, and when warm can move rapidly across the ground. Although they bask openly, they are well camouflaged and difficult to find. The few **pupae** that have been found in the wild have been 1–2 m above the ground, suspended from leaves or twigs.

Silver-washed Fritillary adults are highly mobile and fly rapidly between clearings. However, the butterfly forms discrete colonies within individual woods and, in most regions, rarely seems to travel away from such breeding areas or to move between adjacent woods. The exception to this is in south-west Britain where they are regularly seen flying along wooded lanes.

Silver-washed Fritillary larva basking on a dead leaf.

In hot summers, individuals have been seen considerable distances from known breeding areas in several regions. The recent modest spread of the butterfly also implies that it does have some colonizing ability, though this has yet to be quantified.

Distribution and trends

The Silver-washed Fritillary was once widespread in England and Wales, and there are nineteenth century records from Scotland. It is scattered widely in Ireland, though it is rarely abundant, and is absent from the Isle of Man and the Channel Islands.

The butterfly is comparatively well recorded in Britain where it declined substantially during the twentieth century, especially at the northern and eastern edges of its range. It died out in Scotland during the early 1900s and from much of Yorkshire and East Anglia by the 1940s. The decline has been even more rapid since the 1950s: it became extinct through much of central and eastern England by the late 1950s (for example in Suffolk and Bedfordshire) and became rarer in many other areas, including the New Forest, which had long been regarded as its major stronghold.

Today, the butterfly's main population centres are in the oak woods of south-west England, and the large woodland complexes of the Weald (Surrey and Sussex), Hampshire, Dorset, south Wiltshire, Gloucestershire, and Herefordshire. In Ireland, its distribution is correlated with the presence of oak woods and appears to be fairly stable. It is most numerous in the counties of Wicklow, Waterford, Cork, and Kerry, and in wooded parts of Sligo, Donegal, Down, Fermanagh, and Antrim.

Throughout the twentieth century, the distribution of the Silver-washed Fritillary fluctuated, especially at the edge of its range. It spread during the 1930s and 1940s (for example in Derbyshire and Suffolk), then declined during the 1950s and 1960s, reappearing in several counties during the large population build-up in the hot summers of 1975 and 1976. During the 1990s there was another expansion and it reappeared in Cumbria where it had been absent for almost 20 years. The current survey received records from over 200 new 10 km grid squares in Britain where it was not recorded in the 1970–82 survey, indicating a major re-expansion. Data from the Butterfly Monitoring Scheme show that this has been accompanied by only a slight increase in population size at most monitored sites.

The butterfly's spread is difficult to interpret in some areas as the situation has been clouded by a number of deliberate releases. For example large numbers were released in Suffolk in 1940, about the time of its resurgence, and it was released at several sites in Warwickshire in the 1990s. There are also known to have been releases in Bedfordshire, and clandestine releases cannot be ruled out in both Cumbria and Northamptonshire, where this and several other species have recently reappeared. Nevertheless, the expansion in so many areas appears to represent a genuine recent spread against the backcloth of a general decline during the twentieth century.

European trend

The distribution of the Silver-washed Fritillary appears to be stable in many European countries but it has declined seriously in the Netherlands (now extinct), Austria (>50% decrease in 25 years), Belgium, Denmark, and Latvia (25–50% decrease). In contrast, it seems to be spreading at the northern edge of its range in Finland (>100% increase in 25 years) and Sweden (25–100% increase).

Interpretation and outlook

The decline of the Silver-washed Fritillary is undoubtedly linked to the changing management of woodland during the past century; not just the cessation of coppicing, but the general reduction in active broad-leaved woodland management. However, other factors, such as climatic changes, may have determined the timing of contractions and probably its more recent spread. The butterfly is often regarded as a species of mature woodland but it also requires abundant open spaces where it spends long periods feeding on flowers such as Bramble. It can also be abundant in actively coppiced woods, where it lays its eggs in the shady edges of recently cut areas.

The butterfly seems to do best in well-thinned woods where there is dappled shade within the high forest canopy (around 70–90% canopy cover), or where there are abundant edges such as along broad rides or around glades. For this reason it does particularly well in western oak woods which often grow on steep slopes and where there are numerous small canopy gaps. The butterfly also occurs in several ancient broad-leaved woods that have been cleared and replanted with conifers, although breeding is now mostly confined to the remaining broad-leaved component. Sadly, many modern woods are extremely shady and only capable of supporting small populations. Its future is thus linked closely with the continuation of active woodland management and the maintenance of sunny rides.

The Silver-washed Fritillary appears to be particularly sensitive to climatic changes and tends to spread during warmer periods and contract during cooler ones. It waxed and waned several times during the last 200 years and its recent modest expansion may be temporary. However, there has been a northward movement of the species in Scandinavia, at the northern edge of its range in Europe, and the present shift may be attributable to warming as a result of global climate change. In Britain this has been against the trend in habitat loss and deterioration which caused its earlier decline. The position may become clearer over the next two decades, provided recording continues at the present level.

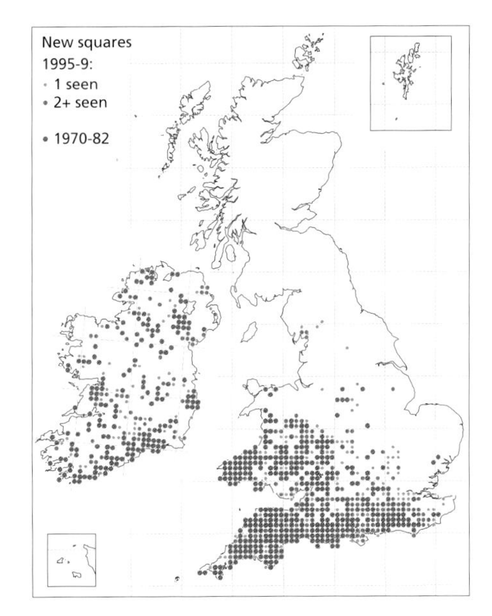

The expansion of range of the Silver-washed Fritillary in southern Britain since 1970–82.

Key references

Thomas and Snazell (1989); Thomas (1991); Warren (1992*b*).

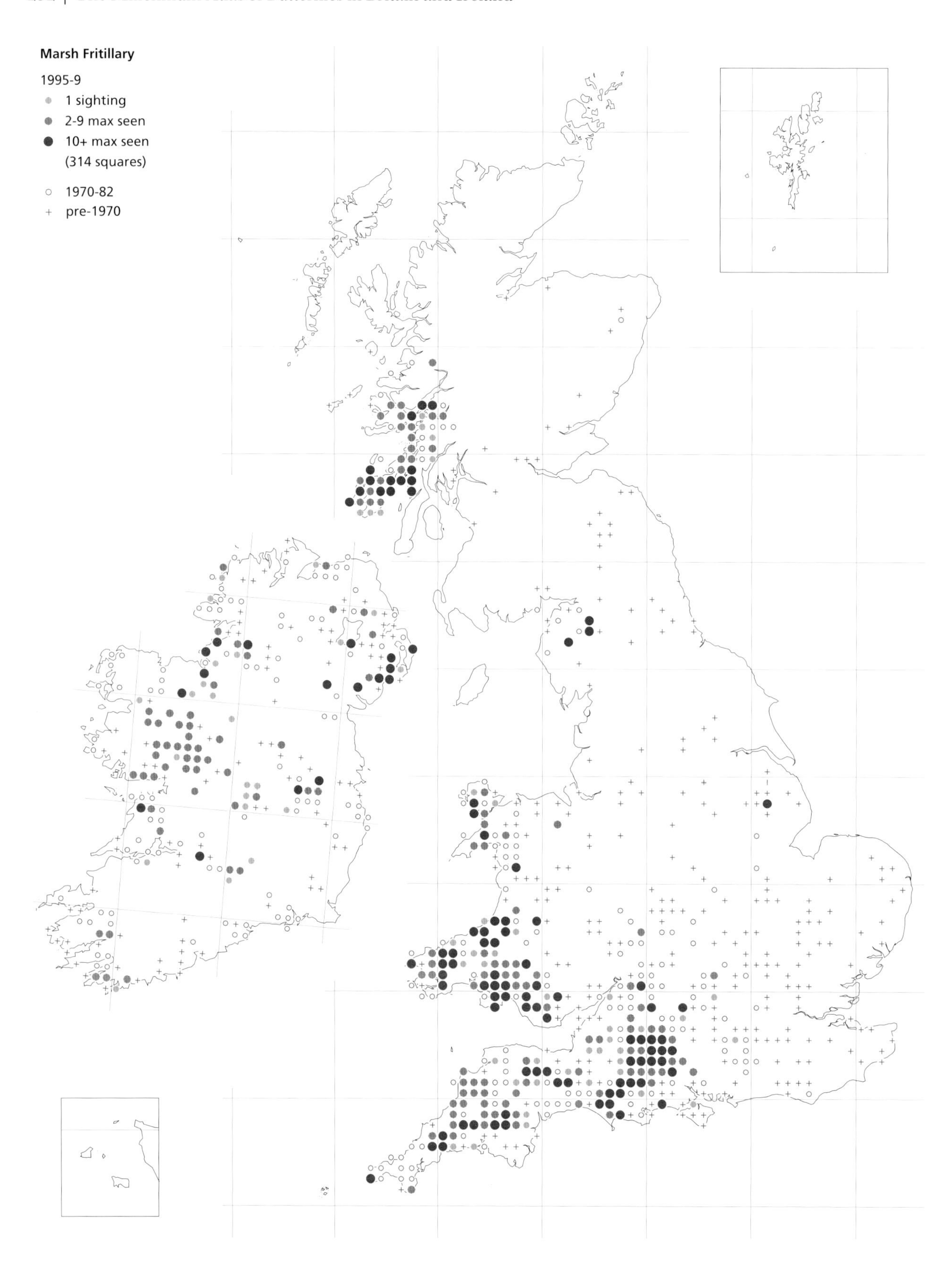
Marsh Fritillary
1995-9
1 sighting
2-9 max seen
10+ max seen
(314 squares)
1970-82
pre-1970

Marsh Fritillary *Euphydryas (Eurodryas) aurinia*

Resident

Range declining.

Conservation status

UK BAP status: Priority Species.
Butterfly Conservation priority: high.
European status: vulnerable.
Fully protected in Great Britain and Northern Ireland. Bern Convention (Annexe II) and EC Habitats and Species Directive (Annexe II).

European/world range

Widely distributed across Europe to Asia as far east as Korea. Declining in most European countries.

The wings of this beautiful butterfly are more brightly patterned than those of other fritillaries, with more heavily marked races being found in Scotland and Ireland. The larvae spin conspicuous webs that can easily be recorded in late summer. The Marsh Fritillary was once widespread in Britain and Ireland but has declined severely over the last century, a decline mirrored throughout Europe. Its populations are highly volatile and the species probably requires extensive habitats or habitat networks for its long term survival.

Foodplants

The main foodplant is Devil's-bit Scabious (*Succisa pratensis*). On calcareous grassland, it occasionally uses Field Scabious (*Knautia arvensis*) and Small Scabious (*Scabiosa columbaria*).

Habitat

The Marsh Fritillary breeds in open grassy habitats, especially damp grassland dominated by tussock-forming grasses; calcareous grassland (usually on west- or south-facing slopes in England or on eskers in Ireland); and heath and mire vegetation with Devil's-bit Scabious. Temporary colonies may also exist in large (>1 ha) woodland clearings and in other grasslands.

Life cycle and colony structure

The butterfly is single brooded with adults flying from mid-May to mid-July, usually with a peak around the end of May to mid-June. The females lay **eggs** in large batches: the first may contain up to 350 but subsequent batches are smaller (*c.* 50–150). Females select larger, more prominent foodplants, or patches of shorter vegetation where the foodplant is abundant. The **larvae** are gregarious and spin a protective web that becomes conspicuous by the end of August. The larval nests occur in intermediate length swards where the turf is 8–20 cm tall, although shorter swards of 5–15 cm can be used where foodplants are abundant (for example on downland and coastal heaths of western Scotland). Many recent surveys have been conducted by counting such autumn webs.

The larvae overwinter in their fourth instar, in a small web close to the ground, usually within a dense grass tussock. They emerge in early spring and can be seen in clusters of up to 150 black larvae, basking in weak sunshine. They eventually become solitary and can disperse widely over the breeding habitat. They **pupate** low in grassy vegetation, either deep within grass tussocks or amongst dead leaves.

The Marsh Fritillary forms compact colonies, often on small patches of habitat: a survey found that over half of sites were less than 2 ha and few were larger than 20 ha.

Marsh Fritillary larvae basking in spring.

Marking studies have shown that most adults rarely fly more than 50–100 m but a proportion seem to diperse further. Initially females rarely move far, but they become more mobile after laying their first egg batch and have been known to colonize sites 10–15 km away. Anecdotal evidence suggests that egg batches may be smaller and colonies more dispersed in parts of Ireland and western Scotland. The butterfly is renowned for the large fluctuations in population size that make it highly prone to local extinction, but in good years it can spread and colonize new sites as well as less suitable habitat. It is therefore thought to exist as metapopulations comprising groups of local populations connected by occasional dispersal.

Distribution and trends

The Marsh Fritillary was once widespread in Britain and Ireland but has declined severely and is now extinct in the eastern half of Britain. There have also been substantial losses in the rest of its range, including Ireland. Its decline began in the late eighteenth century but accelerated during the twentieth century, especially after 1950. Although the general trend has been downward, there is evidence that the butterfly spread onto southern chalk downland in the early twentieth century. There are few records from this well-documented habitat until the 1900s when it seems to have colonized downland, especially in Wiltshire and Dorset.

Overall, the butterfly's British range has contracted by over 60% and a review in 1990 found that colonies were still being lost at over 10% per decade. A total of 432 separate colonies were identified in the UK: 228 in England, 111 in Wales, 58 in Northern Ireland, and 35 in Scotland. Systematic surveys have since been conducted in most of Britain, though it is still under-recorded in some remote areas. Current strongholds are in the Culm grasslands of north Devon and Cornwall; the extensive chalk grassland on the military ranges of Salisbury Plain, the Rhos pastures and heathy commons of south Wales, and the island of Islay in western Scotland.

In Ireland, the situation is probably better, but further surveys are needed and the butterfly is known to be under pressure from drainage and intensive sheep grazing. Its strongholds are in counties Fermanagh, Sligo, Donegal, and the region west of the River Shannon, where many linear colonies exist on road verges across cutaway bogs. There are also good populations on some northern sand dunes and on limestone in the Burren.

The Marsh Fritillary is easy to rear in captivity and has been released in large numbers. These releases are rarely reported and may obscure the butterfly's true status in parts of southern England. The recent record for Lincolnshire, outside its current range, is of an introduced colony.

European trend

The butterfly is declining in most European countries and there have been severe declines in the Netherlands (now extinct); Belgium, Denmark, Germany, Latvia, and Poland (>50% decrease in 25 years); and Austria, Finland, Luxembourg, and Sweden (25–50% decrease).

Interpretation and outlook

The decline of the Marsh Fritillary is linked to three main factors. First, there was a massive loss of unimproved grassland through ploughing and agricultural 'improvement': for example the area of damp pasture in south-west England was reduced by 92% during the twentieth century and over 60% of British chalk downs were lost.

Second, the Marsh Fritillary has suffered greatly from changes in management of the fragments that remain. The species prefers light grazing levels and cannot tolerate heavy grazing, especially during the summer. Although populations can thrive in the absence of grazing, sometimes for even a decade or more, abandoned sites eventually become unsuitable either through loss of foodplant or the spread of rank grasses or scrub. In most of its range, the butterfly is associated with extensive cattle (or pony) grazing rather than sheep grazing as the latter tend to graze the foodplant too tightly. The butterfly's spread on to chalk downland in southern England may even have been linked to the switch from sheep to cattle grazing following the agricultural depression of the late nineteenth century. The only exception to this preference seems to be where sheep graze at low intensity in extensive pastures in western Scotland and Ireland.

Third, Marsh Fritillary populations seem to function at a landscape scale, spreading over large areas in some years and contracting to core breeding patches in others. These cycles are partly driven by severe larval mortalities caused by hymenopteran parasitoids but they may be strongly influenced by changing management. Whatever the cause, the species is extremely prone to local extinctions, which can be balanced by its ability to colonize sites, but only within a range of about 10–15 km. It seems to have survived only in regions where numerous habitat patches still exist. In many regions, potential habitats are too small and isolated to support the species. Although over 80 attempted reintroductions have been documented, all have ultimately failed.

The Marsh Fritillary poses a huge challenge to conservation not only because suitable management is increasingly difficult to maintain under modern farming systems, but because it requires landscape scale measures to maintain extensive habitats, or networks of habitats. Research is now underway to establish how large the patch networks need to be. Recent conservation schemes, such as Environmentally Sensitive Areas, provide support for maintaining low intensity grazing regimes in several important regions for the Marsh Fritillary. Although there are doubts about whether such schemes can ensure suitable grazing systems in the long term, they are promising models that could be applied elsewhere in Europe.

Key references

Barnett and Warren (1995*g*); Hobson (2000*a*,*b*); Lavery (1993); Porter (1982); Warren (1994*b*).

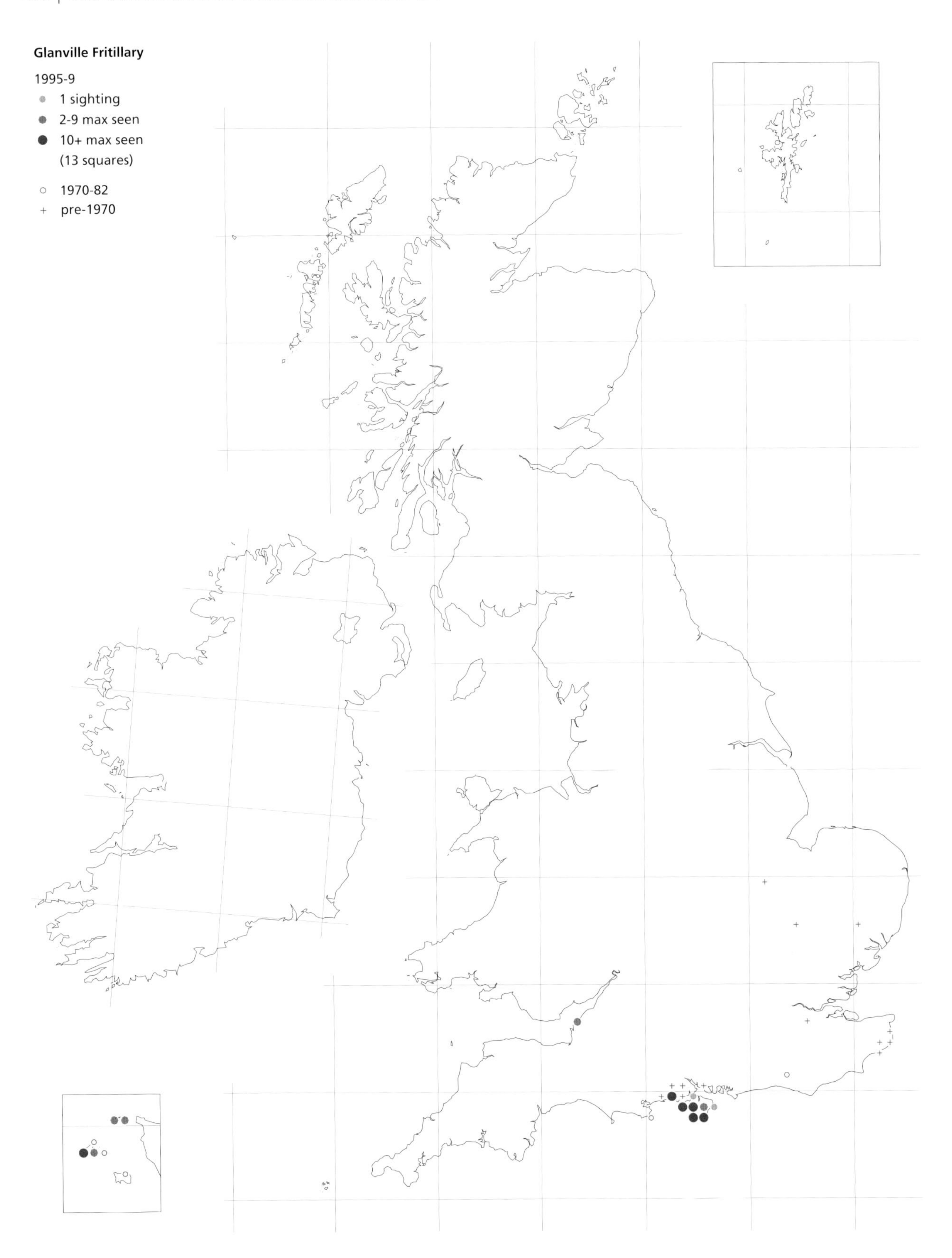
Glanville Fritillary
1995-9
1 sighting
2-9 max seen
10+ max seen
(13 squares)
1970-82
pre-1970

Glanville Fritillary *Melitaea cinxia*

Resident

Range stable.

Conservation status

UK BAP status: Species of Conservation Concern.
Butterfly Conservation priority: medium.
European status: not threatened.
Protected in Great Britain for sale only.

European/world range

Widespread and common through much of Europe to 61 °N in Scandinavia and eastwards to Asia. Severe declines reported in many European countries.

The Glanville Fritillary is virtually restricted to coastal landslips on the southern half of the Isle of Wight and on the Channel Islands. It was named after Lady Eleanor Glanville who was the first to capture British specimens in Lincolnshire during the 1690s. The status of the butterfly appears to have changed little in recent decades, though there has been some loss of habitat due to coastal protection measures. However, there are only a handful of core breeding areas and it remains a vulnerable species.

Foodplants

The main foodplant is Ribwort Plantain (*Plantago lanceolata*). Buck's-horn Plantain (*P. coronopus*) is used occasionally as a secondary foodplant by final instar larvae.

Habitat

Two habitat types are used: coastal grasslands either on undercliffs where there is soil slippage, deeply incised coastal river valleys (chines) with eroding sides, or cliff tops; and south facing chalk downland. The former tends to support the larger, more permanent colonies whereas the latter supports shorter-lived, although sometimes large colonies. There are eighteenth century records from woodland clearings in eastern England.

Life cycle and colony structure

The butterfly is single brooded, with adults flying from late May until early July. In very hot years there can be a small partial second brood in August. Females lay **eggs** in batches of 50–250 on the underside of leaves of Ribwort Plantain. Small plants are selected, generally growing at high densities in short, open vegetation on sheltered south facing slopes. The **larvae** live gregariously in dense webs spun over the foodplant from August to March. These webs are quite conspicuous and web counts are now the main method of population monitoring.

During their fifth instar, the larvae hibernate in large groups within a silken hibernaculum spun in longer grass. They resume feeding in March, basking for much of the time on top of the web in order to raise their body temperature as near as possible to 34 °C, the optimum temperature for development. The larvae become solitary during their sixth and final instar in April when they frequently move tens of metres in search of young, fresh plantains. They **pupate** in dense grass tussocks, hence the species requires a mosaic of short and tall grass.

The Glanville Fritillary is a colonial species that occurs in discrete habitat patches. However, the adults are fairly mobile and are periodically seen several kilometres away from known breeding areas. Larger scale

Larval nest of the Glanville Fritillary.

movements may lead to the establishment of new colonies further inland on the Isle of Wight and even on the mainland. However, large numbers of adults are known to have been released on the south coasts of Dorset and Hampshire and it is impossible to say if any mainland colonies result from natural colonization. Exact breeding areas on the Isle of Wight shift gradually in response to soil slippage and the species has a distinct metapopulation structure with a few more or less permanent colonies that from time to time give rise to smaller, often temporary satellite colonies.

Distribution and trends

The Glanville Fritillary is currently restricted to the Channel Islands and the southern half of the Isle of Wight, though a small colony has been recorded on the mainland coast of Hampshire in most years since 1990. There is a scatter of historical records for the butterfly elsewhere on the mainland, including the original site in Lincolnshire (1690s), a site near London (1760s), and sites on the coast between Folkestone and Sandwich in Kent where it persisted until the mid 1860s.

There have been many attempts, often clandestine, to introduce the Glanville Fritillary outside its known range, sometimes to quite unsuitable sites. The subsequent colonies nearly all died out within a few years and are not shown on the main map so as to avoid confusion with the species true distribution. The only exception is an unauthorized release in 1983, on a limestone promontory in Somerset, which established a small and isolated population that persisted throughout the current survey.

There has been little change in the overall distribution of the Glanville Fritillary on the Isle of Wight in recent years, although a few colonies were lost where the habitat became unsuitable following cliff stabilization and a lack of grazing. The butterfly has colonized three or four sites along the south face of the chalk escarpment since the early 1990s. However, populations are known to fluctuate widely and during the twentieth century it went through several periods of abundance and scarcity. For example, numbers were apparently low at the beginning of the twentieth century, but built up during the 1920s. Peaks were also recorded during the late 1940s, 1989, and in the late 1990s (though numbers were low again in 1999).

On the Channel Islands, the butterfly is currently found only on the south coast of Guernsey and on Alderney. It seems to have become extinct on Jersey following severe storms in 1988 and spells of cold weather during the last ten years.

The status of the butterfly on the Isle of Wight was studied in 1979 by surveying larval nest densities in all known localities. This identified 12 main sites, all but one of them on the west side of the island. Volunteers have resurveyed most of these annually since 1983, revealing huge fluctuations in numbers and a major resurvey in 1996–8 found that nearly all were still occupied. However, the changing pattern revealed by the counts suggests that there may be as few as 5 to 8 core populations that persist through troughs in abundance, indicating that it may be more vulnerable than was previously thought.

European trend

The Glanville Fritillary is apparently stable in the south of its range but has declined severely in several western and northern countries, including Austria, Belgium, Germany, Latvia, Lithuania, and the Netherlands (>50% decrease in 25 years) and the Czech Republic, Finland, and Norway (25–50% decrease). Lesser declines have been reported in Denmark, Poland, Sweden, and parts of north-west Russia (15–25% decrease).

Interpretation and outlook

The Glanville Fritillary requires a warm local microclimate in early spring. This restricts it to early successional habitats, where there is abundant Ribwort Plantain, in the Channel Islands and the extreme south of England. The most important habitats are those created by continuous natural erosion of the coastal zone. On the Isle of Wight, the erosion is due to soil slippage and slumping of the Wealden (Gault) clay. Such habitats are both temporary and variable in extent, but the butterfly should be able to survive provided this natural process is allowed to continue. Although there have been some attempts to stabilize the south coast of the Isle of Wight, only small patches of habitat have been affected. The Isle of Wight Council has now developed an imaginative coastal defence policy that should allow the south-western coastline to continue to erode naturally.

Other threats to the species come from an increase in rabbit grazing, which is reducing habitat suitability at a few sites, and from drought that can lead to sudden population declines. The impact of climatic warming on this species is consequently difficult to predict. The species could spread inland if the climate warms, but populations may be reduced if droughts become more frequent.

Perhaps the biggest difficulty in assessing the future of the Glanville Fritillary is that its true status may be overestimated unless there is accurate monitoring over a period of many years. Recent monitoring evidence, combined with current understanding of metapopulations, suggests that the butterfly is vulnerable to extinction due to natural dynamics of its habitats and populations. Efforts must therefore be made to maintain every possible potential patch of habitat to allow its natural pattern of extinction and colonization. As with many other species, there is no room for complacency amongst conservation bodies for such a localized species.

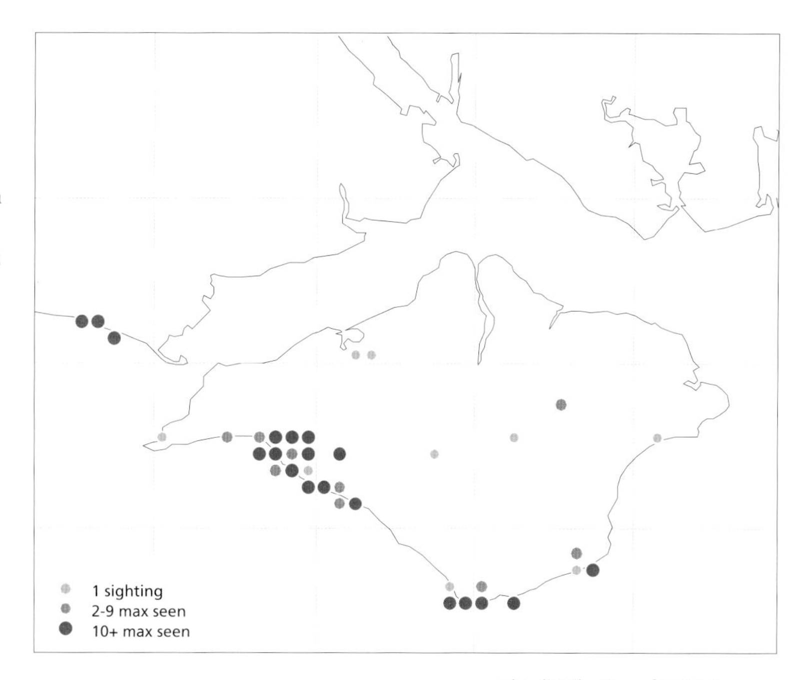

The distribution of 1995–9 records of the Glanville Fritillary at 1km square resolution on the Isle of Wight.

Key references

Bourn and Warren (1997*b*); Clarke (1993); Hanski *et al.* (1994); Kuussari (1998); Pope (1988); Thomas and Simcox (1982).

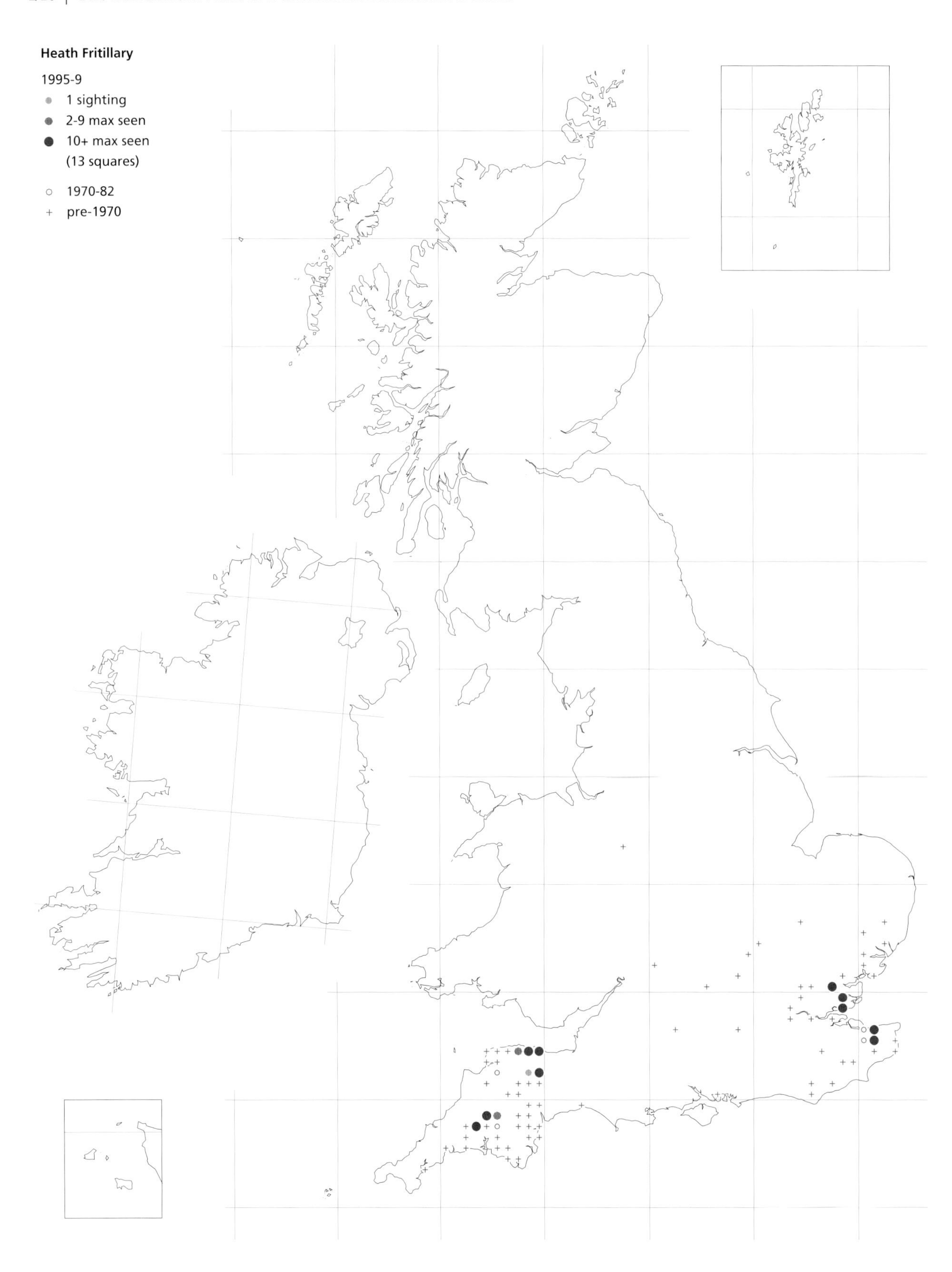
Heath Fritillary
1995-9
1 sighting
2-9 max seen
10+ max seen
(13 squares)
1970-82
pre-1970

Heath Fritillary *Melitaea (Mellicta) athalia*

Resident

Range declining.

Conservation status

UK BAP status: Priority Species.
Butterfly Conservation priority: high.
European status: not threatened.
Fully protected in Great Britain.

European/world range

Widespread and often abundant through most of Europe and across Russia to Asia and Japan. Its range has changed little in southern Europe but it has declined in 12 European countries.

The Heath Fritillary is one of the smaller fritillaries, distinguished by its dusky wing colours. It is restricted to a few specialized habitats where it flies close to the ground with characteristic flits and glides. The butterfly has historically been linked with the traditional practice of woodland coppicing, giving it the local name of the 'Woodman's Follower' as it follows the cycle of cutting around a wood. Sadly it is now one of our rarest butterflies but has been saved from the brink of extinction by the concerted action of conservationists.

Foodplants

The main foodplants are Common Cow-wheat (*Melampyrum pratense*), Ribwort Plantain (*Plantago lanceolata*), Germander Speedwell (*Veronica chamaedrys*), and occasionally other speedwells (*Veronica* spp.). Foxglove (*Digitalis purpurea*) can be a secondary foodplant, especially on Exmoor.

Habitat

The species uses sunny, warm, and sheltered habitats of two main types: coppiced or newly felled woodland on acid soils where Common Cow-wheat is abundant; and sheltered heathland combes (valleys) on Exmoor (up to 200–400 m above sea level) where Common Cow-wheat grows as scattered plants on mineral soils amongst vegetation dominated by Bilberry. On a few sites in south-west England, it also breeds on unimproved grassland with abundant Ribwort Plantain and/or Germander Speedwell growing in short or sparse conditions.

Life cycle and colony structure

The Heath Fritillary is predominantly single brooded with adults flying from the end of May until early July in Cornwall but somewhat later (from mid-June to early August) on Exmoor and in south-east England. There is considerable variation between years and between east and west.

In the south-east there can be a small second generation in hot years during late August and early September (as occurred in 1999).

The **eggs** are typically laid in large batches of between 80–150, though sometimes as few as 15. They are deposited close to the ground on the underside of a leaf immediately next to the foodplant, and only rarely on the foodplant itself. The **larvae** feed gregariously in a small, inconspicuous web, but soon disperse into smaller groups. These groups overwinter during their third instar close to the ground, usually in the leaf litter where they form a hibernaculum by spinning together the edges of a dead, tightly rolled leaf. The larvae emerge again in the first warm, sunny days of March or April and feed sporadically in between lengthy bouts of basking on dead leaves or twigs. They

Heath Fritillary larva on Common Cow-wheat.

pupate within the leaf litter, often within dead leaves.

This species is highly sedentary and forms compact colonies centred on its favoured breeding areas which can be as small as 0.5 ha. Marking experiments have shown that most adults rarely move more than 100 m but a few individuals have been recorded to have dispersed up to 2 km. As many of its habitats are ephemeral, the butterfly has to move regularly to locate new breeding areas. Despite this nomadic lifestyle, it has very limited colonizing ability and studies in south-east England have shown that suitable habitats more than 600 m from a population are colonized only slowly, if at all.

Distribution and trends

Once found in numerous localities scattered across the southern half of England, the Heath Fritillary has always been absent from Wales, Scotland, Ireland, the Channel Islands, and the Isle of Man. Before 1910 it was recorded in at least 58 10 km grid squares but was already undergoing a steep decline that continued throughout the twentieth century. By 1980 it had become highly endangered and the first comprehensive survey found that it was reduced to six 10 km squares. The survey located only 31 colonies: 25 in the Blean Woods complex of Kent and 6 scattered through south-west England (defining a colony as a group of individuals separated from other groups by 300 m or more). Moreover, three-quarters of colonies were small (fewer than 200 adults at peak) and many breeding areas were deteriorating rapidly following the planting of conifers.

Since 1980, further surveys have discovered many new colonies on Exmoor (mostly in the Somerset part) where, intriguingly, there are only one or two historical records. The butterfly has since been recorded at over 30 different localities and Exmoor is now recognized as an important stronghold. However, as in other regions, the butterfly's habitat appears to be ephemeral and breeding areas are constantly shifting, so its status changes almost annually. A thorough survey in 1999 found that numbers had been reduced severely by poor weather in 1997 and 1998 and several colonies had died out.

The Heath Fritillary has had mixed fortunes elsewhere: in Cornwall, two colonies have survived thanks to active conservation management, but in Devon it is almost extinct. In the butterfly's traditional stronghold in the Blean Woods complex, it declined to a low point of just 14 colonies between 1990 and 1995, but has recently recovered to around 18 colonies following increased habitat management.

Over the past hundred years, there have been numerous attempts to re-establish the Heath Fritillary on former sites and outside its historical range. Most of these failed very quickly and have not been mapped. However, two near Southend in Essex lasted for many years during the 1930s and 1940s, only becoming extinct when coppicing was abandoned in the 1950s.

In recent years, a number of reintroductions have been conducted as part of the national conservation plan, to sites where suitable long-term management has been restored. It has so far been reintroduced to two coppiced woods in Essex in 1984 and 1987, along a disused railwayline in Devon in 1993, and to two further woods in Essex in 1997 and 1998.

European trend

The butterfly is common and stable in the south of its range but has declined substantially in Belgium, Luxembourg, and the Netherlands (>50% decrease in 25 years), Denmark and Germany (25–50% decrease), Austria, France, Latvia, Moldova, Slovakia, and Sweden (15–25% decrease).

Interpretation and outlook

The decline of the Heath Fritillary is strongly linked to the cessation of traditional coppicing during the twentieth century. Being confined to the most acidic soils, the butterfly's range was always far more restricted than those of the violet-feeding fritillaries and it became threatened earlier. Fortunately, the butterfly's plight was recognized following the 1980 survey and a substantial conservation effort has saved it from extinction.

The Heath Fritillary would have fared still worse had it not been able to survive in other rare types of habitat, notably dry flower-rich grasslands in Devon and Cornwall, and the sheltered combes of Exmoor. The former have now all but disappeared with the notable exception of two sites where newly planted conifers were removed following the intervention of Prince Charles on his Duchy of Cornwall estate.

Its heathland habitat on Exmoor is also far more secure, thanks largely to the National Trust. It maintains large areas of moorland through extensive sheep and deer grazing, and also burns sections of the moor on rotation, which helps to regenerate short heathland conditions where Common Cow-wheat can flourish. It remains a matter of speculation whether the Heath Fritillary was always so widespread on Exmoor, but it is a remote region and has never been famous amongst entomologists. The species may therefore have been overlooked until the thorough surveys of the 1980s.

Because its habitats are ephemeral, the Heath Fritillary presents an enormous challenge to conservationists, especially as most sites are now relatively small and isolated. Fortunately, the best forms of management in its woodland and grassland habitats are now well understood, enabling populations to be conserved within limited areas. Ideal management of woodland is by continuing an active coppice cycle, where patches are cut close together and interconnected by rides to encourage rapid colonization. This management has led to large population increases on nature reserves in Kent and has allowed successful reintroductions to reserves in Essex.

Despite these successes, the best chances of conserving the butterfly in the long term will be in extensive habitats that allow for natural patterns of colonization and extinction. Appropriate management of large areas of Exmoor and Blean Woods is therefore a top priority. Far more also needs to be learnt about the Heath Fritillary's complex population dynamics on Exmoor and, given its recent decline there, we cannot afford to be complacent.

Heath Fritillary habitat on Exmoor

Key references

Barnett and Warren (1995*h*); Brereton *et al.* (1998*b*); Feber *et al.* (2000); Warren (1987, 1991); Warren *et al.* (1984).

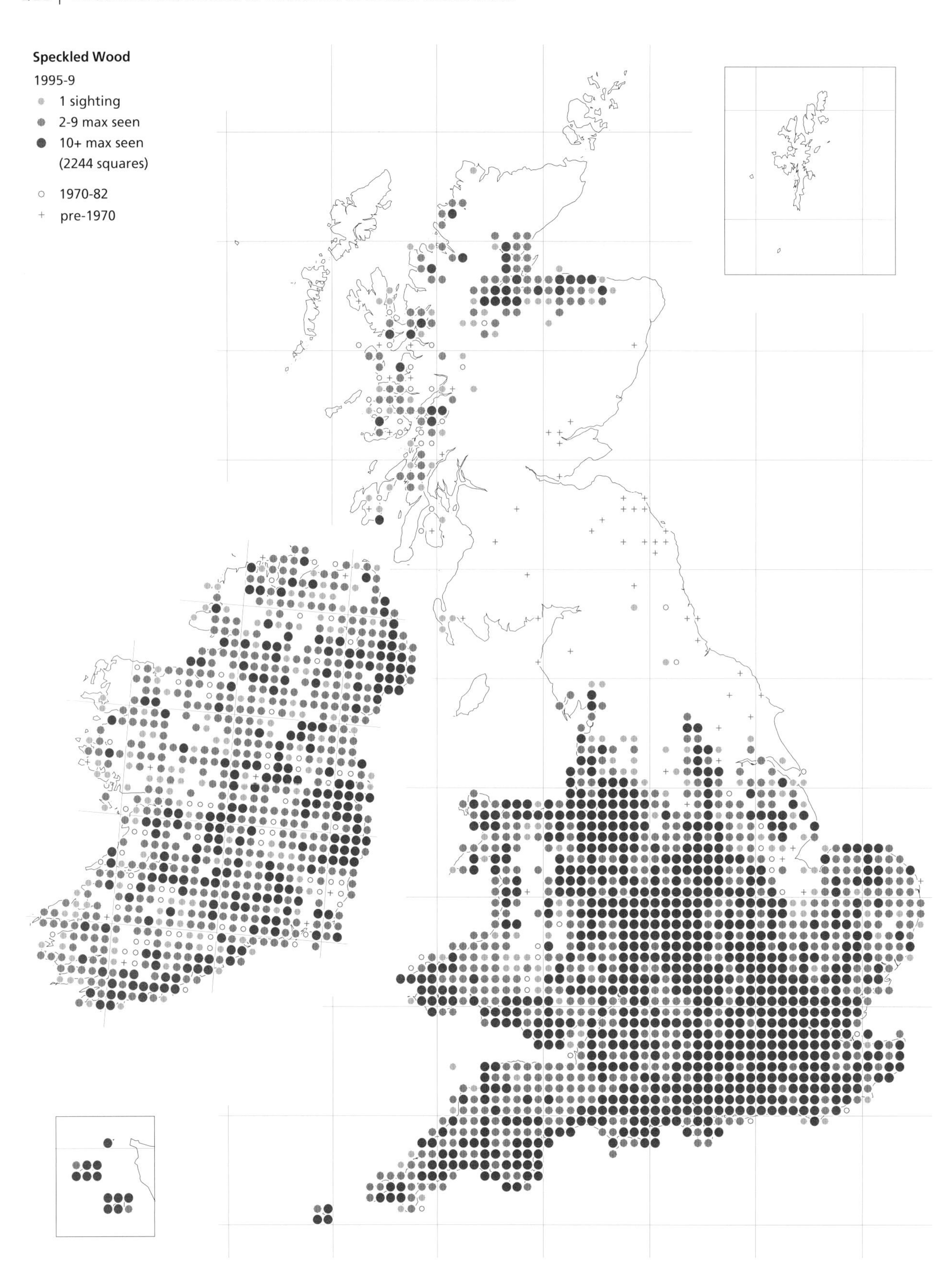
Speckled Wood
1995-9
1 sighting
2-9 max seen
10+ max seen
(2244 squares)
1970-82
pre-1970

Speckled Wood *Pararge aegeria*

Resident

Range expanding.

Conservation status

UK BAP status: not listed.
Butterfly Conservation priority: low.
European status: not threatened.

European/world distribution

Widespread throughout Europe; also in North Africa and eastwards to the Urals. It has expanded in several European countries, and its range has extended northwards.

The aptly named Speckled Wood flies in partially shaded woodland with dappled sunlight. The male usually perches in a small pool of sunlight, from where it rises rapidly to intercept any intruder. Both sexes feed on honeydew in the tree tops and are rarely seen feeding on flowers, except early and late in the year when aphid activity is low. The range of this butterfly contracted during the late nineteenth and early twentieth centuries, but has spread back since the 1920s. It has continued to spread over the past two decades, recolonizing many areas in eastern and northern England and Scotland.

Foodplants

Various grasses are used, including False Brome (*Brachypodium sylvaticum*), Cock's-foot (*Dactylis glomerata*), Yorkshire-fog (*Holcus lanatus*), and Common Couch (*Elytrigia repens*).

Habitat

Towards the northern and eastern margins of its range, the Speckled Wood breeds only in woodland habitats, but elsewhere it also uses lanes and tracks between tall hedgerows, parks, gardens, and scrub. It seems to prefer slightly damp areas where there is tall grass and some shade.

Life cycle and colony structure

There are usually three generations each year in England, Wales, Ireland, and the Channel Islands. In northern Scotland there appear to be only two. However, the flight periods of each generation tend to overlap, making it difficult to separate them precisely. Because of these overlaps, this species can usually be seen in southern areas at almost any time from March to October. Further north, where temperatures are lower, the flight period is reduced (see p. 397).

Eggs are laid singly on grass leaves, sometimes on isolated plants. In spring and autumn, females tend to lay eggs in warm, sheltered spots on the edges of woodland, but in high summer they choose more shady positions within woodland. The **larvae** rest under the grass blades, where their green colour provides excellent camouflage, and feed on the edges of the leaves. The **pupae** are formed on the foodplant or on other vegetation nearby. Unusually among butterflies, this species overwinters at either the larval or pupal stage. It is thought that true hibernation occurs only in the pupal stage, whereas larvae will resume feeding if temperatures rise sufficiently. Larval development rates vary with temperature and, consequently, the emergence of adults is protracted and flight periods overlap.

The behaviour of males makes them more conspicuous than females, which spend more time basking or feeding from aphid honeydew in the tree canopy. Males alternate between two strategies for locating

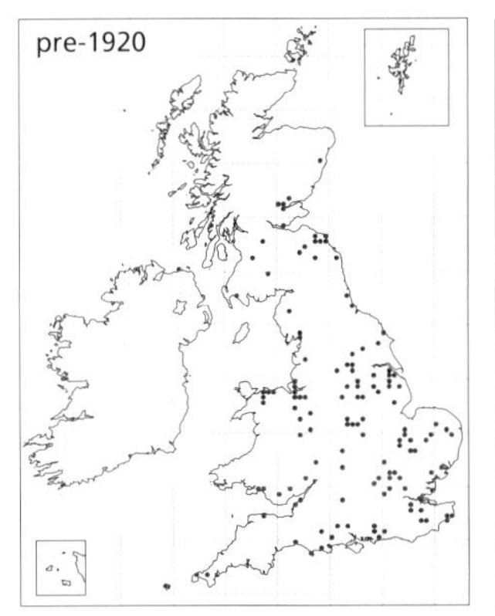

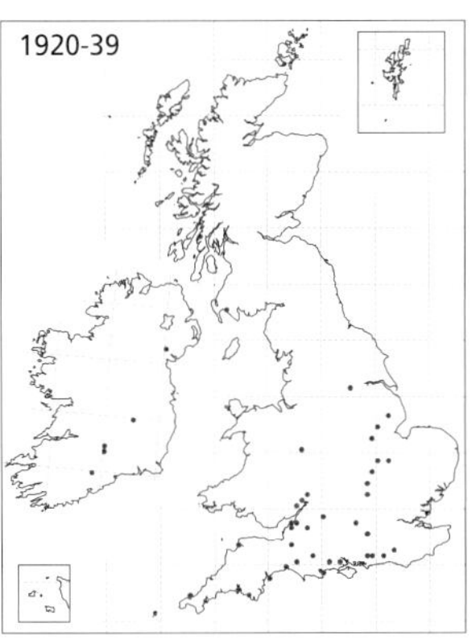

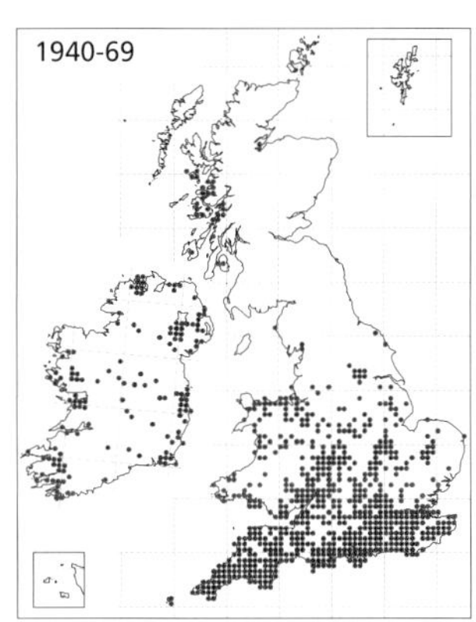

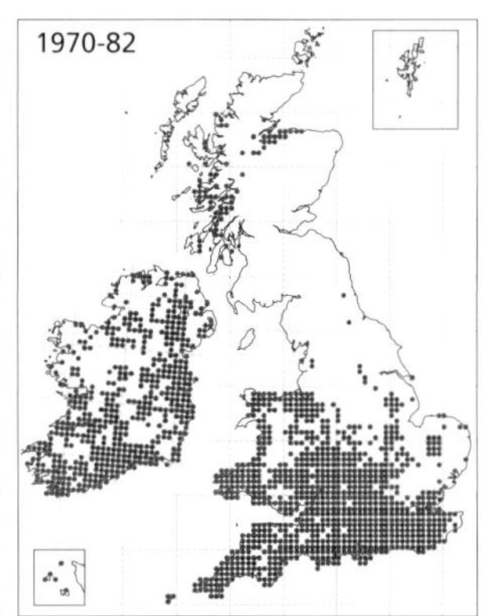

The long-term change in range of the Speckled Wood. Despite a low level of recording, the Speckled Wood was clearly once widespread in Britain. A dramatic contraction of range and subsequent re-expansion followed.

mates, perching in territories and patrolling through the habitat. Research has shown that males with paler wing colour are predominantly territorial and darker males are predominantly patrollers. This is thought to be because darker males warm up more quickly when basking and, therefore, are able to spend longer periods in flight.

The Speckled Wood forms sizeable colonies in blocks of woodland or along wooded tracks and it is normal to see several individuals flying in the same area. However, adults are found at low densities in a wide range of other habitats, such as hedgerows where populations are probably distributed more diffusely. There is little information on the mobility of this butterfly, but the recent range expansion demonstrates considerable ability to colonize new habitat.

Distribution and status

The Speckled Wood has undergone several large changes in range in Britain during the last 200 years. In the second half of the nineteenth century and the early part of the twentieth century, it disappeared from most of its range, becoming restricted to south-west England, lowland Wales, and western Scotland. Starting in the 1920s and gathering pace from the 1940s onwards, the butterfly gradually recolonized southern England and much of Wales. This expansion has continued in recent years, and it has been recorded in over 550 new 10 km squares compared with 1970–82. Overall, the species has expanded its 10 km square distribution by 54% in Britain since the last survey, but it still remains less widely distributed than during its nineteenth century heyday.

In East Anglia, it was virtually confined to Breckland, but it extended further into Suffolk as recently as the 1960s and had become widespread in Norfolk and Suffolk by the end of the twentieth century. Elsewhere in England, the main expansion has been northwards, particularly through the East Midlands and Cheshire and south Lancashire.

In Scotland, the distribution in Argyll seems to be strengthening and that around Inverness has expanded eastwards through Moray into Aberdeenshire, as well as to the north and west over the past 10–20 years. A few recent records from Dumfries and Galloway indicate that there may be small populations in south-west Scotland. The spread has raised great interest amongst recorders and has been well documented.

The history of the species in Ireland is not well known. There are insufficient past records to draw firm conclusions, although Irish literature indicates that it has always been widespread and common. The butterfly is common in the Channel Islands, but it has never been recorded on the Isle of Man.

European trend

The range of the Speckled Wood has extended northwards over the last 30–100 years. More recently, it has spread in several countries, including Denmark, the Netherlands, and Poland (25–100% increase in the past 25 years) and it has colonized the island of Madeira.

Interpretation and outlook

The continuing changes in range of the Speckled Wood seem to be linked to two factors: long-term fluctuations in climate, and habitat change. The Butterfly Monitoring Scheme data show an overall increase in abundance at many sites, especially in the east, and the colonization of several new sites. The data also show that abundance tends to increase following cool wet years and to decrease in drought years. Changes in its distribution may therefore be linked to climatic conditions over the past 150 years.

The other factor influencing its distribution is the substantial change in woodland habitats in Britain. Like the White Admiral, the Speckled Wood is a butterfly of shady woodlands and both species probably benefited from the decline of coppicing during the twentieth century. Many woodlands have since become neglected and it is thought that woodland in England and Wales is generally more shady now than it has been for several hundred years. Although this has led to declines in early successional species, such as fritillaries, which need more direct sunlight, it has increased the habitat available for shade-tolerant species such as the Speckled Wood.

The spread of the Speckled Wood suggests the strong influence of recent climate changes. A recent model of the effects of climate and habitat availability on the expansion of the species has predicted that, if current climate changes continue, most of Britain will become climatically suitable for the butterfly during the twenty-first century. However, the ability of the Speckled Wood to colonize areas as they become climatically suitable is likely to be constrained by the availability of potential habitat. It has not yet spread back into much of northern England and southern Scotland, and it is possible that the fragmentation of woodland is limiting colonization in these areas.

The butterfly's sparseness in the arable 'prairies' of north Cambridgeshire and south Lincolnshire is due simply to the scarcity of suitable woodland. Thus, in our more intensively farmed landscapes, it is important to maintain or plant a network of small woods and hedgerows, with their associated native grasses. The butterfly can be encouraged to breed in parks and gardens, even in urban areas, by leaving patches of tall, shady grassland.

There has been a gradual increase in numbers of the Speckled Wood on sites monitored in the BMS.

Key references

Dyck *et al.* (1997); Hill *et al.* (1999, in press); Mendel and Piotrowski (1986); Shreeve (1986*a*,*b*); Stewart *et al.* (1998); Thomson (1980); Windig and Nylin (1999).

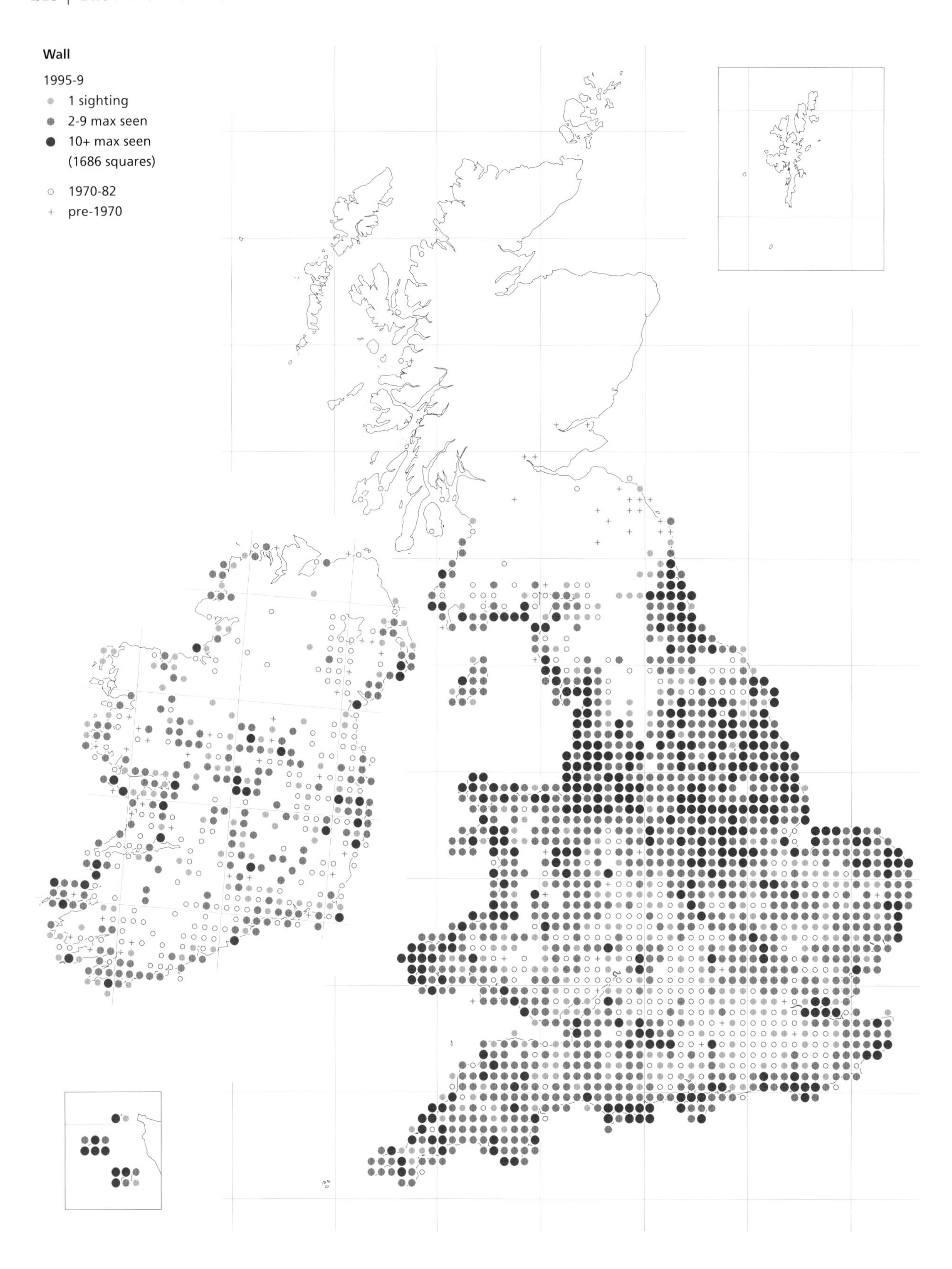
Wall
1995-9
1 sighting
2-9 max seen
10+ max seen
(1686 squares)
1970-82
pre-1970

Wall *Lasiommata megera*

Resident

Declining in some inland areas, slight spread in north of Britain.

Conservation status

UK BAP status: not listed.
Butterfly Conservation priority: low.
European status: not threatened.

European/world range

Extends across most of Europe, North Africa, and the Middle East, between 33 °N and 55 °N, and into southern Scandinavia. The range appears to be expanding northwards in Europe, although declines have been reported in several countries.

The Wall is aptly named after its habit of basking on walls, rocks, and stony places. The delicately patterned light brown undersides provide good camouflage against a stony or sandy surface. In hot weather, males patrol fast and low over the ground, seeking out females. In cooler weather, they will bask in sunny spots and fly up to intercept females, or to drive off other males. The Wall is widely distributed, but rarely occurs in large numbers. Over the last decade, it has declined substantially in many inland areas of central England and Northern Ireland.

Foodplants

Various grasses are used, including Tor-grass (*Brachypodium pinnatum*), False Brome (*B. sylvaticum*), Cock's-foot (*Dactylis glomerata*), bents (*Agrostis* spp.), Wavy Hair-grass (*Deschampsia flexuosa*), and Yorkshire-fog (*Holcus lanatus*).

Habitat

The Wall breeds in short, open grassland where the turf is broken or stony. It is found in dunes and other coastal habitats (including vegetated undercliffs and rocky foreshores) as well as disturbed land (including railway embankments and cuttings), disused quarries, derelict land, and gardens.

Life cycle and colony structure

There are two generations per year, with adults flying from early May to late June and again in late July into September, in the south. Northern populations fly about two weeks later. In warm years, adults emerge earlier and there is an occasional third brood, flying in October in southern England, the Channel Islands and sometimes in southern Ireland (for example in Co. Wexford).

The light green coloured **eggs** are laid singly on grass leaves, amongst broken turf, where there is an exposed vertical edge, such as the top of a cliff, a landslip, or where livestock have poached the turf. Females are also known to lay at the edge of tussocks and where grass grows up a wall, fence, or beneath a hedge. The green **larvae** feed, mainly at night, on the grass leaves. The **pupae** are also green and are formed within the tussocky base of the plants. The larvae of the second brood overwinter whilst still small and resume feeding in early spring.

The Wall rarely occurs in large numbers and the first generation is usually less numerous than the second. It seems to form discrete colonies, but adults are occasionally seen well away from breeding areas and sometimes visit gardens. There is little research evidence about their mobility, but the butterfly's range has varied considerably over the last 100 years. This suggests that

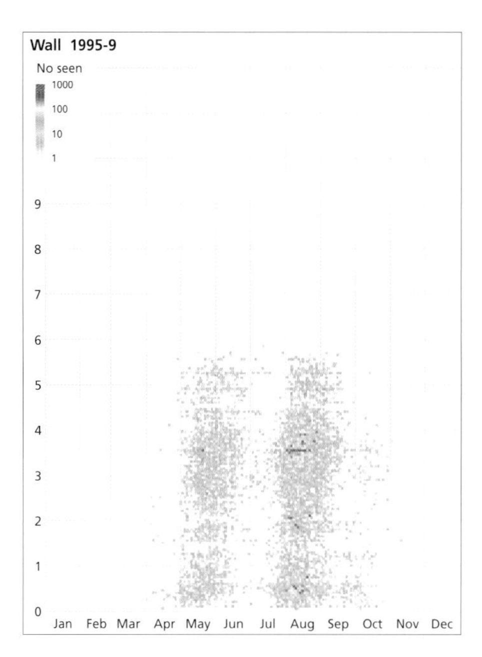

Phenogram of dates of 1995–9 Wall records plotted against latitude, showing two generations throughout the range and a partial third brood. Northern populations fly about two weeks later than those in southern Britain.

adults are capable of colonization over many kilometres, in favourable years. In addition, there is a record from an offshore light vessel, 50 km from the coast.

Distribution and trends

The Wall is a widespread species, occurring across England, Wales and the Channel Islands, but not in mountainous areas. In Scotland, it is found in the coastal areas of Ayrshire and Dumfries and Galloway, with some scattered records from Argyll and Borders. The distribution in Ireland is patchy, and the butterfly is less common towards the north, with colonies located mainly near the coasts.

The range of the Wall has fluctuated considerably in the past. For example, it disappeared from large areas during the cold, wet summers of the 1860s. More recently, it expanded in the 1970s and early 1980s, but has since disappeared from many of its former inland locations across central and southern England, and in Northern Ireland. It was common in these areas during the early 1980s, but had become locally extinct by the late 1990s. The records suggest that there was a slight recovery towards the end of the current survey. In contrast, the Wall seems to have maintained its range in northern England and has spread slightly inland in Lancashire, Northumberland, and Durham.

Data from the Butterfly Monitoring Scheme (BMS) reflect these range changes, showing a large drop in numbers at inland sites, but also that overall numbers increased in 1998 and 1999, after an eight-year decline.

Overall, the results for Britain show that the species has not been re-recorded in 18% (244) of the 10 km squares recorded in the 1970–82 survey, mainly in southern England, although it has been recorded in 260 new squares (mainly in the north). Some of these are due to the increased level of recording within the butterfly's range as well as a slight genuine spread. The identification of trends is more difficult in Ireland, due to patchy recording in both periods. The butterfly has been recorded in many new squares but not re-recorded in many former ones.

This butterfly is a strong flyer with a distinctive appearance, and is easy to spot. It is therefore unlikely to be significantly under-recorded in Britain and the absence of recent records in thoroughly recorded areas of central and southern England represents a genuine decline.

European trend

The distribution of the Wall is stable in most European countries, although there have been small declines in Austria, the Czech Republic, Germany, the Netherlands, Norway, Slovakia, and European Turkey (15–25% decrease in 25 years). Over a longer period (30–100 years), the butterfly's European range has extended northwards.

Interpretation and outlook

Wall butterflies are sensitive to temperature, as is demonstrated by their selection of dry and stony areas that offer a hot microclimate, both for basking and for egg laying. They require body temperatures of about 25–30 °C to fly and, in cooler conditions, will spend long periods basking with wings open on warm bare ground to raise their temperature between brief periods of flight.

Towards the northern edge of the butterfly's range, the development of second brood butterflies is dependent on the weather. If it is not warm enough, the life cycle will be retarded and second generation eggs and larvae will not be able to develop sufficiently to survive the winter. In coastal areas, winter temperatures are higher than inland and the ground probably dries faster in summer, allowing the butterfly to breed further north.

In common with some of the other browns, including the Speckled Wood and Marbled White, the Wall shows long-term changes in its range. Several periods of expansion of range towards the north have been followed by contractions southwards. BMS data show large variations in abundance at monitored sites, with peaks in 1982–3 and 1989–90. The decline in the collated index in 1985–7 coincided with a series of cool, wet summers, and the butterfly has not been recorded at some BMS sites since then. Local studies have indicated a relationship between abundance and summer temperatures. The changes in range in the north are probably accounted for by climate change. In the light of these longer-term cycles, the rarity of the butterfly in central and southern England during the past decade is likely to have been determined by changes in climate rather than habitat.

However, agricultural intensification during the past century has reduced the unimproved grassland habitat of the Wall substantially. In many regions, the butterfly is now abundant only along the coast, where some sparsely vegetated semi-natural habitats have survived, partly due to the effects of coastal erosion. Many inland habitats have become unsuitable for the Wall following the abandonment of traditional grazing in low-productivity grassland. Without grazing animals to break up the sward, good breeding conditions are unlikely to be maintained and colonies may disappear.

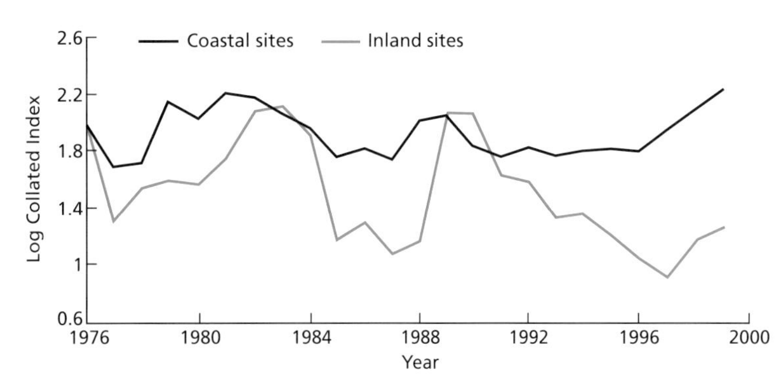

Abundance of the Wall at BMS sites, showing decline and greater variability at inland sites.

Higher stocking levels in fertilized grassland also may have adversely changed habitat conditions. A study in the Netherlands has shown a correlation between larval development and the nitrogen content of grass, indicating that artificial application of nitrogen may disturb the natural life cycle of the butterfly.

Rabbit activity is another factor influencing the Wall, because rabbits create bare patches that are needed for egg-laying. The reduction of rabbit grazing following myxomatosis in the 1950s probably affected the butterfly, but this may have been counter-acted by the warmer weather of the 1970s.

If climate is the main factor causing fluctuations of the Wall and conditions improve again in inland areas, then the core coastal populations are likely to provide an important source for natural recolonization. The ability of the Wall to move significant distances means this could happen quite rapidly, provided that suitable habitats remain available. The major decline in the 1990s, however, remains puzzling and needs both careful monitoring and further research.

Key references

Bink and Siepel (1996); Dennis and Bramley (1985); Ellis (1994); Wickman (1988).

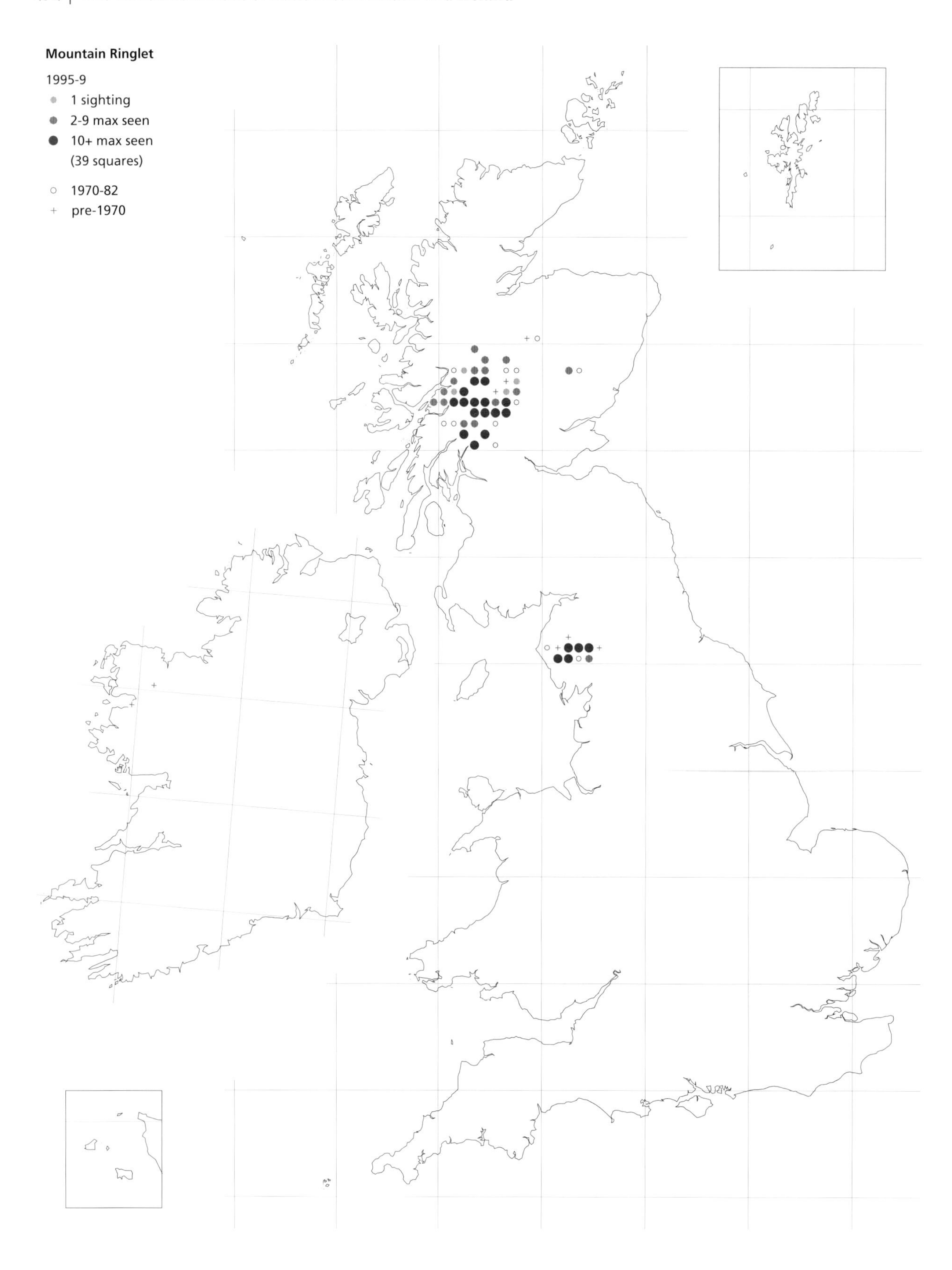
Mountain Ringlet
1995-9
1 sighting
2-9 max seen
10+ max seen
(39 squares)
1970-82
pre-1970

Mountain Ringlet *Erebia epiphron*

Resident
Range stable.

Conservation status
UK BAP status: Species of Conservation Concern.
Butterfly Conservation priority: medium.
European status: not threatened.
Protected in Great Britain for sale only.

European/world range
In mountains across Europe, where it occurs as a series of sub-species from Spain in the south, to the Bavarian Alps in the north, and east to the Carpathians and the Balkans. Its distribution appears stable.

The Mountain Ringlet is our only true montane species and is found on mountains above 350 m amidst the spectacular scenery of the Scottish Highlands and the English Lake District. The adults are highly active only in bright sunshine but can be disturbed from the ground even in quite dull weather. They keep low to the ground in short flights, pausing regularly to bask amongst grass tussocks or feed on the yellow flowers of Tormentil. The butterfly's status is difficult to assess due to the remoteness and unpredictable weather of its mountain habitats, but its range appears stable.

Foodplants
The main foodplant is believed to be Mat-grass (*Nardus stricta*), but the full range is not known. Recent observations suggest a possible association with Sheep's-fescue (*Festuca ovina*).

Habitat
The Mountain Ringlet is found in open mountain grassland dominated by Mat-grass and Heath Bedstraw. The butterfly occurs mainly at altitudes of 500–700 m in the Lake District and 350–900 m in Scotland (in some years it is seen as low as 70 m and as high as 1000 m). Adults tend to concentrate in damper areas and sometimes in flushes dominated by sedges, but they occur also in drier, heathy areas characterized by Bilberry and Wavy Hair-grass. In Scotland, the butterfly favours south facing slopes but a variety of aspects is used in the Lake District, including some north facing slopes.

Life cycle and colony structure
There is a single brood with adults flying for a short period from early June to late July in the Lake District and into early August in Scotland (see p. 397). The flight period varies greatly between years and the peak can vary from the third week of June to the third week of July.

Eggs are laid singly, close to the ground, usually on dead grass blades but also on mosses, sedges, and live grasses. Little is known about the behaviour of **larvae** because they feed by night on the tips of Mat-grass and retreat into grass tussocks by day. They hibernate amongst grass in their third instar during late August or September and emerge in spring to bask and resume feeding. The larvae **pupate** in late May or June, either on the ground or low in grass tussocks. There may be a two-year life cycle on some sites, or in poor seasons, as is the case with many montane moths. This has been demonstrated in captivity but has not been confirmed in the wild.

The Mountain Ringlet forms loose colonies which typically cover large areas of 12–50 ha, although adults often congregate in discrete patches. The location of some colonies also appears to shift slightly from year to year.

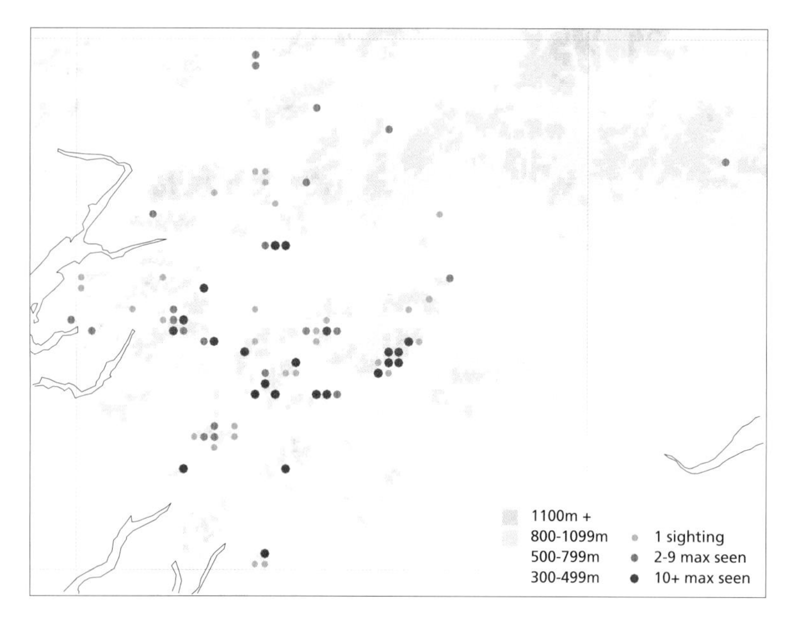

The distribution of 1995–9 records of the Mountain Ringlet plotted at 2 km square resolution in western Scotland, against a background of altitude.

Adults seem to move fairly freely within the wider flight area with males appearing to be more mobile than females. Experiments have shown that males can move up to 200 m in single flights and occasionally further (one male flew 1.3 km in just four hours). In contrast, females are much heavier when they are young and full of eggs, typically moving only 8 m per flight with an observed maximum of 35 m. It seems likely that females fly further once their egg burden has been reduced, but most populations are probably completely isolated from those on adjacent mountains. Adult lifespan is thought to be very short in the wild, with most living for just one or two days when the weather is good.

Distribution and trends

The Mountain Ringlet is found only in the central Lake District (Cumbria) in England, and in central Scotland between Fort William in the west, the Grampians to the east, and Ben Lomond to the south. There are records of the species in Ireland for the nineteenth century, and even museum specimens, but some authors dismiss these and its status in the country remains somewhat of an enigma.

The species is certainly under-recorded due to the remoteness of most colonies and the limited period during which adults may be flying, particularly in poor and variable mountain weather. In Scotland, specific surveys, conducted by the Institute of Terrestrial Ecology and Butterfly Conservation during the early 1990s, revealed new localities and have suggested several more potential areas. Similar surveys were co-ordinated in the Lake District during the late 1990s by Butterfly Conservation, using volunteer surveyors as well as postcards and posters aimed at the many hillwalkers in this popular tourist area. These confirmed colonies at two-thirds of all historical sites, but several colonies still need to be revisited and it is likely that others remain undiscovered.

The BNM survey failed to re-record the Mountain Ringlet in 12 10 km squares recorded in 1970–82, but discovered the butterfly in 22 new 10 km squares. Overall, therefore, the number of recorded squares has increased by 34% since 1970–82.

European trend

The distribution of the Mountain Ringlet is believed to be stable in most of its range, but, as in Britain, it is under-recorded.

Interpretation and outlook

The surveys indicate that the Mountain Ringlet has a stable distribution in Britain but that the butterfly may be more widespread within high mountain areas than is shown in the maps. Despite several recent studies, the ecology of this species remains poorly known and basic facts, such as the larval foodplants and the number of generations per year, need to be confirmed.

The preferred habitat is also poorly known although recent studies provide valuable insights into appropriate management regimes. These should aid its conservation within nature reserves and on National Trust land (which covers significant parts of the butterfly's range in the Lake District). They should also help refine prescriptions for land entered into Environmentally Sensitive Areas and the new Scottish Rural Stewardship Scheme.

Most habitats are grazed by sheep and deer and there is some evidence that this is beneficial, if not essential, to the species' long-term survival. Eggs tend to be laid where the turf height is patchy, between 6–17 cm, and the butterfly often shuns rank, ungrazed grassland. Experiments have been established at two National Nature Reserves (Creag Meagaidh and Ben Lawers) which should shed further light on the impact on the Mountain Ringlet of the exclusion of grazing to promote woodland regeneration.

Future threats to the Mountain Ringlet are difficult to predict, but it could be adversely affected if there are substantial changes to traditional farming patterns and especially the grazing of upland pastures. In some areas it could be threatened by overgrazing, which has been responsible for the loss of at least one Lake District colony since 1990. There are also some threats in the lower parts of its altitudinal range from afforestation. This could have a damaging effect unless sensitive areas are avoided.

Perhaps the most difficult threat to predict is from global warming, which could push the butterfly's habitat zone to higher altitudes and thus smaller areas, as one recent study has suggested. However, this effect could be counteracted if it was consequently able to breed in north-facing aspects in Scotland as it does in the Lake District. Because of these uncertainties, an improved monitoring network is essential for this species which is currently covered by only one regular monitoring transect, at Ben Lawers National Nature Reserve in Perthshire. Much more work is needed before the ecological requirements of this butterfly are fully understood and we are in a position to assess its future.

Mountain Ringlet habitat in the Lake District.

Key references

Boyd-Wallis (1994); McGowan (1997); Ravenscroft and Warren (1996*d*); Redway (1981); Shannon (1995).

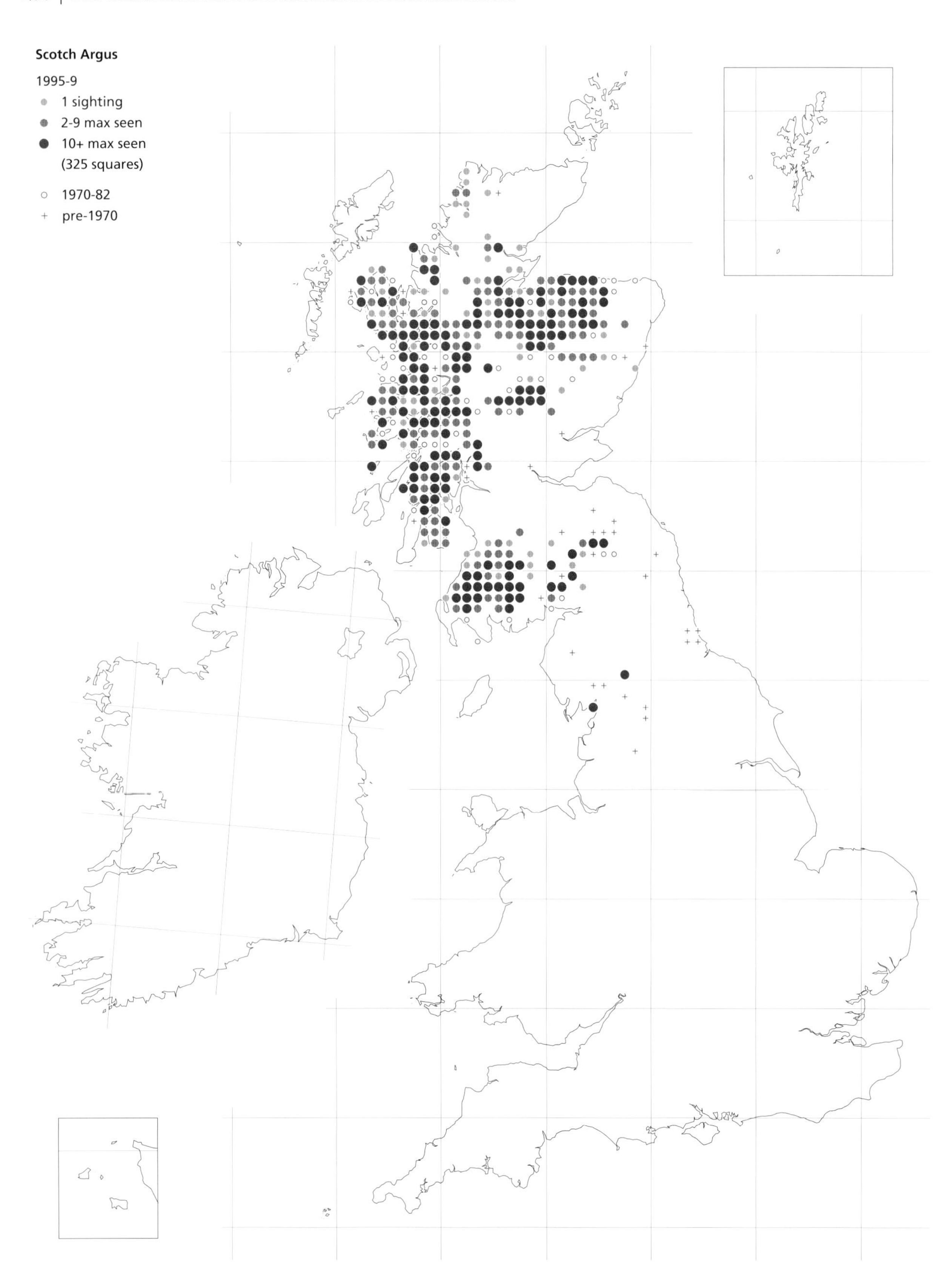
Scotch Argus
1995-9
1 sighting
2-9 max seen
10+ max seen
(325 squares)
1970-82
pre-1970

Scotch Argus *Erebia aethiops*

Resident
Range stable.

Conservation status
UK BAP status: not listed.
Butterfly Conservation priority: low.
European status: near threatened.

European/world range
Mountain areas throughout Europe from central France east to the Urals and western Siberia, but not in Scandinavia. It has declined recently in several European countries.

As its name suggests, this butterfly is found predominantly in Scotland where it flies in tall, damp grassland. In sunshine, males fly almost without rest, weaving low through the grass in search of a mate. In poorer weather they perch on grass clumps, flying out to investigate any passing brown butterflies. The females are far less conspicuous and spend most of their time basking. The Scotch Argus is common and widespread in Scotland but has declined in the southern part of its range, especially in England where it is reduced to just two isolated sites.

Foodplants
The main foodplant in Scotland is thought to be Purple Moor-grass (*Molinia caerulea*) and the populations in northern England use Blue Moor-grass (*Sesleria caerulea*). Though there are few observations from Britain, other grasses may be used, as they are in continental Europe.

Habitat
The Scotch Argus occurs in damp, acid or neutral grassland up to 500 m in montane regions of Scotland, and around the fringes of sheltered bogs, in woodland clearings, and young plantations. In northern England it is now restricted to two sites that contain a mosaic of sheltered limestone grassland, scrub, and woodland. The butterfly is found only in tall grasslands that are lightly grazed or ungrazed.

Life cycle and colony structure
There is a single brood with adults flying from around the last week in July until early September (see p. 397). The life cycle is poorly known as the early stages are difficult to find. The creamy, speckled **eggs** are laid singly on grass blades at heights of 5–20 cm. Although females begin egg-laying flights only in bright sunshine, in captivity they lay eggs readily in the dark. They may therefore be able to climb up grass blades and lay eggs during poor weather, though this still has to be confirmed in the wild.

The **larvae** rest by day within dense grass and emerge only at dusk when they crawl up to feed on the tips of the foodplant. They can be found by searching but great care is needed as they often drop into the vegetation if disturbed. The larvae hibernate when still small, probably deep within grass tussocks. They resume feeding in the spring and **pupate** low down in moss or dense grass.

The Scotch Argus often occurs in very large colonies that can extend over many kilometres of suitable habitat, but its mobility has been examined only in smaller more isolated populations. A mark–recapture study in England found that adults were sedentary and rarely flew more than 100 m; similar observations

Scotch Argus habitat (in foreground) by Loch Lomond, Scotland.

in Switzerland found a maximum distance moved of around 500 m. However, anecdotal evidence in Scotland suggests that they may be more mobile in large areas of unbroken habitat and individuals can be found quite widely, sometimes even coming into gardens.

Distribution and trends

The Scotch Argus is widespread and abundant throughout much of Scotland, especially on the west coast, in the Highland region, and in Dumfries and Galloway. It occurs widely on the Inner Hebridean islands of Skye and Mull, but is probably absent from the Outer Hebrides, although there have been recent unconfirmed reports from the Island of Muldoanich, south of Barra. It is absent from Orkney and Shetland, as well as from Ireland, Wales, and the Isle of Man.

In northern England, the butterfly once occurred in scattered colonies in Cumbria, north Lancashire, western Yorkshire, and the coast of Durham and Northumberland. However most of the colonies died out in the early part of the twentieth century. It is now reduced to two limestone grassland sites in Cumbria which still support colonies of several thousand adults. The large colony at Grassington in western Yorkshire survived until 1955. There have since been several attempts to re-establish it there but with no success, indicating that the habitat is no longer suitable or that the climate has changed.

There is little evidence of any decline in Scotland though a few colonies have been lost from south-east Scotland, including Berwickshire where it was last seen in 1902. In the current survey, the Scotch Argus was found in 130 new 10 km squares since 1970–82, but this is thought to be due to increased recording levels rather than any major spread. The lack of recent records in some other, remote areas of Scotland is probably the result of under-recording rather than declines.

European trend

The Scotch Argus has declined in many countries: it is now extinct in Latvia and Luxembourg, and has declined severely in Belgium (>50% decrease in 25 years), the Czech Republic, France, Germany, and Ukraine (25–50% decrease).

Interpretation and outlook

The distribution of the Scotch Argus appears to be stable in Scotland and it is a very common butterfly in regions where there is tall, damp grassland. The butterfly is sensitive to grazing levels and cannot survive under moderate to heavy grazing. This naturally limits it to habitats that are either not grazed by livestock (e.g. young plantations, woodland rides, and road verges) or hills and moorlands that are grazed only lightly. Fortunately, there are large areas of such habitat in Scotland and it does not seem to be under any immediate threat.

However, a recent review has revealed unexpectedly that the Scotch Argus is declining in many European countries and is in a 'near threatened' category for Europe as a whole. The butterfly's status in Scotland therefore needs to be monitored carefully to ensure that this apparently healthy situation continues.

Unimproved grassland and moorland are being lost at a steady rate due to afforestation and many uplands are becoming increasingly heavily grazed. Habitats are under pressure in the lowlands, as the edges of bogs are drained and improved for agriculture.

Given that levels of recording in the more remote areas of Scotland are generally low, the butterfly may already be becoming less abundant but this has yet to show up in the distribution map. Moreover, if the butterfly is fairly mobile in Scotland, any decline may take decades to become apparent, given present levels of recording and it seems sensible to establish a special sampling scheme as well as conducting further research into the butterfly's precise requirements.

In northern England the Scotch Argus has been reduced to just two sites. The decline occurred over a long period and the underlying causes are poorly understood. As with other species, it may have more specialized requirements at the edge of its range. Fortunately it is still abundant at the two remaining sites which are both on carboniferous limestone hills dominated by Blue Moor-grass and where grazing pressure is light. However, it is a puzzle as to why the butterfly does not occur on other nearby limestone hillsides which appear to have a very similar combination of sheltered grassland and scrub. These neighbouring sites are well recorded yet there have been only occasional sightings and no sign of any establishment. Several attempts were made to introduce the butterfly onto some of these sites during the 1930s and 1940s but they all failed.

Given its steady decline elsewhere in Europe, continued monitoring and further research seem to be important priorities for ensuring the long term future of this intriguing butterfly.

Key references

Beaumont (1995); Kirkland (1996); Loertscher (1991).

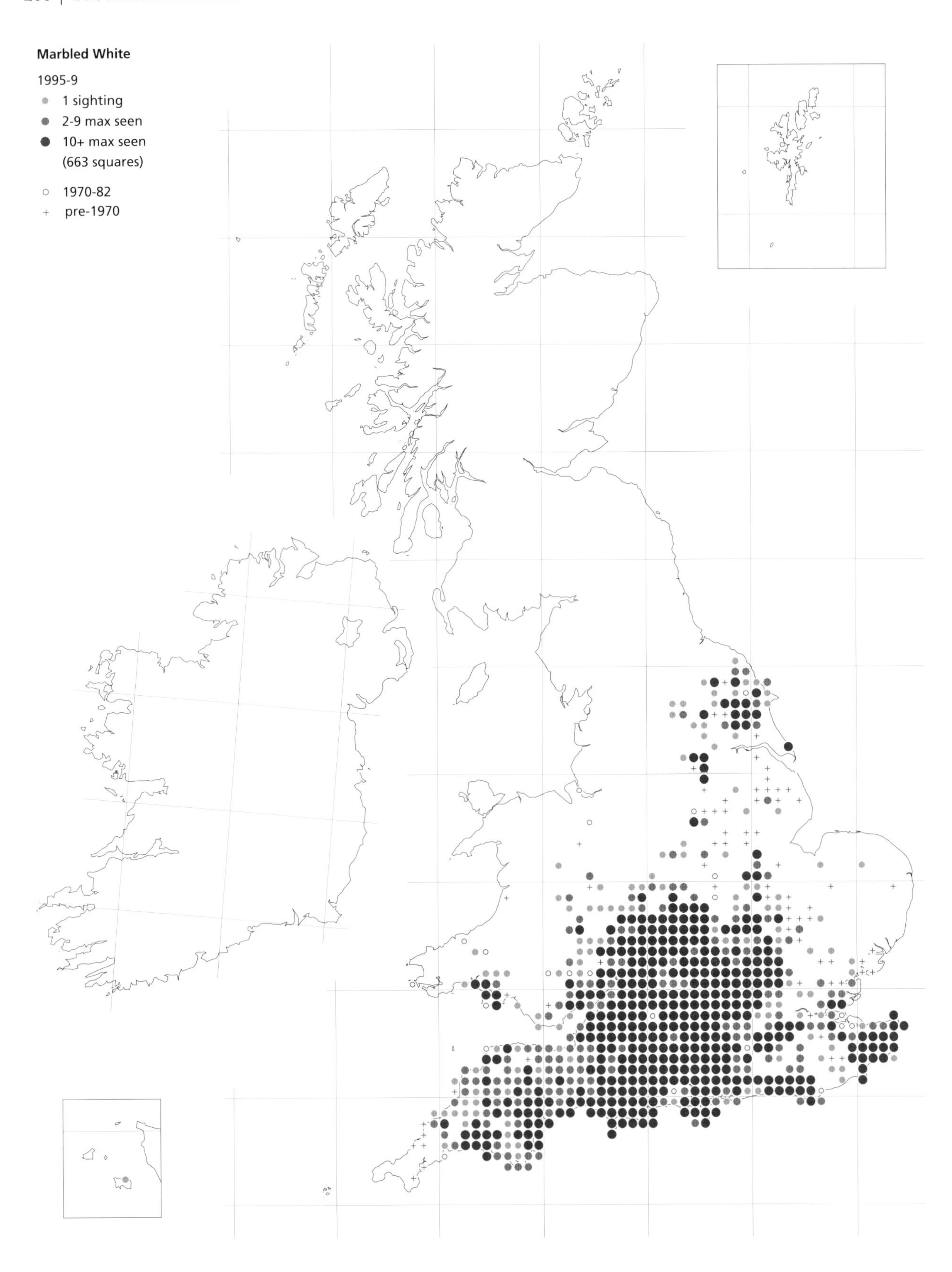
Marbled White
1995-9
1 sighting
2-9 max seen
10+ max seen
(663 squares)
1970-82
pre-1970

Marbled White *Melanargia galathea*

Resident

Range expanding.

Conservation status

UK BAP status: not listed.
Butterfly Conservation priority: low.
European status: not threatened.

European/world range

Most of Europe (except Scandinavia) and North Africa and eastwards to Iran. It is stable in most European countries.

The Marbled White is a distinctive and attractive black and white butterfly, unlikely to be mistaken for any other species. In July it flies in areas of unimproved grassland and can occur in large numbers on southern downland. It shows a marked preference for purple flowers such as Wild Marjoram, Field Scabious, thistles, and knapweeds. Adults may be found roosting halfway down tall grass stems. This species is widespread in southern Britain and has expanded northwards and eastwards over the last twenty years, despite some losses within its range.

Foodplants

Red Fescue (*Festuca rubra*) is thought to be essential in the diet of larvae but Sheep's-fescue (*F. ovina*), Yorkshire-fog (*Holcus lanatus*), and Tor-grass (*Brachypodium pinnatum*) are also eaten. It is thought that several other grasses may be used, but the full range is not known.

Habitat

Colonies occur on unimproved grassland where a range of grass species, including Red Fescue, form a tall sward that is cut or grazed infrequently. The strongest populations are found on chalk or limestone, but a range of habitats is used, including woodland rides and clearings, coastal grassland, waste ground, set-aside, road verges, and railway embankments.

Life cycle and colony structure

There is one generation a year and the flight period is short compared with that of many other species. Adults begin to emerge in mid-June, numbers rise to a peak in July and by mid-August few individuals remain on the wing.

Females do not place **eggs** directly on the foodplant, but perch briefly and eject them as they fly off, leaving them to fall among the grass stems. Newly emerged **larvae** enter hibernation and do not begin to feed until the following spring. Young larvae feed on fine grasses, whereas larger ones are able to eat coarser species. At first, larvae feed during the day, but in the final instar, feeding is nocturnal. They can be found climbing grasses at dusk in late May. When mature, larvae descend to **pupate** low down in grass tussocks or at ground level.

The Marbled White lives in colonies that may contain thousands of adults on suitable grassland. Colonies on small patches of rough grassland, however, may contain only a few dozen adults in a season. There is little detailed information on mobility, but the expansion of range in recent years, and the occurrence of dispersing individuals seen travelling across inhospitable habitats, demonstrate that this species is more mobile than previously thought.

Marbled White larva.

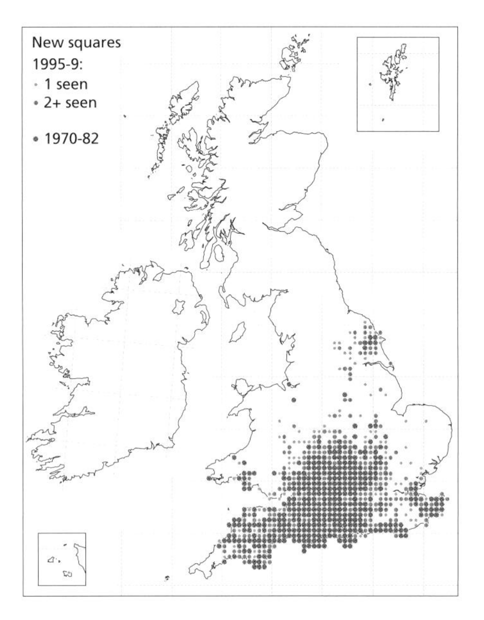

The expansion of range of the Marbled White since 1970–82.

Distribution and trends

Most colonies occur in southern England, but some are found as far north as Yorkshire. From eastern Cornwall in the west, to Hampshire in the east and northwards to the Cotswolds, the Marbled White is a common butterfly. Further north and east it is a more local species with scattered colonies occurring in the Midlands and in the Yorkshire Wolds. In Wales, colonies are restricted to the extreme south, in Monmouthshire, Carmarthenshire, and Glamorgan, and many are coastal. It does not occur in Ireland, Scotland, the Isle of Man, or the Channel Islands.

Since the 1970–82 survey, the Marbled White has spread northwards and eastwards, and has been recorded in 66% more 10 km squares. It has recolonized areas at the edge of its range from which it retreated earlier in the twentieth century and has increased its range slightly, particularly in Monmouthshire, Herefordshire, Worcestershire, Warwickshire, Northamptonshire, further east in Hertfordshire and Essex, and in Yorkshire. On the North Downs in Surrey and Kent, it spread from core sites to form new colonies during the 1980s and 1990s, some of which are monitored and have shown a steady increase in numbers following the first appearance of the butterfly. Since the mid-1990s it has also shown signs of colonizing rough grassland on waste ground and in woodland some distance from the downs. In Cambridgeshire it has spread along railway embankments.

Butterfly Monitoring Scheme data show decreases at a few sites, but these are far outnumbered by increases. There is also a significant upward trend in the collated index for all monitored sites.

Some successful introductions of the Marbled White have been documented, notably in south Yorkshire and Surrey, where it has also been spreading naturally. Other, undocumented, releases may also have taken place. The scale of the recent spread, however, accompanied by reports of individuals flying purposefully until out of sight, indicates genuine and widespread dispersal. The ease with which this species may be seen and identified suggests that it has not been overlooked except perhaps in some under-recorded parts of Wales.

European trend

The Marbled White is stable in most European countries but it has declined in Austria (25–50% decrease in 25 years), Belgium, Germany, and Luxembourg (15–25% decrease).

Interpretation and outlook

The reasons for the expansion in range of the Marbled White are not entirely clear, but the most recent colonizations have coincided with years in which the weather during the flight period has been warm and sunny. It is known to increase in numbers following a warm, dry summer and such conditions also appear to increase the likelihood of dispersal in a species that is normally sedentary.

In most of the areas where expansion has taken place, the amount of suitable grassland has decreased with agricultural intensification in recent decades. However, the butterfly has made use of patches of habitat which have survived or have recently become suitable, such as road verges and embankments, railway tracks (used and disused), disused pits, waste ground, and ungrazed downland.

There are also records of Marbled White colonizing (and presumably breeding in) fields that have been set-aside from arable cultivation for a few years following changes in EC agricultural subsidies in 1992. Such non-rotational set-aside develops a long grassy sward very quickly and can become suitable after two or three years. The extent to which the butterfly uses set-aside and its importance in aiding the spread of the species are not known.

Although the Marbled White is associated with tall grassland, some management is required to ensure that this does not become too overgrown. The butterfly flourishes where grazing is light enough to allow some grass to grow long and form tussocks, yet heavy enough to ensure a sward sufficiently open to encourage the growth of both fine and coarse grasses, and plenty of flowers to provide nectar. Tall plants, such as knapweeds and scabious, are ideal basking sites in the early morning and late afternoon when temperatures are low, and on south-facing chalk slopes, where it may become very hot during the middle of the day, they may provide shade. Shelter is important and large populations may flourish in broad, sunny rides in young plantations.

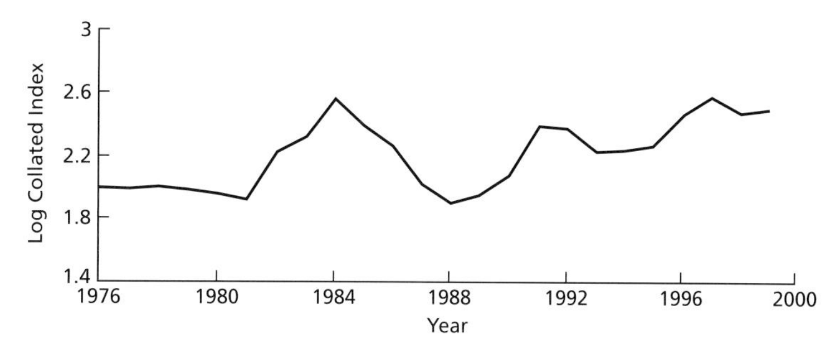

The Marbled White has increased in abundance at BMS sites during the recent period of range expansion.

Although the Marbled White has spread recently, there were losses of colonies throughout its range during the twentieth century. It disappeared from many sites by the late 1970s due to the destruction of its grassland habitats. On the other hand, patches of suitable habitat occur well within the range of the butterfly, but remain unoccupied. There appears to be scope for future expansion both within and beyond the current distribution range and it remains to be seen whether the current increases will continue, or whether a series of poor summers will cause a retreat from recently occupied areas.

Key references

Loertscher *et al.* (1997); Rafe and Jefferson (1983).

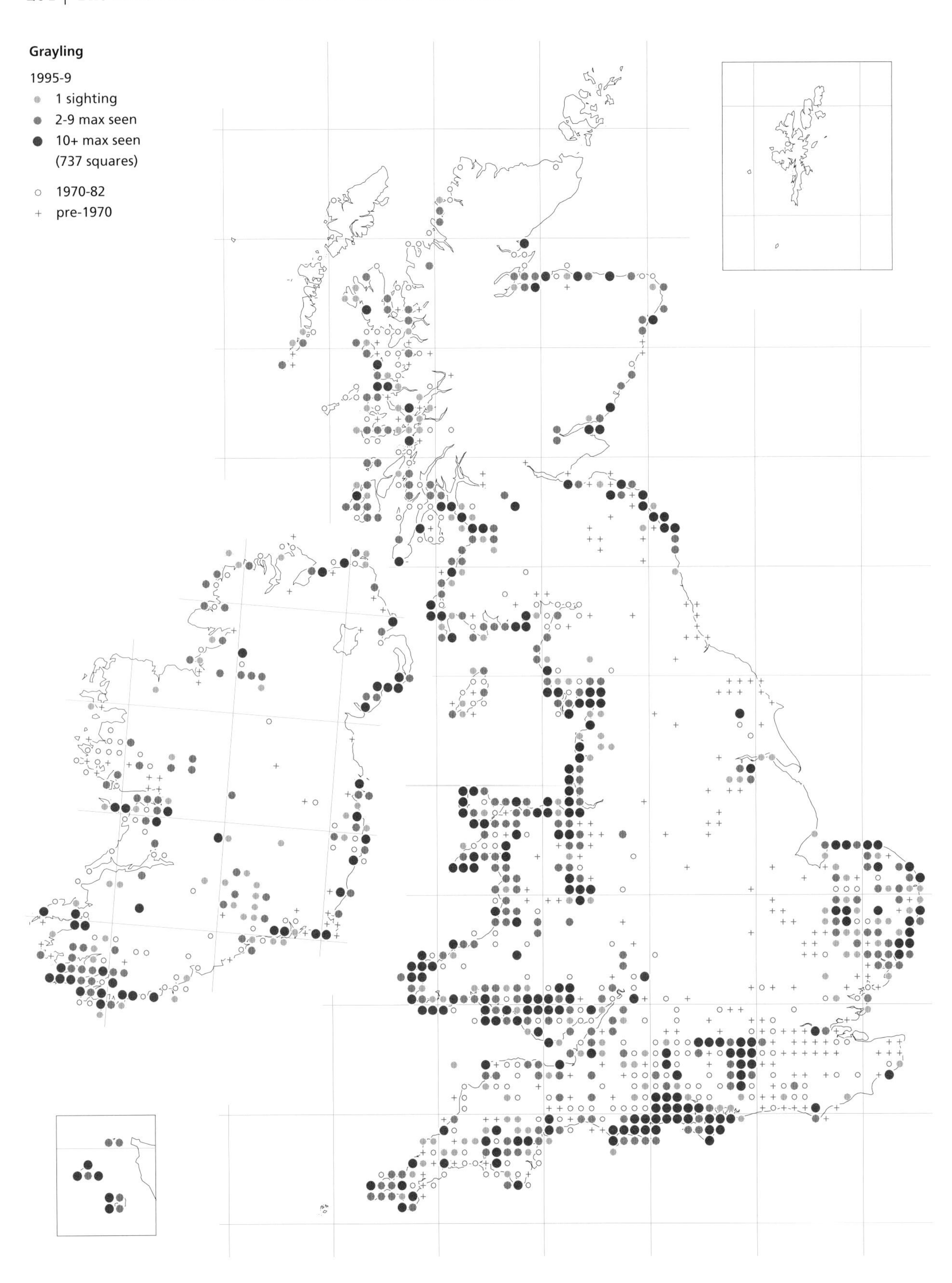
Grayling
1995-9
1 sighting
2-9 max seen
10+ max seen
(737 squares)
1970-82
pre-1970

Grayling *Hipparchia semele*

Resident

Range declining in some areas.

Conservation status

UK BAP status: not listed.
Butterfly Conservation priority: low.
European status: not threatened.

European/world range

Through Europe as far north as 63 °N, but absent from parts of south-east Europe, and extending into western and northern Asia. It is declining in many European countries.

Cryptic colouring provides the Grayling with excellent camouflage, making it difficult to see when at rest on bare ground, tree trunks, or stones. The wings are kept closed when not in flight and the fore wings are usually tucked behind the hind wings, concealing the eyespots and making the butterfly appear smaller. In flight this is a distinctive, large butterfly with a looping and gliding flight, during which the paler bands on the upperwings are visible. The Grayling is widespread on the coast and southern heaths, but is declining in many areas, particularly inland.

Foodplants

The main species used include Sheep's-fescue (*Festuca ovina*), Red Fescue (*F. rubra*), Bristle Bent (*Agrostis curtisii*), and Early Hair-grass (*Aira praecox*). Coarser grasses such as Tufted Hair-grass (*Deschampsia cespitosa*) and Marram (*Ammophila arenaria*) are occasionally used.

Habitat

Many colonies are coastal, on dunes, saltmarsh, undercliffs, and clifftops. Inland, colonies are found on dry heathland, calcareous grassland, old quarries, earthworks, derelict industrial sites such as old spoil heaps and, in a few areas, in open woodland on stony ground. It occurs on a wide range of soil types, but all are dry and well-drained, with sparse vegetation and plenty of bare ground in open positions.

Life cycle and colony structure

This is a very variable species of butterfly and distinct races have been named in northern Scotland, Ireland, and on Great Ormes Head in north Wales. It is single brooded, with adults emerging in early July and peaking in August, but with a few still flying in mid-September (see p. 397). The race on Great Ormes Head is markedly smaller and emerges a month earlier, flying in June.

Graylings spend less time nectaring than many butterflies, but are attracted to feed from muddy puddles and sap oozing from tree trunks. Males are territorial and perch in sunshine, usually on a patch of bare ground but also on boulders or tree trunks, to wait for passing females. Any approaching insect of similar size is investigated by the male and, when a receptive female is located, there is a characteristic courtship display. The male flaps his wings and folds them so that the female's antennae are drawn across the scent glands on his forewings before pairing takes place.

Females are seen less often, except when flying between scattered tufts of grass to lay **eggs**. They lay these singly on plants growing in full sun, usually surrounded by bare ground. The **larvae** feed on the grass

Grayling habitat on heathland.

leaves at night. They hibernate in tussocks during the winter and resume feeding in spring, becoming full-grown in June. **Pupation** takes place in a silk-lined cavity just below the surface of the ground.

Graylings live in colonies that range in size from fewer than 50 adults to thousands where there are large stretches of suitable habitat. Little is known about their mobility, but adults are rarely seen away from breeding areas. However, their presence on many offshore islands implies that they may be able to spread in favourable years.

Distribution and trends

This is a widespread coastal species in Britain and Ireland, although it is absent from much of the east coast between Norfolk and Northumberland, and sparse in the extreme north of Scotland. It is present on many offshore islands, including the Channel Islands, Lundy, the Isle of Man, and the Western Isles of Scotland, but absent from Orkney and Shetland.

In England, there are some inland colonies on dry lowland heath, especially in Dorset, the New Forest in Hampshire, in Surrey, and in the Breckland of Suffolk and Norfolk. Others are found on old mine spoil-heaps, for example in Shropshire, and a few remain on chalk grassland in the south. In Wales, inland populations are scattered and occur mainly on old mine spoil; in Ireland they occur mainly on limestone grassland.

The distinctive appearance and behaviour of the Grayling ensure that it is unlikely to be confused with any other butterfly. It has been seen in over 220 new 10 km squares in Britain since the 1970–82 survey, and over 80 in Ireland, probably due to past under-recording rather than expansion. It is thought to remain under-recorded in parts of Ireland, Scotland and Wales. There has been a distribution decline in recent decades in parts of southern England on all habitats and the current survey shows that it has not been re-recorded in 38% (214) of the 10 km squares where it was found in Britain in 1970–82.

There has also been a decline in abundance at many sites monitored by the Butterfly Monitoring Scheme (BMS), although there have been no extinctions. The BMS collated index of abundance declined during the 1990s, remaining at a low level until 1999. These declines have occurred despite warm summers, which should have benefited this species.

European trend

The Grayling has declined in many countries, especially Austria, Belgium and Luxembourg (50–75% decrease in 25 years), Germany, Latvia, the Netherlands, and Slovakia (25–50% decrease).

Interpretation and outlook

In coastal areas, the Grayling remains widespread and locally abundant in many parts of Britain and Ireland. However, many colonies currently face a range of threats and there have been losses due to building development and the construction of golf courses and car parks on sand dunes. Agricultural intensification has affected land right up to the cliff edge in many places, restricting suitable habitat to the extreme edge of the coast where land cannot be worked. However, dune systems undergo changes as a result of tide and wind and, in places, new breeding areas are created, providing the Grayling with opportunities to increase.

Colonies on lowland heaths have been in decline for many years, as the amount of suitable habitat has decreased through factors such as: the complete destruction of heathland for housing and road development; losses through agricultural improvement and ploughing; and successional changes following the cessation of grazing and invasion by coarse grasses, scrub and trees. The overall loss of lowland heathland has been estimated at over 60% during the last hundred years. The remaining fragments often contain very small breeding areas suitable for the Grayling and many populations have reduced in size in recent years.

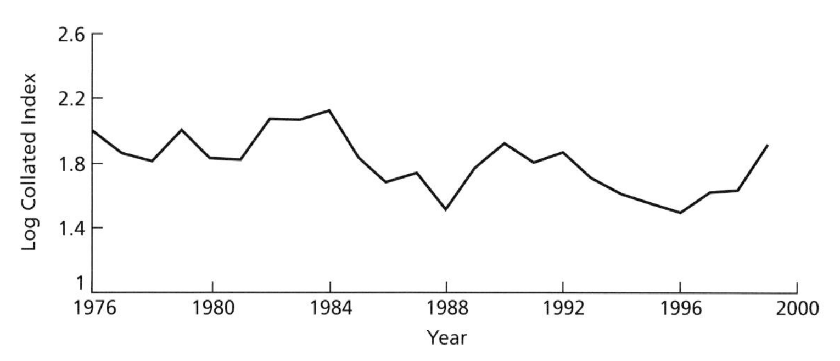

Grayling abundance on BMS sites, showing general decline and increase since 1997.

Hopefully, better management of heathland, with removal of invasive species and reinstatement of traditional grazing and cutting regimes, will prevent further losses. The importance of lowland heath is now recognized internationally and restoration programmes are taking place in several regions, helped by schemes such as Countryside Stewardship and the Wildlife Enhancement Scheme.

This species was never common on southern chalk downland and declined in the 1950s when myxomatosis led to a dramatic fall in numbers of rabbits, and a consequent increase in the height and density of vegetation. Few colonies now survive in this habitat, most being restricted to old quarries where there is very short, sparse vegetation on steep slopes. The Grayling was lost from the North Downs in the 1950s and from the Berkshire Downs in the 1990s. Although the introduction of grazing specifically for conservation may mean that suitable habitat is again available on some sites, distance from surviving colonies prohibits recolonization. Other inland colonies, such as those on spoil heaps and other old industrial sites, are being lost due to reclamation schemes or development. Where colonies survive, management should be aimed at maintaining sparse vegetation with plenty of bare ground.

The decline in numbers may continue as habitat is lost through destruction and successional changes. More knowledge of the mobility and habitat requirements of this species is therefore needed. Warm summers, such as those experienced during the 1990s, might have been expected to have helped the species, but the BMS data show declines in abundance in most years. Given these recent changes, greater attention should now be paid to the conservation of this species, and its priority rating should be reviewed.

Key references

Dennis and Bardell (1996); Dennis *et al.* (1998*a*,*b*); Findlay *et al.* (1983); Tinbergen (1972).

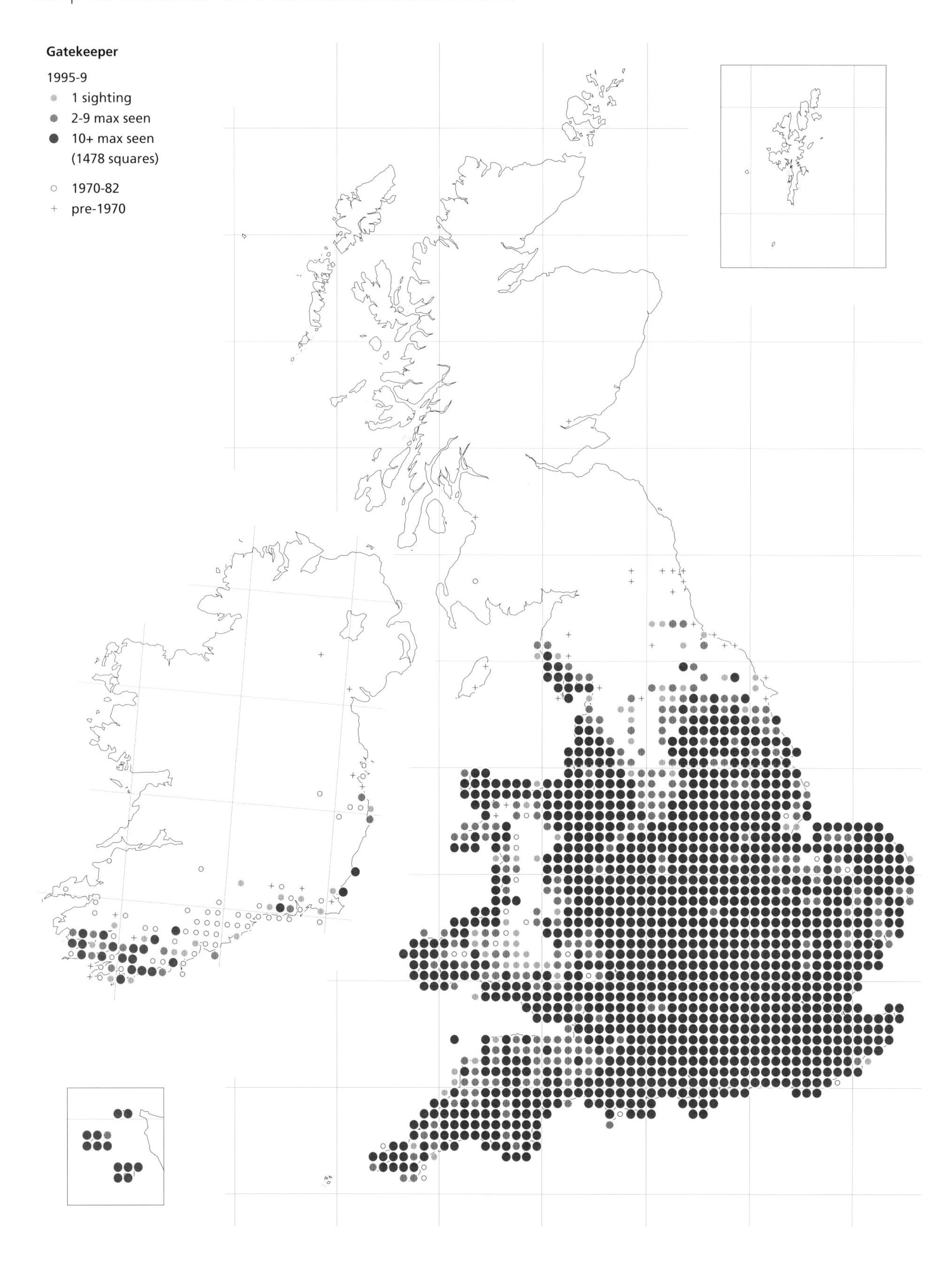
Gatekeeper
1995-9
1 sighting
2-9 max seen
10+ max seen
(1478 squares)
1970-82
pre-1970

Gatekeeper *Pyronia tithonus*

Resident
Range expanding in Britain.

Conservation status
UK BAP status: not listed.
Butterfly Conservation priority: low.
European status: not threatened.

European/world range
It occurs widely across Europe south of 53 °N, to Asia Minor and the Caucasus, and also occurs locally in Morocco. It is declining in several European countries.

As its English names suggest, the Gatekeeper (also known as the Hedge Brown) is often encountered where clumps of flowers grow in gateways and along hedgerows and field edges. It is often seen together with the Meadow Brown and Ringlet, from which it is easily distinguished when basking or nectaring with open wings. The colour and patterning of the wings are very variable and about a dozen aberrations have been named. Favourite nectar sources include Wild Marjoram, Common Fleabane, ragworts, and Bramble. It is widespread in southern Britain and its range has extended northwards in recent years. Its range is far more localized in southern Ireland.

Foodplants
Various grasses are used, with a preference for fine grasses such as bents (*Agrostis* spp.), fescues (*Festuca* spp.), and meadow-grasses (*Poa* spp.). Common Couch (*Elytrigia repens*) is also used. The full range of other species used is not known.

Habitat
This species is found in grassland where tall grasses grow close to hedges, trees, or scrub, particularly along hedgerows and woodland rides and also in habitats such as undercliffs, heathland, and downland where there are patches of scrub. Open grassland with short vegetation is avoided.

Life cycle and colony structure
There is one generation a year with a short flight period, beginning in mid-July, reaching peak numbers in early August and finishing by the end of August (see p. 398). Adult longevity is reduced in hot dry summers.

Males perch on bushes and taller herbs to await passing females, investigating any passing butterfly. Bramble is a particularly favoured nectar source and Gatekeepers spend long periods nectaring and basking on Bramble bushes in a sunny position. Females choose taller grassy vegetation, close to shrubs, on which to lay **eggs**, which are either placed on foodplants or deposited on nearby vegetation. The young **larvae** hibernate during the second instar, hidden at the base of grass clumps. Although closely related species, such as the Meadow Brown and Ringlet, continue to feed during warm spells in the winter, the Gatekeeper is thought to cease feeding completely and enter an inactive state. After winter, larvae feed nocturnally until fully grown in June, resting during the day at the base of tall grass stems. The **pupae** are suspended among vegetation at the base of shrubs.

Gatekeepers live in colonies that vary greatly in size, the smallest producing fewer than a hundred individuals in a season, the largest producing thousands. They are sedentary butterflies, rarely flying beyond the

Gatekeeper larva.

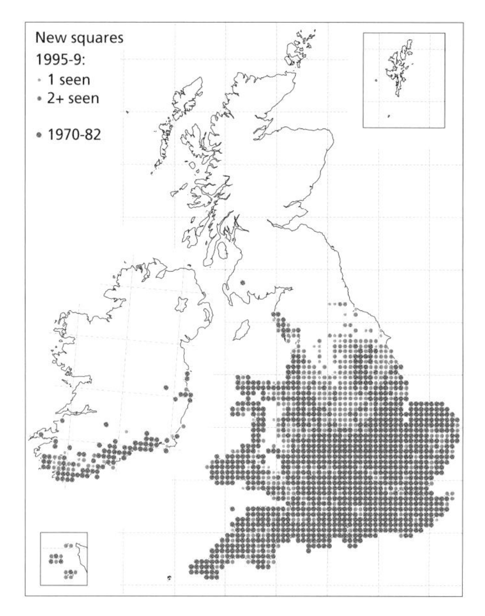

The expansion of range of the Gatekeeper since 1970–82.

boundary of their colony. A few longer flights are made, however, and the spread of the species in recent years demonstrates that it has some colonizing ability. Adults may be seen nectaring in gardens, but it is not clear whether they breed in this habitat.

Distribution and trends

In southern England, the Gatekeeper is a common and widespread butterfly. It occurs on the Channel Islands and other southern offshore islands. In the Midlands it is less widespread, occurring mainly in woodland. It is found as far north as Yorkshire, Lancashire, and Cumbria, where it is confined to the warmest, most sheltered locations. The northern edge of its range on the west side of England, in Cumbria, does not appear to have changed in recent decades, but in the east it has extended northwards and westwards in Yorkshire. It has not yet reached Co. Durham and parts of south Northumberland from which it disappeared at the end of the nineteenth century. The current map also shows considerable northwards expansion in the central counties of Leicestershire, Nottinghamshire, and Derbyshire into areas occupied in the nineteenth century.

This expansion began in the 1970s and has continued, particularly during the warm summers of the early 1980s and since 1989. On the western side of England, where colonies at the edge of the range are mainly coastal, there has been both an increase in numbers at traditional sites and a spread eastwards to new areas away from the coast. The current survey recorded the Gatekeeper in 33% more 10 km squares in Britain than in 1970–82. The spread northwards has not been accompanied by a general increase in abundance at Butterfly Monitoring Scheme sites.

In Wales, populations are especially strong in coastal areas, but also occur inland, though not in mountainous parts. In Ireland the distribution is limited to an area within about 50 km of the south coast, extending northwards along the east coast towards Dublin. The Gatekeeper has not extended its range northwards there in recent years. It is not clear whether there has been a decline in Ireland since the 1970–82 survey, or whether the species is under-recorded in the 1995–9 period. This butterfly does not occur in Scotland (where a few historic records are considered to be erroneous or of introduced individuals) or on the Isle of Man.

The Gatekeeper is not generally a species of urban habitats, but in recent years has been recorded at new sites in the London area. On Hampstead Heath, where it first appeared on the transect in 1991, numbers have risen steadily since. It has also been recorded recently on Wimbledon and Mitcham Commons; transect figures show an increase in numbers on commons in the London Borough of Croydon which exceeds the regional trend. It has also spread southwards in Essex to recolonize Epping Forest and urban parts of north-east London from which it disappeared in the nineteenth century.

European trend

The Gatekeeper's range is stable in most countries, but the species has become extinct in Poland and Slovakia, and has declined seriously in Romania and Turkey (>50% decrease in 25 years). In contrast, it has expanded recently in Hungary and the Netherlands.

Interpretation and outlook

Although this species is common and widespread across most of England and

Wales, colonies have been lost as fields have become larger with the intensification of agriculture. This butterfly has been affected more than most by the scale of hedgerow removal and the decrease in availability of field edge habitat, especially in the arable areas of eastern England. There have also been substantial declines throughout its range due to loss of unimproved grassland. The Gatekeeper will colonize even small patches of suitable habitat, so there is scope for populations to be increased by the planting of hedges and the growth of strips of native grasses alongside them.

Much remains to be learned about the mobility, temperature requirements, and larval foodplants of this species. Numbers fluctuate considerably, with close synchrony of fluctuations across its range, suggesting that the cause is climatic. Increases in abundance are associated with warm weather in the previous and current summers, and June temperatures are considered to be a major determining factor in date of emergence. The recent northwards range expansion is thought to be associated with the higher temperatures experienced in recent summers. Although the Gatekeeper is generally a sedentary species, it seems to be more mobile in warm summers, and more likely to disperse away from breeding areas.

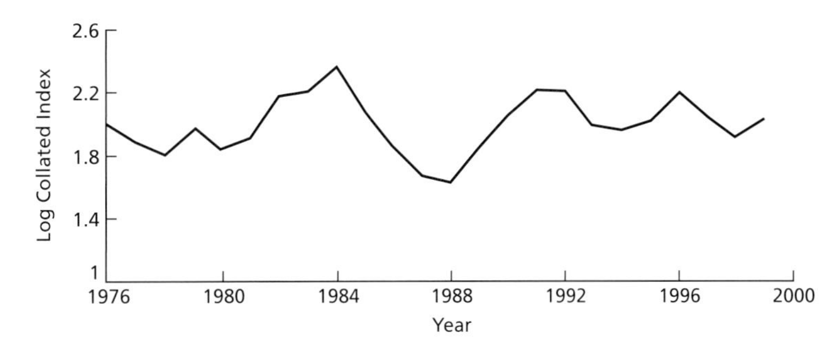

There has been no overall increase in Gatekeeper abundance at BMS sites despite its range expansion.

The colonizations in the London area, however, are unlikely to be due to climatic changes, as the species occurs so much further north. Air pollution may have been responsible for the loss of the species in the London area in the nineteenth century and its recolonization may be due to the recent improvements in air quality. This theory has not been proved and increases may be due, in part, to the creation of wildflower areas and less frequent mowing regimes in many parks, gardens, and cemeteries.

The Gatekeeper is not able to use as wide a range of grassland habitats as the Meadow Brown and this may also be a limiting factor in its spread northwards. It may continue to spread and recolonize areas in north-east England that it occupied in the nineteenth century, and continued survey and monitoring at the northern edge of its range are required.

Key references

Brakefield (1987); Hill *et al.* (in press); Pollard (1991).

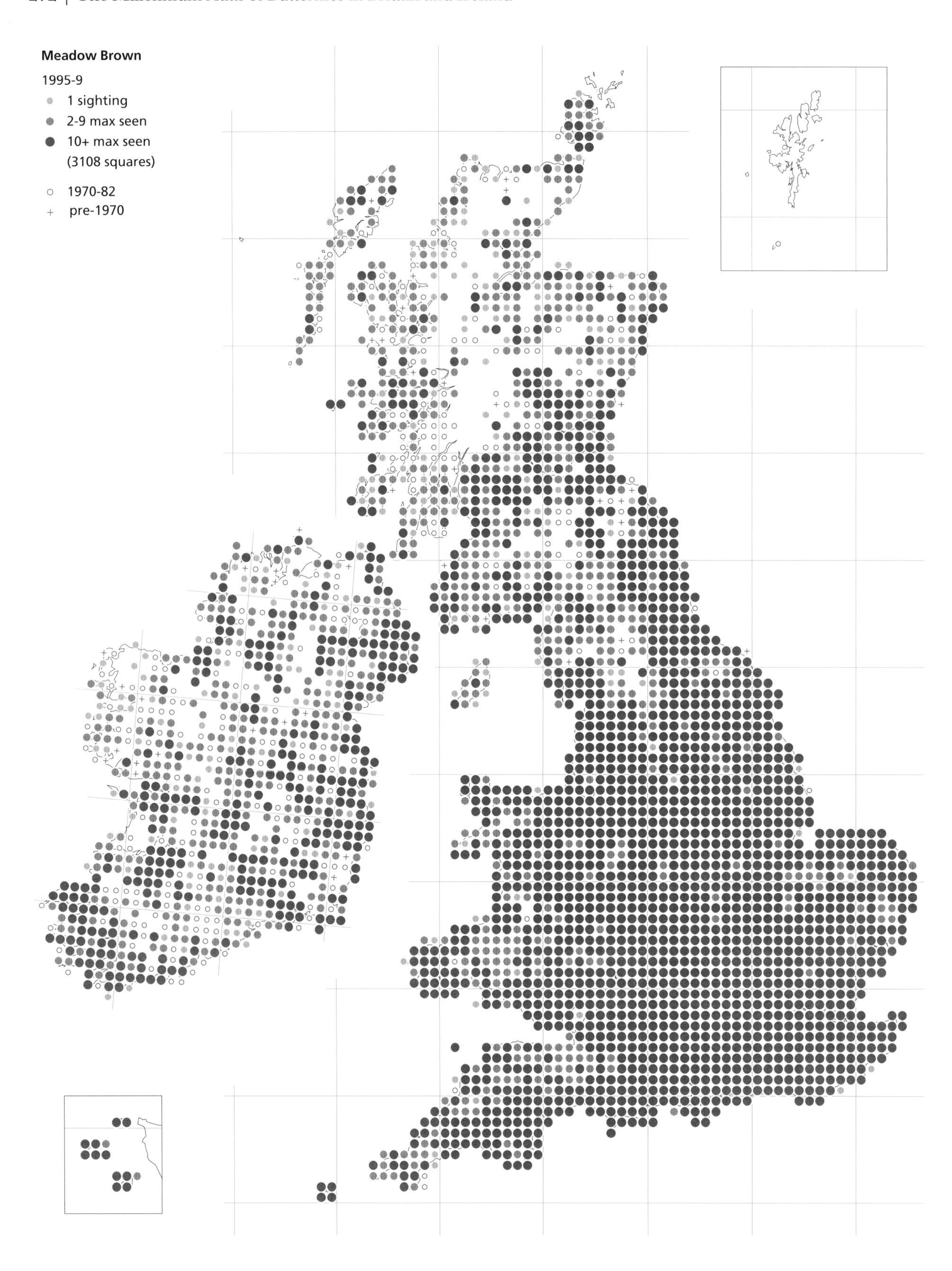
Meadow Brown
1995-9
1 sighting
2-9 max seen
10+ max seen
(3108 squares)
1970-82
pre-1970

Meadow Brown *Maniola jurtina*

Resident

Range stable.

Conservation status

UK BAP status: not listed.
Butterfly Conservation priority: low.
European status: not threatened.

European/world range

Its range extends across Europe south of 62 °N, and eastwards to the Urals, Asia Minor and Iran. It is stable in most European countries.

The Meadow Brown is the most abundant butterfly species in many habitats. Hundreds may be seen together at some sites, flying low over the vegetation. Adults fly even in dull weather when most other butterflies are inactive. Regional variations in the spotting pattern on the wings have led to it being studied extensively by geneticists over many years. Larger forms occur in Ireland and the north of Scotland. It is one of our most widespread species, but many colonies have been lost due to agricultural intensification.

Foodplants

A wide range of grasses is used. Those with finer leaves such as fescues (*Festuca* spp.), bents (*Agrostis* spp.), and meadow-grasses (*Poa* spp.) are preferred, but some coarser species such as Cock's-foot (*Dactylis glomerata*), Downy Oat-grass (*Helictotrichon pubescens*), and False Brome (*Brachypodium sylvaticum*) are also eaten by larger larvae. Other species of grass are also believed to be used.

Habitat

Open grasslands, including downland, heathland, coastal dunes and undercliffs, hay meadows, roadside verges, hedgerows, woodland rides and clearings, and waste ground. It also occurs in urban habitats such as parks, larger gardens, and cemeteries.

Life cycle and colony structure

There is one generation a year (see p. 398), the first adults emerging at the beginning of June on downland, where the flight period tends to be longer than in other habitats. The first adults to emerge are those in particularly warm localities, for example, sheltered south-facing banks in full sun. Males emerge several days before the first females. Fresh individuals may be seen as late as the end of October and it is sometimes suggested that there may be more than one generation. Late sightings of freshly emerged individuals are, however, thought to be due to the very varied length of time taken by larvae to reach maturity.

Females lay **eggs** on or near grasses, sometimes placing them carefully on a foodplant, at other times depositing them on other vegetation nearby. The young **larvae** feed during the day. They overwinter among grass stems, continuing to feed in mild weather. In the spring they feed nocturnally, and can be found climbing grass stems after dusk. **Pupation** takes place among grass stems or litter at ground level. The colour and patterning of both larvae and pupae are very variable.

Meadow Brown larva.

This butterfly breeds in suitable patches of habitat over wide areas of the countryside, sometimes occurring at high densities. Colonies in the most suitable habitats, such as downland, coastal grassland, and

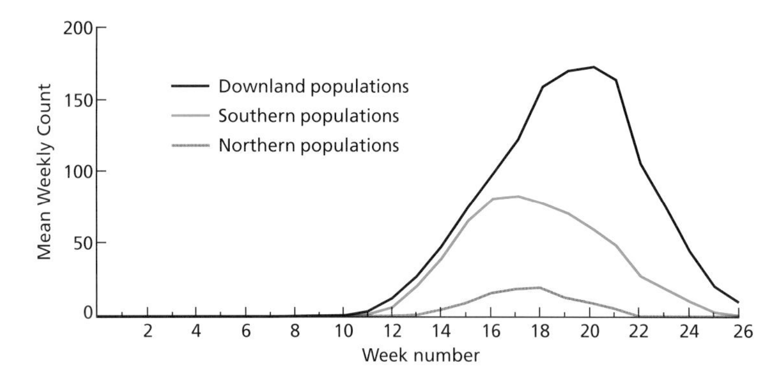

Average flight period of the Meadow Brown on BMS sites (1976–99), showing larger populations and later and longer flight period at downland sites.

unfertilized meadows, can reach high densities, estimated to be as high as 2000 adults per hectare at some sites. Marking studies have shown that adults regularly travel more than 100 m, even flying across crops and open fields.

Distribution and trends

The Meadow Brown is widespread in Britain and Ireland, and is absent only from Shetland. In Orkney, it is at the extreme edge of its range and is usually found only in the warmest, south-facing situations, although in warm summers it becomes more widely distributed over the islands. In mountainous areas, it becomes increasingly scarce above 200–300 m and it is absent from the tops of higher mountains. In the north of Scotland, this butterfly occurs more commonly in coastal areas, although it is thought to be under-recorded in parts of Scotland.

There has been no change in range in recent years but populations have become much reduced in areas of intensive agriculture, where arable crops and rye-grass monocultures have replaced native grasses, leaving only small fragments of habitat suitable for breeding, for example field margins and hedgerows. In London and other cities, colonies of the Meadow Brown are more common than those of closely related species such as the Gatekeeper and Ringlet.

European trend

The Meadow Brown has undergone a severe decline in Finland (50–75% decrease in 25 years) and Malta (25–50% decrease) and to a lesser extent in Luxembourg, Sweden, and parts of Russia. There have been increases in the Netherlands and Romania over the same period.

Interpretation and outlook

This remains a common and widespread butterfly throughout Britain and Ireland. At one time the Meadow Brown, as its name suggests, was strongly associated with hay meadows, where females laid eggs after the crop was cut. The enormous area occupied by hay meadows in earlier centuries probably supported numbers many times greater than those remaining today. Across Britain such expanses of native grasses have been replaced by arable crops and rye-grass leys, which will not support colonies.

Except for nature reserves and land used for recreation, the Meadow Brown is now restricted to smaller populations in fragments of suitable habitat such as roadside verges, small corners of waste ground, and hedgerows. It does not tolerate very heavily grazed swards and increased stocking rates in upland areas have led to widespread declines in numbers. Increased shade due to lack of management in many woods has also resulted in losses.

Numbers are likely to continue to decline in farmland if intensification continues. However, populations may be increased by encouraging greater provision of conservation headlands, field margins, and other patches of suitable habitat. Where more extensive tracts of native grasses remain, higher numbers can be encouraged by providing an open sward of medium height in a sheltered situation, with a good mixture of fine grasses and plenty of flowers to provide nectar.

This is the most widely recorded butterfly in the Butterfly Monitoring Scheme and numbers recorded annually in the scheme are far higher than those of any other species. There have been significant increases in population levels in recent years at several monitored sites. However, most monitored sites are nature reserves and the pattern of increase does not necessarily reflect the situation in the wider countryside.

The wide variability in the length of the flight period of this species has long been recognized. Higher summer temperatures result in earlier emergence and increased abundance. The flight period is longest on short turf, especially on chalk grassland, and is markedly shorter in the north of Scotland. The wide range of habitats used by the Meadow Brown probably results in greater variation in development time and emergence than is the case with the Gatekeeper. The use of warmer, short turf habitat may also enable the Meadow Brown to survive further north than the Gatekeeper.

Overall, the Meadow Brown does not seem to be at threat either in the wider countryside, or in urban areas, provided that some areas of semi-natural grassland remain available.

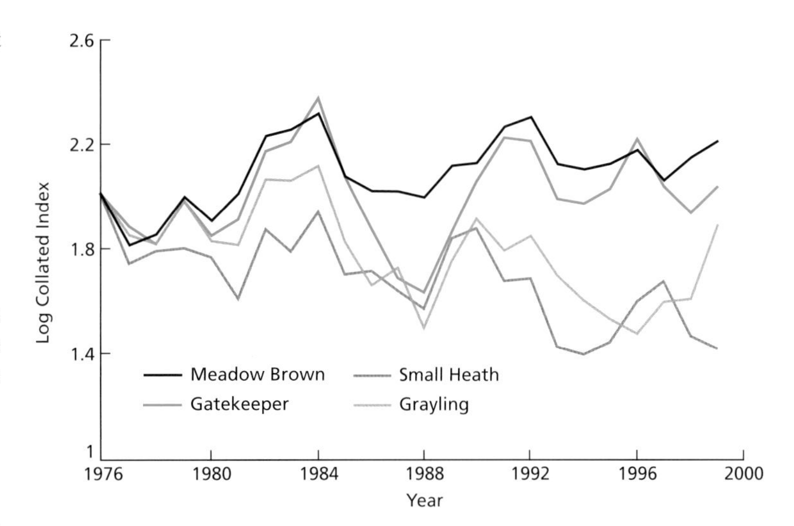

Abundance of four species of browns at BMS sites, showing synchrony of fluctuations (and the recent decline of the Small Heath).

Key references

Brakefield (1982*a*,*b*, 1987); Cribb (1975); Dowdeswell (1981); Pollard (1981).

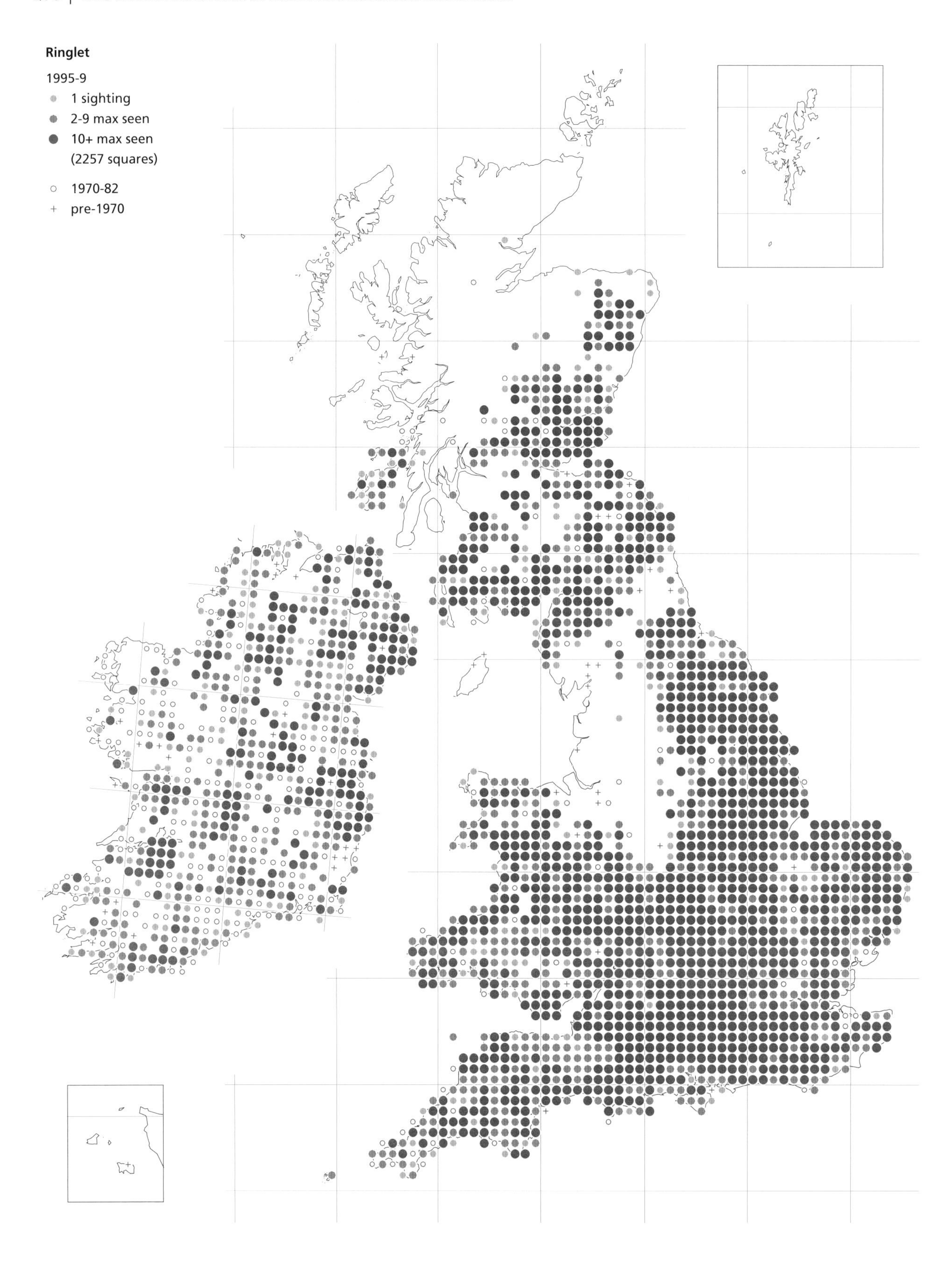
Ringlet
1995-9
1 sighting
2-9 max seen
10+ max seen
(2257 squares)
1970-82
pre-1970

Ringlet *Aphantopus hyperantus*

Resident

Range expanding.

Conservation status

UK BAP status: not listed.
Butterfly Conservation priority: low.
European status: not threatened.

European/world range

Across Europe except in central and southern Spain, Portugal, peninsular Italy, and northern Scandinavia, and extends eastwards across Asia to Japan. It is stable in most European countries, but its overall range appears to be shifting northwards.

When newly emerged, the Ringlet has a velvety appearance and is almost black, with a white fringe to the wings. The small circles on the underwings, which give the butterfly its name, vary in number and size and may be enlarged and elongated or reduced to small white spots; occasionally they lack the black ring. Bramble and Wild Privet flowers are favourite nectar sources, and adults continue to fly with a characteristic bobbing flight in dull, cloudy conditions when most other butterflies are inactive. This widespread butterfly has extended it range in England and Scotland in recent years.

Foodplants

Coarser grasses are used, including Cock's-foot (*Dactylis glomerata*), False Brome (*Brachypodium sylvaticum*), Tufted Hair-grass (*Deschampsia cespitosa*), Common Couch (*Elytrigia repens*), and meadow-grasses (*Poa* spp.). Other species of grass may also be used.

Habitat

Tall grassland is used, mainly in damp situations in partial shade on heavy soils where grasses are lush, especially in woodland rides and glades. The butterfly also occurs on commons, verges, and riverbanks, especially on clay soils. In northern areas, it is found in more open, less shady, situations.

Life cycle and colony structure

There is one generation a year. Adult emergence begins at the end of June and the first males are on the wing up to ten days before the first females. Peak numbers occur in the third week of July and during early August numbers decline rapidly, the flight period being over by mid-August (see p. 398). Males can be seen weaving through long grass as they patrol in search of females. **Eggs** are dropped as the female perches at the top of tall grass stems, rather than attached to the plant. **Larvae** feed nocturnally on tall, lush grasses. Like the larvae of other browns, they can be found climbing grass stems after dusk. They spend the winter hidden low down among the vegetation, feeding during warm spells. They are fully grown in May. **Pupation** takes place in a cocoon of silk at the base of a grass clump.

The Ringlet, like other browns, lives in colonies. These can range in size from a few individuals in a corner of a field to many hundreds in woodland rides or areas with scrub and rough grassland. Adults are generally sedentary and rarely fly beyond the boundary of the colony, although if a large area of suitable habitat is available they will wander freely through it, travelling distances of several hundred metres. Its recent spread also suggests that adults are capable of dispersing over large distances, though this has not been quantified.

Ringlet larva.

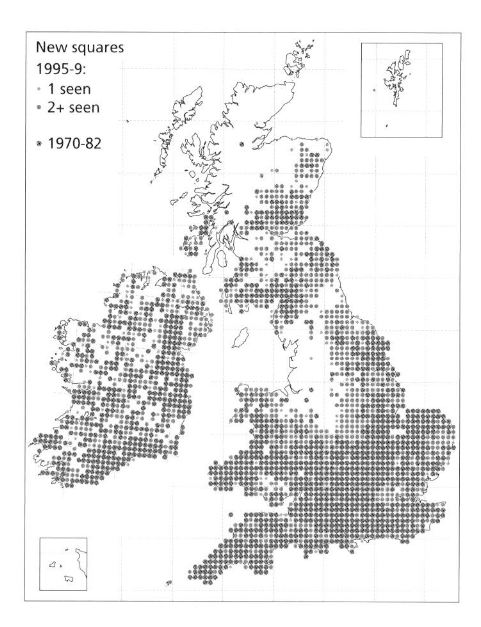

The expansion of range of the Ringlet since 1970–82.

Distribution and trends

The Ringlet is widely distributed across southern and eastern England, Wales, Ireland and parts of lowland Scotland. It also occurs on some of the Inner Hebrides, including Islay and Jura. It does not occur on the Channel Islands, Isle of Man, Outer Hebrides, Orkney, or Shetland. There is a large gap in its distribution, in the Midlands and north of England, parts of southern and central Scotland, and north-east Wales. It occurred at one time in parts of this large area but disappeared in the nineteenth century, when it also disappeared from the London area. In Wales, it began to spread across Gwynedd in the 1930s, colonizing Anglesey before 1950. The recolonization of parts of the Midlands, Yorkshire and north-east England began later, and much of it took place during the 1990s. Across Britain it has been recorded in over 50% more squares during the current survey than in 1970–82.

During the recent period of expansion it has moved northwards in Shropshire and westwards across the counties of Leicestershire, Nottinghamshire, and Derbyshire. It has also increased considerably in West Yorkshire and County Durham, reappearing in areas from which it retreated in the late nineteenth century. In eastern Scotland it has moved northwards in the border area and southwards in Fife to close the gap in distribution that existed in the 1970s between these areas, colonizing new parts of Berwick and Lothian. A few recent records from the extreme north-east of Scotland indicate that it may have colonized here, though it does not occur at altitudes above 200–300 m. As much of the recent expansion has taken place in Scotland, where there are fewer recorders, the current map may not show the full distribution there. In recent years the Ringlet has also shown signs of colonizing some parks and commons in London.

Some colonizations have taken place within the range of the species in habitats, such as chalk downland, where scrub encroachment has led to the development of patches of suitable breeding vegetation. However, it is thought that many colonies were lost in the twentieth century throughout its range, especially in intensively farmed parts of eastern England.

There may be a small recording bias as older individuals, faded to a dark brown and losing the white edge to the wings, can be mistaken in flight for Meadow Browns. When at rest, however, the distinctive eyespots are clearly visible and the two species are easily separated.

European trend

Its distribution is stable in most countries. There have been increases in the Netherlands and Romania in the past 25 years, but small decreases in Luxembourg and Moldova. Over the last 30–100 years its range appears to have shifted northwards in Europe, extending in the north but contracting in the south.

Interpretation and outlook

The reasons for the decline of the Ringlet from parts of its range in the nineteenth century, and for its subsequent re-expansion, are not clear. Climatic factors may be partly responsible, but temperature alone cannot explain why the species is absent from large parts of northern England and southern Scotland, when it survives much further north. Suitable habitat and foodplants are available throughout Britain, and do not appear to limit the distribution of the species, though drainage and land management affect populations at a local level.

The Ringlet may be adversely affected by atmospheric pollution and that would explain its disappearance from industrial areas of Scotland and central and northern England, and also from the London area, during the nineteenth century. If this is the case, cleaner air may at least in part explain its recovery in these areas. Although this theory has not been thoroughly investigated, the distribution of the Ringlet coincides with the distribution of lichens, which are adversely affected by sulphur dioxide pollution.

Numbers fell dramatically in 1977 on monitored sites following the extreme drought in 1976. At the Butterfly Monitoring Scheme (BMS) site at Monks Wood, Cambridgeshire, Ringlets disappeared from open areas and areas with dense shade for several years after this drought, remaining only in rides in moderate shade, which provide the most suitable habitat. After 1977 there were significant increases at a number of BMS sites, mainly in the south and east of England, as the species recovered. These increases continued until 1984. Since then there has been little change in abundance, although some decreases occurred in 1996 following the 1995 drought. These were not as severe as in 1977.

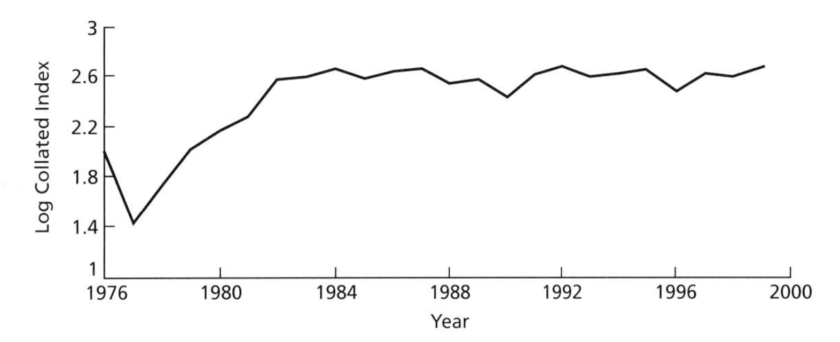

Ringlet abundance declined at BMS sites after the 1976 drought and subsequently increased.

Many colonies have disappeared within the species' range due to loss of native grasses and drainage of damp habitats, and in some areas overgrazing has also resulted in losses. Overzealous cutting of vegetation and tidying up of hedgerows and commons also destroys habitat. Grasses are only used for egg-laying when they are in a vigorous, ungrazed state. Where colonies are known to exist, retention of uncut grassy rides, verges, and headlands, especially in damp situations, will benefit the species, while provision of shade and shelter from wind are also important.

A recent model of the effects of climate on the expansion of the Ringlet has predicted that, if current climate changes continue, further areas of Britain are likely to be colonized.

Key references

Corke (1999); Hill *et al.* (in press); Sutcliffe and Thomas (1996); Sutcliffe *et al.* (1997*a*,*b*).

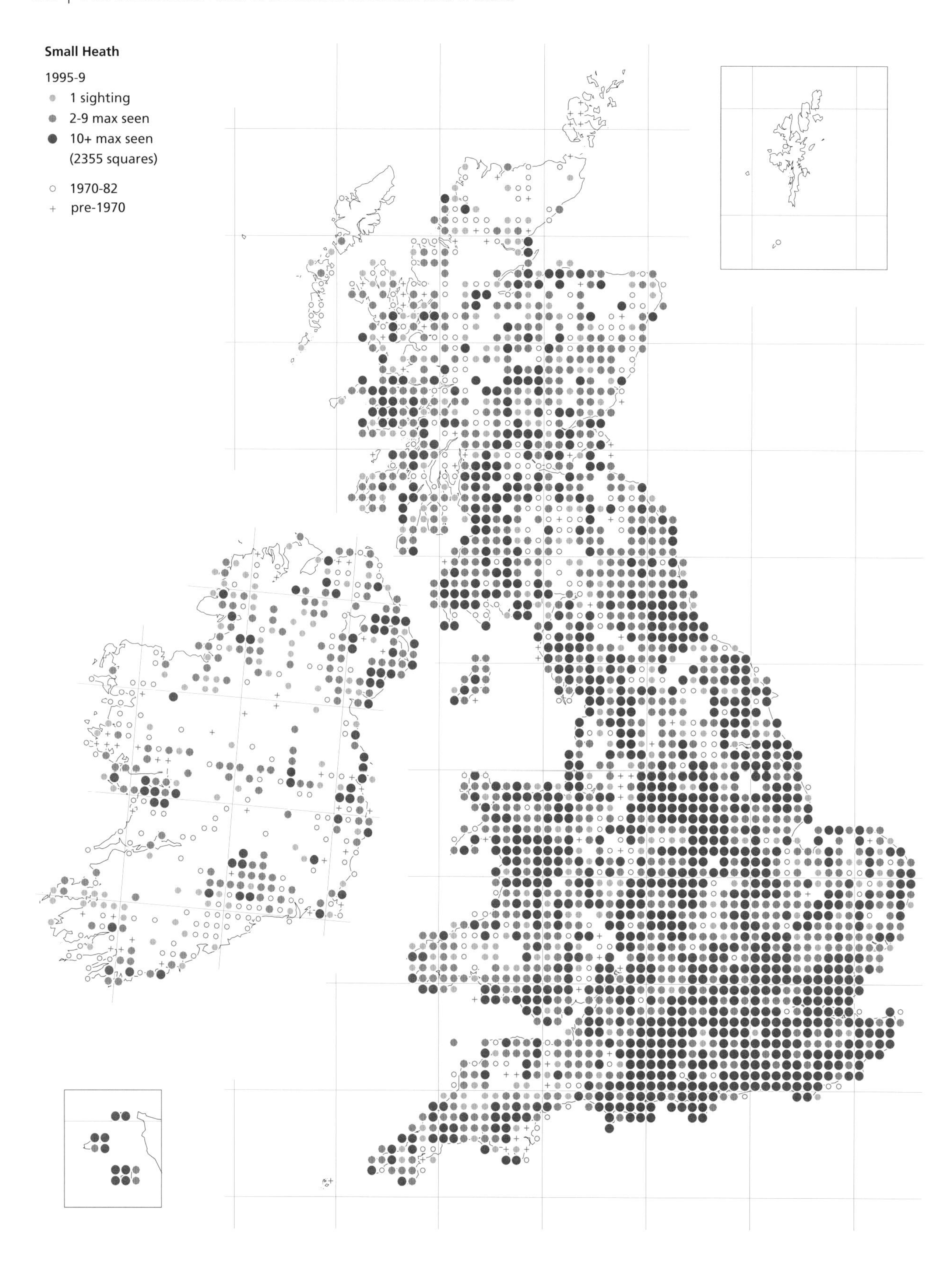
Small Heath
1995-9
1 sighting
2-9 max seen
10+ max seen
(2355 squares)
1970-82
pre-1970

Small Heath *Coenonympha pamphilus*

Resident

Range stable.

Conservation status

UK BAP status: not listed.
Butterfly Conservation priority: low.
European status: not threatened.

European/world range

Throughout Europe to 69 °N, in north Africa and eastwards to Mongolia. It is stable in most European countries.

The Small Heath is an inconspicuous butterfly that flies only in sunshine and rarely settles more than a metre above the ground. Its wings are always kept closed when at rest. The number of broods and the flight periods are variable and adults may be seen continuously from late April to September on some sites in southern England. This relatively widespread butterfly can occupy a range of habitat types and, although its range has changed little, many colonies have disappeared in recent decades.

Foodplants

Fine grasses, especially fescues (*Festuca* spp.), meadow-grasses (*Poa* spp.), and bents (*Agrostis* spp.).

Habitat

This species occurs on grassland where there are fine grasses, especially in dry, well-drained situations where the sward is short and sparse. The largest colonies occur on downland, heathland, and coastal dunes. Smaller populations occur in many other locations including roadside verges, waste ground, woodland rides and glades, moorland, and parkland.

Life cycle and colony structure

Adults are usually seen when nectaring from yellow flowers such as dandelions and hawkbits. The number of broods and timing of flight periods vary in different areas. In northern England and in Scotland there is just one generation of adult butterflies a year, emerging as late as July in the far north. In most of southern Britain and in Ireland there are two or three generations a year following a pattern identified by Richard South in 1906:

May and June butterflies from May and June eggs (twelve months' cycle), July butterflies from August eggs (eleven months' cycle), August and September butterflies (partial second brood) from May and June eggs (four month cycle)

Little needs to be added to this description of a pattern that has been shown clearly by Butterfly Monitoring Scheme (BMS) data. In the single brood pattern in the north of Britain, **eggs** are laid on the leaves of fine grasses by the spring emergence of adults and the resulting **larvae** develop slowly over the summer and following winter to **pupate**

Small Heath larva.

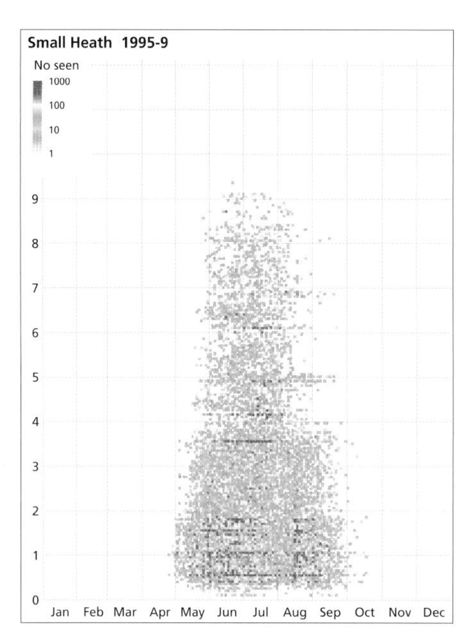

Phenogram of dates of 1995–9 Small Heath records plotted against latitude, showing the switch to a single brood in the north.

and then emerge as adults a year later. Larvae feed by day, and overwinter at different stages of development with feeding being resumed during warm spells. The pupae are suspended from grass stems.

In the warmer conditions further south some of the larvae develop more rapidly to give rise to a late summer (August–September) emergence, a partial second brood. At the warmest locations, for example on southern chalk downland, this second late summer brood will lay eggs that may give rise to larvae that overwinter and emerge the following July, thereby overlapping with the last of the spring emergence. Thus adults may be seen continuously from late April to early September where populations are strong on southern downland sites.

The Small Heath occurs in colonies. Males congregate to compete for dominance and to attract females in breeding areas, known as leks, which are usually situated close to landmarks such as trees and bushes in open grassland. Females visit these areas only to mate, there being no resources such as nectar or larval foodplants. Once mated, female butterflies avoid males and disperse to lay eggs. Mobility in this species has not been studied in detail. The Small Heath is capable of moving some distance and individuals have been found several kilometres from known populations.

Distribution and trends

This is one of the most widely distributed butterflies in Britain and Ireland, being absent only from Shetland and Orkney. Records on Orkney published in the 1984 atlas are now known to be of misidentified Large Heaths. In Ireland, populations are widespread but scattered and the species is thought to be under-recorded. The Small Heath breeds at higher altitudes than other species except for the Large Heath and Mountain Ringlet and colonies occur in upland areas of northern England, Wales, and Scotland where most other butterflies are absent or encountered only as vagrants. It is almost certainly under-recorded in some upland areas.

The range of this species is not thought to have changed in recent years, though the current survey shows that it has not been re-recorded in 14% (255) of the 10 km squares where it was found in Britain in 1970–82. This could represent a genuine decline within its range, especially as recording effort has been greater during the recent survey and the species has been found in over 480 new squares in Britain.

There have undoubtedly been widespread local losses and populations have decreased on a variety of habitats at monitored sites in England and Wales. Four extinctions have occurred at sites in the BMS; all of them in woodland. In southern England numbers fell in the summers following 1990 to a low point in 1994, when this was a scarce butterfly on downland and was rarely seen in other habitats. The reasons for this decline are not known. It remains a less common butterfly than in 1990.

European trend

The Small Heath is stable in most European countries, but there has been a severe decline in the Netherlands (50–75% decrease in 25 years) and smaller declines in Finland and Malta (15–25% decrease). There have been slight increases in Romania and parts of Russia.

Interpretation and outlook

Many colonies have been lost throughout the range of this species where changes in agriculture have resulted in the replacement of native grasses by arable crops or improved grassland. Other sites have become unsuitable due to successional changes following the cessation of grazing. However there have also been declines in numbers of Small Heath in habitats that appear to remain suitable; the reasons for this are not clear but may be due to changes in weather patterns.

This species is most abundant on grassland that is short and sparse and contains a mixture of fine grasses, with a variety of nectar sources available throughout the flight period. Thin, well-drained soils such as chalk and sand appear to be most suitable, whereas cultivated areas enriched with fertilizers are not. Some grazing is usually required to maintain the short sward necessary for this species.

Though the Small Heath remains widespread in distribution it seems to have become much less abundant at a local scale, for example, in comparison with the Meadow Brown (see p. 336). This may be due to habitat fragmentation, overgrazing, or weather conditions, but much remains to be learned about the ecology and requirements of the species before the reasons for its current decline are understood and a prognosis can be given for its future. Like that of the House Sparrow and Song Thrush, the decline of such a common species is a cause for serious concern.

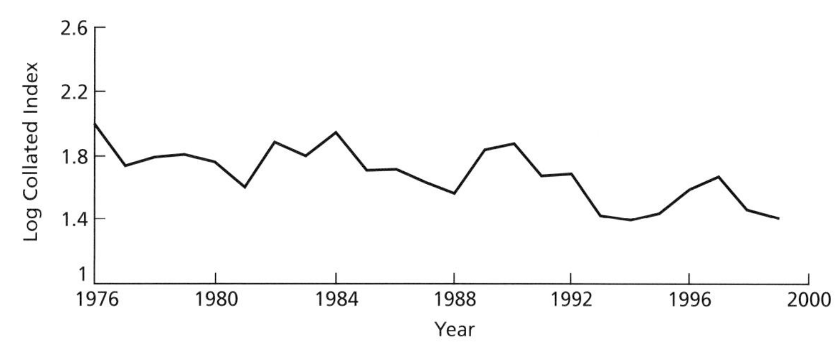

Small Heath abundance is steadily declining at BMS sites.

Key references

Pollard and Greatorex-Davies (1997); Wickman (1992); Wickman *et al.* (1994).

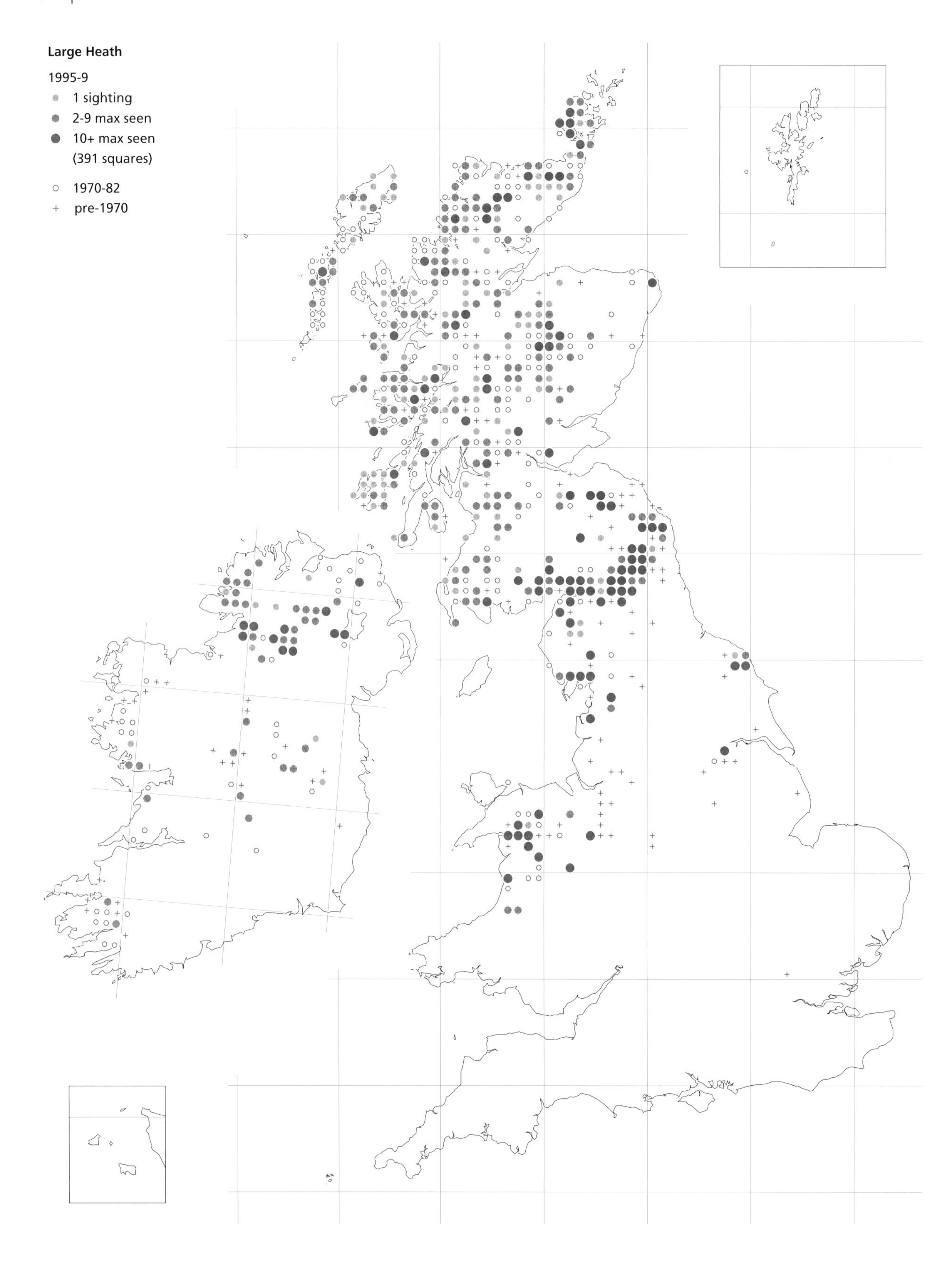
Large Heath
1995-9
1 sighting
2-9 max seen
10+ max seen
(391 squares)
1970-82
pre-1970

Large Heath *Coenonympha tullia*

Resident

Range declining.

Conservation status

UK BAP status: Species of Conservation Concern.
Butterfly Conservation priority: medium.
European status: vulnerable.
Protected in Great Britain for sale only.

European/world range

Occurs across Europe and Asia, and in Canada and western USA. It has declined seriously in many European countries.

The Large Heath is restricted to wet boggy habitats in northern Britain, Ireland, and a few isolated sites in Wales and central England. The adults always sit with their wings closed and can fly even in quite dull weather provided the air temperature is higher than 14 °C. The size of the underwing spots varies across its range; a heavily spotted form (*davus*) is found in lowland England, a virtually spotless race (*scotica*) in northern Scotland, and a range of intermediate races elsewhere (referred to as *polydama*). The butterfly has declined seriously in England and Wales, but is still widespread in parts of Ireland and Scotland.

Foodplants

The main foodplant is Hare's-tail Cottongrass (*Eriophorum vaginatum*) but larvae have been found occasionally on Common Cottongrass (*E. angustifolium*) and Jointed Rush (*Juncus articulatus*). Early literature references to White Beak-sedge, *Rhyncospora alba*, are probably erroneous.

Habitat

The butterflies breed in open, wet areas where the foodplant grows: lowland raised bogs, upland blanket bogs; and damp acidic moorland. Sites are usually below around 500 m (600 m in the far north) and have a base of *Sphagnum* moss, interspersed with dense tussocks of Hare's-tail Cottongrass and abundant Cross-leaved Heath, the main adult nectar source. In Ireland, the butterfly can be found where manual peat extraction has lowered the surface of the bog, creating damp areas with local concentrations of foodplant.

Life cycle and colony structure

There is one generation per year with adults generally flying between mid- to late June and early August, with a peak in mid-July. However, the precise dates vary considerably with altitude and latitude, and the butterfly can fly up to a month later in the mountains of north Scotland (see p. 398).

The **eggs** are laid singly on dead stems at the base of tussocks of the larval foodplant and the **larvae** feed on the stems of the foodplant from late July to late September. They hibernate low down in tussocks in their second or third instar and emerge in March to continue feeding during the day. The larvae **pupate** in late April or early May suspended either from the foodplant or surrounding vegetation. In the far north, a small proportion of larvae have a two-year life cycle and remain as third instar larvae throughout their second summer. This flexibility may help the species to cope with a more extreme climate and the unpredictable weather during the adult flight period.

Large Heath larva on Hare's-tail Cottongrass.

The Large Heath is a highly colonial and sedentary butterfly, with adults rarely seen

The distribution of Hare's-tail Cottongrass (1950 onwards records), the larval foodplant of the Large Heath.

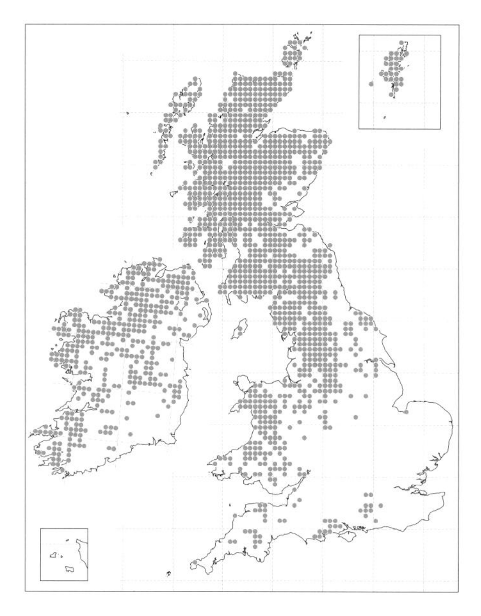

outside their boggy habitats. Marking studies have shown that most adults move less than 100 m and the maximum recorded movement was 450 m. Some populations are very large, numbering up to 15 000 adults, but there is likely to be little interchange of adults unless populations occur very close together.

Distribution and trends

The Large Heath is widespread and locally common in Ireland, Scotland, and the far north of England. In central and eastern England the butterfly and its habitat have always been more restricted but it has declined severely and is now confined to a small number of localities. The butterfly has also declined in Wales (to around 10 known localities), though upland areas remain under-recorded. The butterfly is similarly under-recorded in parts of Ireland and Scotland that are difficult to access, and may be more widespread than shown.

In Ireland, the Large Heath is found on bogs and moors on northern and western hills as far west as Co. Kerry, and on lowland bogs in Northern Ireland and the midlands (though it appears to be absent from the mountains of Leinster and east Munster). In Scotland, it is still widespread but there have undoubtedly been huge losses due to large-scale drainage and afforestation of its moorland and bog habitats. Changes in habitat management, particularly overgrazing, may also be causing declines in many regions and it is possible that many colonies are being lost before they have been recorded.

In England, intensive surveys have located a total of around 140 colonies. The butterfly has died out from most of central and eastern England, apart from a few well known sites in Lincolnshire, Shropshire, and Yorkshire (1, 1, and 6 colonies respectively). The main English strongholds are in Northumberland and Cumbria, with around 100 and 30 colonies respectively, but even here colonies are still being lost at a high rate (for example 20% lost in Northumberland in 12 years prior to 1996).

Overall, the Large Heath was recorded in 187 new 10 km grid squares in Britain during the current survey, but was not recorded in 47% (137) of the squares where it was observed in 1970–82. This implies a large loss of colonies, especially given the generally increased level of recording. The latter figure is even higher for Ireland (88%) but patchy recording during the short flight period, and inaccessibility of many boggy sites, makes the loss more difficult to interpret.

European trend

The Large Heath has become extinct in Croatia and Hungary and has declined severely in many countries, notably Austria, Belgium, the Czech Republic, and the Netherlands (>75% decrease in 25 years); Germany and Romania (50–75% decrease); and Denmark, Latvia, and Slovenia (25–50% decrease).

Interpretation and outlook

The distribution of the Large Heath is determined primarily by its dependence on wet moorland and bog habitats. In recent decades, these habitats have been severely reduced by drainage, afforestation, overgrazing, and peat extraction. Over 90% of lowland raised bogs have been destroyed in both Britain and Ireland during the last 200–300 years, as well almost 80% of the more extensive Irish blanket bogs. The extraction of peat, combined with soil shrinkage in surrounding areas following the intensification of agriculture, has reduced water levels in many surviving patches. This has led to a loss of the characteristic peatland vegetation (including Hare's-tail Cottongrass) and invasion by birches and pines.

Similar factors have depleted lowland raised bogs and blanket bogs across Europe, leading to the decline of the Large Heath in many countries. Both types of habitat are now listed under the EC Habitats and Species Directive and this should encourage better protection of some large tracts, although many smaller areas remain vulnerable. Peatland habitats are recognized as being of international importance for their wildlife and have been the subject of a number of high profile conservation campaigns, both by the Peatlands Campaign Consortium and the Irish Peatland Conservation Council. Numerous projects are also underway to conserve peatland habitats in Ireland and Britain. However, peat continues to be cut extensively for use in horticulture and gardens, and is still used as a fuel in several Irish power stations. Other threats are from open cast coal mining and continuing afforestation.

The future of the Large Heath will depend on the effective safeguarding of remaining peatland habitats and the continuation of suitable management. Recent research has shown that the butterfly cannot tolerate heavy grazing, as this reduces the grass tussocks used for breeding. Recent overgrazing of upland habitats by sheep, encouraged by EC subsidies, has been particularly damaging. Light and extensive grazing regimes appear to be the most suitable. The peatland habitat itself also requires the maintenance of a high water table, but care must be taken when restoring sites that this does not lead to flooding as Large Heath larvae cannot tolerate prolonged submersion.

The bog habitat of the Large Heath in Cumbria.

Although colonies can survive for many years on small isolated sites, the Large Heath has poor colonizing ability and is highly susceptible to local extinction. Wherever possible, large areas of habitat and large populations must be maintained, as these will be most resistant to extinction in the long term.

The spot frequency and wing colouring of the Large Heath vary considerably across Britain and Ireland, producing a gradient of genetic variation that has been the subject of many studies. The conservation of the butterfly throughout its geographic range is important to maintain this genetic diversity. The Large Heath also provides an important indicator of boggy habitats in these islands, which are special for many other animals and plants.

Key references

Bourn and Warren (1997*c*); Dennis and Eales (1997, 1999); Eales (1999*b*); Joy and Pullin (1997, 1999); Melling (1987).

Large Chequered Skipper *Heteropterus morpheus*

Probably introduced
Extinct since 1996.

Conservation status
UK BAP status: not assessed.
Butterfly Conservation priority: not assessed.
European status: not threatened.

European/world range
Through central Europe from north Spain and western France eastwards across Russia and Asia. Stable in parts of Europe but declines reported in 12 countries and expansions in 3.

This delicate butterfly was first seen on Jersey in 1946 but has been restricted to one small site in the north of the Island. It seems likely that it was introduced accidentally in hay brought over from France during the wartime occupation. The site has gradually become more overgrown and numbers have been very low since 1985. None have been seen since 1996 and it may now be extinct. In Jersey, the butterfly has always been known by its French name *Le Miroir*.

Foodplants

The only recorded foodplant on Jersey is Purple Moor-grass (*Molinia caerulea*). In continental Europe it uses a variety of other grasses including Purple Small-reed (*Calamagrostis canescens*), False Brome (*Brachypodium sylvaticum*), and Common Reed (*Phragmites australis*).

Habitat

On Jersey it was found in sheltered damp grassland dominated by tall Purple Moor-grass. The main site is a flushed area on the edge of a *Sphagnum* bog at the edge of a wood, on a north facing slope. In continental Europe, it breeds in damp meadows and sheltered woodland clearings, usually with damp or peaty soils.

Life cycle and colony structure

The butterfly is single brooded throughout its range and on Jersey adults have been recorded from mid-July to mid-August. The **eggs** are laid singly on the foodplant and the **larvae** live within a tube formed by spinning together the edges of the leaf blade. They feed slowly for around 70 days until early October, by which time most Purple Moor-grass plants have died back. They can survive only where this foodplant grows in wet flushes and dies back later in the year. In Europe, they prefer Purple Small-reed if the two foodplants grow together.

When half-grown, they hibernate within the tubes and resume feeding the following spring, moving to create new shelters as they grow. They **pupate** within loose shelters formed by spinning together several grass blades. As the butterfly lives in tall grassland for most of the year, it is highly susceptible to grazing or cutting.

The Large Chequered Skipper forms discrete colonies in suitable breeding habitat and adults and larvae can occur at high density. On Jersey, it has been almost entirely restricted to one small breeding area since its discovery, with only a few nearby records. However, studies in the Netherlands have shown that adults can be quite mobile and are regularly recorded up to 5 km from known populations, and sometimes even 100 km away.

Distribution and trends

The Large Chequered Skipper has never been recorded from Britain or Ireland and was not recorded on Jersey until 1946 when it was seen in the north-east corner of the island near Bouley Bay. It was originally recorded at three separate small localities in this area, but was soon reduced to a single site of only 0.1 ha, in a damp flush at the edge of a wood. Numbers seemed to reach a peak in the 1960s when a maximum of 50 adults could be seen at one time. It has declined steadily and, since 1985, the maximum seen per day has been just six adults. None has been seen since 1996 and it seems likely that the colony is now extinct.

European trend

The butterfly has declined in several countries, including the Asian part of Turkey (extinct), Austria, Latvia, and Romania (>50% decrease in 25 years), Hungary, Slovakia, and Ukraine (25–50% decrease), Belgium, Germany, Lithuania, the Netherlands, and Switzerland (15–25% decrease). However, its range has expanded in Slovenia (>100% increase in 25 years) and the Czech Republic and Poland (25–100% increase).

Interpretation and outlook

The Large Chequered Skipper was probably introduced to Jersey from France in hay imported as fodder for horses, which were being used to replace mechanized transport because of the wartime petrol shortages. The butterfly is found commonly on the nearby mainland of France but even this quite mobile species is unlikely to have colonized naturally over a linear distance of more than 40 km, 25 km of which is across the sea.

Large Chequered Skipper larva.

The habitat on the main Jersey site has always been exceptionally small and it is perhaps surprising that the butterfly survived as long as it did. The site was reported to have deteriorated due to encroaching bracken and scrub, and some experimental clearance was conducted by Jersey Environmental Services during the early 1990s. However, there was also some nutrient enrichment that was suspected to have altered the vegetation. The main effort must now be focused on conserving populations of the species in more extensive habitats available in France, and in other European countries where it is becoming threatened.

Key references

Baker (1992); Clarke (1993); Le Quesne (1946); Long (1977); Raemakers and van der Made (1991); Swaay (1998).

Pale Clouded Yellow *Colias hyale*

Rare migrant

Conservation status

UK BAP status: not assessed.
Butterfly Conservation priority: not assessed.
European status: not threatened.

European/world range

Much of western and central Europe, but reaching Scandinavia only as a migrant, and eastwards through northern Turkey and central Asia as far as north-east China. Range uncertain because of identification difficulties. The European distribution is considered stable.

The Pale Clouded Yellow is a rare visitor to Britain and Ireland. It is probably very rare or absent in most years, but occasionally arrives in some numbers and breeds here, although it is unlikely to survive our winters. Close similarities with the Berger's Clouded Yellow and females of the *helice* form of the Clouded Yellow make identification of this migrant extremely difficult in the field.

Foodplants

Several leguminous plants have been recorded, notably clovers (*Trifolium* spp.) and Lucerne (*Medicago sativa*).

Habitat

Migrant Pale Clouded Yellows tend to locate fields with abundant foodplants and thereafter range little. However they may be encountered in almost any habitat. In continental Europe this species breeds in a wide variety of flowery habitats including cultivated fields of its foodplants.

Life cycle and colony structure

The species is thought to be able to complete at least two generations in Britain and Ireland in favourable years. Most sightings have been in August and September, and many of these individuals may be the progeny of earlier immigrants.

Eggs are laid singly on the leaves of foodplants and in good conditions many **larvae** develop rapidly, **pupate** attached to vegetation, and produce a new generation of adults. Alternatively the larvae enter diapause, the overwintering stage in continental Europe. In most years, it appears that winters in Britain and Ireland are too damp for survival. However, there are convincing reports of overwintering and subsequent adult emergence in Kent and Essex following a major immigration in 1947.

Distribution and trends

Records of Pale Clouded Yellows were not uncommon prior to 1950, but they undoubt-edly included some Berger's Clouded Yellows (which was not recognized here as a separate species until 1947) and misidentified pale forms of the Clouded Yellow. The best periods were 1900 and the late 1940s, but only a few have been reported in most years since then.

Sightings tend to occur in south-east England and, less frequently, in other counties on the south coast. Irish sightings are also largely restricted to the southern coast. There have been reports of this species in the current survey period, including one in Hampshire in September 1998, but none has been confirmed.

Reliable identification is a problem with this species and its distribution is unclear due to confusion with the other two migrant species of *Colias*. The ground colour of the male Pale Clouded Yellow's upperwing is a pale primrose yellow and is easily distinguished from the deep orange-yellow of the Clouded Yellow. However, separation of females is more difficult and both sexes are virtually indistinguishable from the Berger's Clouded Yellow. Positive identification of adults requires examination of their genitalia, but the full-grown larvae are easily distinguished.

European trend

The butterfly is thought to be stable in most countries, but distribution declines have occurred in Luxembourg (50–75% decrease in the past 25 years) and Italy (25–50% decrease).

Interpretation and outlook

Most Pale Clouded Yellow immigrants are thought to arrive from northern France or central Europe and the species is considered far less migratory than the Clouded Yellow. Significant immigrations of all of the *Colias* species are so irregular that it is impossible to make predictions about their future occurrence in Britain and Ireland.

Berger's Clouded Yellow *Colias alfacariensis*

Rare migrant

Conservation status

UK BAP status: not assessed.
Butterfly Conservation priority: not assessed.
European status: not threatened.

European/world range

Much of southern and central Europe northwards to Belgium, the Netherlands, and southern Germany. Extends eastwards to Turkey, but the limits are uncertain because of identification problems. The distribution in Europe is stable.

In recent decades the Berger's Clouded Yellow has been the rarest of the three *Colias* species that migrate to Britain and Ireland, and it has been recorded only from England. It was recognized as a species separate from the Pale Clouded Yellow only in 1947 and has since been identified in English butterfly collections dating back to the nineteenth century. Although separation of the adults of the two species remains extremely difficult in the field, full-grown larvae and the choice of foodplant can greatly aid identification.

Foodplants

The only native plant used by the Berger's Clouded Yellow in Britain is Horseshoe Vetch (*Hippocrepis comosa*), although the naturalized Crown Vetch (*Securigera varia*) may be used, as it is in continental Europe.

Habitat

Breeding is generally restricted to dry, calcareous habitats where its larval foodplant, Horseshoe Vetch, grows. However, it may breed occasionally in other habitats if Crown Vetch is present.

Life cycle and colony structure

Such is the rarity of this butterfly that the early stages of the life cycle have been recorded in the wild in Britain only recently. It is not clear how many generations of the Berger's Clouded Yellow might occur here, but in continental Europe two or three per year is normal depending on the locality.

In captivity, females lay **eggs** singly on the leaves of the foodplant and the **larvae** feed initially low down on plants. Older larvae feed on the upper leaves, often in strong sunshine. At this stage they are easily distinguished from those of the other two *Colias* species by the presence of a bright yellow lateral stripe, flanked by rows of black dots. In its permanent range, it overwinters in the larval stage, but there is little evidence to suggest that this has occurred successfully in Britain. **Pupae** are attached to low vegetation.

Although it is considered to be a migratory species, little is known of the dispersal behaviour of the Berger's Clouded Yellow. Recent genetic studies show a surprising level of geographical variation in the species, suggesting that it may not disperse to the same extent as the other two *Colias* migrants.

Distribution and trends

The Berger's Clouded Yellow is even less frequently recorded than the Pale Clouded Yellow. There are a few records from May and June, with most in August and September. The later years of the 1940s saw many sighting of both species, although the Berger's Clouded Yellow records were largely confined to Kent and Sussex. Most sightings have been on chalk downs and cliff tops.

There have been very few confirmed records since then, including an adult in Cornwall in 1960 and larvae on the Isle of Portland, Dorset in the early 1990s. A female was reported laying eggs on Horseshoe Vetch in Dorset in 1996. Identification difficulties have probably resulted in this species being under-recorded, but it is undoubtedly a very rare migrant to Britain.

European trend

The Berger's Clouded Yellow has declined in a few countries, notably Belgium (50–75% decrease in 25 years) and also Germany and Luxembourg (15–25% decrease). In contrast, its distribution has expanded in Bosnia, Macedonia, Romania, and Yugoslavia over the same period.

Interpretation and outlook

The rarity and identification difficulties associated with the Berger's Clouded Yellow make any interpretation of records in Britain extremely difficult. Although unimproved calcareous habitats have declined as a result of the intensification of agriculture, the butterfly's foodplant remains relatively common in places. The continued sporadic occurrence of the Berger's Clouded Yellow is entirely dependent on the maintenance of continental populations.

Key reference

Cleary *et al.* (1995).

Black-veined White *Aporia crataegi*

Extinct (1920s)

Conservation status

UK BAP status: not assessed.
Butterfly Conservation priority: not assessed.
European status: not threatened.

European/world range

Most of Europe including southern Scandinavia and eastwards through Turkey and the Middle East, across Asia to Japan. It also occurs in North Africa. It is declining in many European countries.

The mysterious decline of the Black-veined White has been the subject of discussion amongst entomologists for over a century. This impressive butterfly became extinct in Britain in the 1920s. Since then, many reintroductions have been attempted, all of which have failed. It is a large species with a powerful, soaring flight. When they become worn, the wings look like parchment patterned with distinctive black veins.

Foodplants

Both wild and cultivated fruit trees are used. In Britain, Blackthorn (*Prunus spinosa*) and hawthorns (*Crataegus* spp.) were the main foodplants, although fruit trees in orchards were used in certain areas. In continental Europe, the larvae are also recorded as feeding on many *Prunus* spp. (e.g. Plum, Peach, Wild and Bird Cherry), apples (*Malus* spp.), and Pear (*Pyrus communis*).

Habitat

Breeding occurs in scrub and woodland edge habitats, hedgerows, and cultivated areas, particularly orchards. The species may be a significant orchard pest in some parts of its range, but appears never to have attained this status in Britain. In continental Europe, the adult butterflies are seen in a diverse range of habitat types, often favouring warm, sunny, nectar-rich meadows.

Life cycle and colony structure

The species has a single generation each year and overwinters in the larval stage. In Britain, adult emergence normally peaked in late June and July. **Eggs** are laid in batches on the leaves of foodplants, and the **larvae** live in groups. When young, they spin light silken webs from which they emerge to feed. Later on, they construct a more substantial silk nest in which to hibernate. In spring, feeding recommences with the larvae eventually becoming solitary. **Pupae** are attached to stems of foodplants or other vegetation.

The Black-veined White tends to occur in discrete colonies, although there is some evidence of dispersive or even migratory tendencies in Europe. In southern England, strongholds were often surrounded by areas with few or no colonies, suggesting restricted dispersal and limited colonization. Throughout its range, the species is associated with large fluctuations in abundance.

Distribution and trends

The former distribution and the decline of the Black-veined White have been well documented. It has been recorded from most counties in England south of Yorkshire and Lancashire, as well as in southern Wales and Merionethshire (southern Gwynedd). There are no accepted records for Scotland, the Isle of Man, or Ireland, but records for the Channel Islands appear genuine.

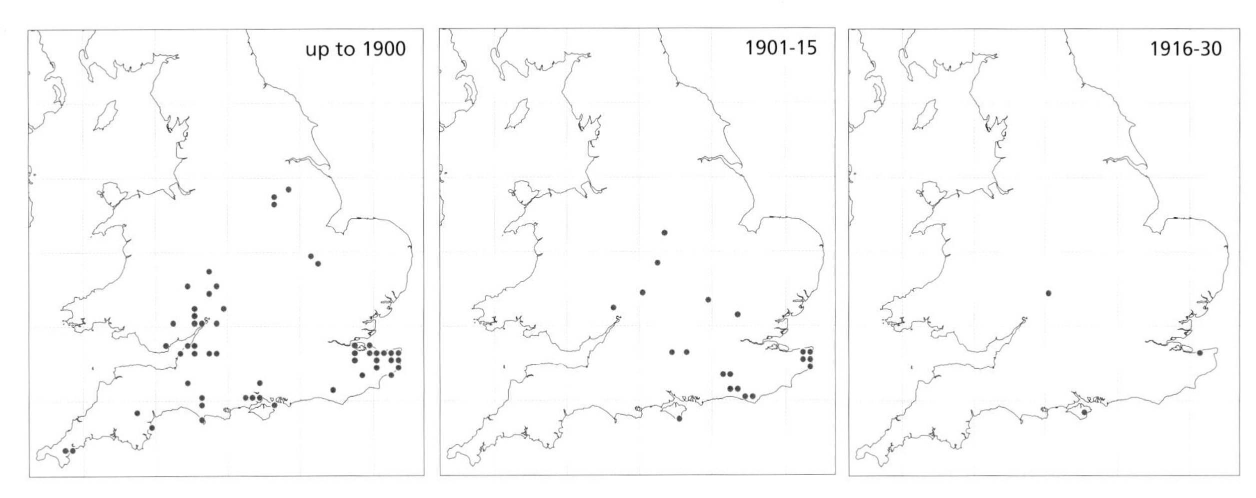

The distribution of historic records of the Black-veined White at 10 km square resolution in southern Britain.

The decline of the species began at least 200 years ago, with extinctions at the county level being noted early in the nineteenth century. A rapid decline during the first half of the nineteenth century reduced the species to three core areas: the New Forest in Hampshire, eastern Kent, and the Welsh borders (incorporating parts of Monmouthshire, Glamorgan, and west Midland counties). Despite periods of local abundance in the second half of the nineteenth century, the loss of colonies continued. The last records of colonies were in north-east Kent and Worcestershire and extinction is believed to have occurred by 1925.

There have been rare sightings of the Black-veined White in Britain since its extinction. Although the possibility of vagrants from continental Europe cannot be excluded, it seems likely that many of these sightings are either misidentifications or of individuals originating from captive stock. In addition, it is possible that the immature stages may be imported accidentally with fruit.

Attempts to reintroduce the Black-veined White were made prior to its extinction and have continued since. A few temporary colonies have been established using stock from continental Europe, but there has been no lasting success. As well as apparently being futile, it is now illegal under the Wildlife and Countryside Act (1981) to release this species in Britain.

The long history of captive breeding and escapes and releases of this species makes it difficult to interpret both the recent and historical records. It has even been suggested that the last colonies in north-east Kent resulted from reintroduction.

European trend

Serious declines have occurred in many countries. It has become extinct in the Czech Republic and the Netherlands, and has suffered declines in Austria and Slovakia (>75% decrease in 25 years), Croatia, Luxembourg, and Poland (50–75% decrease), Belgium and France (25–50% decrease), and Albania, Germany, and Latvia (15–25% decrease). Expansions have been reported at the northern edge of its distribution, in Denmark and Finland.

Interpretation and outlook

There has been much debate about the decline of the Black-veined White. Several factors were probably involved but, in this instance, habitat loss was not implicated. There is evidence that above average September rainfall may have contributed to its demise, perhaps by increasing susceptibility to disease. Overall, the explanation of the decline is incomplete. Similarly, the dramatic fluctuations in abundance, for which this species is renowned, remain unexplained.

The status of this species in Britain is unlikely to change in the near future. All attempts at reintroduction have failed, perhaps because stocks taken from continental Europe lack the necessary adaptation to the British climate. Although the Black-veined White appears to be expanding at the northern edge of its European distribution, it is declining in the countries closest to Britain making recolonization less likely.

Key references

Oates and Warren (1990); Pratt (1983).

Bath White *Pontia daplidice*

Rare migrant

Conservation status

UK BAP status: not assessed.
Butterfly Conservation priority: not assessed.
European status: not threatened.

European/world range

Much of Europe, except Britain, Ireland, and most of Scandinavia. Also occurs in North Africa, the Middle East, and central and eastern Asia. East of the Alps, and in parts of the Middle East and Asia, it is replaced by a distinct race, the Eastern Bath White. The species is thought to have a stable distribution in most European countries.

The Bath White is a scarce immigrant to Britain and Ireland, although there have been some years when it was more abundant. In common with many other Whites, the species migrates northwards across Europe each year, breeding for a few generations but usually failing to survive winters in central and northern Europe. Recent molecular research has shown the existence of two distinct races, the Bath White and the Eastern Bath White.

Foodplants

Mignonettes (*Reseda* spp.) and various crucifers are used. In Britain and Ireland, Wild Mignonette (*Reseda lutea*), Hedge Mustard (*Sisymbrium officinale*), and Sea Radish (*Raphanus raphanistrum*) have all been recorded as foodplants.

Habitat

Open and disturbed habitats, such as roadside verges, cultivated and derelict ground, are often used for breeding in continental Europe. Adults are highly mobile and can be seen in almost any type of habitat.

Life cycle and colony structure

The Bath White has several generations during the year in southern Europe, but appears unlikely to have ever achieved more than one in Britain and Ireland. **Eggs** are laid singly on foodplants and the **larvae** feed on leaves. Overwintering occurs in the **pupal** stage and there is some evidence to suggest that the species has occasionally overwintered in Britain. Parasitism and predation are believed to result in extremely high levels of mortality of the immature stages in the wild.

This is a mobile species that seems to colonize suitable habitat rapidly, as well as occurring as a rare migrant well outside its normal breeding range. A high dispersal rate (over 4 km per generation) was calculated from molecular studies on populations within the permanent breeding range.

Despite this high level of mobility, substantial genetic differences remain between populations and at one time it was proposed that the Bath White and Eastern Bath White were separate species. More recent evidence suggests that they are only races of the same species.

Distribution and trends

Few Bath Whites have been recorded in Britain and Ireland in recent decades. In the current survey period single individuals were recorded in West Sussex (1995), Kent (1996), and Cornwall and Somerset (1998).

The only well-recorded episode of mass immigration occurred in 1945 and was centred on the Channel Islands and

Cornwall. Bath Whites were recorded in all of the south-coast counties of England in that year (often in some numbers), as well as in other southern and Midland counties, and in southern Wales and Ireland. It seems certain that local breeding occurred during this season, and some records from late summer and autumn, were probably of homebred individuals.

Most sightings of immigrants occur in late summer (late July and August) and are strongly biased to the south of England. A few early (April and May) individuals were recorded in 1946 providing circumstantial evidence that overwintering had occurred.

The identification of Bath Whites is not easy. In flight the butterfly may resemble common native species, and even good views might lead an inexperienced observer to record a female Orange-tip as a Bath White, or vice versa. Although the respective flight periods should help, it is possible that Bath Whites have been under-recorded. Nevertheless, the status of the species as an extremely rare migrant is undoubtedly correct.

European trend

The butterfly's distribution is stable in many countries, although there has been a serious decline in Sweden (>75% decrease in the past 25 years).

Interpretation and outlook

With the division of the Bath White into two races, it would be of interest to know the identity of immigrants to Britain and Ireland. The great immigration of 1945 may have come directly from Spain or even northern Africa, suggesting that these individuals were Bath Whites rather than Eastern Bath Whites. However, the identity of the regular trickle of individuals sighted in south-eastern England is less clear, as the two races are impossible to distinguish in the field and both are thought to be present as migrants in countries along the east coast of the North Sea.

The Bath White is a very rare visitor and its continued occurrence in Britain and Ireland depends on the survival of resident populations in continental Europe and beyond.

Key references

Dannreuther (1946); Geiger *et al.* (1988); Porter *et al.* (1997).

Long-tailed Blue *Lampides boeticus*

Rare migrant

Conservation status

UK BAP status: not assessed.
Butterfly Conservation priority: not assessed.
European status: not threatened.

European/world range

Throughout Africa, southern Asia, Australia, Hawaii, and southern Europe. In Europe, probably only resident in Mediterranean countries, but occurs as a regular migrant and summer breeding species across Spain, France, Germany, and much of central Europe. Its distribution in most European countries is stable.

The Long-tailed Blue is a fast-flying, migratory butterfly with a very widespread global distribution. It reaches Britain only rarely and in small numbers. Although it is known to breed here occasionally, it appears unable to survive our winter climate. The tail on the trailing edge of each hindwing and the white striping distinguish this butterfly from any resident blues in Britain and Ireland.

Foodplants

Over 45 foodplants have been recorded worldwide, most being wild and cultivated legumes. In Britain larvae have been recorded on Broad-leaved and Narrow-leaved Everlasting-pea (*Lathyrus latifolius* and *L. sylvestris* respectively), Broom (*Cytisus scoparius*), and Bladder-senna (*Colutea arborescens*). In addition, larvae are found regularly in imported produce such as mange-tout peas (*Pisum sativum*).

Habitat

A diverse range of hot, flowery habitats is used for breeding in continental Europe. In Britain, Long-tailed Blues have generally been recorded in gardens or with other blues on open downland.

Life cycle and colony structure

The species appears to breed all year round in its southern permanent range. Both eggs and larvae have been discovered in the wild in the Channel Islands and southern England, showing that breeding is occasionally possible. However, it appears extremely unlikely that the Long-tailed Blue has ever overwintered here.

Eggs are laid on foodplants, and the young **larvae** feed initially on flowers. Later they bore into seed pods, where they feed on both the pod tissue and the developing seeds. In common with many other blue butterflies, the larvae are attended by a range of ant species. **Pupation** occurs amongst dead leaves.

Migration is a regular feature of Long-tailed Blue populations and mass movements have been observed in Europe. The species is a strong flyer and has been recorded at high altitudes in mountain ranges. It is somewhat surprising that the butterfly has been recorded so infrequently in Britain.

Distribution and trends

This rare migrant is most often recorded in the Channel Islands and southern counties of England. There appear to be no records from Ireland or the Isle of Man, and none recently from Scotland. Most sightings have been of single butterflies and have generally occurred in late summer (August and September), although one was recorded in Devon in late June 1998.

In the 1995–9 survey period, Long-tailed Blues have been reported in every year, although some of the records refer to accidentally imported individuals and others may have been deliberately released. Adults have emerged successfully from immature stages discovered in imported foodstuffs, for example from two separate batches of mange-tout peas from Kenya in 1998 and 1999. These have not been mapped. An adult was also recorded during 1998 in a greengrocer's shop in Essex.

Since the species was first observed in Britain in the 1850s, there have been many years when none has been recorded at all, and the only year of notable abundance was in 1945.

Records of this rare immigrant are sometimes difficult to interpret. Genuine immigrants can be confused with specimens resulting from deliberate or accidental introduction. Furthermore, the species has probably been under-recorded because of its superficial resemblance to more common species (e.g. the Common Blue).

European trend

The butterfly appears to be stable in most countries, although small declines in distribution have been reported from Albania, Romania, and Ukraine (15–25% decrease in 25 years).

Interpretation and outlook

Sightings of the Long-tailed Blue are still very rare and the apparent recent increase may simply reflect a greater intensity of recording, or more importation with fresh foodstuffs. Although foodplants are widely available and breeding is possible, it seems that the climate of Britain is unsuitable for colonization in the near future.

As is the case with other southern European migrants, the continued (albeit rare) occurrence of the Long-tailed Blue in Britain will depend on the survival of populations in its permanent range.

The distribution of records of the Long-tailed Blue at 10 km square resolution.

Key reference

Mavi (1992).

Short-tailed Blue *Cupido (Everes) argiades*

Rare migrant

Conservation status

UK BAP status: not assessed.
Butterfly Conservation priority: not assessed.
European status: not threatened.

European/world range

From northern Spain through France, Germany, and much of central Europe to the Baltic States and Turkey. In northern Europe and Scandinavia as a migrant. The species extends across Asia to Japan. Its distribution is stable in many European countries, but its range appears to be shifting northwards.

The Short-tailed Blue is one of the rarest visitors to Britain. Although populations are present in France and neighbouring countries in continental Europe, the species does not appear to have strong migratory habits and its rare occurrences in the Channel Islands and southern England may owe more to chance wanderings. This butterfly is still occasionally referred to as the Bloxworth Blue, after the heath in Dorset where the first British specimens were identified.

Foodplants

A range of legumes is used, many of which are common in Britain and Ireland. These include Common Bird's-foot-trefoil (*Lotus corniculatus*), Red Clover (*Trifolium pratense*), Tufted Vetch (*Vicia cracca*), and Lucerne (*Medicago sativa*).

Habitat

In continental Europe the butterfly is often associated with woodland edges and clearings, scrubby grasslands, and heaths.

Life cycle and colony structure

The Short-tailed Blue has two broods in continental Europe, with flight periods in May and June and July to September. All British records have been made during the latter period.

The **eggs** are laid singly on foodplant leaves, but the **larvae** feed mainly on flower buds and developing seed pods (and are also cannibalistic). Both hibernation, which occurs in the larval stage, and **pupation** also take place on foodplants.

There is no direct evidence that the Short-tailed Blue has ever bred successfully in Britain. However, the inland locations and numbers of butterflies involved in some of the early records provide circumstantial evidence of breeding. If this has ever taken place here, it must have been an extremely rare event.

Little information is available on the mobility and colony structure of the Short-tailed Blue. It does not appear to be a powerful flier and is not thought to have strong migratory tendencies.

Distribution and trends

Almost all records are from southern counties of England, in particular, Dorset, Somerset, Hampshire, Sussex, and Kent. There are also records from the Channel Islands. It has not been recorded from Scotland, Wales, the Isle of Man, or Ireland.

Records of the Short-tailed Blue are very infrequent, with 1885 (when the species was found at Bloxworth, Dorset) and 1945 accounting for many of the sightings. In recent years there have been several sightings, for example in Kent in 1994 and in

Derbyshire and Warwickshire in 1998. However, the origins of these individuals remain unclear and they may have escaped from captivity rather than being genuine immigrants.

The superficial similarity of this species to other small blues such as the Silver-studded Blue and Small Blue may have resulted in it being under-recorded. Nevertheless this is an extremely rare visitor.

European trend

The butterfly's distribution is stable in many countries, but serious declines have been reported in the Czech Republic and Poland (50–75% decrease in 25 years) and lesser declines in Italy, Moldova, and Switzerland (15–25% decrease). In contrast, the species appears to be expanding at the northern edge of its range (for example in parts of Russia). During the past 30–100 years its range has been shifting northwards, with declines in the extreme south and expansions at the northern edge.

Interpretation and outlook

It has been suggested that the occurrence of the Short-tailed Blue in Dorset, Hampshire, and south Somerset may be due to immigration from the heaths of Brittany. With so few and such irregular sightings though, it is extremely difficult to draw any conclusions from the distribution of records in Britain. For the present, the butterfly's occasional appearances are reliant on the continued survival of populations in continental Europe. The apparent northwards spread of the species in continental Europe, perhaps in response to global climate change, raises the possibility of eventual colonization.

Mazarine Blue *Polyommatus (Cyaniris) semiargus*

Extinct (1904)

Conservation Status

UK BAP status: not assessed.
Butterfly Conservation priority: not assessed.
European status: not threatened.

European/world range

Widespread across much of Europe, northwards into the Arctic Circle and across the Middle East and central Asia to China. In North Africa it is restricted to Morocco. Declining in many European countries.

Little is known about the British distribution of this former resident species. It became extinct in Britain at the beginning of the twentieth century, but remains widespread in continental Europe. The adults are similar in size to the Common Blue, but have darker upperwings and no orange markings on the undersides. The reasons for its decline are not known.

Foodplants

The main foodplant is Red Clover (*Trifolium pratense*), although several other leguminous plants have been suggested. It is assumed that Red Clover was used in Britain although there are no known records of larval stages in the wild.

Habitat

A range of flowery habitats, particularly hay meadows and unimproved pasture at a variety of altitudes, is used in continental Europe.

Life cycle and colony structure

The species has one generation each year in northern Europe with adults on the wing from mid-June until mid-July. **Eggs** are laid in small numbers, on clover flowers, upon which the young **larvae** feed. After hibernation the larvae resume feeding, concentrating on young leaves and shoots. As with many other blues, ants are attracted to the larvae and **pupae**, and the latter may be buried by them.

The Mazarine Blue forms discrete colonies in continental Europe, but little is known about the mobility of individuals.

Distribution and trends

Although the species was recorded from many counties in southern Britain (as far north as Yorkshire) in the eighteenth and nineteenth centuries, it appears always to have been scarce. However, it is difficult to judge the true extent of its distribution because of the lack of records from the nineteenth century and confusion arising from the fraudulent sale of specimens from continental Europe as native individuals. There are no records from Ireland, the Isle of Man, or Scotland.

Colonies were well documented in only three main areas: Dorset, Gloucestershire, and Glamorgan, and these are believed to have become extinct in 1841, 1865, and 1877 respectively. The last records of a colony were on the Lincolnshire/Yorkshire border in 1904.

There have been very few records of this species in the past 50 years and none is known from the current survey period. Some were recorded in coastal locations in southern England and in Jersey, and may be genuine immigrants, but others are assumed to have been accidental or deliberate releases of captive stock.

European trend

The distribution of the Mazarine Blue is stable in many countries, but serious recent declines have occurred in Belgium and the Netherlands (>75% decrease in 25 years) and Luxembourg (25–50% decrease). Lesser declines have also been reported in the Czech Republic, Denmark, Germany, Latvia, Moldova, Slovakia, Sweden, European Turkey, and Ukraine (15–25% decrease). In contrast, increases have been reported in Bosnia, Macedonia, and parts of Russia.

Interpretation and outlook

The lack of knowledge of the previous distribution and decline of the Mazarine Blue in Britain makes interpretation difficult. Extinction occurred before the widespread destruction of its grassland habitats, and the larval foodplant remains widespread.

The explanation put forward nearly a century ago implicated changes in the timing of hay cuts and the use of Red Clover as a crop. The theory suggested that more widespread cultivation of Red Clover at the end of the nineteenth century caused females to be attracted to the fields, where their progeny were then destroyed by cutting. Little progress has been made in elucidating the story of this former resident since that time.

There is little chance of encountering a wild Mazarine Blue in Britain or Ireland. The species is not regarded as migratory and no concerted effort has been made to reintroduce it. Although it is not currently considered threatened in Europe, the outlook is gloomy, following population declines in many intensively-developed countries in northern Europe. This makes the species a worthy candidate for research in continental Europe.

The distribution of historical records of the Mazarine Blue in southern Britain at a 10 km square distribution.

Key reference

Bretherton (1951*a*).

Geranium Bronze *Cacyreus marshalli*

Introduced

Conservation status

UK BAP status: not assessed.
Butterfly Conservation priority: not assessed.
European status: not threatened.

European/world range

Southern Africa, but recently established in the Balearic Islands and recorded in the wild from mainland Spain, France, Italy, Portugal, the Canary Islands, and Britain. Also recorded in Morocco. Distribution expanding rapidly in southern France.

The Geranium Bronze is the only butterfly covered by this atlas whose occurrence in Britain is believed to have resulted entirely from accidental human introduction. It has become established in several Mediterranean countries in recent years and wild adults were recorded for the first time in Britain in 1997. The species breeds on cultivated *Pelargonium* and immature stages may be transported inadvertently with pot plants.

Foodplants

Species in the genera *Pelargonium* and *Geranium* are used. In Europe, only cultivated species of *Pelargonium* have been recorded as foodplants to date, although the butterfly uses a wide variety of wild plants in these genera in southern Africa.

Habitat

In its European range, the Geranium Bronze can occur wherever cultivated *Pelargonium* is planted (including gardens, window boxes, and municipal plantings along roads). If the butterfly were to adapt to use wild foodplants in Europe a wide range of habitats would be available, as is the situation in southern Africa. It has been recorded at altitudes up to 1500 m in France, and in areas well away from human habitation.

Life cycle and colony structure

The species is continuously brooded in warm climates. In southern France, adults were recorded from March until early November in 1998 and reports from Spain indicate that they can be seen in every month of the year.

Eggs are laid singly (although often several on any one plant) on the flower buds or undersides of foodplant leaves. Initially **larvae** may reside inside plant tissues, mining buds, leaves and stems, or feed externally as do older individuals. **Pupae** occur either attached to the host plant or in leaf litter on the ground.

The butterfly is not considered to be very mobile in its native range and its arrival in Europe was undoubtedly due to accidental importation with its foodplants. It is unclear, though, how much of the subsequent expansion has been due to natural expansion from points of introduction.

Distribution and trends

The first observations of breeding in Britain and Ireland were made in Sussex in 1997, almost 20 years after the Ministry of Agriculture, Fisheries, and Food (MAFF) first discovered imported larvae in Britain. Adults were recorded between September and November 1997 and egg-laying was observed on *Pelargonium* plants in two private gardens near Lewes (East Sussex). There were further sightings in the same area in 1998. Three adults emerged in February from a *Pelargonium* that had been in a conservatory during the winter, and a single individual was recorded in May in a garden. It was suspected that this butterfly had also emerged from plants overwintered in a glasshouse.

This species is easy to identify from the plain, bronze-brown wings with white and brown chequered fringes, distinctive underwing pattern, and the presence of a tail on the rear edge of each hindwing. Nevertheless, the widespread press coverage and appeal for records generated a number of erroneous identifications from inexperienced observers. There have been no other confirmed wild sightings from Britain and Ireland.

European trend

The overall trend has been one of expansion in south-western Europe since its introduction.

Interpretation and outlook

The events of 1997 showed that the summer climate in southern England can be suitable for breeding and development of the Geranium Bronze. However, the ecological requirements of both the butterfly and its known foodplants indicate that overwinter survival in the wild is unlikely at present. A continued trend towards warmer winters, in which *Pelargonium* plants can survive outdoors, or adaptations by the butterfly to use wild foodplants, may enable it to overwinter in the future.

Geranium Bronze larva and feeding damage on *Pelargonium*.

The long-range dispersal of the species, including the occurrence in England, is thought to be due to the accidental introduction of pre-adult stages with imported foodplants. The Geranium Bronze has proved to be a significant pest of the horticultural trade in the Balearic Islands and has been designated as such by the European and Mediterranean Plant Protection Organization. The status of the species in Britain is being closely monitored by MAFF and any occurrences should be reported immediately to the Plant Health Service. Such surveillance is necessary to reduce the probability of future accidental introductions.

Key references

Clark and Dickson (1971); Delmas and Maechler (1999); Holloway (1998, 1999); Honey (1993).

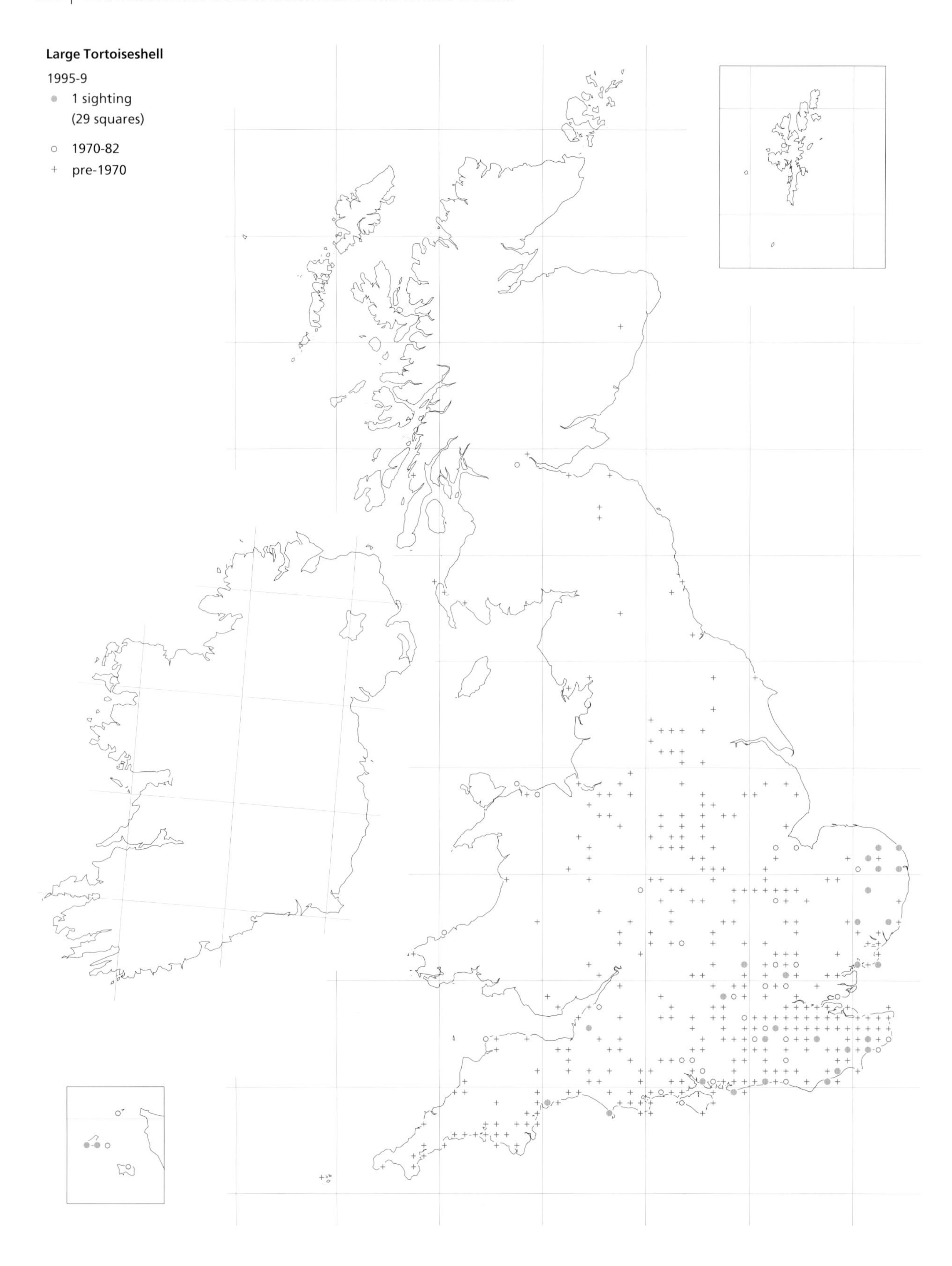
Large Tortoiseshell
1995-9
1 sighting
(29 squares)
1970-82
pre-1970

Large Tortoiseshell *Nymphalis polychloros*

Extinct (1980s)

Conservation status

UK BAP status: not assessed.
Butterfly Conservation priority: not assessed.
European status: not threatened.
Protected in Great Britain for sale only.

European/world range

Across Europe, except for northern Scandinavia, though it is increasingly scarce in northern Europe. It also occurs in North Africa and eastwards into Asia. There has been a marked decline in a number of European countries.

This large, mobile butterfly is now seen very rarely and is thought to be extinct in Britain. Sightings in recent years are thought to be of released or escaped individuals, or migrants from continental Europe. Adults are recorded mainly in spring when they emerge from hibernation and feed on willow flowers, from which they soar rapidly into the treetops at the least disturbance. They may also be seen in late summer feeding before hibernation. In common with other nymphalids, the undersides of the wings are cryptically coloured.

Foodplants

Elms (*Ulmus* spp.) are the main foodplants, especially Wych Elm (*Ulmus glabra*). Other trees may be used in the wild, including willows (*Salix* spp.), Aspen (*Populus tremula*), poplars (*Populus* spp.), and birches (*Betula* spp.).

Habitat

In mainland Europe (and in the past in Britain), clearings and the edges of mature deciduous woodland are favoured, as are lines of trees, for example along hedgerows, avenues in parkland, and wooded lanes. Willows are usually abundant at breeding localities and their flowers are the main nectar source for adults emerging from hibernation.

Life cycle and colony structure

There is one generation a year. Adults emerge in July and August and feed for a short period before entering hibernation in piles of wood, hollow trees, sheds, and other outbuildings.

In early April, they briefly become active again, feeding, basking and searching for mates, before the females lay their **eggs** in batches of up to 200. These large batches are laid in neat bands and encircle young twigs in sunny situations, usually at the top of mature elms. Younger, suckering growth may also be used. **Larvae** spin webs of silk in which they live gregariously, feeding on young leaves until they are fully grown in June, when they drop to the ground and crawl away to pupate singly. **Pupae** are suspended from woody stems or trunks.

The adults are strong flyers and very mobile. Individuals range widely rather than forming colonies, and sightings are almost always of single butterflies. They remain close to woods and other locations where there are large, mature trees.

Distribution and trends

In the past, numbers of records of the Large Tortoiseshell have varied greatly in Britain, with years of abundance and years of scarcity. In the nineteenth century it was numerous in some areas, but the last period of abundance was from 1945–8, when high numbers were also recorded in other

Large Tortoiseshell larva.

European countries. Since then there have been no documented sightings of immature stages in the wild and records of adults have been very infrequent.

Records shown on the current map have been carefully checked to exclude, as far as possible, cases of mistaken identity, which may arise, particularly with inexperienced recorders who confuse the Large and Small Tortoiseshells. Some of the observations are thought to have been of released individuals, though the number of sightings near the coast suggest that many are of migrants from mainland Europe.

In the past the Large Tortoiseshell occurred mainly in southern England with concentrations in south Devon, the New Forest, Sussex, Kent, Essex, and Suffolk. In the nineteenth century it was also common in the Midlands and there were occasional records from Scotland and Wales. The species has been recorded in the Channel Islands, but has never been confirmed as occurring in Ireland or on the Isle of Man.

The period of time that has elapsed since the Large Tortoiseshell was recorded frequently in Britain is now longer than any previous period of scarcity and it is feared that the butterfly is extinct as a breeding species. However, there were few sites where this wide-ranging butterfly bred regularly in the past, making it difficult to be sure of its status. The current interest in butterfly recording is such that, if breeding were to take place, there is a high probability that either the conspicuous larval webs or the adult butterflies would be seen and reported.

European trend

There have been severe declines in many countries, especially Austria, Denmark, and the Netherlands (>75% decrease in 25 years), in Belgium and Romania (50–75% decrease) and the Czech Republic, Germany, and Slovakia (25–50% decrease).

Interpretation and outlook

The reasons for the fluctuations in numbers of the Large Tortoiseshell in the past are not clearly understood, though many theories have been put forward. It probably occurred mainly as a migrant from continental Europe that, unlike most migrant species, was able to survive our winters and establish populations for a few years before dying out. However, since the 1945–8 period of abundance, which coincided with large migrations of other butterflies from the continent, records have not coincided with migrations of other species.

The few genuine sightings in recent years are thought to be mainly of individuals that have been reared in captivity from larvae obtained abroad and that have either escaped or been released into the wild. The loss of mature elms through Dutch Elm Disease in recent decades may have contributed to the decline of the Large Tortoiseshell, although it had become rare before the disease became widespread.

The species might become established in Britain again in the future, though the declines in Belgium, Germany, and the Netherlands suggest that this is increasingly unlikely. All genuine sightings in the wild should be reported to enable the status of the species to be monitored.

Little ecological information is available on this species, but is needed urgently to help reverse the Large Tortoiseshell's decline in continental Europe.

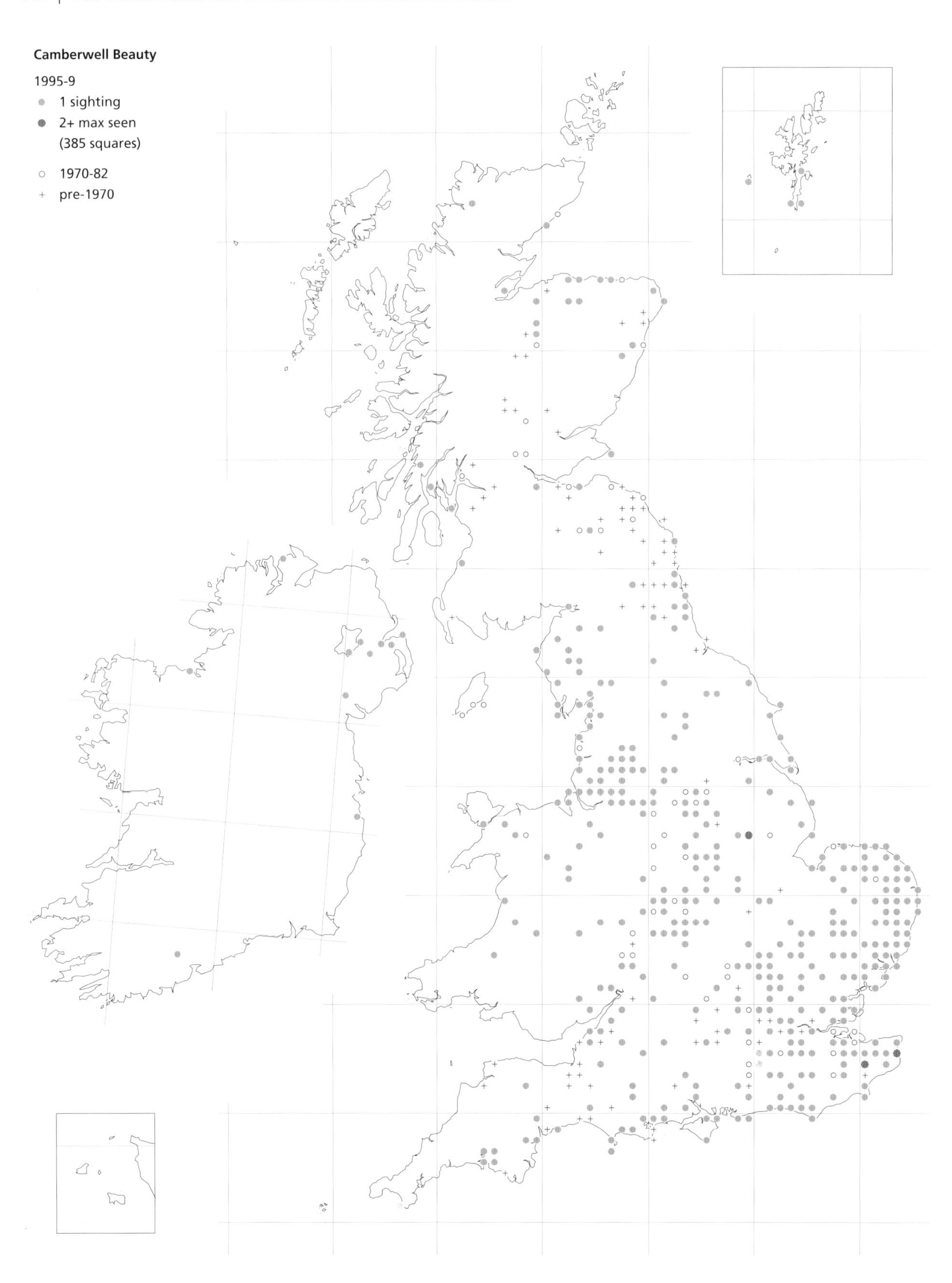
Camberwell Beauty
1995-9
1 sighting
2+ max seen
(385 squares)
1970-82
pre-1970

Camberwell Beauty *Nymphalis antiopa*

Rare migrant

Conservation status

UK BAP status: not assessed.
Butterfly Conservation priority: not assessed.
European status: not threatened.

European/world range

Widespread across central Europe to Norway and Sweden, and found across Asia and North America. The overall range in Europe appears stable, although the distribution fluctuates and recent severe declines have been reported in many countries.

The Camberwell Beauty is a large and striking butterfly rarely seen in Britain and Ireland, but which occasionally arrives in large numbers. Most of the individuals seen here probably migrate from Scandinavia where it is widespread and sometimes locally abundant. When they arrive, the adults are known to visit gardens and nectar on buddleias. It is named after its first reported sighting in Britain, near Camberwell in south-east London, in 1748.

Foodplants

A range of tree species is used, mainly willows (*Salix* spp.), elms (*Ulmus* spp.), and poplars (*Populus* spp.).

Habitat

The butterfly is normally found in woodland habitats in mainland Europe. It also breeds in a range of other habitats such as river valleys, dunes, parks, and gardens.

Life cycle and colony structure

In continental Europe, the Camberwell Beauty has one generation per year, flying from mid-June to July in southern countries and August to September in Scandinavia. Migrants are seen in Britain and Ireland usually in late summer, although there have been records from every month of the year. There are no known records of Camberwell Beauties breeding here.

In continental Europe, large batches of **eggs** are laid in tight spiralling clusters around the stems of its foodplants. The **larvae** feed gregariously on young leaves but disperse prior to forming **pupae** that hatch within about three weeks. The adults hibernate and are seen flying again in the following spring. The temperate winter conditions of Britain and Ireland are considered to be too mild and damp for successful hibernation in most years, although there is evidence for very occasional overwintering, leading to a few spring sightings.

In continental Europe, Camberwell Beauties are highly mobile and range widely through the countryside, especially after hibernation.

Distribution and trends

Sightings of the Camberwell Beauty are relatively common, in parts of England at

Camberwell Beauty larva.

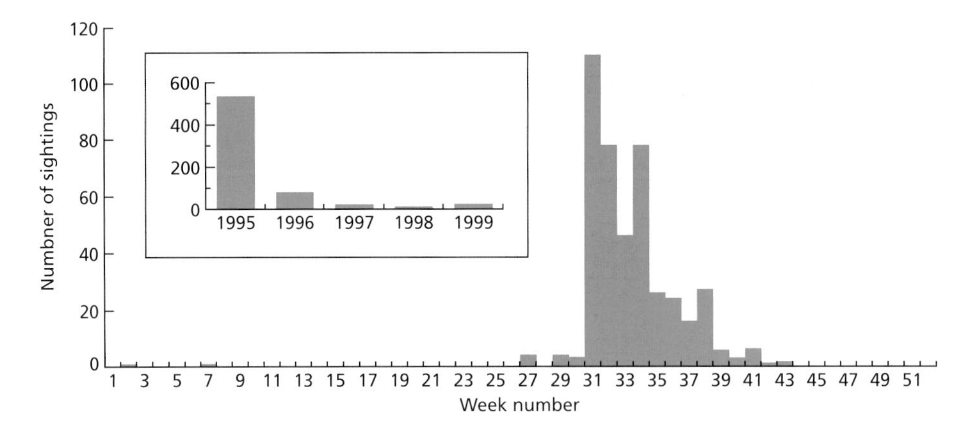

The number of Camberwell Beauty records by week during 1995, showing the timing of the major influx of migrants. The inset graph shows the total number of records of the Camberwell Beauty received in each year (1995–9).

least, in comparison with the other rare migrants included in this atlas. Almost all sightings are thought to be of immigrants, although some may be escapes or releases of captive-bred stock. Historical records show an easterly distribution, with many records in south-east England and East Anglia, and this pattern is also clear in the current survey. Nevertheless, the species has been recorded in each of the countries covered by this atlas.

As with most migrant butterflies, large influxes of Camberwell Beauties have occurred in particular years, notably 1846, 1872, 1947, and 1976. The species has been recorded in every year of the current survey, but there was an unusually large influx into Britain in 1995, with around 350 sightings reported. As well as being much more widespread in England than in a typical year, individuals were recorded in Ireland and as far north as Sutherland and Shetland.

The large size and distinct appearance of this butterfly make it hard to overlook or mis-identify. Although many sightings were reported by people who were not particularly knowledgeable about butterflies, their descriptions were clear and several were backed up by photographs. During the same year, the Netherlands and Denmark recorded a much larger influx of Camberwell Beauties. It seems likely, therefore, that Britain experienced the edge of a substantial long-distance migration.

In the following year, 1996, there were again more sightings than usual, but far fewer than in 1995. Several were recorded in early spring, some of them well away from East Anglia, suggesting that some individuals might have overwintered here in 1995–6. A few were even found in places typical for hibernation, such as garden sheds.

European trend

Large annual fluctuations make it difficult to assess the trend of the Camberwell Beauty. Its range appears to be stable in many European countries and has expanded in Sweden in the past 25 years. However, it has declined seriously in the Netherlands, where it is now extinct, in Belgium, Luxembourg, and Romania (>75% decrease in 25 years), Austria and European Turkey (50–75% decrease), and in several other countries.

Interpretation and outlook

Prevailing weather conditions over northern Europe at the time of the 1976 and 1995 influxes to Britain and Ireland were very similar: a stable high-pressure area was centred over the northern North Sea for about four weeks in August. This led to a persistently strong airflow from southern Scandinavia across Denmark and the Netherlands to south-east England. The timing of the earliest records of the main influx in both 1976 and 1995 was similar, the beginning of August. Studies of wind patterns at the time indicate that Scandinavia was a likely origin for the migrant butterflies in both years. The large numbers recorded in Britain and Ireland probably resulted from the combination of a population explosion in Scandinavia (where numbers vary dramatically) and a mass emigration carried on the prevailing airflow.

There is an indication from the distribution of records from both 1976 and 1995 that there were two main areas of entry, across East Anglia and along the Humber estuary and its tributaries. The phenogram for 1995 records of the Camberwell Beauty shows the near simultaneous arrival of butterflies at all latitudes, which is consistent with migration from the east.

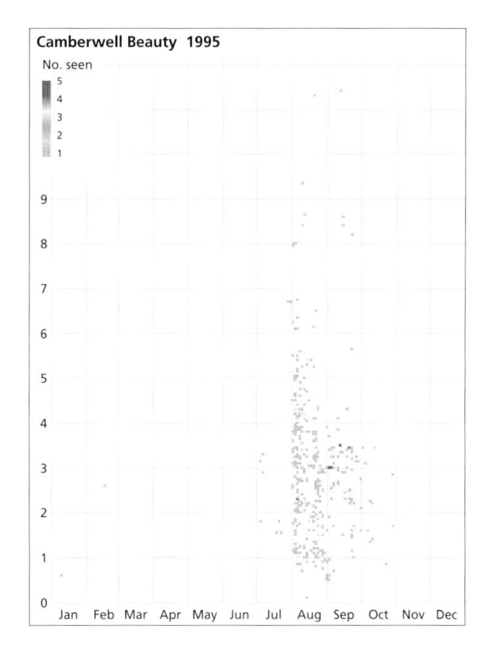

Phenogram of dates of 1995 records of Camberwell Beauty plotted against latitude.

Numbers of Camberwell Beauties seen across Europe fluctuate widely from year to year. In some years few are seen, even in its Scandinavian strongholds. The arrival of migrants in Britain and Ireland clearly depends on the status of these populations and there is growing concern about its decrease in many countries. The cause of these declines is poorly known but may be related to changing woodland habitats or possibly loss of its foodplants. However, the butterfly is reported to be expanding in Sweden and further research is needed to identify its conservation requirements and status in other key parts of its range. With no evidence for breeding in Britain and Ireland, the Camberwell Beauty seems likely to remain at best a rare visitor in most years.

Key references

Chalmers-Hunt (1977); Tunmore (1996); Vos and Rutten (1996).

Map *Araschnia levana*

Rare migrant

Conservation status

UK BAP status: not assessed.
Butterfly Conservation priority: not assessed.
European status: not threatened.

European/world range

Occurs across central Europe from northern Spain to southern Scandinavia, through eastern Europe and across central Asia as far as Japan. Its European distribution appears to be expanding both to the north and the south.

The Map is a very unusual butterfly, in that adults of the first generation have a quite different appearance from those of the second. The spring brood has colouring somewhat like the Comma, predominantly orange with black markings, whereas the summer brood is dark brown with white markings, not unlike a small White Admiral. It is a very rare visitor to Britain, although there have been several unsuccessful introductions in the past 100 years.

Foodplants

Eggs are laid on Common Nettle (*Urtica dioica*) and Small Nettle (*U. urens*).

Habitat

In mainland Europe, the butterfly breeds mainly in sheltered woodland clearings, scrub margins, wet meadows, and hedgerows.

Life cycle and colony structure

There are two broods per year, with adults flying in mainland Europe in April/June and in July/August. The **eggs** are laid in long strings that resemble nettle flowers, and are fixed to the undersides of the leaves of its foodplants. The **larvae** feed gregariously on nettle leaves, finally dispersing to form **pupae**. They overwinter in the pupal stage.

In continental Europe, the Map seems to occur in distinct colonies, although individuals range widely. A recent study has suggested that individuals of the spring generation may be more sedentary than those of the summer brood, which are more likely to disperse.

The recent spread of the species in many European countries indicates considerable mobility. The Netherlands, for example, was colonized entirely in only about 20 years, during the middle of the twentieth century. This spread from south to north equates to an average rate of expansion of over 15 km per year, crossing many areas of unsuitable habitat.

Distribution and trends

There have been few British records of the Map during the last two centuries, and all have been in England or Wales. Introductions are thought to account for most or all of these butterflies. A few sightings in recent years at a locality in the south Midlands of England are strongly suspected to be of individuals released from captive-bred stock (which is illegal under the 1981 Wildlife and Countryside Act) and these have not been mapped. There is one recent record from Kent in 1995, which may have been of a vagrant from north-east France.

There have been several failed attempts to introduce this species into Britain over the past century: in the Wye valley around 1912, the Wyre Forest in the 1920s, south Devon in 1942, near Worcester in the late 1960s, and Cheshire in the early 1970s. There is only limited documentary evidence about these introductions, but it appears that this species is not suited to conditions in Britain and colonies have not persisted for long. In fact there are no known records of immature stages from the wild in Britain.

European trend

The Map has spread extensively in Europe over the past 50 years, expanding its range both to the north and south. It has continued to increase in several countries at the edge of its range over the past 25 years, notably Denmark, Estonia, Finland, Latvia, north-east Russia, and Sweden in the north and Spain to the south. However, it has declined significantly in Austria (50–75% decrease over 25 years) and Slovenia (15–25% decrease).

Interpretation and outlook

Because of its recent expansion in Europe, the Map butterfly could be considered a likely candidate for colonizing Britain and perhaps Ireland. However, the failure of past introductions suggests that climatic conditions may not be suitable here at present. In particular, the British climate may encourage the emergence of adults too early in the spring, when weather conditions are unreliable for survival. Climate change causing drier and sunnier spring weather in south-eastern England might improve the prospects for the Map to colonize from France.

However, the rarity of sightings in Britain over the last few decades suggests that the English Channel represents a substantial barrier to this species. Until sightings of vagrant or immigrant individuals are a more regular occurrence, it seems unlikely that the Map will establish itself in Britain.

Summer brood Map.

Key references

Fric and Konvička (2000); Oates and Warren (1990); Swaay (1995).

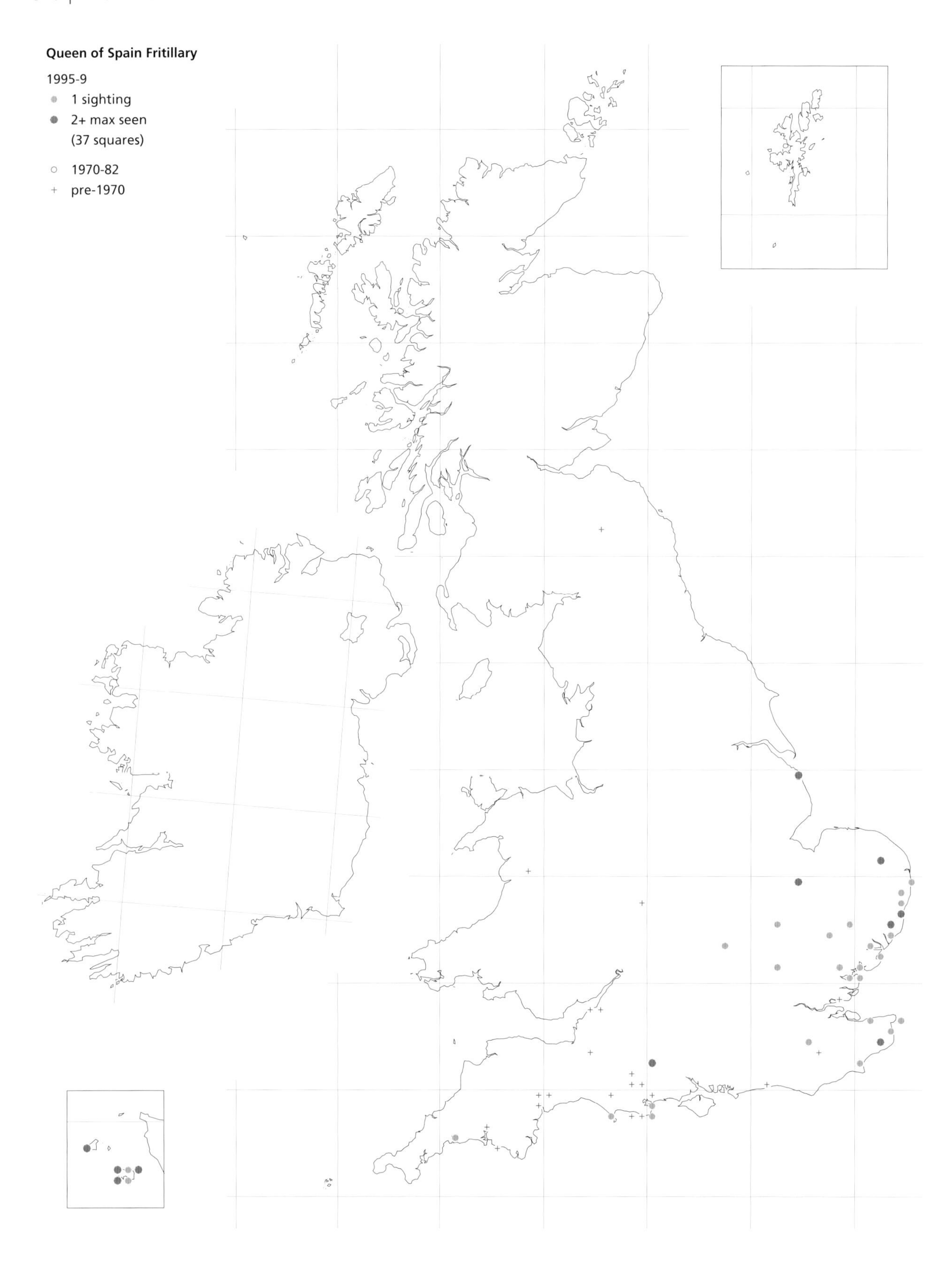
Queen of Spain Fritillary
1995-9
1 sighting
2+ max seen
(37 squares)
1970-82
pre-1970

Queen of Spain Fritillary *Issoria (Argynnis) lathonia*

Rare migrant

Conservation status

UK BAP status: not assessed.
Butterfly Conservation priority: not assessed.
European status: not threatened.

European/world range

Throughout continental Europe as far north as southern Scandinavia, with migrants occurring almost to the Arctic Circle. Resident in North Africa and eastwards through Turkey and the Middle East into Asia. Declining in north-west Europe, but stable or expanding in other European countries.

The beautiful Queen of Spain Fritillary is an uncommon visitor to Britain and Ireland, despite having a migratory habit and the presence of permanent colonies just across the English Channel. In flight, it may be confused with the Dark Green Fritillary, but it is smaller, its wings have a more angular shape, and the large silver patches on the undersides of the hindwings are unmistakable. Sightings have increased in recent years and it is possible that a breeding colony became established in Suffolk.

Foodplants

Species in the genus *Viola* (violets and pansies) are used in continental Europe. Northern populations rely mainly on Wild Pansy (*Viola tricolor*) and Field Pansy (*V. arvensis*) and it is assumed that the butterfly is restricted to these two species in Britain.

Habitat

The butterfly is found in a diverse range of habitats in continental Europe, but in the north of its range is associated predominantly with dry, open habitats such as limestone pavement, heathland, and dunes.

Life cycle and colony structure

The Queen of Spain Fritillary has several generations each year in continental Europe and appears to have a flexible life cycle. Overwintering is reputed to occur in larval, pupal, or adult stages.

Eggs are normally laid singly on the leaves of foodplants, although there are reports of females laying eggs on other low-growing vegetation. **Larvae** and adults spend much time basking. **Pupae** are located near ground level, suspended from vegetation.

Migration is a regular feature of populations in continental Europe, although movements are generally much shorter than those undertaken by species such as the Red Admiral and the Painted Lady. Most records from Britain and Ireland have been in September and October, although there have been some in June, July, and August. After arrival here, there appears to be little evidence of dispersal inland.

Distribution and trends

The Queen of Spain Fritillary is a rare visitor to England and the Channel Islands, and very few records exist for Ireland, Scotland and Wales. The butterfly was more common in the latter half of the nineteenth century (with notable influxes in 1868 and 1872) and in the 1940s (particularly 1945). In the first period, most sightings were from south-east England (particularly Kent), but latterly the butterfly was recorded in all of the counties on the south coast of England. Unfortunately, many of these historical records are not in the BNM database and,

The Queen of Spain Fritillary.

therefore, do not appear on the distribution map.

Larvae and pupae were found on Jersey in several years during the 1950s and mating and egg laying have been observed in mainland Britain. Therefore it appears that early immigrants may give rise to a summer brood. A sighting of 25 individuals flying together in Cornwall in September 1945 added some credence to this possibility, but there is no historical evidence of successful overwintering.

Sightings of the Queen of Spain Fritillary were very rare between 1950 and 1989, but the 1990s saw a dramatic increase in records. Notable observations included five individuals at one site in the Channel Islands in 1991 and several at Spurn Head, Lincolnshire in 1993.

In the current survey period there have been many sightings, particularly in East Anglia where the butterfly was very rare during most of the twentieth century. The recent records were centred mainly on a small area of the Suffolk coast. There were two sightings at the Royal Society for the Protection of Birds' Minsmere Reserve in August 1995, several others between July and September 1996, and many more from a wider area of the Suffolk coast in 1997. Most notable was one in late June, again at Minsmere. Only a few individuals were recorded in this area in 1998 and 1999. In the same period there were also several sightings of individuals in Essex.

European trend

Serious distribution declines have occurred in parts of north-west Europe, particularly in Belgium (>75% decrease in 25 years) and Luxembourg and the Netherlands (50–75% decrease). In other areas, however, it is expanding (notably in Finland, Romania, Russia, and Ukraine) or stable.

Interpretation and outlook

The relatively large numbers of Queen of Spain Fritillaries in consecutive years, and in a small geographical area, prompted speculation that breeding was occurring at Minsmere. The evidence is circumstantial, but the increase in sightings was not mirrored in other counties or, indeed, elsewhere in Suffolk. Deliberate release appears unlikely to account for all of the sightings, as many of the sites have no public access and are regularly wardened.

The timing of the 1995 and 1996 Minsmere records is probably consistent with July immigrants giving rise to a 'native' late-summer brood using Field Pansy as a foodplant. Moreover, the early (June) sighting in 1997 suggested that the species may have overwintered in Suffolk, especially as weather conditions at this time appeared unfavourable for immigration. The low number of sightings in both 1998 and 1999, however, suggest that any breeding colony was short-lived.

The outlook for this species appears brighter than in recent decades. In the Netherlands, populations on coastal dunes have responded favourably to grazing and, elsewhere in the country, the species is reversing previous losses in range. Reports also indicate recent increases in sightings in Belgium and northern France. These trends may lead to increasing numbers of immigrants in Britain (and Ireland) in the future. The evidence strongly suggests that summer breeding is possible here, but reports of overwintering are still speculative.

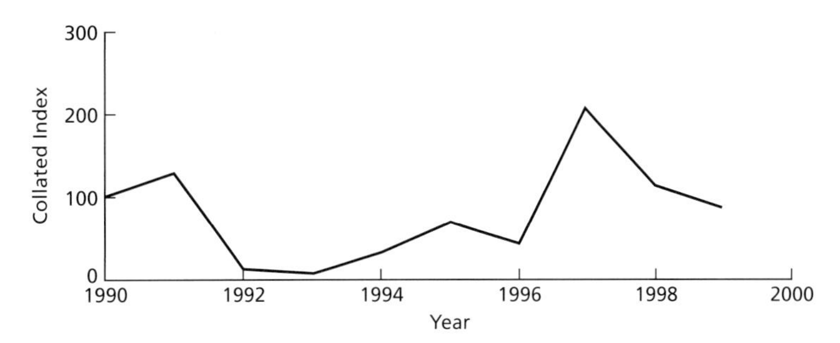

Dutch Butterfly Conservation collated index of abundance of the Queen of Spain Fritillary at monitored sites in the Netherlands, showing high numbers in 1997 (the same year that activity peaked at Minsmere, Suffolk).

The Queen of Spain Fritillary has had short periods of more frequent occurrence in the past. Only time will tell whether the spate of modern records is a temporary phenomenon or marks a long-term change in the status of this species, perhaps in response to climatic warming.

Key references

Wallis DeVries and Raemakers (1998); Willmott (1978); Wilson (1998).

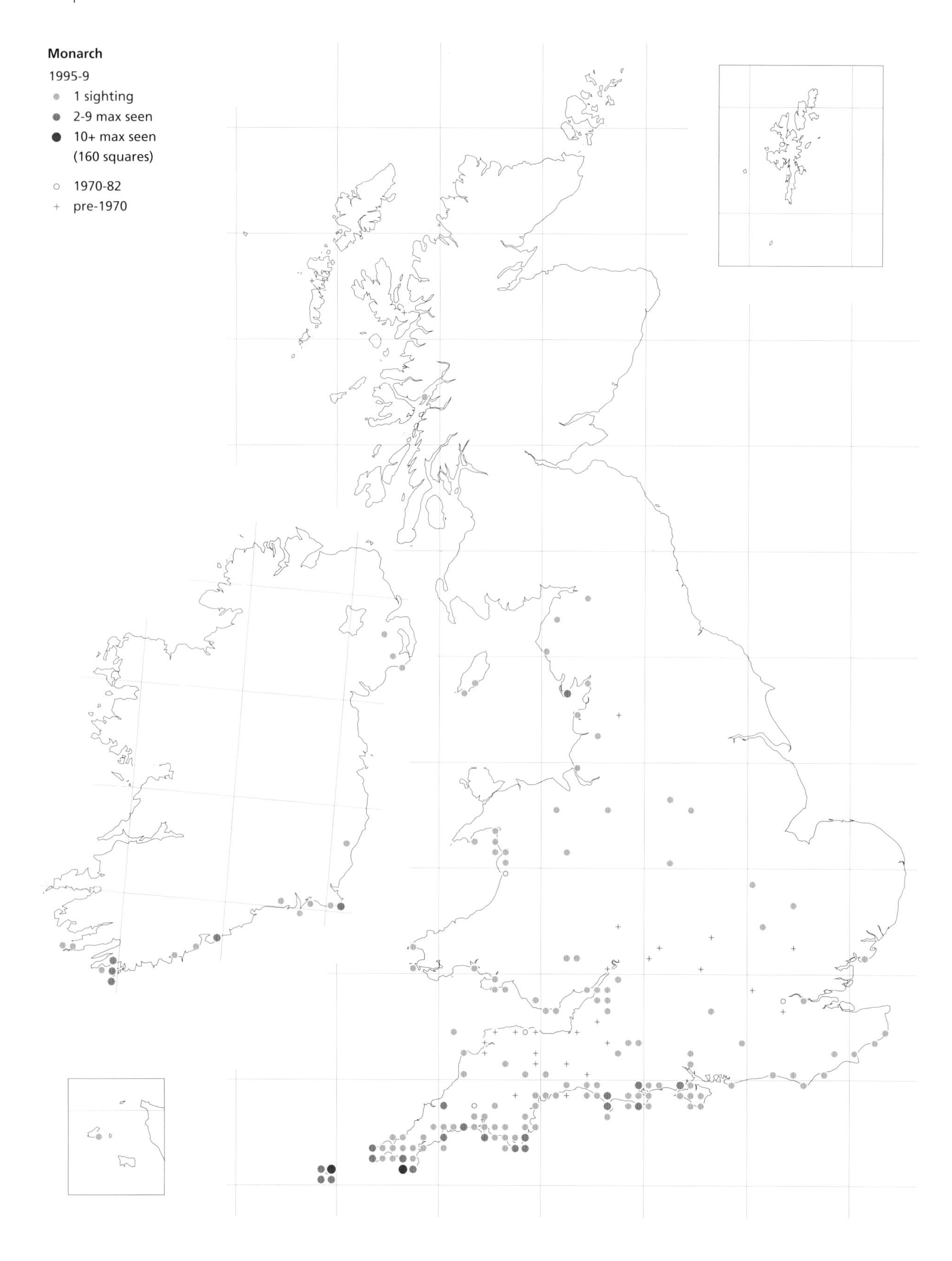
Monarch
1995-9
1 sighting
2-9 max seen
10+ max seen
(160 squares)
1970-82
pre-1970

Monarch *Danaus plexippus*

Rare migrant

Conservation status

UK BAP status: not assessed.
Butterfly Conservation priority: not assessed.
European status: not threatened.

European/world range

Originally North and South America, but now includes several Pacific Islands, Australia, New Zealand, and parts of Asia. Easterly expansion led to colonisation of the Azores, Canary Islands, Madeira, southern Spain, and Gibraltar.

This large and distinctive butterfly is famed for its trans-continental migrations in North America and the discovery of its communal overwintering sites in Mexico. It also occasionally crosses the Atlantic to reach Europe. In common with many other migrants, it has particular years of abundance in Britain and Ireland, and sightings appear to be increasing in frequency.

Foodplants

Milkweeds in the genus *Asclepias* are the main foodplants, giving rise to the Monarch's alternative vernacular name, the Milkweed. There are no native *Asclepias* species in Britain and Ireland, and these plants are not widely cultivated. The Monarch populations in south-western Europe and the Canary Islands commonly utilize *Asclepias curassavica*, which has become naturalized in some areas.

Habitat

In North America the butterfly breeds in open, flowery habitats including meadows, dry and damp pasture, and waste ground. In Britain and Ireland there is no natural breeding habitat and most adults are seen near the coast, and in areas rich in nectar, including gardens.

Life cycle and colony structure

Populations in the Canary Islands and Spain are continuously brooded, but in North America the adults enter reproductive diapause at the end of the summer.

Eggs are laid singly on foodplant leaves and the brightly coloured **larvae** feed voraciously, often stripping whole plants. Plant toxins are absorbed by the larvae and passed onto the adults, making them unpalatable to most predators. **Pupae** are attached to stems or leaves of the foodplants or other nearby vegetation. When conditions are suitable, complete development from egg to adult may take less than a month.

In North America, Monarchs spend the winter months in aggregations in Mexico or coastal California and migrate inland north and eastwards in spring stopping to breed on the way. Those migrating back from the Mexican overwintering sites lay their eggs on milkweeds along the Gulf Coast and the new generation reaches Canada by late May or early June. In the autumn, individuals make an extraordinary migration to the south, covering distances of more than 3600 km. During the later stages of this migration, enormous swarms develop, and overwintering occurs in spectacular numbers, particularly in Mexico.

Distribution and trends

Historical records of the Monarch in Britain and Ireland show a strong south-westerly and coastal distribution, and the 1995–9

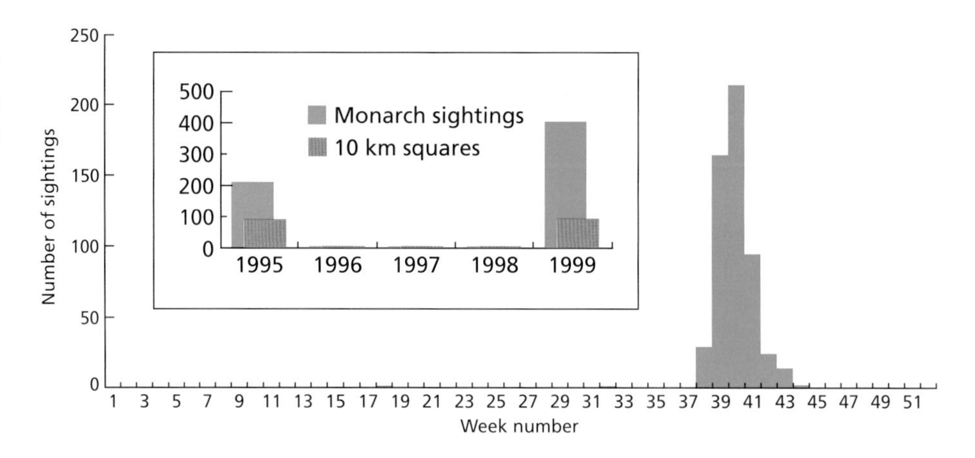

The number of Monarch records by week during 1999, showing the timing of the major influx of migrants. The inset graph shows the total number of records received and 10 km squares recorded for the Monarch in each year (1995–9).

survey reflects this pattern. Most sightings have occurred in Cornwall, the Isles of Scilly, Devon, Dorset, Hampshire, and the Isle of Wight. Monarchs have also been reported regularly from the less well-recorded areas of south-west Wales and Ireland.

Sightings have occurred in most areas during the last 150 years, from the Channel Islands to Shetland. Despite their strong flight and migratory habits, Monarchs often remain close to the coast on arrival. The only breeding record came from Kew Gardens, London, in 1981, following an escape from a nearby butterfly house.

The majority of Monarch sightings have been in September and October and, as with other migrants, high numbers of individuals have occurred in particular years. Notable years were 1968 and 1981, which together represent over 40% of the records between 1876 and 1988. The two largest immigrations recorded to date both occurred during the current survey, in 1995 and 1999. It can be difficult to determine the number of individual butterflies involved. Such is the excitement that sightings are often reported several times! However, there were approximately 150–200 records in 1995 and a conservative estimate of 300 individuals has been put forward for September and October 1999. These two years account for most of the records in the current survey, but Monarchs have been seen in every year.

Phenogram of dates of 1999 records of the Monarch plotted against latitude. The simultaneous arrival of Monarchs in south-west England and south-west Ireland and the delay before sightings were recorded elsewhere, are consistent with the influx arriving from the south-west.

The Monarch is unlikely to be misidentified, but the origin of individuals is difficult to ascertain and records of captive bred Monarchs may slightly distort the distribution of true migrants.

European trend

The Monarch's distribution in the Canary Islands has declined recently (15–25% decrease in 25 years), but colonization of parts of southern Spain and, most recently, Gibraltar has occurred over the same period.

Interpretation and outlook

There is strong evidence that the Monarchs that reach Britain and Ireland come from North America. The dates of most records correlate well with major migrations on the other side of the Atlantic and with favourable meteorological conditions. In addition, several of the larger immigrations (including those of 1995 and 1999) have coincided with the arrival of North American bird species. There are also records of Monarchs seen at sea in the North Atlantic.

The 1995 immigration occurred shortly after reports of mass movements of Monarchs on the east coast of the United States and during a strong easterly flow of warm air. Similar meteorological conditions occurred during the 1999 influx, as two storm systems moved eastwards across the Atlantic. A study of wind speeds suggested that it may have taken as little as four days for the butterflies to cross the Atlantic.

Alternative explanations for Monarch records in Britain and Ireland include the accidental or deliberate release of captive bred individuals, and Monarchs arriving on ships. Migration from the European populations is another potential source. Indeed, some immigrations have been accompanied by moths from these south-western parts of Europe (for example in 1995 and 1983).

On balance, none of these alternative explanations seems able to account for the number and distribution of sightings in major immigration events such as those in the autumns of 1995 and 1999.

Irrespective of the Monarch's fame, its occurrence in Britain and Ireland is currently little more than a delightful curiosity. There are no realistic prospects of breeding or colonization. Nevertheless, recent decades have seen a substantial increase in sightings, which may be expected to continue. In the long-term, the immigration of Monarchs is dependent on their survival in America and, perhaps increasingly, in Europe.

Key references

Brower (1995); Davey (2000); Coombes and Tucker (1996); Martin and Gurrea (1988); Nelson (1996); Skinner and Parsons (1998); Tunmore (2000); Vanholder (1996).

6 The pattern and cause of change

The maps collated in this atlas show that butterfly distributions in Britain and Ireland have changed substantially since recording began in the nineteenth century. Some of these changes have taken place gradually over the last two centuries, but others have occurred more rapidly, especially during the last few decades. The Butterflies for the New Millennium (BNM) survey (1995–9) demonstrates that major changes are still occurring.

Habitat specialists and wider countryside species

Butterflies in Britain and Ireland have been divided previously into two categories: 'matrix' (common and widespread) species, which occur throughout the farmed countryside and in urban habitats; and 'island' (rarer and more localized) species, which are usually restricted to discrete patches of semi-natural habitat.[1] This concept is useful in evaluating changes in distribution and predicting future trends, and has been developed here in the light of recent ecological studies, using the terms 'wider countryside species' and 'habitat specialists' (see summary below).

Wider countryside species use habitats and foodplants that are still relatively common in the modern landscape (e.g. hedgerows, field margins, road verges, and nettle beds). They also tend to be mobile species that are able to colonize suitable habitat as it becomes available. As a result, many have extensive and continuous distributions. In contrast, the habitats used by specialist butterflies are more localized and often occur as small, isolated patches (e.g. lowland heath, coppice woodland, and calcareous grassland). Often habitat specialists are less able to colonize suitable habitat patches, and their distributions tend to be more restricted and localized.

Most species fit well into one or other category, but some species have different requirements near the edge of their range. For example, the Speckled Wood is restricted to woodland habitats in northern Britain and might be considered a habitat specialist there, but breeds in a wider range of habitats in the south. In such cases the species have been classified according to the characteristics displayed in the

[1] Pollard and Yates (1993*a*); Pollard and Eversham (1995).

Comparison of the attributes of habitat specialists and wider countryside species.*

Habitat specialists	*Wider countryside species*
• Confined to specific, discrete habitat 'islands' that are localized or patchy in the modern landscape (i.e. butterfly species of downland, heathland, woodland clearings etc.)	• Broad habitat requirements or use habitats that are widely distributed in the farmed countryside (e.g. generalist grassland, woodland, or hedgerow species)
• Rarely or never use linear habitats such as hedgerows or road verges	• Can use linear habitats such as hedgerows and road verges
• Usually only one or two species of larval foodplant	• Often several species of larval foodplants
• Relatively sedentary	• Relatively mobile
• Mostly have a single generation per year	• Often multi-brooded

* Developed from Pollard and Eversham (1995).

majority of their range. In addition, species that are restricted to a certain region, but which use habitats that are widespread and common within that area, are considered to be wider countryside species.

Our classification differs from those of previous authors in a few aspects. The Purple Hairstreak, White-letter Hairstreak, and Scotch Argus have been placed with the wider countryside species because their habitats are widespread in the landscape within their range. The Brown Argus and Marbled White have been placed in this category too, because they have shown the ability to colonize new sites and a wide range of habitats in recent years.

Distribution change since the 1970–82 survey

Coverage and possible bias

An important difference between the 1970–82 and 1995–9 data sets is that, in Britain, the latter contains over 12 times as many records in total and records from almost 7% (179) more 10 km grid squares. The increased coverage has been even greater in Ireland where 15% (126) more squares have been recorded and the number of records has increased fourfold since the 1970–82 survey. The number of squares with records of common species is therefore greater in the new survey, and declines of other species are less apparent. The changes must therefore be interpreted with caution as the data in each period contain some elements of bias.

Although the 10 km square resolution maps presented in the species accounts reveal broad changes in range, past coverage may have been too patchy to measure the degree of change with precision. The limitations of mapping schemes have been explored in several recent studies and the rate of decline at the population level may be under-represented by such coarse-scale distribution maps.[2] Moreover, declines in the abundance of some widespread species may be completely obscured. The main analysis presented is a simple comparison between the number of 10 km squares recorded for each species in the BNM survey (1995–9) and the data gathered in the 1970–82 survey. Separate analyses are shown for Great Britain (and the Isle of Man) and Ireland (see pp. 326–7 and 330).

In addition to the overall increase in recording intensity, considerable efforts were made in the BNM survey to cover rare or declining species and to target previously under-recorded areas (p. 34). Some species are difficult to record, for example canopy-dwellers such as the Purple Hairstreak, or the Green Hairstreak whose adult

Types of change in the status and distribution of butterflies referred to in the text

- Decreases in the total number of squares recorded for a species (contraction of range) due to:
  1) genuine decline
  2) under-recording
- Increase in the total number of squares recorded for a species due to:
  1) genuine range expansion (extension of the edge of range)
  2) genuine infilling (increased distribution within range)
  3) increased survey effort

[2] Dennis (2000); Dennis and Thomas (2000); León-Cortés *et al.* (1999); Thomas and Abery (1995).

behaviour makes it difficult to observe. These and other factors affecting survey coverage in both data sets vary from species to species, but need to be borne in mind when evaluating changes and trends.

Such differences in recording effort occur for many, if not all, other taxa, and several methods have been developed to make allowances in analysis. One method, which was applied in the analysis of changes in butterfly distribution in Flanders (Belgium) and the Netherlands during the twentieth century, used 'reference species' to correct for different recording intensity in different periods.[3] However, no suitable reference species could be selected from the data available in Britain and Ireland, because of the under-representation of records of widespread and common species in the early decades of the twentieth century. A method that takes into account the complex differences in recording effort between the data sets available for butterflies in Britain and Ireland requires further development.

The tables give the total number of recorded squares in each data set and the

[3] Maes and Swaay (1997).

Britain and the Isle of Man: analysis of butterfly distribution changes between 1970–82 and in 1995–9.
Figures refer to the number of 10 km grid squares with records within each recording period, dividing species into habitat specialists and wider countryside species. Data are included for the three migrant species that regularly breed in Britain.

Species	*Recorded squares 1970–82*	*Recorded squares 1995–9*	*Overall change*	*New squares added*	*% New squares added*	*Previous squares lost*	*% Previous squares lost*
Habitat specialists							
Chequered Skipper	30	27	−10%	6	**20%**	9	30%
Lulworth Skipper	13	12	−8%	2	**15%**	3	23%
Silver-spotted Skipper	21	31	+48%	16	**76%**	6	29%
Dingy Skipper	584	571	−2%	216	**37%**	229	39%
Grizzled Skipper	386	385	0%	137	**35%**	138	36%
Swallowtail†	6	6	0%	0	**0%**	0	0%
Wood White	109	82	−24%	41	**38%**	68	62%
Green Hairstreak	603	911	+51%	506	**84%**	198	33%
Brown Hairstreak	160	137	−14%	32	**20%**	55	34%
Black Hairstreak	24	25	+4%	9	**38%**	8	33%
Small Blue	194	245	+26%	109	**56%**	58	30%
Silver-studded Blue	104	97	−7%	39	**38%**	46	44%
Northern Brown Argus*	60	121	+102%	79	**132%**	18	30%
Chalkhill Blue	200	204	+2%	54	**27%**	50	25%
Adonis Blue	79	95	+20%	40	**51%**	24	30%
Duke of Burgundy	113	106	−6%	37	**33%**	44	39%
White Admiral	241	375	+56%	183	**76%**	49	20%
Purple Emperor	67	89	+33%	44	**66%**	22	33%
Small Pearl-bordered Fritillary	643	768	+19%	373	**58%**	248	39%
Pearl-bordered Fritillary	328	262	−20%	132	**40%**	198	60%
High Brown Fritillary	105	57	−46%	33	**31%**	81	77%
Dark Green Fritillary	645	925	+43%	514	**80%**	234	36%
Silver-washed Fritillary	381	495	+30%	207	**54%**	93	24%
Marsh Fritillary	252	224	−11%	110	**44%**	138	55%
Glanville Fritillary	7	9	+29%	4	**57%**	2	29%
Heath Fritillary	13	13	0%	4	**31%**	4	31%
Mountain Ringlet	29	39	+34%	22	**76%**	12	41%
Grayling	570	582	+2%	226	**40%**	214	38%
Large Heath	290	340	+17%	187	**64%**	137	47%

percentage changes between them. Because of the greater intensity of recording in 1995–9 these totals obscure the true picture. A more accurate measure of loss of range for declining species is drawn from the number of squares recorded in 1970–82 from which species have been 'lost' (not re-recorded in 1995–9). This is shown in the final column of each table and forms the basis for much of the following discussion. As an artefact of the analysis, all species show a small 'decline' (at best zero) because not all squares were re-surveyed exhaustively in 1995–9. This is evident for most of the wider countryside species in Britain, which show small reductions (<10%) in relation to the 1970–82 squares. They also show higher percentages of new squares, reflecting more intensive recording in 1995–9.

In Ireland, although the overall coverage is good, many previously recorded 10 km squares have not been re-surveyed in sufficient detail or during the flight periods of all species. Therefore, in some squares, even widespread and common species recorded in 1970–82 have not been re-recorded in the 1995–9 survey, making

Species	*Recorded squares 1970–82*	*Recorded squares 1995–9*	*Overall change*	*New squares added*	*% New squares added*	*Previous squares lost*	*% Previous squares lost*
Wider countryside species							
Small Skipper	1068	1463	+37%	439	**41%**	44	4%
Essex Skipper	267	638	+139%	389	**146%**	18	7%
Large Skipper	1202	1578	+31%	454	**38%**	78	6%
Brimstone	907	1240	+37%	378	**42%**	45	5%
Large White	1880	2191	+17%	459	**24%**	148	8%
Small White	1805	2062	+14%	400	**22%**	143	8%
Green-veined White	2132	2540	+19%	471	**22%**	63	3%
Orange-tip	1428	2048	+43%	692	**48%**	72	5%
Purple Hairstreak	562	999	+78%	547	**97%**	110	20%
White-letter Hairstreak	443	575	+30%	310	**70%**	178	40%
Small Copper	1614	1958	+21%	501	**31%**	157	10%
Brown Argus*	346	718	+108%	423	**122%**	51	15%
Common Blue	1840	2220	+21%	576	**31%**	196	11%
Holly Blue	770	1456	+89%	711	**92%**	25	3%
Small Tortoiseshell	2073	2416	+17%	456	**22%**	113	5%
Peacock	1497	2010	+34%	573	**38%**	60	4%
Comma	838	1501	+79%	691	**82%**	28	3%
Speckled Wood	948	1462	+54%	556	**59%**	42	4%
Wall	1362	1378	+1%	260	**19%**	244	18%
Scotch Argus	243	325	+34%	130	**53%**	48	20%
Marbled White	398	662	+66%	286	**72%**	22	6%
Gatekeeper	1068	1416	+33%	370	**35%**	22	2%
Meadow Brown	2042	2388	+17%	453	**22%**	107	5%
Ringlet	1108	1694	+53%	649	**59%**	63	6%
Small Heath	1825	2058	+13%	488	**27%**	255	14%
Regular migrants							
Clouded Yellow	211	1118	+430%	936	**444%**	29	14%
Red Admiral	1466	2331	+59%	927	**63%**	62	4%
Painted Lady	1219	2114	+73%	991	**81%**	96	8%

† Excluding records of migrants.

*The squares recorded for the Northern Brown Argus and Brown Argus during 1970–82 have been reclassified in light of the current interpretation of these species.

numerical comparison inconclusive. The percentage of new squares is higher for many species than in Britain and only the Green-veined White shows a reduction that is less than 10%. Such a comparison should become appropriate for Ireland in a few years time, if the current recording effort continues as planned.

Changes affecting habitat specialists since 1970–82: Britain

The results of the BNM survey show that most habitat specialists have continued to decline in Britain. They have not been re-recorded in a substantial proportion of 10 km squares where they were found in 1970–82, despite the increased intensity of the 1995–9 survey. For most species this is believed to represent a genuine decline, but there are exceptions, and the reader is referred to the species accounts for case by case interpretation.

The habitat specialist species that have undergone the most serious declines in Britain since 1970–82 based on the percentage of 1970–82 squares in which each has not been re-recorded in 1995–9 are: the High Brown Fritillary (–77%), Wood White (–62%), Pearl-bordered Fritillary (–60%), Marsh Fritillary (–55%), Large Heath (–47%), Silver-studded Blue (–44%), Duke of Burgundy (–39%), and Dingy Skipper (–39%). Most other species have shown genuine declines, reflected in a loss of more than 20% of their 1970–82 squares. These losses are all the more dramatic considering that the level of recording in 1995–9 has been far higher than in 1970–82 and that weather conditions have been generally favourable for butterflies. Note that the number of squares in which the Black Hairstreak and Mountain Ringlet were not re-recorded may be due partly to difficulties in observing these species.

Some species remain under-recorded and are characterized by both large percentage declines of 1970–82 squares and the recent discovery of many new squares. Percentage declines of extremely restricted species are also probably exaggerated due to sporadic records outside their core range and attempts at reintroduction. For example, the 29% decline in recorded squares for the Glanville Fritillary since 1970–82 is due to the failure to re-record the species in only two 10 km squares, both of which represented temporary colonies away from core habitats. The butterfly has also been recorded in four new squares in the BNM survey, giving an overall increase of 29%. In fact, detailed surveys have shown that the core range of this species and that of the Lulworth Skipper have remained very similar over the last 20 years.[4]

Despite this background of severe decline, new squares have been recorded for many habitat specialist species during 1995–9, due to improvements in survey coverage. These include threatened species that have been surveyed comprehensively for the first time, such as the Chequered Skipper, Pearl-bordered Fritillary, High Brown Fritillary, and Heath Fritillary. For each species, additional colonies have been discovered in areas such as Exmoor, Dartmoor, and parts of Wales and Scotland. There is no evidence that these newly discovered colonies are the result of recent colonization, whereas there are many documented cases of recent extinctions.

Seven other habitat specialists were recorded in many more 10 km grid squares during the more recent survey, but this is believed to be due to improved recording rather than expansion. The Northern Brown Argus has shown the largest of these increases, followed (in decreasing order) by the Green Hairstreak, Dark Green Fritillary, Mountain Ringlet, Small Blue, Small Pearl-bordered Fritillary, and Large Heath. Most of the new squares that have been added for these species are in

[4] Bourn and Warren (1997*b*); Bourn *et al.* (2000*b*).

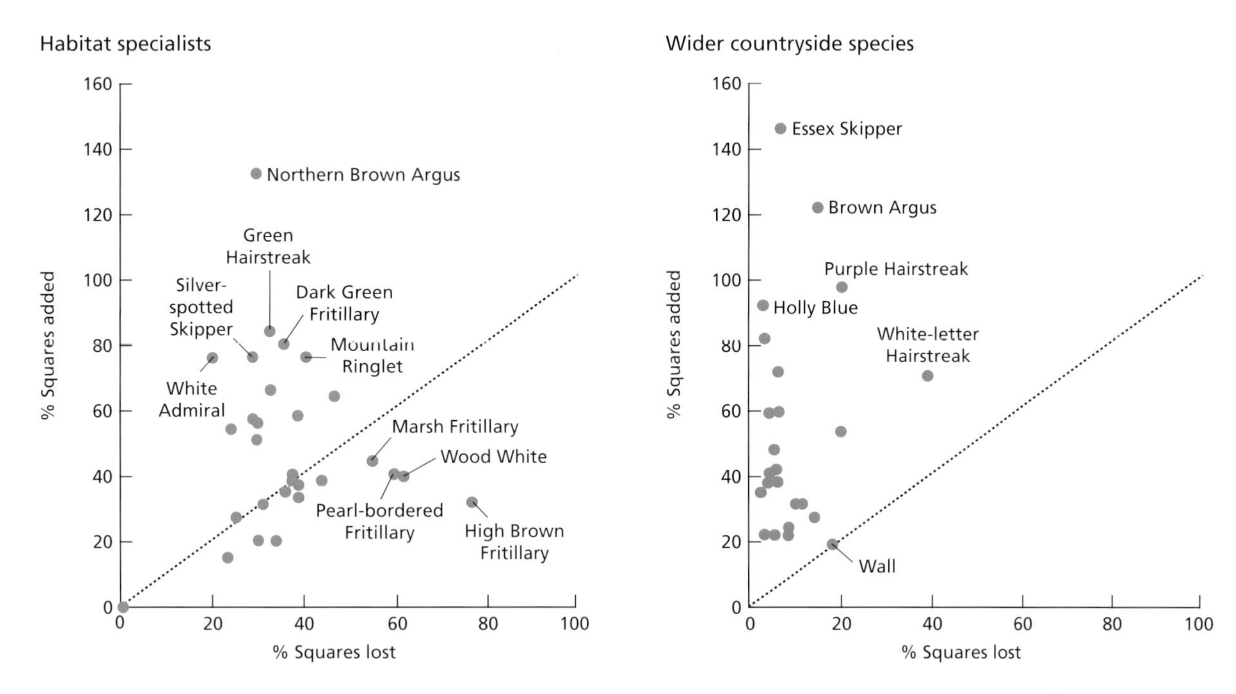

Gains and losses amongst butterfly species in Britain between 1970–82 and 1995–9, based on the number of 10 km squares recorded in each period (see Table p. 326–7). Species lying below the diagonal line show a net decrease in recorded squares between these periods and those above the line show a net increase.

previously under-recorded areas, particularly parts of Scotland, where our knowledge of their distributions has now improved markedly. With the exception of the Green Hairstreak and Mountain Ringlet (which are believed to be under-recorded), all of these species have shown genuine declines in parts of their former range. The Small Pearl-bordered Fritillary, for example, has undergone a substantial decline (−39%) from its 1970–82 range, with 248 former squares being lost, mainly in England. This decline has been offset by BNM records of the species from 373 new 10 km squares, mainly in Scotland. Again there is no evidence to suggest that any of the new squares are due to colonization (genuine range expansion). Furthermore, this and other species have probably undergone substantial local declines within their range due to habitat loss, which have not yet become apparent at a 10 km square level.

Although the increased level of recording effort in the BNM survey makes the interpretation of change since 1970–82 complex, this improved coverage is of great significance to conservation. Thousands of new colonies of habitat specialist butterflies have been discovered as a result of both the BNM project and detailed species surveys, representing an important step towards conserving such species in Britain and Ireland.

In contrast to the declining species, four habitat specialists that had declined rapidly over the period 1950–80 have recently undergone a partial recovery, as shown by the number of new squares recorded by the BNM survey. The Silver-spotted Skipper has returned to parts of its former range on chalk downland in south-east England while the Adonis Blue has spread in Dorset and Wiltshire. Two woodland species, the Purple Emperor and Silver-washed Fritillary, have also returned to parts of their former range in southern England, although both remain more restricted than in the early twentieth century. These recoveries have occurred in well-recorded areas and are not thought to be biased greatly by increased recording effort. The largest expansion among the habitat specialists is that of the White

Ireland: analysis of butterfly distribution changes between 1970-82 and in 1995-9.
Figures refer to the number of 10 km grid squares with records within each recording period, dividing species into habitat specialists and wider countryside species. Data are included for the three migrant species that regularly breed in Ireland.

Species	*Recorded squares 1970–82*	*Recorded squares 1995–9*	*New squares added*	*Previous squares lost*
Habitat specialists				
Dingy Skipper	36	64	47	19
Wood White	205	241	151	115
Green Hairstreak	81	118	98	61
Brown Hairstreak	19	16	6	9
Small Blue	44	58	40	26
Pearl-bordered Fritillary	5	6	3	2
Dark Green Fritillary	138	99	55	94
Silver-washed Fritillary	178	166	103	115
Marsh Fritillary	126	90	61	97
Grayling	126	145	83	64
Large Heath	40	51	46	35
Wider countryside species				
Brimstone	83	116	84	51
Large White	469	589	258	138
Small White	458	532	232	158
Green-veined White	578	884	353	47
Orange-tip	361	686	393	68
Purple Hairstreak	14	19	15	10
Small Copper	339	372	206	173
Common Blue	386	431	205	160
Holly Blue	77	106	74	45
Small Tortoiseshell	541	629	241	153
Peacock	336	645	404	95
Speckled Wood	515	769	318	64
Wall	305	294	162	173
Gatekeeper	80	49	23	54
Meadow Brown	615	707	236	144
Ringlet	441	563	285	163
Small Heath	239	285	180	134
Regular migrants				
Clouded Yellow	84	108	76	52
Red Admiral	296	550	334	80
Painted Lady	138	353	271	56

Admiral (56% increase since 1970–82), whose range has been expanding since the 1920s following an earlier contraction. The latest results show that it continued to spread and expand its range margin northwards in the 1980s and 90s.

Changes affecting habitat specialists since 1970–82: Ireland

In Ireland, the picture is less clear due to lower levels of recording than in Britain in both survey periods. The Large Heath and Marsh Fritillary show particularly large numbers of 1970–82 squares in which they have not been re-recorded. This may reflect the genuine declines witnessed in some parts of Britain, but may partly be the result of under-recording. The Large Heath, for example, was recorded in more new 10 km squares during the current survey, than were known in 1970–82. Although the interpretation of changes in Ireland is difficult with this analysis, certain conclusions can be drawn. The ranges of a few important species, such as the Wood White, Brown Hairstreak, and Pearl-bordered Fritillary, have not suffered such serious contractions at the 10 km square level as they have in Britain. In addition, many new colonies have been located for other habitat specialists, notably the Dingy Skipper and Small Blue, as well as the Large Heath.

Changes affecting wider countryside species since 1970–82: Britain

In contrast to habitat specialists, most wider countryside species in Britain have been re-recorded in a high proportion of the 10 km squares that were recorded in 1970–82. Moreover, nearly all have been recorded in many additional squares in the recent survey. Despite the overall increase in this group, a few species have undergone a significant loss of 1970–82 squares. Apparent losses of the Purple Hairstreak may be attributable to difficulties in recording this canopy-dwelling species. The same recording problem may also have played a part in the recorded decline of the White-letter Hairstreak. This latter species has suffered some genuine decline as a result of the initial effects of Dutch Elm Disease on its foodplants, but is now recovering in many areas.

The Wall is the only wider countryside species for which the BNM survey has not recorded a substantial increase in the number of occupied squares. Improved recording and, possibly, some local infilling and expansion in parts of the Wall's range have offset a major loss of colonies across a large area of the West Midlands and southern England. Many other wider countryside species (e.g. the Small Heath) may be declining within their range, but the true extent of changes have not been shown at the 10 km square level of resolution.

The increase for 11 species has been either relatively small or is thought to be due largely to increased recording, rather than any real change in distribution. The remaining 14 species have been recorded in over 30% more 10 km squares than in 1970–82 and all appear to have undergone genuine expansions of range in Britain. They now occupy regions where they were absent or rare at the time of the last survey, but a proportion of the increase for many of these species can also be attributed to 'infilling' (i.e. local increases in distribution that do not extend the range margins of the species). The biggest expansions have been for the Essex Skipper and Brown Argus, both of which have more than doubled their known range in the last 20 years. Major expansions have also been recorded for (in descending order of magnitude) the Holly Blue, Comma, Purple Hairstreak,

Marbled White, Speckled Wood, and Ringlet, which have increased their range by more than 50% since 1970–82. Other species with expanding ranges are the Orange-tip, Small Skipper, Brimstone, Peacock, Gatekeeper, and Large Skipper.

The recent expansions of the Orange-tip, Peacock, Comma, and Speckled Wood represent the continuation of trends identified by the previous survey.[5] However, the rate of spread has increased in the past two decades. The only wider countryside species that had suffered a major contraction of range at the time of the 1970–82 survey was the Marbled White. This trend has since been reversed and the butterfly has re-colonized almost all of its pre-1940 range. Many of the other range expansions of wider countryside species have commenced or only become apparent since 1970–82, further highlighting the rapid and dramatic nature of fluctuations in the distributions of butterflies. All of these expansions are thought to be real and not unduly biased by improvements in recording effort, although species such as the Essex Skipper and Purple Hairstreak were (and remain) somewhat under-recorded.

Most of the species that are expanding in Britain are spreading northwards, although some are spreading to the east or west at the same time. The pattern of change varies with each species. Some have spread away from core areas that have been occupied for many decades (e.g. the Essex Skipper and Brown Argus). Others, including the Comma and Speckled Wood, have re-expanded following an earlier contraction during the late nineteenth and early twentieth century.

Overall, the distribution of 14 wider countryside species has expanded in recent years compared with just five habitat specialists (the Silver-spotted Skipper, Adonis Blue, White Admiral, Purple Emperor, and Silver-washed Fritillary). On the other hand, over 20 habitat specialists have undergone substantial losses at the 10 km square level in the last 20 years, compared with just two wider countryside species (the White-letter Hairstreak and Wall).

Changes affecting wider countryside species since 1970–82: Ireland

The Orange-tip, Peacock, and Speckled Wood appear to be faring particularly well, as are the Purple Hairstreak and Holly Blue, species that could be considered as habitat specialists in Ireland. The Gatekeeper is the only wider countryside species to show a major reduction in the number of 10 km squares recorded in Ireland, but we remain unsure whether this represents a genuine decline. Although the current and historical distribution data are insufficient to draw firm conclusions, it is likely that many wider countryside species have suffered widespread local losses.

Changes in distributions since the nineteenth century

The distribution maps presented in this atlas include records of butterflies back to the nineteenth century. The most striking trend over this longer period is that five of the 59 resident species have become extinct and a large proportion have undergone substantial declines (p. 334). Of these, 15 have undergone major contractions in range (defined as more than 50% range loss at the 10 km square level) and a further 14 have disappeared from parts of their range (i.e. they have experienced range losses of at least 20%). Thus the range of half of our resident species has declined and 8% have become extinct. Moreover, the extent of these losses may have been underestimated due to the comparatively low level of past recording, especially prior to 1970 when recording was very patchy.

[5] Heath *et al.* (1984).

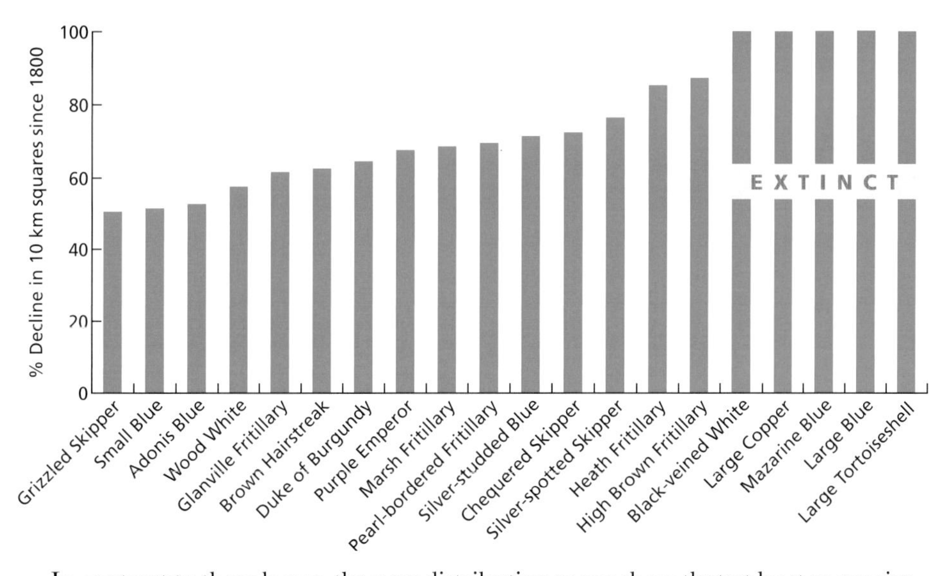

Most serious losses of range amongst butterflies in Britain.

In contrast to these losses, the new distribution maps show that at least 15 species have expanded significantly in recent years, and a further five have shown small recoveries of their former range. A few species seem to show a more complex pattern of change. The Holly Blue, for example, undergoes cycles in abundance, with peaks every 6–10 years, so that in some periods it is widespread and common while in others it is scarce.[6] The BNM recording period included the peak of one such cycle. The fluctuations of the Wall may also be cyclical as it has been through two or three periods of scarcity over the last 150 years, including one during the current survey. However, such long-term cycles are difficult to identify and the scarcity of the Wall during 1995–9 may be only a temporary response to a short-term factor (e.g. a weather event or the outbreak of disease) or may mark the beginning of a long-term decline. Further monitoring will be necessary before an interpretation can be given.

Only 10 species show no obvious change in range. Two of these, the Swallowtail and Mountain Ringlet, are highly restricted species whose habitats have remained fairly stable; the other eight are widespread species, with little scope for range expansion.

Butterfly recording over the past 150 years has revealed a complex pattern of change, but the over-riding picture is one of widespread losses. These have sometimes been part of long-term declines but there has been a noticeable acceleration in the last 50 years.

Changes in local distribution and abundance

The main problem with summarizing distribution by mapping at a 10 km square resolution is that it does not reveal changes in butterfly numbers or flight areas[7] at the local level. Most butterflies have flight areas or populations that are far smaller than 10×10 km and local changes will not be reflected in distribution maps plotted at this scale, unless the species has completely disappeared. The relationship between butterfly distributions and abundance is complex and depends on a variety of factors including the population structure of the species concerned, the distribution of suitable habitats, and the mapping resolution.[8] Because of these questions of scale, distribution maps at the 10 km square level have been shown to underestimate the overall rate of population decline. For example, one study showed that maps at this

[6] Pollard and Yates (1993*b*).

[7] Flight areas represent land units that support populations of butterflies and contain all the resources necessary for the butterfly's life cycle. They thus represent a natural population unit (after Cowley *et al.* 1999).

[8] e.g. Gaston and Kunin (1997); Quinn *et al.* (1997).

Changes in status of the 59 resident butterfly species over the last 200 years. This is based on the total number of 10 km grid squares with records since 1800 and the number recorded in the latest recording period (1995–9). Most changes have occurred within the last 50–100 years.

Extinct (year)	*Major range loss (>50%)*	*Local range loss (c. 20–50%)*	*Small (<20%) or no change in distribution but major decline in abundance within range*[†]	*Small (<20%) or no change in distribution and no evidence for decline in abundance within range*	*Recent range expansion (>20%)*[‡] *or re-expansion to most of former range*
Black-veined White (1920s)	Chequered Skipper	Lulworth Skipper	Swallowtail	Large White	Small Skipper
Large Copper (1851)	Silver-spotted Skipper*	Dingy Skipper	Small Copper	Small White	Essex Skipper
Mazarine Blue (1904)	Grizzled Skipper	Green Hairstreak	Common Blue	Green-veined White	Large Skipper
Large Blue (1979)	Wood White	White-letter Hairstreak*	Meadow Brown	Small Tortoiseshell	Brimstone
Large Tortoiseshell (1980s)	Brown Hairstreak	Black Hairstreak	Small Heath	Mountain Ringlet[¶]	Orange-tip
	Small Blue	Northern Brown Argus			Purple Hairstreak
	Silver-studded Blue	Chalkhill Blue			Brown Argus
	Adonis Blue*	Small Pearl-bordered Fritillary			Holly Blue
	Duke of Burgundy	Dark Green Fritillary			White Admiral
	Purple Emperor*	Silver-washed Fritillary*			Peacock
	Pearl-bordered Fritillary	Wall			Comma
	High Brown Fritillary	Scotch Argus			Speckled Wood
	Marsh Fritillary	Grayling			Marbled White
	Glanville Fritillary	Large Heath			Gatekeeper
	Heath Fritillary				Ringlet
Totals: 5	15	14	5	5	15

* Partial recovery in the last 25 years

[†] Based on flight area data in Cowley *et al.* 1999 (see table opposite), except for Swallowtail which is inferred from loss of open fen habitat during the twentieth century.

[‡] Note that several of these species have almost certainly declined in abundance within their range while still expanding in range.

[¶] Lack of recent records in some previously recorded squares is believed to be due to under-recording.

Loss of butterfly flight areas since 1904 in part of north Wales.* Migratory species are excluded. The following species became extinct between 1904 and 1997 in the study area of 35 km^2: Green Hairstreak, Small Blue, Duke of Burgundy, Small Pearl-bordered Fritillary, Pearl-bordered Fritillary, High Brown Fritillary, and Silver-washed Fritillary.

Species	*500 m grid squares occupied in 1997(%)*	*Total current land surface occupied by flight areas (%)*	*Estimated loss of flight area since 1904 (%)*
Small Copper	66	1.0	91
Brown Argus	53	10.0	77
Common Blue	92	13.0	75
Large Skipper	46	2.0	75
Gatekeeper	96	16.0	73
Speckled Wood	72	14.0	70
Grayling	41	5.0	60
Meadow Brown	99	41.0	54
Small Heath	48	11.0	49
Dark Green Fritillary	26	0.3	41
Silver–studded Blue	16	3.0	17
Wall	41	2.0	12
Dingy Skipper	7	1.4	9

* from Cowley *et al.* (1999)

resolution probably reflect real declines in only the rarest species, whereas declines of species of intermediate rarity may be underestimated by as much as 85%, and population losses of common species are unlikely to be detected at all.[9] Studies in north Wales have developed this theme and shown that many wider countryside species have undergone serious declines in total flight areas[10] although their overall distribution has remained more or less constant.

The local scarcity of some wider countryside species can be seen clearly in the BNM data. For example, even in the well recorded county of Warwickshire, very different numbers of local colonies of two species are not reflected at 10 km square resolution (see figure overleaf). One wider countryside species (the Small Heath) is locally scarce in comparison to another (the Meadow Brown). Both these species have ranges that cover the whole of Britain and Ireland, and neither has shown any significant range contraction at the 10 km square level since 1970–82. It would be desirable to compare the changing distribution of such species at a more detailed level of resolution, but there are insufficient data at high precision in the 1970–82 data set. However, the new data provide a far more detailed baseline against which to measure future change.

The overall ranges of five species have shown little change, but they have probably undergone significant declines in numbers and local distribution within their range (see p. 334, column 4). It is also possible, even likely, that some of the expanding species are becoming less numerous within their range than formerly, due to factors such as habitat loss.

Other evidence for changes in abundance at individual sites comes from the Butterfly Monitoring Scheme (BMS), which was developed in 1976 specifically to monitor changes from a network of over 100 sites around the UK. The data gathered show that six species have undergone significant decreases in abundance at

[9] Thomas and Abery (1995).

[10] Cowley *et al.* (1999); León-Cortés *et al.* (1999).

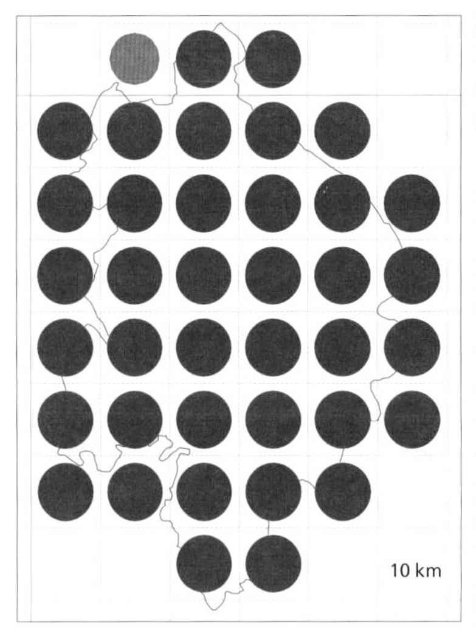

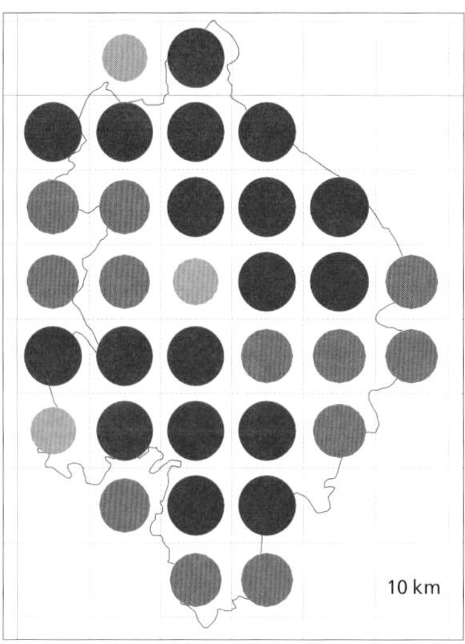

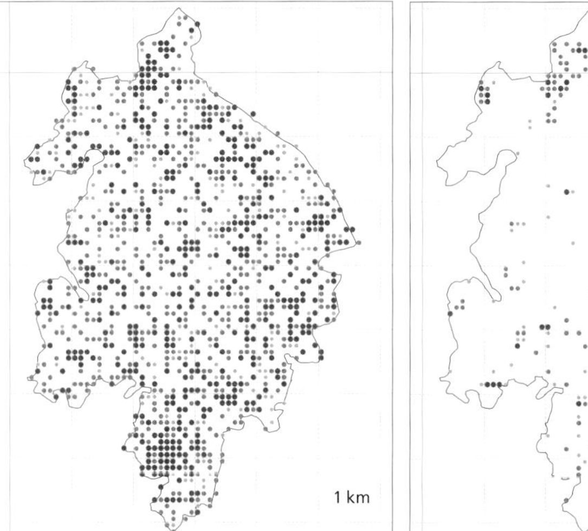

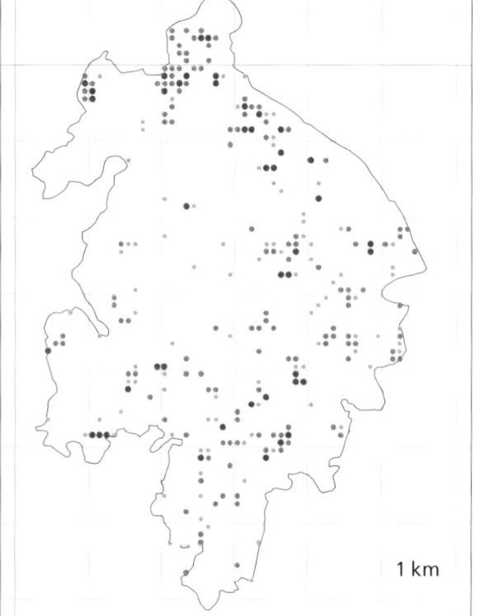

The 1995–9 distribution of two wider countryside species, the Meadow Brown and Small Heath, shown at 1 km square level in Warwickshire (vice-county 38), illustrating the comparative scarcity of some wider countryside species in the modern landscape. At 10 km square resolution, both maps show almost complete coverage.

these sites, 14 have increased significantly, and a further 14 have shown no significant trend.[11] Most declining species are habitat specialists, but they include two wider countryside species: the Wall and Small Heath. In contrast, the species increasing in abundance at the BMS sites are mostly wider countryside species. These data may not be representative of the countryside as a whole because most monitored sites are nature reserves or are specially managed; however, this makes the number of declining species at BMS sites even more alarming.

Causes of decline

Previous studies[12] have examined the causes of decline of butterflies and are summarised in the species accounts. Most species have declined primarily due to habitat loss, such as the destruction of semi-natural grasslands, fenland, hedgerows, or woodland. However, many butterflies have very precise habitat requirements and have been affected as much by changes in land management. Together these have led to increasing fragmentation and isolation of habitats, causing further losses as populations have become less viable, and the natural pattern of extinction and re-colonization has been disrupted. The impact of these factors is discussed below.

Habitat loss

Land use in Britain and Ireland has changed and evolved gradually over the centuries. Until the nineteenth century, butterflies probably thrived in the variety of habitats that were created by traditional agriculture and forestry. However, the pace of change in land use quickened during the twentieth century. Farming became more intensive, while traditional systems were largely abandoned. Writing in the 1940s, E.B. Ford commented 'since the beginning of the nineteenth century the English countryside has become steadily less favourable for our butterflies'.[13] Changes over the last 50 years have accelerated and the effect on semi-natural habitats, and the wildlife that inhabits them, has been devastating.[14] The massive and widespread destruction of semi-natural vegetation and consequent loss of breeding habitats has severely affected butterflies and their distributions.

One of the earliest habitats to suffer was fenland, which provided valuable

[11] Greatorex-Davies and Roy (2000).

[12] e.g. Feltwell (1995); Fox (in press); Heath (1974); Heath *et al.* (1984); Thomas (1984, 1991); Warren (1992*a*, 1993*a*,*b*).

[13] Ford (1945).

[14] Rackham (1986).

The main causes of decline of butterflies in Britain and Ireland.

Species category	*Main causes of decline*
Extinct species	
Black-veined white	Not known (possibly climatic change or disease)
Large Copper	Drainage of fenland
Mazarine Blue	Not known (possibly changes in agricultural practice)
Large Blue	Loss of semi-natural grassland and reduction in grazing
Large Tortoiseshell	Not known (possibly due to loss of suitable woodland and reduced levels of immigration)
Major range loss	
Chequered Skipper	Probably shading of woodland in England and decline in management
Silver-spotted Skipper	Loss/fragmentation of calcareous grassland and reduction in grazing
Grizzled Skipper	Loss of calcareous grassland and reduction in grazing
Wood White*	Shading of woodland rides and decline in coppicing
Brown Hairstreak	Hedgerow removal and annual hedge trimming
Small Blue	Loss of semi-natural grassland and reduction in grazing
Silver-studded Blue	Loss/fragmentation of semi-natural grassland and heathland; reduction in grazing (and burning)
Adonis Blue	Loss/fragmentation of calcareous grassland and reduction in grazing
Duke of Burgundy	Shading of woodland rides and decline in coppicing; loss of calcareous grassland
Purple Emperor	Loss/fragmentation of woodland; changes in woodland management
Pearl-bordered Fritillary	Decline in wood management, especially coppicing; loss and abandonment of bracken habitats
High Brown Fritillary	Decline in wood management, especially coppicing; loss and abandonment of bracken habitats
Marsh Fritillary	Loss/fragmentation of semi-natural grassland and changing grazing patterns
Glanville Fritillary	Loss of semi-natural grassland and cliff stabilisation
Heath Fritillary	Decline in coppicing and loss of semi-natural grassland
Local range loss	
Lulworth Skipper	Not known
Dingy Skipper	Loss of semi-natural grassland
Green Hairstreak	Loss of semi-natural grassland and heathland
White-letter Hairstreak	Loss of elms through Dutch Elm Disease
Black Hairstreak	Clear-felling/removal of Blackthorn stands
Northern Brown Argus	Loss/fragmentation of semi-natural grassland?
Chalkhill Blue	Loss/fragmentation of calcareous grassland
Small Pearl-bordered Fritillary	Decline in coppicing; loss of semi-natural grassland
Dark Green Fritillary	Decline in coppicing; loss of semi-natural grassland
Silver-washed Fritillary	Changing woodland management and fragmentation?
Wall	Not known (possibly loss of semi-natural grasslands and climatic changes)
Scotch Argus	Not known
Grayling	Loss of heathland and semi-natural grassland, and reduction in grazing.
Large Heath	Loss of bogs (especially lowland raised bogs)
Stable range but decline within range	
Swallowtail	Loss of fenland
Small Copper	Loss and 'improvement' of semi-natural grassland, hedges, and ditches
Common Blue	Loss and 'improvement' of semi-natural grassland, hedges, and ditches
Meadow Brown	Loss and 'improvement' of semi-natural grassland, hedges, and ditches
Small Heath	Loss and 'improvement' of semi-natural grassland, and reduction in grazing

* Britain only. Has spread in Ireland.

Information is taken from the species accounts in Chapter 5 and species categories from table on p. 334.

Loss of important butterfly habitats in Britain and Ireland.

Habitat	*Approximate loss since the 1940s*
Britain	
Lowland flower-rich grassland*	97%
Fenland (East Anglia)†	90% (99% since 1700)
Chalk and limestone grassland†	80%
Actively coppiced woodland‡	75% (95% since 1905)
Lowland raised bogs†	60% (94% since 1800)
Ancient broad-leaved woodland†	50%
Hedgerows (England)¶	67%
Hedgerows (Scotland)**	54%
Lowland heathland†	40% (83% since 1800)
Ireland	
Raised bogs††	89%
Blanket bogs††	79%

*Department of the Environment (1995).
†Nature Conservancy Council (1984).
‡Peterken and Allison (1989).
¶Council for the Protection of Rural England (1999).
**Mackey *et al.* (1998).
††Department of Arts, Heritage, Gaeltacht and the Islands (1998).

farmland when drained. The vast fenlands of East Anglia, the main habitat of the Swallowtail and Large Copper, once covered over 3000 square kilometres and included the largest inland lake in lowland England (Whittlesey Mere, measuring 9×5 km). The fens remained more or less intact until the seventeenth century when they were progressively drained, so that by the early 1900s less than 3% remained. This destruction accounted for the first known butterfly extinction, that of the Large Copper, last seen in 1851. The Swallowtail was able to survive in the Norfolk Broads, now the largest remaining fenland fragment in Britain.

The widespread drainage and destruction of bog vegetation has also seriously reduced the habitat of the Large Heath, especially in lowland areas of Britain and Ireland. Another crucial factor affecting butterflies has been the reduction in the extent of managed ancient woodland, an important habitat for at least three-quarters of resident species and the sole habitat of 16 species.

Perhaps the biggest change affecting butterflies has been the agricultural 'improvement' of semi-natural grasslands through ploughing, fertilizing, and cultivation. These grasslands are important habitats for over 43 species and, until the twentieth century, would have been composed mainly of native grasses and herbs, including a range of larval foodplants. However, most modern grasslands typically contain a narrow range of grasses and very few herbs, and a large proportion have been reseeded with Perennial Rye-grass. These 'improved' grasslands are of little value to butterflies because they contain few or no larval foodplants, or they are managed intensively by grazing or repeated cutting. Only small areas of semi-natural grassland remain in lowland Britain and these collectively represent less than 2% of the permanent grassland area.[15] Far larger areas of semi-natural grassland still exist in the uplands, but these support only a small proportion of native butterfly species.

For the last two centuries, most of the lowlands of Britain and Ireland have been divided up into small fields, often bounded by hedges or walls and associated strips of grassland (field margins). These enclosures provided a network of linear habitats across the landscape where many butterflies could breed. We have described these

[15] The total area of semi-natural lowland grassland is thought to be reduced to between 50 000 and 100 000 ha in England and Wales (Blackstock *et al.* 1999).

butterflies as wider countryside species partly because of this feature. However, this vital network has been severely reduced over the last 50 years. More than half our hedgerows have disappeared as field sizes have been increased to accommodate larger machinery and greater productivity. Where hedgerows remain in farmland, their value for wildlife has often been damaged or destroyed by management practices such as the cultivation of former field margins, the use of herbicides, and annual flailing of hedges.

Loss of hedgerows and the increased frequency and mechanisation of cutting have caused the decline of many butterflies.

Changing habitat management

The rapid intensification of farming and forestry has had a profound effect on the management of remaining semi-natural habitats as these generally represent small fragments of marginal and low productivity land. As explained in the species accounts, most habitat specialist butterflies are very sensitive to the way their habitats are managed and can disappear rapidly from sites if management regimes become unsuitable.

In grasslands, many species require precise grazing regimes to maintain a suitable sward structure where the larval stages can develop.[16] One of the most serious factors affecting butterflies that need short turf has been the abandonment of low productivity grassland. Fewer domestic stock and the impact of myxomatosis on rabbit populations, have led to increased sward height and gradual invasion by scrub. One casualty was the Large Blue, which declined rapidly following the introduction of myxomatosis in the 1950s and became extinct in 1979. Reduction in grazing pressure also caused severe declines in species that require short turf on calcareous grasslands, notably the Adonis Blue and Silver-spotted Skipper. The Silver-studded Blue and Grayling have also died out from most inland chalk downlands, possibly because of the absence of the short, sparse vegetation needed for breeding. These species now survive mainly on heathland and coastal habitats. In the damp grasslands of south-west England, the Marsh Fritillary has been affected similarly by the long-term reduction in grazing pressure.

In the last two decades, grazing levels have increased on some lowland grasslands, especially chalk downs, due in part to the recovery of rabbit populations and partly to the reintroduction of grazing regimes for conservation purposes. As a result, both the Adonis Blue and Silver-spotted Skipper have shown some re-expansion of their range, though they remain far more restricted than in the 1940s. It has also been possible to reintroduce the Large Blue, following a programme that has restored grazing regimes and recreated suitable habitat on some former sites.[17]

Because butterflies have different requirements, the effects of changing grassland management have varied from species to species. The effects of abandonment have been temporarily beneficial to some species. The Marsh Fritillary is thought to have spread onto southern chalk downland during the early twentieth century, probably as a result of abandonment of grazing. This was particularly noticeable on Salisbury

[16] Turf height summaries are given in BUTT (1986).

[17] Thomas (1999*a*,*b*).

Remaining semi-natural habitats, such as this coastal grassland in Cornwall, have frequently suffered from neglect, in this case leading to scrub invasion.

Plain after it was requisitioned for military use. The Duke of Burgundy prefers taller grass and some scrub, and survived comparatively well on chalk downland, at least until the 1990s. There is evidence that other species associated with tall grasses, such as the Marbled White and Lulworth Skipper, have benefited from reductions in grazing pressure.[18]

However, reduced grazing pressure has allowed coarse grasses and scrub to spread on many sites, further reducing the extent of floristically-rich grassland habitat available. Abandonment has had a particularly severe effect on coastal habitats and many species have declined locally on sites that have been invaded by dense scrub.

Abandonment of low-productivity grazing land has led to declines in several fritillary species that breed in bracken-dominated habitats: notably the High Brown Fritillary and Pearl-bordered Fritillary. These species require grazing to maintain open conditions where the larval foodplants, violets, can thrive amongst patchy bracken litter. The declines of these fritillaries have been most noticeable on the large expanses of common land in lowland areas. Traditionally, these commons were grazed, but most of those that remain have been abandoned or developed.

At the other extreme, overgrazing has adversely affected some butterfly habitats, especially in the uplands, often as a direct result of subsidies available under the EC Common Agricultural Policy (p. 355). The Pearl-bordered Fritillary has disappeared following overgrazing of many of its traditional bracken habitats in Wales,[19] and the Marsh Fritillary has declined or been lost from numerous sites because of increased grazing by sheep (p. 350), or a switch from cattle to sheep.[20]

One of the biggest causes of decline amongst woodland butterflies has been the cessation of active broad-leaved woodland management, especially the decline of coppicing. Until the twentieth century, the majority of woods in lowland Britain were managed by regular cutting, coppicing, of patches of woodland, typically on rotations of 5–20 years.[21] This created a constant supply of clearings and early successional habitats in which a variety of species could flourish. During the twentieth century this practice declined so that only 3% of woodland is still actively coppiced (see figure opposite). The High Brown Fritillary, Pearl-bordered Fritillary, Small Pearl-bordered Fritillary, Heath Fritillary, Wood White, and Duke of Burgundy have all been badly affected. Although overall woodland cover has increased during the last 200 years, most of the former localities for these species now lack the clearings and rides needed to support populations.

Habitat fragmentation and isolation

In addition to the reduction in available habitat, fragmentation and isolation have a profound effect on butterfly distributions. Smaller patches of habitat tend to support small populations that are susceptible to extinction due to chance events (such as unfavourable weather, fires, or disease) or demographic effects[22] (such as difficulty in

[18] Pearman *et al.* (1998); Thomas (1990*a*).

[19] Brereton *et al.* (1999).

[20] Barnett and Warren (1995*g*).

[21] e.g. Warren and Key (1991).

[22] Kuussaari *et al.* (1998) found strong evidence for reduced population growth rates in low-density colonies of the Glanville Fritillary. This is called the Allee effect.

finding a mate when population densities are very low). The increasing isolation of habitats reduces the chances of colonization by butterflies and makes it more likely that local extinction becomes permanent. Genetic effects such as inbreeding, which occur in small, isolated populations, may also play a significant role in local extinction.[23]

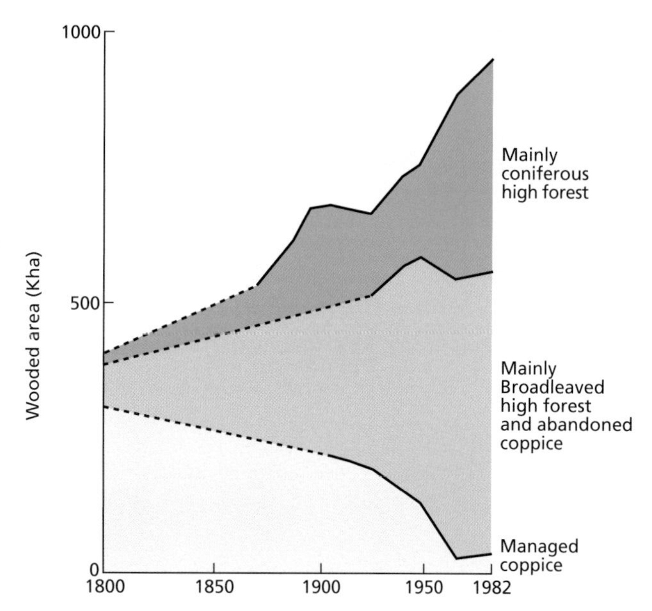

Changes in the extent and composition of woodland in England since 1800. Dotted lines indicate estimated changes where census data do not exist (after Warren and Key 1991).

The extent to which fragmentation has directly caused the decline of butterflies is unclear, but it has undoubtedly exacerbated the effects of habitat loss and changing management. Studies have shown that apparently suitable sites for habitat specialist species may remain unoccupied because they are either too small or too isolated to be colonized.[24] Although a few attempts at reintroducing species in isolated habitats have been successful (e.g. the Black Hairstreak and Large Blue), the majority have failed, usually because the sites are too small to support viable populations by themselves.[25]

The total flight area occupied by many rarer butterflies is very small and detailed site surveys have shown that at least four species now occupy less than 3 square km (0.001% of the entire British land surface).[26] The average habitat patch size occupied by these species is also very small, typically 3–6 ha. The population structures of many butterflies (perhaps most of the habitat specialist species)[27] are now influenced strongly by metapopulation dynamics.[28] The survival of species occupying a few small habitat patches may depend not only on the management of each patch,[29] but also on movement between the populations. Such movement allows the periodic recolonization of patches where populations have become extinct, or may prevent local extinction by boosting numbers and genetic diversity. Species may occupy only some of the available habitat at any one time, but are able to move to different patches as they become suitable. In effect, local colonies are linked and dependent upon each other for the long-term survival of the species. However, the number and spatial arrangement of habitat patches necessary to support a metapopulation are not known and will vary from species to species. Current theoretical estimates suggest a minimum of twenty interconnected populations may be needed to ensure metapopulation persistence for some habitat specialists, but the larger each patch, the fewer will be needed.[30] Habitat quality is also a crucial factor.

Habitat fragmentation is most severe in lowland areas, especially following the reduction in semi-natural grassland. Detailed surveys suggest that the average size of remaining fragments of acid and neutral grassland in the lowlands is only 2.2–2.6 ha, and calcareous grassland patches average 7.7 ha.[31] This fragmentation is an important limiting factor in the recovery of species such as the Silver-spotted Skipper and Adonis Blue, and will reduce the chances of these and other species surviving except in extensive habitat networks. The Marsh Fritillary is particularly susceptible to habitat fragmentation because it undergoes huge fluctuations in population size, making local extinction commonplace. It is thus of great concern that the butterfly

[23] Saccheri *et al.* (1998) showed that small, isolated populations of the Glanville Fritillary produce smaller egg batches, and suffer higher egg and larval mortality, due to inbreeding.

[24] C. D. Thomas *et al.* (1992, 1998).

[25] Oates and Warren (1990).

[26] Cowley *et al.* (1999).

[27] Examples include the Marsh Fritillary (Warren 1994*b*), Silver-studded Blue (Lewis *et al.* 1997), and Dingy Skipper (Gutiérrez *et al.* 1999).

[28] Gilpin and Hanski (1991); Hanski (1998, 1999); Hanski and Gilpin (1991); Thomas and Kunin (1999).

[29] Dennis and Eales (1997, 1999).

[30] C. D. Thomas (1994, 1995).

[31] Blackstock *et al.* (1999).

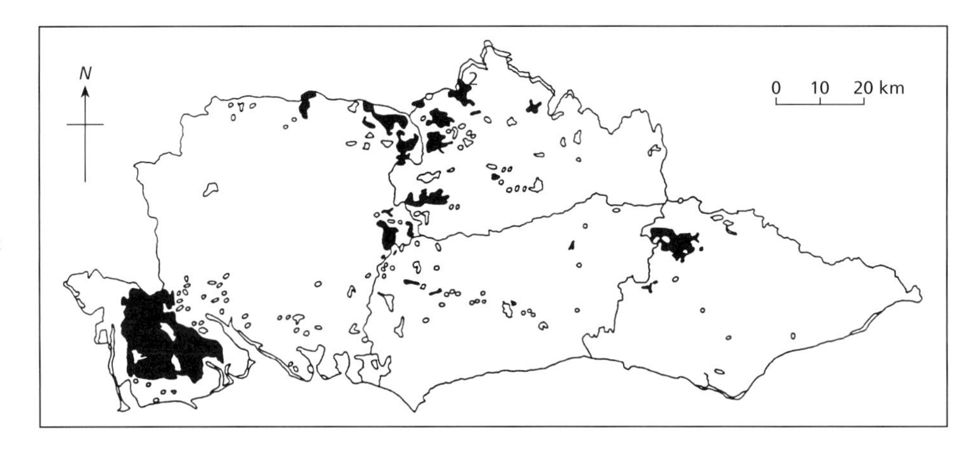

Distribution of Silver-studded Blue on heathland fragments in southern England (Hampshire, Surrey, West Sussex, and East Sussex). Shaded areas contain the butterfly, unshaded areas are unoccupied habitat patches. Sizes of the smallest fragments are slightly exaggerated, so that they can be seen. After C.D. Thomas *et al.* (1998).

has recently died out from several regions where many suitable, but small, habitat patches exist (e.g. in Hampshire, and parts of north Wales and central England).

Habitat fragmentation may even be having a negative effect on wider countryside species. For example the availability of habitat may be constraining the current expansion of species such as the Speckled Wood and Ringlet.[32]

Other causes of decline

During the twentieth century, the use of agricultural chemicals on the land increased dramatically. The greatest impact on butterflies has been the indirect effect of fertilizers and herbicides, which have allowed the agricultural intensification of semi-natural habitats. Artificial fertilizers encourage a few nutrient-demanding, vigorous plants to flourish at the expense of others, including the larval foodplants of most butterflies.

The decline of butterflies is also commonly blamed on the use of pesticides, but there is little evidence for this assertion. Few important butterfly habitats are deliberately sprayed with pesticides, most of which are directed at crop monocultures that contain no larval foodplants. The exceptions are field margins, which are often the refuge of many wider countryside species in intensively farmed landscapes. Lethal effects have been recorded on Large White larvae at quite low doses, and at higher doses on the Green-veined White, Common Blue, and Gatekeeper.[33] In contrast, spray drifts of two persistent chemicals (Dieldrin and DDT, both now banned) had little effect on larval survival of the Small Tortoiseshell.[34] Field margin habitats are also at risk from herbicide drift, which can cause loss of butterfly foodplants at distances of up to 10 m from the end of the spray boom.[35]

Airborne pollution, such as acid rain and nitrogen deposition, can have a serious impact on wildlife and habitats, but there is no direct evidence that they have affected butterflies.[36] They have been implicated in the decline of some butterflies in continental Europe[37] and

[32] Hill *et al.* (1999, in press).

[33] Davis *et al.* (1991); Sinha *et al.* (1990).

[34] Moriarty (1969*a,b*).

[35] Feber and Smith (1995); Marrs *et al.* (1989).

[36] See discussions in Barbour (1986), Fry and Cooke (1984), and Warren (1992*a*).

[37] Heath (1981).

Intensive cereal farm

nitrogen deposition is suspected to be encouraging the spread of coarse grasses on chalk grassland in the Netherlands and, possibly, in woodland rides in eastern England.[38] The latter has resulted in an increase in butterflies that breed on coarse grasses at the expense of species feeding on low-growing herbs. More research is needed to determine whether such pollutants have any impact on butterfly distributions.

Collecting has been cited as a cause of butterfly declines in the past, but there is little evidence to support this. Collecting butterflies was a common hobby in the past but, although large numbers of individuals may have been killed on some sites, this probably had a minimal impact on population levels. Moreover, nearly all documented local extinctions have occurred on sites where there has been little or no collecting. Nevertheless, collecting may have had an impact on a few highly prized species, especially once populations had been reduced to low levels due to habitat loss. For example, collecting is usually cited as causing the final demise of the Large Copper and some colonies of the Large Blue, but it seems likely that these would have become extinct anyway due to habitat degradation.

Causes of range expansion

The range expansion of 15 butterfly species over the last two decades has been particularly noteworthy, especially as it runs counter to the downward trend of most butterfly habitats. Most of the currently expanding species are classed as wider countryside species, which can breed in linear habitats such as field margins, hedges, and roadside verges. This suggests that their populations can remain viable in the modern landscape. The larvae of almost half of the expanding species are grass feeders, whereas the remainder breed on herbs, trees, and shrubs that are widespread and common. The exception is the White Admiral, a habitat specialist of shady woodland. Weather factors and increasing availability of particular habitats have been proposed to account for these spreads.

Weather and climate

The recent expansion of so many butterflies, whose habitats apparently have not increased (and have probably decreased), strongly suggests a response to climatic change. The climate has fluctuated in the past, and with it the distribution of butterflies, but the current changes are unique because they are almost certainly driven by human activity (e.g. the increasing level of greenhouse gases, such as carbon dioxide, in the atmosphere). Records maintained in central England since the seventeenth century show that average annual temperatures have increased by approximately 1.5 °C over the last 25 years, and summer temperatures by 1 °C.[39]

The impact of climate on butterfly populations is complex and depends on events occurring at critical periods of the life cycle. Analysis of the BMS data has shown that populations of many species increase if the weather is warm and sunny during the flight period or immediately before it.[40] This encourages the survival of larval and pupal stages, and allows good numbers of eggs to be laid. For other species (e.g. the Grizzled Skipper and Brimstone) increases are related to warm dry weather during the egg-laying period of the preceding year. In contrast, increases of eight species are associated with high rainfall during the early months of the previous year. This relationship is particularly strong for the Speckled Wood and Ringlet, suggesting that they favour warm, moist conditions. An additional factor is that the

[38] Bobbink and Willems (1987); Pollard *et al.* (1998*b*).

[39] Roy and Sparks (2000).

[40] Pollard (1988); Pollard and Yates (1993*a*); Pollard *et al.* (1995); Roy *et al.* (in press).

The increased burning of fossil fuels may be causing climatic change and the expansion of some butterfly species. However, it may threaten others.

first emergence dates of many species have become earlier and their flight periods longer in recent years.[41] The data collected by the BNM survey suggest that some species are producing more frequent or more sizeable second or third generations (e.g. single brooded Brown Argus, Peacock, and Heath Fritillary).

In some cases, species are recolonizing their historical range after periods of contraction (e.g. the Orange-tip, White Admiral, Comma, and Speckled Wood). Some of these began to spread during the early part of the twentieth century: both the Comma and White Admiral started to spread during the 1920s and continued more or less throughout the century. The timing of these expansions has been related to periods of favourable weather, which allowed populations to build up and spread. For example, the expansion of the White Admiral has been related to a period of warmer than average summer temperatures (especially in June) during the 1920s and 30s, conditions that are thought to favour larval survival.[42]

The distribution of the Speckled Wood is closely correlated with climatic factors that are thought to limit the butterfly's growth and survival. These factors include the cumulative annual temperature above 5 °C (the threshold for larval growth); the coldest month mean temperature (related to overwintering survival); and moisture availability (related to foodplant quality).[43] Similar models have also been developed for the Gatekeeper and Ringlet, and all three species are expected to continue their northward range expansions under predicted climate change.[44] However, habitat fragmentation may constrain this expansion by preventing the species from recolonizing new areas when they become climatically suitable.

The impact of climate change on butterflies is not confined to Britain and Ireland, and many species are known to be spreading northwards in mainland Europe. In a sample of 35 non-migratory species that are not constrained by habitat availability, the ranges of 63% have shifted northwards (and only 3% have moved southwards).[45] A number of these species are shifting their whole range northwards, with movement at both the southern and northern edges. The large-scale movement of so many species, spread across a range of butterfly families, strongly suggests the impact of a continent-wide phenomenon such as global warming. The average annual temperature of Europe increased by 0.8 °C during the twentieth century, shifting climatic isotherms northwards by 120 km, roughly equivalent to the range shift observed in these butterflies.

Land use change

Climate change is not the only factor implicated in butterfly expansions and several species have undoubtedly benefited from habitat changes. For example, the spread of the White Admiral is almost certainly connected with the increased availability of shady woodland during the twentieth century, as coppicing was abandoned and many ancient woods became overgrown. The butterfly has even continued to spread north during periods of less favourable climate, indicating that a combination of the two factors is involved. The spread of the Brown Argus has been linked to the availability of a new type of habitat, set-aside fields, which support a range of annual

[41] Roy and Sparks (2000); Sparks and Yates (1997).

[42] Pollard (1979).

[43] Hill *et al.* (1999).

[44] Hill *et al.* (in press).

[45] Parmesan *et al.* (1999).

foodplants rarely used previously. The full extent of the use of this new habitat, which became available following reforms in agricultural policy in the early 1990s, is not clear and there are many records from other habitats such as road verges and woodland clearings.

Although most butterflies suffered from widespread agricultural intensification during the twentieth century, those that feed on nettles may have benefited. The Common Nettle is one of the few butterfly foodplants that may have increased due to the application of fertilizers in the farmed countryside. This could have helped the spread of species such as the Peacock and Comma.

Habitat factors may also be involved in the expansion of other species. The spread of coarse grasses in Monks Wood National Nature Reserve, Cambridgeshire has favoured several grass-feeding species (in particular the Large Skipper, Speckled Wood, and Ringlet)[46] and could be assisting their spread in other areas. Many of the expanding species breed in linear habitats and may be strongly influenced by changing management practices. Species favouring longer grass (e.g. the Essex Skipper and Marbled White), may have benefited from the reduced frequency of cutting and spraying of road and railway verges since the 1960s and 70s and, possibly, from set-aside. Furthermore, areas of long grass are increasingly common in parks and may have assisted the spread of some species into urban areas, including central London. Butterflies are also known to benefit from unsprayed 'conservation' headlands that are now encouraged around field margins for the benefit of game and wildlife.[47] Although the total area of such headlands is still small, it may have helped to offset some of the enormous impact of hedgerow loss.

More research is needed before we can understand the causes of butterfly range expansions. If the climate continues to warm as predicted (at least in the medium-term), we can expect several butterflies to spread further. Wider countryside species are well placed to take advantage of such change, because they are able to move through the landscape to colonize new habitat, as it becomes available. However, many habitat specialist species are constrained by habitat and ecological factors. For example, the Black Hairstreak is known to be very slow to spread and the large stands of Blackthorn needed for breeding are scattered in the modern landscape of eastern England. Most other habitat specialists are also constrained by the fragmentation of suitable habitat.

Other habitat specialists may be directly threatened by the predicted climate change, as their habitats shift northwards and to higher elevations. Species such as the Mountain Ringlet, which are dependent on habitat that is already restricted to high altitude and northern latitude, may be most at risk.

A few habitat specialists (e.g. the Dingy Skipper, Purple Emperor, and Silver-washed Fritillary) that appear better able to colonize new habitat patches may be expected to spread in future years if warming continues. In addition, butterflies that currently occur only as migrants in Britain and Ireland may become resident species. During the current survey period both the Clouded Yellow and Painted Lady have successfully overwintered, and a few of the rarer migrants have been recorded more frequently than in the past few decades (e.g. the Queen of Spain Fritillary).

Butterflies will be valuable bio-indicators of future climatic change, but many species are not expected to be capable of responding because their habitats are already so greatly reduced and fragmented. Their survival will depend on the success of conservation measures and future land-use change.

[46] Pollard *et al.* (1998*b*).

[47] Feber and Smith (1995); Feber *et al.* (1999).

7 Conserving butterflies in the new millennium

Butterflies in crisis

The preceding chapters have documented how the distributions of butterfly species in Britain and Ireland have changed significantly over the last 150–200 years. Some species have spread, some have fluctuated, but most have declined and five have become extinct. The scale of loss identified in the current survey indicates that a serious crisis exists for the majority of resident butterflies. Even amongst the 15 species that have recently expanded their range, some may be declining in abundance within their range as habitats continue to deteriorate or disappear.

Concern about the loss of butterfly species began during the nineteenth century and was brought into sharp focus by the extinction of the Large Copper in the 1860s. This gained the attention of entomologists such as Charles Rothschild, who helped establish the first nature reserves in Britain during the late nineteenth and early twentieth centuries.[1] Later, in 1925, a Committee for the Protection of British Lepidoptera was formed. At its first meeting, the most threatened species were listed as the Black-veined White, Wood White, Large Copper, Mazarine Blue, Large Blue, Heath Fritillary, and Glanville Fritillary.[2] Sadly, this list proved to be accurate as four of the species became extinct, and the others declined seriously. From today's perspective, several species are conspicuous by their absence from the list, notably violet-feeding fritillaries, which were still common and widespread at that time, but are now among our most endangered butterflies.

Even set against earlier changes, the rate of loss appeared to increase during the latter half of the twentieth century and an even greater proportion of species became threatened. This led to a growing concern about the plight of butterflies and stimulated numerous ecological studies of rare and declining species. It also led to specific conservation measures being taken by a range of voluntary and statutory conservation organizations,[3] and to the foundation, in 1968, of the British Butterfly Conservation Society (now Butterfly Conservation).

Numerous interacting factors are responsible for the declines of butterflies, but the main ones have been the unprecedented rate of loss of semi-natural habitats

[1] Early reserves included Wicken Fen, a famous locality for the Swallowtail, which was designated in 1889, and Woodwalton Fen, the site where the Large Copper was reintroduced, which was designated in 1910 (see Rothschild and Marren 1997).

[2] Feltwell (1995).

[3] e.g. Feltwell (1995); Thomas (1984); Warren (1992*a*).

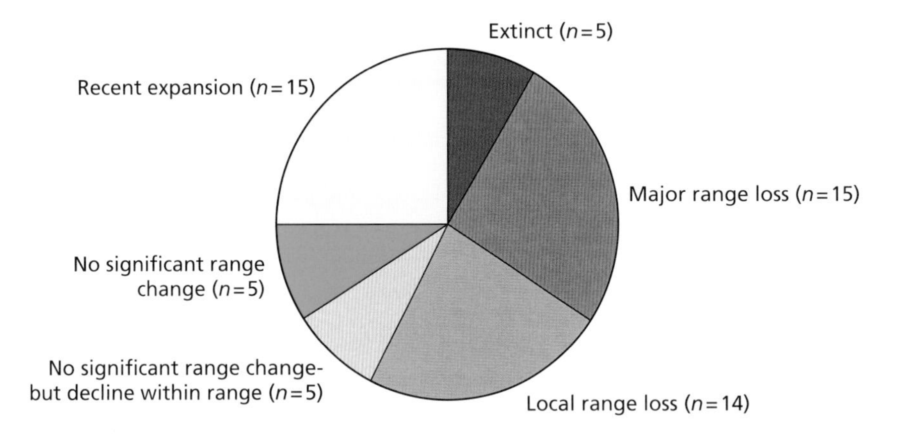

Overall changes in the status of butterflies in Britain and Ireland (data from Table, p. 334).

during the last 50 years and profound changes in farming and forestry. Similar factors are affecting butterflies across Europe and a recent review has shown that 71 species are threatened, 12% of all resident European species.[4] Many more are declining rapidly in several countries, but have not yet reached the thresholds that define threat under international criteria.[5] We are thus facing a pan-European crisis that requires urgent action at international, national, and local levels.

This chapter describes some of the recent conservation initiatives and examines the main issues that need to be addressed.

Developing a strategic approach to conservation

The UK Biodiversity Action Plan

At the Earth Summit at Rio de Janeiro in 1992, the UK Government made a commitment to wildlife conservation by signing the Convention on the Conservation of Biological Diversity. This led to a new approach to conservation and the production of the UK Biodiversity Action Plan (UK BAP).[6] The plan recognized that despite efforts to conserve wildlife during the twentieth century, many species continued to decline and new action was needed urgently.

The plan also recommended that biodiversity conservation should be placed at the heart of Government policy and that it should be the responsibility of all Government departments that have an impact on the environment. Moreover, it envisaged a new partnership, with Government departments and voluntary conservation organizations working together on shared objectives that have since been laid out in detailed Habitat Action Plans and Species Action Plans.[7] The UK BAP provides an important new framework on which to build a comprehensive and co-ordinated plan for butterflies, as an integral part of an overall strategy for conserving biodiversity. Butterfly Conservation was involved closely with this process and with the work of the Biodiversity Challenge Group of voluntary organizations.[8] It has also taken on the role of Lead Partner to oversee the implementation of plans for the priority butterflies and moths, primarily under the Action for Butterflies project.[9]

The Republic of Ireland Biodiversity Plan

The Republic of Ireland Government is also a signatory to the Convention on Biological Diversity and is developing its own National Biodiversity Plan.[10] This is expected to include the designation of extensive Natural Heritage Areas, which will encompass Special Areas of Conservation designated under the EC Habitats and Species Directive. It may also include measures for threatened species.

Identifying priorities for action

Resources available for nature conservation are always limited and clear priorities are required to target effort where it is most needed. Over the years, several different methods have been used, the most popular being the production of Red Data Books or Red Lists. The first official Red Data Book covering British insects was published in 1987,[11] based on criteria developed by the International Union for the Conservation of Nature (IUCN). The criteria have subsequently been improved and made more objective by IUCN, and have been used as the basis of designating

[4] Swaay and Warren (1999).

[5] The criteria have been revised recently by the International Union for the Conservation of Nature (IUCN 1994).

[6] Department of the Environment (1994).

[7] Department of the Environment (1995); UK Biodiversity Group (1998*a*,*b*, 1999*a*,*b*,*c*).

[8] Wynne *et al.* (1995).

[9] Barnett (1995); Bourn and Warren (1997*d*). Species Action Plans have been written for all high and most medium priority species listed on p. 348. References are given in species accounts.

[10] Department of Arts, Heritage, Gaeltacht, and the Islands (1998).

[11] Shirt (1987).

Priority status and legal protection of threatened butterflies in Britain and Ireland. Species not covered by legislation and considered low priority in previous analyses are not listed.

Species	*European threat status*[1]	*Butterfly Conservation Priority*[2]	*UK BAP status*[3]	*Legal protection*[4]
Chequered Skipper		High	Priority	Sale only (GB)
Lulworth Skipper	Vulnerable	Medium	Conservation Concern	Sale only (GB)
Silver-spotted Skipper		High	Priority	Sale only (GB)
Dingy Skipper				NI
Grizzled Skipper		Medium		
Swallowtail		Medium	Conservation Concern	Full (GB)
Wood White		Medium	Conservation Concern	Sale only (GB)
Brimstone				NI
Brown Hairstreak		Medium	Conservation Concern	Sale only (GB)
Purple Hairstreak				NI
White-letter Hairstreak		Medium		Sale only (GB)
Black Hairstreak		Medium	Conservation Concern	Sale only (GB)
Large Copper		High	Priority	Full (GB)
Small Blue		Medium	Conservation Concern	Sale only (GB) + NI
Silver-studded Blue		Medium	Priority	Sale only (GB)
Northern Brown Argus		High	Priority	Sale only (GB)
Chalkhill Blue			Conservation Concern	Sale only (GB)
Adonis Blue		Medium	Priority	Sale only (GB)
Holly Blue				NI
Large Blue	Endangered	High	Priority	Full (GB)
Duke of Burgundy	Near Threatened	Medium	Conservation Concern	Sale only (GB)
Purple Emperor		Medium	Conservation Concern	Sale only (GB)
Large Tortoiseshell		(Extinct)		Sale only (GB)
Small Pearl-bordered Fritillary		Medium	Conservation Concern	
Pearl-bordered Fritillary		High	Priority	Sale only (GB)
High Brown Fritillary		High	Priority	Full (GB)
Silver-washed Fritillary			Conservation Concern	
Marsh Fritillary	Vulnerable	High	Priority	Full (GB) + NI
Glanville Fritillary		Medium	Conservation Concern	Sale only (GB)
Heath Fritillary		High	Priority	Full (GB)
Mountain Ringlet		Medium	Conservation Concern	Sale only (GB)
Scotch Argus	Near Threatened			
Large Heath	Vulnerable	Medium	Conservation Concern	NI

Notes :

1. After Swaay and Warren (1999). Red Data Book categories follow new IUCN criteria, as applied to Europe as a whole.
2. Species priority as identified by Butterfly Conservation for targeting conservation action in the UK (after Warren *et al.* 1997). These priorities are being reviewed using data from the current survey.
3. Species listed in the UK Biodiversity Action Plan (UK Biodiversity Group, 1998*a*)
4. Legal status in Great Britain (England, Scotland, and Wales) refers to Schedule 5 of the Wildlife and Countryside Act 1981 (as amended). Full protection covers taking and harming, while species listed for sale only require a licence for trading wild-caught livestock or specimens. In Northern Ireland (NI), legal protection is granted under the Wildlife (Northern Ireland) Order 1985 and is broadly equivalent to the full protection category in Britain. The Republic of Ireland, Isle of Man, and the Channel Islands currently have no legislation that lists butterfly species.

conservation priority to species within the UK BAP.[12] This defines Priority Species as globally threatened species, or species where numbers or range in the UK have declined by more than 50% in the last 25 years. Species of Conservation Concern include threatened endemic or other globally threatened species, and species where numbers or range have declined by more than 25% in the last 25 years.

UK butterfly species were assessed by analysing distribution changes shown at the 10 km square level in the 1984 atlas[13] and drawing on more recent local information.[14] The UK BAP lists 25 species as conservation priorities: 11 Priority Species and 14 Species of Conservation Concern (see opposite).

Unfortunately, this approach summarizes data at a coarse scale (10 km squares) that tends to underestimate the true rate of population decline. Although this was taken into account in the analyses, subsequent research has shown that such summaries can mask population declines in all but the rarest species (see p. 325).

Despite these limitations, distribution maps plotted at 10 km square resolution remain helpful in identifying priorities. They are particularly useful in identifying declines of localized species that are most susceptible to extinction. Such species are usually habitat specialists with particular threats that can be tackled within Species Action Plans. Widespread species tend to have more general problems, which are catered for under habitat conservation measures laid down within Habitat Action Plans. The combination of Species and Habitat Action Plans within the UK BAP thus represents a coherent approach to conserving all butterfly species, both the habitat specialists and those occurring in the wider countryside.

Identifying rates of species decline remains a vital, though complex, issue and there is a pressing need for a systematic and scientifically robust method for identifying butterfly population changes throughout Britain and Ireland.[15] The development of such a method is a logical next step following the BNM project. A review of priorities will also be a key task for the next few years. The results presented in this atlas suggest that several species should be considered as having increased priority, especially the Wood White, Large Heath, Duke of Burgundy, Dingy Skipper, Brown Hairstreak, Dark Green Fritillary, Small Pearl-bordered Fritillary, Grayling, and Grizzled Skipper. The priority of the Lulworth Skipper may also need to be reviewed to reflect its threatened status in Europe.

Conserving genetic diversity

Although most butterflies in Britain and Ireland also occur widely across Europe, many have unique races that have become adapted to local conditions in these islands. For example, small-sized races of the Silver-studded Blue and Grayling occur on Great Ormes Head in north Wales, and heavily-marked forms of several species occur in Scotland and Ireland. Some of these races have evolved different foodplant preferences and, in the case of the Silver-studded Blue, even adaptations to different ant species.[16] Until recently, most of our understanding of genetic variation has come from studies of wing patterns, but modern molecular-level techniques are revealing fascinating new features in the genetic composition and inter-relationships of butterfly populations.[17]

The conservation of variety within and between populations is important:[18]

- Each race contains unique genes that confer particular advantages and may be vital to future evolution and survival.

[12] Department of the Environment (1995).

[13] Heath *et al.* (1984).

[14] Warren *et al.* (1997).

[15] e.g. Dennis *et al.* (1999); Thomas and Abery (1995).

[16] Thomas *et al.* (1999).

[17] e.g. Hoole *et al.* (1999); Pullin *et al.* (1997).

[18] Brakefield (1991).

- Endemic races are likely to indicate habitats where comparable divergence has evolved in other invertebrate groups.
- Genetic variation is essential to ensure robust populations that are able to adapt and survive in the long term.[19]
- It preserves variation that enables us to understand how evolution works and how species adapt to local conditions.

As yet, there is no statutory framework for the conservation of genetic variation, but useful principles are built into Species Action Plans and Butterfly Conservation's Regional Action Plans,[20] and should be covered by local biodiversity initiatives. These principles include conserving populations of each species throughout its range and maintaining large populations, which can hold a reservoir of genetic variation to act as a buffer against change.

Threats to butterflies of conservation concern

The current threats to butterflies of conservation concern have been listed in the species accounts and within Species Action Plans.[21] In many cases, the threats are the same factors that caused the historical decline of the species, namely changing habitat management and habitat fragmentation and isolation, but in others these are becoming less important (e.g. due to policy change) and new threats have become apparent (e.g. climate change).

Changing habitat management

Because threatened butterflies have highly specific ecological requirements, their most serious and immediate future threat is often from changing habitat management. The maintenance of sustainable management regimes on semi-natural habitats is one of the biggest practical challenges currently facing conservationists, especially now that most traditional management is uneconomic and has largely been abandoned (see p. 339).

The problem of achieving suitable management is exacerbated by the fact that the breeding areas used by habitat specialists tend to be small and isolated. Such sites can be expensive to manage and are frequently a minor part of modern farming enterprises. Increasingly, they will be managed effectively only if financial support is available, either through grants to support low intensity, and environmentally sensitive, farming, or through management agreements with conservation agencies.

Habitat fragmentation and isolation

The process of fragmentation seems certain to continue and lead to further local extinctions over coming decades. An especially worrying possibility is that there may be a time lag between habitat

Livestock numbers in the UK have increased from 39.6 million head in 1960 to 54.5 million head in 1995. Numbers of sheep almost doubled between 1980 and 1995, which had a serious effect on upland birds and other wildlife (RSPB 1999). Sheep numbers in Ireland trebled between 1980 and 1991, devastating many peatland, sand dune, and machair habitats (Cabot 1999).

[19] Saccheri *et al.* (1998).

[20] Butterfly Conservation has produced Regional Action Plans for the whole of the UK in order to bring together local information for butterflies and moths, and to guide local conservation action (e.g. Bourn *et al.* 1996; S. Ellis 1999*b*; Joy 1997, 1998*b*; Nelson 1998).

[21] Warren *et al.* (1997).

loss and fragmentation and the eventual extinction of a species from a region.[22] Good habitat management may be able to reduce the chances of extinction, but most habitat specialists require conservation measures that operate at the landscape level as well as on individual sites (p. 355).

Agricultural intensification and chemical pollution

Agricultural intensification continues to be one of the biggest threats to wider countryside butterflies, as well as to a wide variety of other farmland wildlife.[23] Intensification is also a serious threat to habitat specialists like the Marsh Fritillary and Small Pearl-bordered Fritillary whose damp grassland habitats are still being lost due to the application of artificial fertilizers or as a consequence of conversion to arable crops. Drainage and agricultural 'improvement' threaten the bracken-dominated habitats of the Pearl-bordered Fritillary and Small Pearl-bordered Fritillary, and the bog habitats of the Large Heath. Specific agricultural practices continue to threaten other species, such as the Brown Hairstreak whose hedgerow habitats are still being destroyed or rendered unsuitable by annual flailing.

For some Species of Conservation Concern, further agricultural intensification is not regarded as the main threat. In part, this is because some remaining habitats are unlikely to be improved for agriculture (e.g. most remaining chalk downland is too steep to plough with current technology) or the economic stimulus for intensification has been reduced (e.g. subsidies for land drainage). Similarly, herbicides and pesticides are not thought to be a great threat to most habitat specialists, because now they generally breed in areas separated from intensive agricultural land.

It seems likely that new threats will arise in the future as agriculture continues to be mechanized and industrialized. For example, new chemicals will undoubtedly be developed with possible direct and indirect effects on butterflies and their habitats. Another insidious, progressive change that could affect butterfly habitats is the increasing level of nitrogen in soils, caused by increased burning of fossil fuels, vehicle emissions, and application of fertilizers. This has already caused problems in countries such as the Netherlands and is suspected to be steadily altering many habitats in Britain.[24] Other serious impacts could arise from the cultivation of genetically modified crops, both directly through the growing of insect resistant crops and indirectly through the further intensification of agriculture.[25]

[22] Hanski (1998).

[23] e.g. Pain and Pienkowski (1997); Siriwardena *et al.* (1998).

[24] Effects of increasing nitrogen deposition during the 1980s have been noticed in a variety of vegetation types, but especially unimproved grasslands, heaths, and bogs (Firbank *et al.* 1999). Conservation problems identified in the Netherlands include the spread of coarse grasses on chalk downland and enrichment of heathlands (Bobbink and Willems 1987).

[25] Jeffcoate (1999).

Afforestation

Although the area of ancient semi-natural woodland was reduced greatly during the twentieth century, the overall woodland cover of both Britain and Ireland more than doubled. Most of this expansion has been through the afforestation of upland areas, a process that is planned to continue at least for the next few decades. The potential impact on butterflies is poorly known, but threats are posed for several species, especially as planting is generally

Large areas of Britain and Ireland have been planted with conifers over recent decades and continuing afforestation could threaten some butterfly habitats.

targeted at low productivity land and has been dominated by non-native conifers. A possible threat to the Pearl-bordered Fritillary has been identified recently in Scotland, where native woodland regeneration schemes can lead to the loss of bracken edge habitats that are used for breeding.[26]

Urban and industrial development

Urban development may not pose an immediate threat to most habitat specialists, partly because these species are already rare due to extensive habitat loss and inhabit remote regions away from large-scale development. Many of their sites are also protected as nature reserves or designated Sites of Special Scientific Interest (SSSI). However, new roads and other developments continue to be a potential threat and some important sites have been destroyed.[27] Urban encroachment and industrial developments continue to erode habitats for wider countryside butterflies and much other wildlife, often in the very areas where they could be enjoyed close to towns and cities. Urban expansion is a particular threat in southern England where hundreds of thousands of new houses are planned.

Natural ecological change

Natural ecological change is a significant threat to many species, especially those breeding in early successional habitats that will become unsuitable without substantial management input. For example, in central and northern England, the Grizzled Skipper relies almost entirely on post-industrial habitats created by quarrying and mining. These will eventually develop into tall grassland and scrub and will lose the sparse conditions that the butterfly needs. Similar problems face many woodland butterflies whose breeding habitats in open clearings disappear rapidly without regular management. Some sites have been acquired as nature reserves, but will need constant management, and a certain amount of experimentation, to maintain their present richness. Other butterflies, such as the Silver-spotted Skipper and Adonis Blue, could be affected by natural ecological change as they rely heavily on rabbit grazing to maintain the short turf required for breeding. They would be vulnerable if rabbit numbers are reduced again by new strains of myxomatosis, or other diseases such as viral haemorrhagic disease.

Collecting

Collecting is thought likely to be a minor threat in the future, not least because most butterfly enthusiasts now prefer to 'collect' photographs instead. Nevertheless, some small populations may be at risk from even low-level collecting and the current laws protecting a few threatened species remain a sensible precaution.

Climate change

The impact of climate change on butterflies is difficult to predict, partly because of uncertainties about how the climate of Britain and Ireland will be affected and partly because the impact will be different for each species and each habitat. Current forecasts are for an average temperature increase across Europe of around 2 °C by 2050,[28] but some climate models predict significant differences with latitude, with northern areas becoming cooler while southern areas become hotter and drier.

[26] See species account and Brereton and Warren (1999).

[27] For example Selar Farm SSSI, a Marsh Fritillary site in south Wales, which became an open-cast coal mine during the early 1990s (Kirkland 1995).

[28] Alcamo and Kreilemann (1996); Hulme and Jenkins (1998).

Changes in average annual temperature (relative to the 1961–90 average) for 30-year periods centred on the 2020s, 2050s, and 2080s, and for four of the scenarios projected by the UK Climate Impacts Programme (Hulme and Jenkins 1998).

Changes in average annual rainfall (relative to the 1961–90 average) for 30-year periods centred on the 2020s, 2050s, and 2080s, and for four of the scenarios projected by the UK Climate Impacts Programme (Hulme and Jenkins 1998).

However, other studies suggest that the climate may cool dramatically if the North Atlantic Drift ocean current is reversed, bringing cooler surface water to the seas around Britain and Ireland.[29]

Several studies have examined the possible impact on butterflies; their general conclusion is that many species will benefit if the climate warms while a few habitat specialists may be harmed.[30] The information gathered in the current survey shows that at least 15 species are spreading northwards, probably in response to warmer temperatures. However, despite this background of favourable climatic conditions, most habitat specialists have continued to decline, possibly because any benefit has been overridden by habitat loss and change. Recent research also shows that butterflies living in fragmented habitats may be unable to keep pace with large scale environmental changes and may become more prone to local extinction.[31] In addition, populations of some short-turf species may be reduced if droughts become more frequent.

The most serious effect of climate change is likely to be on species breeding in habitats with thin or damp soils, or montane habitats. Such habitats are highly sensitive to climatic conditions and their vegetation communities could alter rapidly in response to rainfall or temperature changes. An analysis of threatened butterflies indicates that climate change might threaten eight species over the next 25 years, although many more might be affected over a longer timescale.[32]

Conserving butterflies in the wider countryside

Fundamental issues and major environmental change

One of the main underlying causes of butterfly declines during the twentieth century was a succession of agricultural policies aimed at increasing food production. In recent decades, the driving force in western Europe has been the EC Common Agricultural Policy (CAP), which has led to further intensification and industrialization of agriculture, at the expense of wildlife and semi-natural habitats. Equally damaging to many woodland butterflies have been changes in forestry and the abandonment of traditional practices such as coppicing. Tackling these fundamental causes of butterfly decline will continue to be of vital importance.

Some environmental problems, such as global climate change, may cause serious additional threats to butterflies and their habitats. Discussion of these wider issues is beyond the scope of this book, but they will have to be addressed if they are not to undermine the effectiveness of other conservation measures.

The need for a landscape approach to conserving butterflies

Throughout the twentieth century, nature conservation in Britain and Ireland was based largely around the protection of key sites, either good examples of particular habitat types or important localities for rare species. A series of National Nature Reserves and National Parks has been declared, and in Britain thousands of SSSIs have been designated.[33] Although these measures may have served the prime purpose of protecting some key sites, they have not proved very effective in conserving threatened butterflies. Indeed, studies have shown that some butterflies have declined as fast on SSSIs as on unprotected land and that the butterfly fauna of some famous nature reserves has continued to decline.[34] The main problems are

[29] New Scientist, 27 November 1999.

[30] Dennis (1993); Elmes and Free (1994).

[31] Thomas (1994).

[32] Warren *et al.* (1997).

[33] For details see Rothschild and Marren (1997) and Sheail (1998). The equivalent designation in Northern Ireland is Area of Special Scientific Interest (ASSI).

[34] Warren (1992*a*, 1993*a*,*b*).

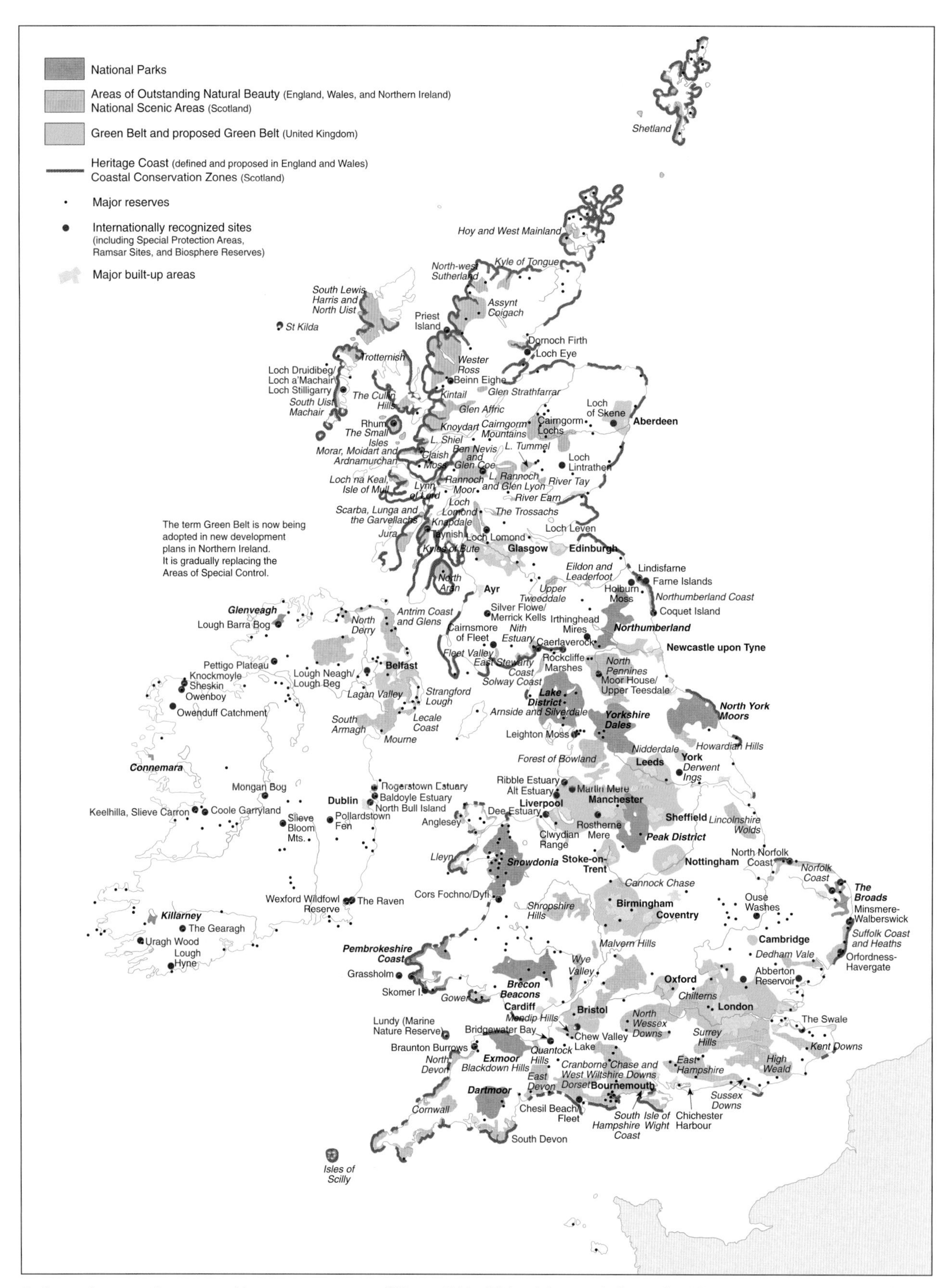

Designated conservation areas and larger nature reserves. (NB over 6000 additional sites are designated as SSSIs or ASSIs, comprising around 10% of the UK).

thought to be insufficient or incorrect habitat management, a lack of knowledge of species requirements, and the small size of many protected areas. Despite this disappointing record, some nature reserves have been very successful (see p. 359).

Many wider countryside butterflies in Britain and Ireland appear to have declined almost as rapidly as the habitat specialist species that have been the focus of recent attention.[35] It is clear that efforts to protect individual sites, however laudable, have been insufficient to prevent the decline of most butterfly species. A more integrated approach, such as described in the UK BAP, is therefore needed to protect the countryside as a whole, as well as the special areas within it. This should include the maintenance of extensive habitats for butterflies within the context of a rich and diverse countryside. If we fail to achieve this, today's common species may become tomorrow's threatened ones.

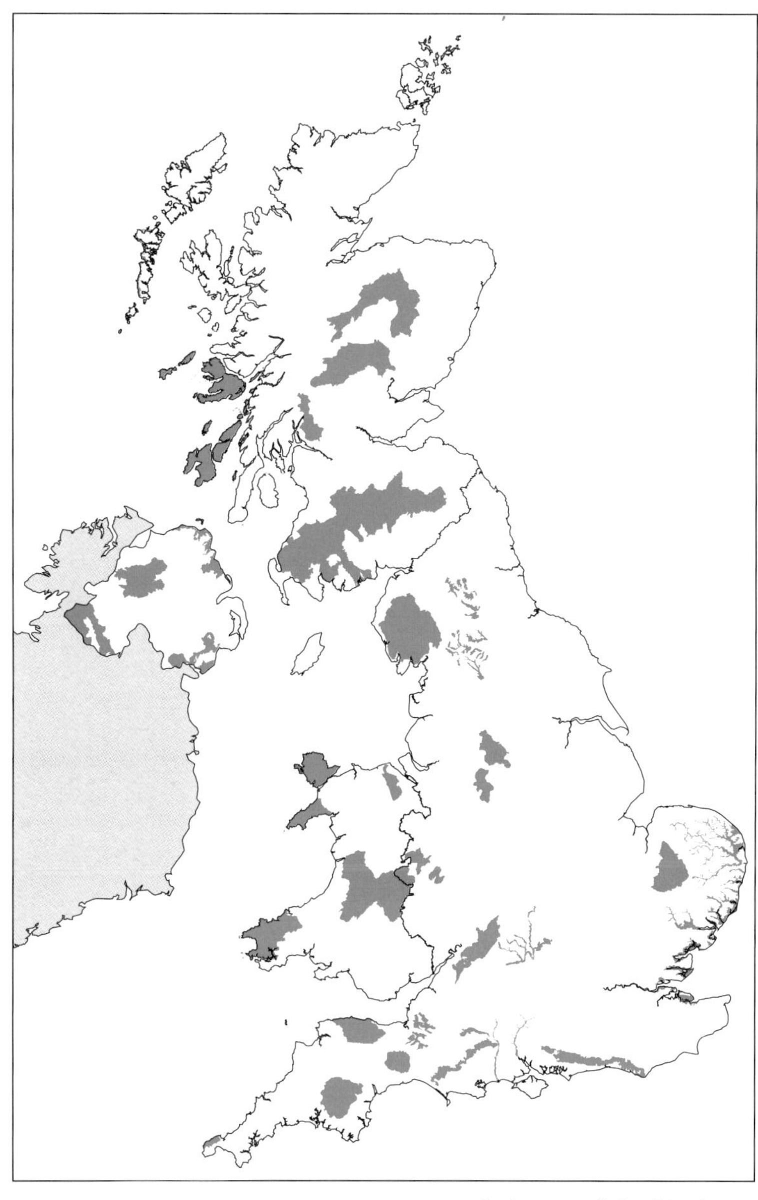

Environmentally Sensitive Areas in the UK. In Ireland, a similar Rural Environment Protection Scheme has been developed.

The development of new landscape schemes

The development of landscape-scale conservation measures is in its infancy, but there have been some promising new initiatives, notably the development of agri-environment schemes.[36] These provide grant incentives to farmers to manage land in a more environmentally sensitive way, in order to maintain valued landscapes and to enhance or maintain nature conservation interest. Within the UK, Environmentally Sensitive Areas (ESAs) now cover over 10% of the land area and include key butterfly regions such as Dartmoor, Exmoor, the South Wessex Downs, the South Downs, the Cotswolds, Cumbria, and the Argyll Islands. In the Republic of Ireland, the Rural Environment Protection Scheme has been developed to help farmers to maintain traditional low intensity farming and address ecological objectives in designated areas of high conservation value.

Other schemes have been designed to cover specific habitats, such as the Countryside Stewardship Scheme in England whose targets include key butterfly habitats such as chalk downland, wet pastures, and field margins.[37] With devolution, new schemes are being developed to replace ESAs; for example Tir Gofal ('Land Care') in Wales and the Rural Stewardship Scheme in Scotland.

[35] Cowley *et al.* (1999); León-Cortés *et al.* (1999).

[36] These schemes were first introduced in the UK in 1987 and were expanded under an EC Regulation in 1992. See Baldock and Mitchell (1998); RSPB (1995); Sheldrick (1997).

[37] See reviews by Dover (1996); Feber *et al.* (1996, 1999).

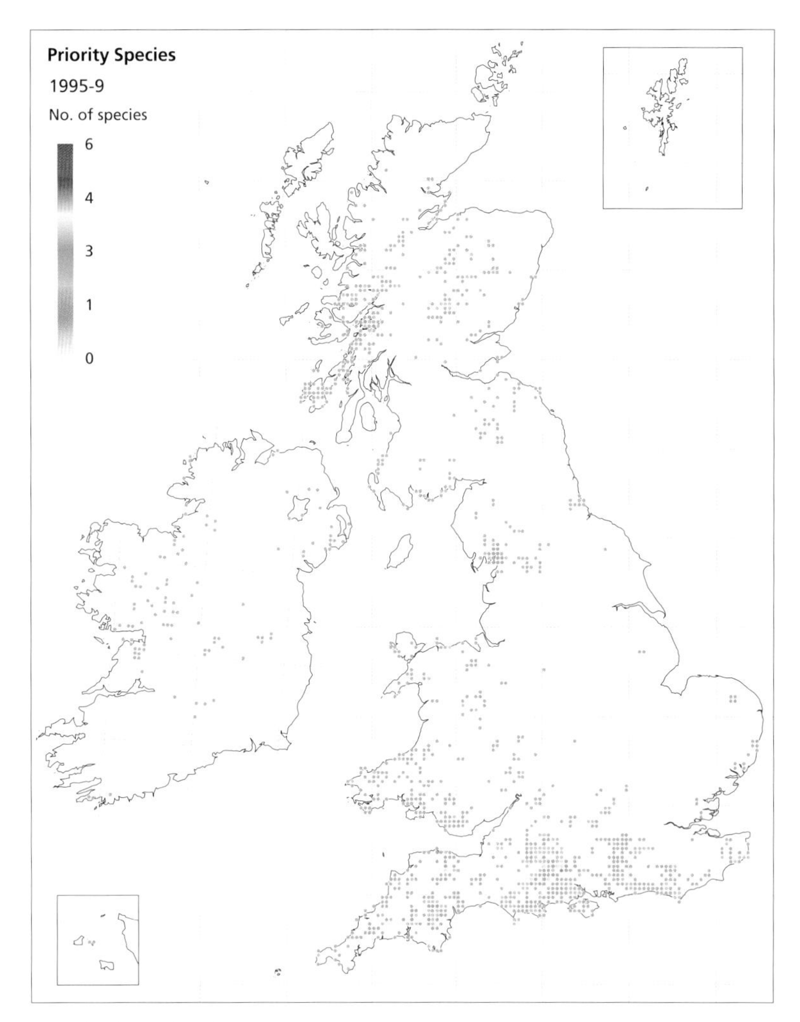

Butterfly 'hotspots': nine Priority Species listed in the UK BAP (p. 348) plotted at 5 km square resolution (excluding the extinct Large Copper and reintroduced Large Blue).

Although entry into agri-environment schemes is voluntary, the uptake has been high. However, only 4% of the agricultural support budget is currently spent on promoting environmentally sensitive farming; the remaining 96% provides price support that encourages intensification elsewhere. Moreover, the payments available often cannot compete with grants to improve land under other EC schemes.[38] Most conservation organizations are pressing for urgent reform of the CAP, including a major expansion of agri-environment schemes and the attachment of environmental conditions to all payments.[39]

The most recent CAP reforms, known as Agenda 2000, were agreed in March 1999 and included a new Rural Development Regulation. Though poorly funded, it is seen as a significant first step towards a more integrated rural policy, which could protect the environment as well as maintaining rural communities.[40] The UK Government has since taken up an option to reduce subsidies for production (by 5%) and has used some of the funds to expand agri-environment schemes. This is a welcome initiative which will almost double expenditure on environmentally sensitive farming by 2006 (to over £200 million p.a.) and shows the direction that needs to be taken in future reforms.

The impact of schemes on butterflies

Schemes such as ESAs and Countryside Stewardship are now making a significant contribution to the protection and enhancement of important butterfly areas.[41] A large proportion of remaining chalk grassland is now being managed under these schemes. In the last few years, the management prescriptions have also been revised to take account of biodiversity priorities. Some ESAs now include measures to encourage extensive grazing specifically for butterflies such as the Marsh Fritillary and bracken management to benefit the High Brown Fritillary and Pearl-bordered Fritillary.

However, the schemes are not without their problems and the grazing regimes supported have not always been suitable for habitat specialists such as the Marsh Fritillary.[42] Efforts are now being made to identify the requirements of such species within conservation plans, but this requires trained and knowledgeable project officers who can tailor management to the priorities on individual sites.

[38] Baldock and Mitchell (1998); Harvey (1997).

[39] e.g. Baldock and Mitchell (1998); RSPB (1998*b*).

[40] English Nature (1999*a*).

[41] Warren and Bourn (1997).

[42] Bourn and Warren (1996).

Natural Areas and Heritage Zones

Another approach to landscape conservation is being promoted by English Nature under their Natural Areas initiative.[43] This divides England into natural landscape areas that can be used to frame land use policies and maintain distinct local features. Scottish Natural Heritage is developing a similar concept known as Natural Heritage Zones. These approaches are intended to help conserve habitats at the landscape level and promote local distinctiveness.

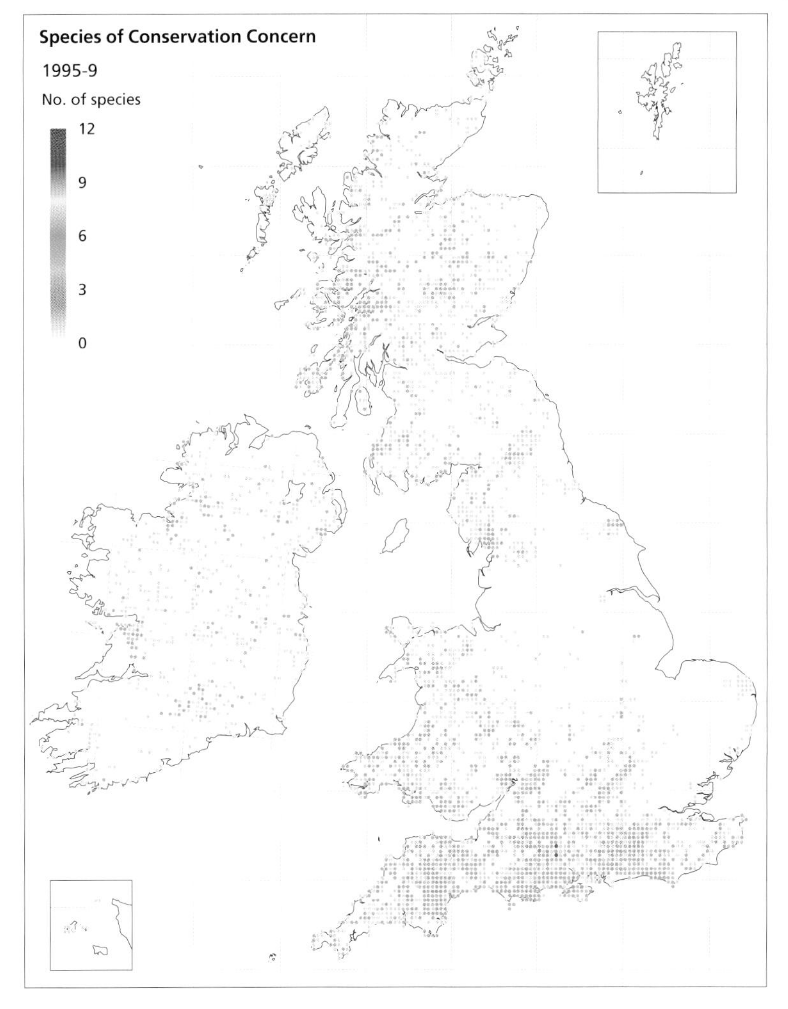

Butterfly 'hotspots': 23 Species of Conservation Concern listed in the UK BAP (p. 360 348) plotted at 5 km square resolution (excluding the Large Copper and Large Blue).

Key regions for butterflies

Data from the current survey help to identify some of the 'hotspots' for threatened butterflies that should be targeted for immediate conservation action. Important concentrations of Priority Species occur in Argyll and the uplands of Scotland, parts of north and south-west Wales, and the Burren in Ireland. In England, key areas are the limestone region around Morecambe Bay (north Lancashire and south Cumbria), east Kent, the chalk and limestone downs of central-southern England, and Devon and east Cornwall (especially Dartmoor and Exmoor). Species in the lower category of Conservation Concern are more widely distributed around Britain and Ireland, though there is a notable paucity in the more intensive arable regions of central and eastern England.

Important areas for conserving individual butterflies in the UK are described in the species accounts and Species Action Plans, and key areas are listed within Butterfly Conservation's series of Regional Action Plans. The Forestry Commission has used these areas to target grants within their Coppice for Butterflies Challenge scheme, developed as part of a range of new biodiversity grants in 1996.[44]

The role of nature reserves

Since its inception, the conservation movement has focused on the acquisition of important sites as nature reserves. In England and Wales, over 40% of SSSIs are now owned or managed by voluntary and statutory conservation organizations.[45]

However, the main problem with reserves is that many are small and may not be viable ecological units, especially if they become isolated through continuing loss of surrounding habitats. Reserves need to be considered within a wider landscape context and, wherever possible, used to influence neighbouring habitats and land use. Reserves can still play a central role, especially if they are large enough to

[43] English Nature (1999*b*).

[44] Forestry Authority (1996); Warren (1996).

[45] DETR (1998).

support core populations or if they occupy strategically important positions in the landscape.

Despite some past failures, there are reserves where habitat specialists have thrived (e.g. the Black Hairstreak, High Brown Fritillary, and Heath Fritillary).[46] In these cases, reserve management has been based on sound ecological research and populations have been able to thrive while elsewhere others have declined. Our knowledge of the ecological requirements of most rare and declining species has improved greatly over recent years, and with it our ability to manage reserves more effectively. We can therefore anticipate a better success rate in the future. Nature reserves also serve other important purposes of course: to evaluate different forms of management; for education, recreation, and leisure; to act as demonstration areas; and for scientific research.

Managing habitats for butterflies

Butterflies, notably the habitat specialists, are highly sensitive to management and even short periods of unsuitability can lead to local extinction. Most butterfly habitats in Britain and Ireland have been created or heavily modified by human activity for thousands of years, but the pattern of land use has changed rapidly in recent decades. The survival of many butterflies will thus depend to a large extent on how well their habitats can be managed in a changing world.

Recommended management regimes are frequently based on traditional systems under which species are known to have thrived in the past. However, such systems may have to be modified to fit in with modern practicalities and economics, and in some cases completely new management regimes have to be developed. Many previous publications have covered habitat conservation,[47] and only a brief resumé is given here for two of the main butterfly habitats: grasslands and woodlands.

Grasslands

The suitability of grassland habitats for butterflies is strongly influenced by the height and composition of the turf, and the availability of shelter, all of which depend on cutting and/or grazing. Recent experience has shown that both the timing of grazing and the type of grazing stock are important.[48] On chalk grassland, autumn or winter grazing has been suggested to be best for most butterflies, but spring and summer grazing may be essential for species that prefer short turf (e.g. the Adonis Blue).[49] Most species can survive under a range of other grazing regimes, many of which are described in the species accounts.

An increasing problem for grassland management is the limited availability of livestock breeds that are suited to grazing unimproved pastures. Traditional breeds are often preferable for grazing such low productivity grasslands when compared to modern breeds, which have been selected to grow quickly on improved grassland and processed feed. However, traditional livestock breeds are becoming increasingly rare and conservation bodies have had to develop their own grazing networks.[50]

Woodlands

In woodlands, butterflies breed in a range of habitats, from sunny rides and clearings to shady woodland. Most of these features can be built into the design of modern woodlands and can be covered by grants from the Forestry Commission.[51] Many of the fritillaries require the regular creation of clearings, either coppiced or recently

[46] Thomas (1991); Warren (1992*a*, 1995*a*).

[47] Reviews of habitat management for butterflies are given by Dunbar (1993), Pullin (1995), Thomas (1984, 1991), and Warren (1992*a*). Other key references are given in the species accounts.

[48] Oates (1993); Thomas (1991); Warren (1994*a*,*b*); Warren and Thomas (1993).

[49] Butterflies Under Threat Team (1986).

[50] A Grazing Animals Project (GAP) and a Forum for the Application of Conservation Techniques (FACT) have recently been initiated to share knowledge about grazing and to develop specialist machinery for conservation sites (Appendix 7).

[51] The Forestry Commission runs a Woodland Grant Scheme that gives grants for planting and restocking, and for maintenance of rides and glades. Biodiversity grants are available for management (e.g. coppicing) aimed at conserving wildlife.

felled areas, close enough to each other to allow colonization. The requirements of woodland butterflies and guidelines for managing woods for butterflies are covered in several publications.[52] These principles have been put into practice with great success on some nature reserves.[53]

Suitable management can also be incorporated into commercial woodlands, for example by modifying felling cycles, and introducing interconnecting rides and glades. The Forestry Commission has recently drawn up a Species Action Plan for Butterflies to address the key issues within its own woods, including special measures to conserve Priority Species listed in the UK BAP. The chief mechanism for ensuring suitable management is through Forest Design Plans. These can identify the location of key habitats and plan long term management such as the expansion and cutting of rides or the break up of large, even-aged plantations.

Active woodland management such as coppicing can encourage butterfly larval foodplants, including violets, and provide the sunny conditions required by many species.

Grazing with traditional livestock breeds is often the best method of maintaining grassland habitats for butterflies.

Rebuilding the countryside: habitat and species restoration

The progressive and continuing loss of semi-natural vegetation has led to a growing recognition of the need to restore habitats and reverse the effects of fragmentation and isolation. There is also a strong desire among some conservationists to reintroduce extinct species, particularly if habitat restoration has been successful.

Although new techniques have been developed that can reproduce many elements of natural habitats, restored habitats are rarely, if ever, as diverse as ancient ones. The restoration of a complete community may take hundreds of years[54] and may never occur if sources of constituent plants and animals are no longer available in the vicinity. The conservation of surviving semi-natural habitats must therefore remain the top priority.

Habitat restoration has already been incorporated into schemes such as Environmentally Sensitive Areas (ESAs). For example, over 1000 ha is being restored to chalk downland within the South Wessex Downs ESA, though most of this has been planted with only a simple mix of native grasses and few herbs. There has been some recolonization by common butterflies, notably the Meadow Brown, but very little by downland specialists.[55] One of the main constraints on grassland restoration

[52] Barker *et al.* (1998); Fuller and Warren (1991, 1993); Joy (1998*a*); Robertson *et al.* (1995); Warren and Fuller (1993); Warren and Key (1991); Warren *et al.* (1994).

[53] e.g. Pollard and Yates (1993*a*); Wilson (1997).

[54] Techniques for habitat restoration are reviewed by Dunbar (1993), English Nature (1993), Fry and Lonsdale (1991), Gibson (1995), Gilbert and Anderson (1998), Kirby (1992), and Wells *et al.* (1981, 1986).

[55] Hoare (1999).

The maintenance of extensive cattle grazing is vital to the survival of many butterflies, especially the Marsh Fritillary, as shown here on one of its Culm grassland sites in Devon.

has been the difficulty of obtaining appropriate seed mixes, derived from native stock. This problem is now being addressed and new guidelines have been produced.[56]

The deliberate reintroduction of locally extinct species can be more controversial. Most documented releases involve the movement of resident species from site to site. There have been a few attempts to introduce non-native species such as the Map, but these are now prohibited by law. A review conducted in 1990 discovered that at least 300 butterfly (re)introductions had been attempted in Britain and Ireland, covering 43 species.[57] The outcome of these attempts is variable; a few have succeeded in establishing thriving populations while the majority have failed after a few years. Many hundreds, possibly thousands, more releases have probably been conducted covertly and have made it difficult to interpret distribution changes of several species (e.g. see accounts for the Adonis Blue, 365Marsh Fritillary, and Marbled White).

Although reintroductions have frequently failed in the past, they can have a role within an overall conservation strategy for threatened species. The Large Blue has been re-established successfully at several sites since its extinction in 1979, and the Heath Fritillary has been re-established at several sites in Essex following detailed research.

A code on species restoration is available from Butterfly Conservation.[58] The main guidelines are that:

- reintroductions should be conducted within an overall conservation strategy for the species that places the emphasis on conserving existing populations first.
- attempts should also concentrate on large potential habitats, or networks of habitats, capable of supporting populations in the long term.
- all attempts should be fully documented at the time and monitored subsequently to assess their success.

Recommendations for future action

Research, surveying, and monitoring

Experience has shown repeatedly that measures to conserve butterflies are likely to be effective only if they are based on sound scientific and ecological knowledge.[59] Many species have complex habitat requirements that have come to be understood only following painstaking research. Although our understanding of butterfly ecology has improved greatly in recent decades, more research is needed urgently on a range of topics including the practicalities of habitat management and the implications of habitat fragmentation.

Similarly, effective conservation relies on detailed knowledge about species distributions and the location of populations, which have formed the chief rationale

[56] Swash and Belding (1999).
[57] Oates and Warren (1990).
[58] Butterfly Conservation (1995).
[59] See reviews by New (1997), Thomas (1984), and Warren (1992*a*).

for the present survey. The main changes observed highlight the need to update information about species distributions regularly and to monitor butterfly populations at both the site and landscape levels. The Butterfly Monitoring Scheme (BMS) has also revealed a wealth of information about butterflies and will be vital in helping to understand the impact of future environmental change, as well as the effectiveness of conservation action.

There is now scope for a major expansion of butterfly monitoring, not least because individual volunteers and members of the voluntary conservation organizations are conducting regular monitoring transects on hundreds of sites outside the BMS. Butterfly Conservation has recently established a system to collate these additional data, which cover a larger number and broader range of sites than the BMS, initially to examine the impact of agri-environment schemes.[60] These large data sets will enable a wider range of issues to be examined, including site-level conservation, land use policies, and overall environmental changes.

The role of legislation

Few butterfly species are fully protected by law in Britain and Ireland. Full protection makes it illegal to kill or injure individuals, or to possess them, without a licence. Sale or trading in a further 22 species is prohibited in Britain without a licence (see p. 348). The drawback with this legislation is that it is designed mainly to protect butterflies from collectors and does little to address the serious problems of habitat destruction or management.

Far more significant aspects of current legislation are those designed to protect habitats and special areas. In Britain, the most important legislation was the 1981 Wildlife and Countryside Act, which provides for the designation of SSSIs. However, the legislation has many weaknesses and there have been numerous cases where SSSIs, including some important butterfly sites, have been damaged or destroyed.[61] Many other sites have deteriorated through neglect. The UK Government introduced a new wildlife bill in 2000 to strengthen the legislation and a new Wildlife Amendment Bill is planned in the Republic of Ireland.[62]

Another key piece of legislation that affects many important butterfly habitats is the EC Habitats and Species Directive (approved in 1992). This aims to establish a network of sites across Member States, known as Natura 2000 sites, which will maintain listed habitats and species 'at a favourable conservation status'. Both Britain and Ireland have put forward numerous sites to cover the listed habitats as well as smaller numbers for the listed species. The only resident butterfly species listed under the Directive is the Marsh Fritillary[63] and several important habitat networks have been proposed for this species in the UK and Ireland.[64]

The European law covering Natura 2000 sites is potentially far stronger than that covering SSSIs, but has yet to be tested in practice. The main drawback is that many important butterfly sites will not be covered and the legislation does little to address many of the fundamental problems mentioned above.

Engaging the public

Nature conservation does not (and should not) operate in isolation and has to involve both public perception and politics. The future care of the urban landscape and the countryside, and the wildlife that inhabit them, will be strongly influenced by public and thus political opinion. As our society becomes ever more urbanized, it will

[60] Brereton and Goodhand (2000).

[61] Friends of the Earth (1994); RSPB (1998*a*).

[62] Buckley (1998).

[63] The Large Blue is also listed under a different Annexe (IV) that requires legislation to protect the butterfly, but not necessarily its habitat.

[64] Warren (1995*b*).

A ten-point plan for the conservation of butterflies

Agricultural reform A major reform of the EC Common Agricultural Policy (CAP) is needed which makes biodiversity conservation an integral part of agricultural policy. Key reforms should include: a) The attachment of strong environmental conditions and cross-compliance to all price support payments under CAP. These would require maintenance of valuable existing features such as hedgerows and semi-natural habitats. b) A major expansion of agri-environment schemes so that they account for at least one-third of the CAP budget and provide support for suitable management of semi-natural habitats.[65]

Implementation of national Biodiversity Action Plans The biodiversity plans of the UK and Irish Governments are important new initiatives and should be implemented fully as a matter of urgency.

Strengthening of wildlife legislation Wildlife legislation should afford maximum possible safeguards to designated sites of special wildlife interest (e.g. SSSIs in Britain, ASSIs in Northern Ireland, and Natura 2000 sites throughout Britain and Ireland) and underpin national biodiversity action plans.

Encouragement of woodland biodiversity Forestry grants should be restructured to give greater priority to wildlife conservation. Schemes to encourage natural woodland regeneration and new afforestation should be designed to avoid damaging important butterfly habitats.

Promotion of sustainable development Local and national planning policies should ensure that development is sustainable[66] and that biodiversity is maintained in both town and countryside. Wherever possible, planning policies should seek to restore wildlife habitats or create new ones to help reverse habitat loss and fragmentation. Measures are also needed to reduce the pressures from urban and suburban expansion.

Protection of nationally important butterfly sites Only a small proportion of nationally important butterfly sites (i.e. sites containing good populations of priority species) is currently given any protection under existing legislation. It is crucial that all such sites are given stronger protection.

Restoration of the landscape Measures are needed to encourage the restoration of fragmented habitats and to rebuild the countryside by restoring links between habitats (e.g. hedgerows and field margins). For butterflies, such measures should be concentrated initially in core regions of threatened species.

Development of research, survey, and monitoring Further research is needed urgently to underpin conservation programmes. Key topics are identified within the UK BAP and Butterfly Conservation's Species Action Plans. A comprehensive survey and monitoring programme is needed to enable changes in butterfly populations to be identified in the future and to assess the effectiveness of conservation measures.

Increasing awareness of the need for conservation The conservation of butterflies is likely to be effective in the long run only if policy makers and the public are aware of the problems. Continual effort must be made to raise awareness of butterflies and the wider conservation issues that affect them.

Obtaining adequate funds for the above actions Funds are needed to stem the loss of habitats and address the many issues outlined above. The popularity of voluntary conservation organizations (combined membership of over three million) shows that there is wide support for improved protection of indigenous wildlife, both for its own sake and to provide enjoyment for present and future generations.

[65] Detailed proposals for CAP reform, elaborating these points, have been produced by several non Governmental organizations (e.g. RSPB 1998*b*) and by the statutory conservation agencies (Anon 1999; Tilzey 1998).

[66] Biodiversity is a seen as a key test of sustainability (English Nature 1999*c*). A strategy for sustainable development in the UK has been produced by the Department of the Environment, Transport and the Regions (1999*a*).

become increasingly important to engage the public in wildlife issues that may otherwise become remote from most people's everyday lives. Recent opinion polls still put the environment high on the public's list of priorities and the UK Government has even placed a wildlife indicator, an index of bird abundance, as one of ten 'quality of life' indicators.[67] However, we cannot be complacent and unless we continue to persuade the people of today that conserving nature is important, there will continue to be less for the people of tomorrow to enjoy.

Conclusions

The distributions of butterflies in Britain and Ireland have undergone rapid changes during the last two centuries, with particularly heavy losses over the last 50 years. At the start of the new century and millennium, butterflies continue to be threatened by many factors, ranging from agricultural intensification and land use change, to climate change and nitrogen deposition. There is little doubt that butterfly distribution and status will continue to change in the future and we must be prepared to re-assess the situation constantly and base our actions on sound, up to date information. The Butterflies for the New Millennium project has yielded an unprecedented amount of information on how distributions have changed and a unique new database that can be used to target conservation. Further recording will be vital in the future and anyone interested in participating is urged to contact Butterfly Conservation (see p. 431).

Conserving butterflies is undoubtedly a large and difficult task, but we have never been better equipped to face the daunting challenge. We have the benefit of decades of pioneering research into butterfly ecology and conservation, and a wealth of experience with practical habitat management. Measures to conserve threatened butterflies are now enshrined within the UK BAP and many new conservation initiatives are underway both in the UK and Ireland. Butterfly Conservation has also grown to become an influential wildlife organization with an active and effective volunteer force, backed by a core of professional staff. The Biological Records Centre is continuing to improve recording as part of the National Biodiversity Network, and the Dublin Naturalists' Field Club has harnessed the energy of butterfly enthusiasts in Ireland.

Some of the prominent issues that need to be addressed over the coming decades have been outlined, but it is of necessity a partial review based on incomplete knowledge. There is no doubt that new problems will appear in the future and conservationists will need to be forward-looking and adaptable. For much of the last century, conservationists fought a rearguard action that was unable to prevent the progressive loss of habitats and species, except in a few notable cases. A more positive agenda for conservation must now be developed, to resolve underlying causes of decline and try to tackle problems before they become insurmountable. The ten-point plan (opposite) lists the most pressing actions required.

The loss of butterflies must be seen in the context of the decline in many other wildlife groups. In almost all cases, measures to protect one species or habitat will serve others and unified policies are needed to avoid duplication of effort. The Biodiversity Action Plan process shows how this integration might be achieved and it is to be hoped that it will ensure that wildlife conservation is given greater emphasis, locally and nationally, in the future.

[67] Department of the Environment Transport and the Regions (1999*b*).

Appendix 1

Vernacular and scientific names of butterfly species

The vernacular names and taxonomic order of species detailed in this atlas follow Emmet and Heath (1990), the paperback edition (revised with minor corrections) of their 1989 book. All scientific names follow Karsholt and Razowski (1996). Where the genus has been changed in relation to Emmet and Heath, the former genus has been given in brackets. Note that both of these authorities place the Duke of Burgundy within the family Lycaenidae and the Monarch within the family Nymphalidae.

VERNACULAR NAME	SCIENTIFIC NAME
Hesperiidae (the 'skippers')	
Chequered Skipper	*Carterocephalus palaemon*
Large Chequered Skipper	*Heteropterus morpheus*
Small Skipper	*Thymelicus sylvestris*
Essex Skipper	*Thymelicus lineola*
Lulworth Skipper	*Thymelicus acteon*
Silver-spotted Skipper	*Hesperia comma*
Large Skipper	*Ochlodes venata*
Dingy Skipper	*Erynnis tages*
Grizzled Skipper	*Pyrgus malvae*
Papilionidae (the 'swallowtails')	
Swallowtail	*Papilio machaon*
Pieridae (the 'whites')	
Wood White	*Leptidea sinapis*
Pale Clouded Yellow	*Colias hyale*
Berger's Clouded Yellow	*Colias alfacariensis*
Clouded Yellow	*Colias croceus*
Brimstone	*Gonepteryx rhamni*
Black-veined White	*Aporia crataegi*
Large White	*Pieris brassicae*
Small White	*Pieris rapae*
Green-veined White	*Pieris napi*
Bath White	*Pontia daplidice*
Orange-tip	*Anthocharis cardamines*
Lycaenidae (the 'hairstreaks', 'coppers', and 'blues')	
Green Hairstreak	*Callophrys rubi*
Brown Hairstreak	*Thecla betulae*
Purple Hairstreak	*Neozephyrus* (*Quercusia*) *quercus*
White-letter Hairstreak	*Satyrium w-album*
Black Hairstreak	*Satyrium pruni*
Small Copper	*Lycaena phlaeas*
Large Copper	*Lycaena dispar*
Long-tailed Blue	*Lampides boeticus*
Small Blue	*Cupido minimus*
Short-tailed Blue	*Cupido* (*Everes*) *argiades*
Silver-studded Blue	*Plebeius* (*Plebejus*) *argus*

VERNACULAR NAME	SCIENTIFIC NAME
Continued	
Brown Argus	*Aricia agestis*
Northern Brown Argus	*Aricia artaxerxes*
Common Blue	*Polyommatus icarus*
Chalkhill Blue	*Polyommatus (Lysandra) coridon*
Adonis Blue	*Polyommatus (Lysandra) bellargus*
Mazarine Blue	*Polyommatus (Cyaniris) semiargus*
Holly Blue	*Celastrina argiolus*
Large Blue	*Maculinea arion*
Geranium Bronze	*Cacyreus marshalli*
Duke of Burgundy	*Hamearis lucina*
Nymphalidae (the 'nymphalids' and 'fritillaries')	
White Admiral	*Limenitis (Ladoga) camilla*
Purple Emperor	*Apatura iris*
Red Admiral	*Vanessa atalanta*
Painted Lady	*Vanessa (Cynthia) cardui*
Small Tortoiseshell	*Aglais urticae*
Large Tortoiseshell	*Nymphalis polychloros*
Camberwell Beauty	*Nymphalis antiopa*
Peacock	*Inachis io*
Comma	*Polygonia c-album*
Map	*Araschnia levana*
Small Pearl-bordered Fritillary	*Boloria selene*
Pearl-bordered Fritillary	*Boloria euphrosyne*
Queen of Spain Fritillary	*Issoria (Argynnis) lathonia*
High Brown Fritillary	*Argynnis adippe*
Dark Green Fritillary	*Argynnis aglaja*
Silver-washed Fritillary	*Argynnis paphia*
Marsh Fritillary	*Euphydryas (Eurodryas) aurinia*
Glanville Fritillary	*Melitaea cinxia*
Heath Fritillary	*Melitaea (Mellicta) athalia*
Nymphalidae (the 'browns')	
Speckled Wood	*Pararge aegeria*
Wall	*Lasiommata megera*
Mountain Ringlet	*Erebia epiphron*
Scotch Argus	*Erebia aethiops*
Marbled White	*Melanargia galathea*
Grayling	*Hipparchia semele*
Gatekeeper	*Pyronia tithonus*
Meadow Brown	*Maniola jurtina*
Ringlet	*Aphantopus hyperantus*
Small Heath	*Coenonympha pamphilus*
Large Heath	*Coenonympha tullia*
Nymphalidae (the 'milkweeds')	
Monarch	*Danaus plexippus*

Appendix 2

Vernacular and scientific names of plants and other organisms

The vernacular and scientific names of plants follow Stace (1997) except that, for simplicity, sub-species and variety names have not been included here.

VERNACULAR NAME	SCIENTIFIC NAME
Agrimony	*Agrimonia eupatoria*
Alder	*Alnus glutinosa*
Alder Buckthorn	*Frangula alnus*
apples	*Malus* spp.
Ash	*Fraxinus excelsior*
Aspen	*Populus tremula*
Barren Strawberry	*Potentilla sterilis*
Bell Heather	*Erica cinerea*
bents	*Agrostis* spp.
Bilberry	*Vaccinium myrtillus*
birches	*Betula* spp.
Bird Cherry	*Prunus padus*
Bitter-vetch	*Lathyrus linifolius*
Black Medick	*Medicago lupulina*
Blackthorn	*Prunus spinosa*
Bladder-senna	*Colutea arborescens*
Blue Moor-grass	*Sesleria caerulea*
Bog-myrtle	*Myrica gale*
Bracken	*Pteridium aquilinum*
Bramble	*Rubus fruticosus* agg.
brassicas	*Brassicaceae*
Bristle Bent	*Agrostis curtisii*
Broad-leaved Dock	*Rumex obtusifolius*
Broad-leaved Everlasting-pea	*Lathyrus latifolius*
Broom	*Cytisus scoparius*
Brussels-sprout	*Brassica oleracea*
Buck's-horn Plantain	*Plantago coronopus*
Buckthorn	*Rhamnus cathartica*
buckthorns	*Rhamnaceae*
buddleias	*Buddleja* spp.
Bugle	*Ajuga reptans*
Bullace	*Prunus domestica*
Bush Vetch	*Vicia sepium*
Cabbage	*Brassica oleracea*
Charlock	*Sinapis arvensis*
clovers	*Trifolium* spp.
Cock's-foot	*Dactylis glomerata*
Common Bird's-foot-trefoil	*Lotus corniculatus*
Common Cottongrass	*Eriophorum angustifolium*
Common Couch	*Elytrigia repens*
Common Cow-wheat	*Melampyrum pratense*
Common Dog-violet	*Viola riviniana*
Common Fleabane	*Pulicaria dysenterica*
Common Nettle	*Urtica dioica*
Common Reed	*Phragmites australis*

Common Restharrow	*Ononis repens*
Common Rock-rose	*Helianthemum nummularium*
Common Sorrel	*Rumex acetosa*
Common Stork's-bill	*Erodium cicutarium*
Common Vetch	*Vicia sativa*
Cow Parsley	*Anthriscus sylvestris*
Cowslip	*Primula veris*
Crack-willow	*Salix fragilis*
crane's-bills	*Geranium* spp.
Creeping Cinquefoil	*Potentilla reptans*
Creeping Soft-grass	*Holcus mollis*
Creeping Thistle	*Cirsium arvense*
Cross-leaved Heath	*Erica tetralix*
Crown Vetch	*Securigera varia*
crucifers	*Cruciferae*
Cuckooflower	*Cardamine pratensis*
currants	*Ribes* spp.
Cut-leaved Crane's-bill	*Geranium dissectum*
Dame's-violet	*Hesperis matronalis*
dandelions	*Taraxacum* spp.
Devil's-bit Scabious	*Succisa pratensis*
Dog-rose	*Rosa canina*
Dogwood	*Cornus sanguinea*
dogwoods	*Cornus* spp.
Dove's-foot Crane's-bill	*Geranium molle*
Downy Oat-grass	*Helictotrichon pubescens*
Dwarf Thistle	*Cirsium acaule*
Dyer's Greenweed	*Genista tinctoria*
Early Hair-grass	*Aira praecox*
elms	*Ulmus* spp.
English Elm	*Ulmus procera*
Evergreen Oak	*Quercus ilex*
False Brome	*Brachypodium sylvaticum*
False Oxlip	*Primula × polyantha (P. veris × vulgaris)*
fescues	*Festuca* spp.
Field Maple	*Acer campestre*
Field Pansy	*Viola arvensis*
Field Scabious	*Knautia arvensis*
Foxglove	*Digitalis purpurea*
Garlic Mustard	*Alliaria petiolata*
Germander Speedwell	*Veronica chamaedrys*
Goat Willow	*Salix caprea*
Gorse	*Ulex europaeus*
gorses	*Ulex* spp.
grasses	*Poaceae*
Greater Bird's-foot-trefoil	*Lotus pedunculatus*
Grey Willow	*Salix cinerea*
Hairy Rock-cress	*Arabis hirsuta*
Hairy Violet	*Viola hirta*
Hare's-tail Cottongrass	*Eriophorum vaginatum*
hawkbits	*Leontodon* spp.
hawthorns	*Crataegus* spp.
Heath Bedstraw	*Galium saxatile*
Heath Dog-violet	*Viola canina*
Heather	*Calluna vulgaris*

VERNACULAR NAME	SCIENTIFIC NAME
Continued	
heathers	*Calluna vulgaris* and *Erica* spp.
Hedge Mustard	*Sisymbrium officinale*
Hedgerow Crane's-bill	*Geranium pyrenaicum*
Hemp-agrimony	*Eupatorium cannabinum*
Hoary cress	*Lepidium draba*
Holly	*Ilex aquifolium*
Honesty	*Lunaria annua*
Honeysuckle	*Lonicera periclymenum*
Hop	*Humulus lupulus*
Horseshoe Vetch	*Hippocrepis comosa*
Ivy	*Hedera helix*
Jointed Rush	*Juncus articulatus*
Kidney Vetch	*Anthyllis vulneraria*
knapweeds	*Centaurea* spp. and *Acroptilon* spp.
Large Bitter-cress	*Cardamine amara*
Lesser Trefoil	*Trifolium dubium*
Lucerne	*Medicago sativa*
mallows	*Malva* spp.
Mange-tout pea	*Pisum sativum*
Marram	*Ammophila arenaria*
Marsh Violet	*Viola palustris*
Mat-grass	*Nardus stricta*
Meadow Crane's-bill	*Geranium pratense*
Meadow Foxtail	*Alopecurus pratensis*
Meadow Vetchling	*Lathyrus pratensis*
meadow-grasses	*Poa* spp.
Michaelmas-daisy	*Aster novo-belgii*
mignonettes	*Reseda* spp.
Milk-parsley	*Peucedanum palustre*
milkweeds	*Asclepias* spp.
Narrow-leaved Everlasting-pea	*Lathyrus sylvestris*
Nasturtium	*Tropaeolum majus*
nettles	*Urtica* spp.
oaks	*Quercus* spp.
Pale Dog-violet	*Viola lactea*
pansies	*Viola* spp.
Peach	*Prunus persica*
Pear	*Pyrus communis*
Pellitory-of-the-wall	*Parietaria judaica*
Pedunculate Oak	*Quercus robur*
Perennial Rye-grass	*Lolium perenne*
pines	*Pinus* spp.
Plum	*Prunus domestica*
poplars	*Populus* spp.
Primrose	*Primula vulgaris*
Purple Moor-grass	*Molinia caerulea*
Purple Small-reed	*Calamagrostis canescens*
Purple-loosestrife	*Lythrum salicaria*
Ragged-Robin	*Lychnis flos-cuculi*
ragworts	*Senecio* spp.
Rape (including Oil-seed Rape)	*Brassica napus*
Red Clover	*Trifolium pratense*

Red Fescue	*Festuca rubra*
Restharrow	*Ononis repens*
Ribwort Plantain	*Plantago lanceolata*
Runner Bean	*Phaseolus coccineus*
rye-grass	*Lolium* spp.
Salad Burnet	*Sanguisorba minor*
Sea Radish	*Raphanus raphanistrum*
Sea-kale	*Crambe maritima*
Sessile Oak	*Quercus petraea*
Sheep's-fescue	*Festuca ovina*
Sheep's Sorrel	*Rumex acetosella*
Small-leaved Elm	*Ulmus minor*
Small Nettle	*Urtica urens*
Small Scabious	*Scabiosa columbaria*
snowberries	*Symphoricarpos* spp.
speedwells	*Veronica* spp.
sphagnum moss	*Sphagnum* spp.
Spindle	*Euonymus europaeus*
stonecrops	*Sedum* spp.
stork's-bills	*Erodium* spp.
teasels	*Dipsacus* spp.
thistles	*Cirsium* spp. and *Carduus* spp.
Timothy	*Phleum pratense*
Tor-grass	*Brachypodium pinnatum*
Tormentil	*Potentilla erecta*
Tufted Hair-grass	*Deschampsia cespitosa*
Tufted Vetch	*Vicia cracca*
Turkey Oak	*Quercus cerris*
Turnip	*Brassica rapa*
vetches	*Anthyllis* spp., *Hippocrepis* spp., and *Vicia* spp.
violets	*Viola* spp.
Viper's-bugloss	*Echium vulgare*
Water Dock	*Rumex hydrolapathum*
Water-cress	*Rorippa nasturtium-aquaticum*
Wavy Hair-grass	*Deschampsia flexuosa*
White Beak-sedge	*Rhynchospora alba*
White Clover	*Trifolium repens*
Wild Angelica	*Angelica sylvestris*
Wild Cabbage	*Brassica oleracea*
Wild Carrot	*Daucus carota*
Wild Cherry	*Prunus avium*
Wild Marjoram	*Origanum vulgare*
Wild Mignonette	*Reseda lutea*
Wild Pansy	*Viola tricolor*
Wild Plum	*Prunus domestica*
Wild Privet	*Ligustrum vulgare*
Wild Radish	*Raphanus raphanistrum*
Wild Strawberry	*Fragaria vesca*
Wild Thyme	*Thymus polytrichus*
willows	*Salix* spp.
Winter-cress	*Barbarea vulgaris*
Wood Avens	*Geum urbanum*
Wood Small-reed	*Calamagrostis epigejos*
Wych Elm	*Ulmus glabra*
Yorkshire-fog	*Holcus lanatus*

Other organisms

VERNACULAR NAME	SCIENTIFIC NAME
fungus causing Dutch Elm Disease	*Ophiostoma novo-ulmi*
harvestmen	*Opiliones*
beetles	*Coleoptera*
black ant spp.	*Lasius alienus*
	Lasius niger
Burnet Companion moth	*Euclidia glyphica*
elm bark beetles	*Scolytus scolytus*
	Scolytus multistriatus
Mother Shipton moth	*Callistege mi*
red ant spp.	*Myrmica sabuleti*
	Myrmica scabrinodis
	Myrmica ruginodis
wasp, parasitic on Large White larvae	*Apanteles glomeratus*
wasp, parasitic on Holly Blue larvae	*Listrodomus nycthemerus*
Wood Ant	*Formica rufa*
House Sparrow	*Passer domesticus*
Song Thrush	*Turdus philomelos*
Fallow Deer	*Dama dama*
Mole	*Talpa europaea*
Muntjac	*Muntiacus reevesi*
Rabbit	*Oryctolagus cuniculus*

Appendix 3

Local butterfly atlases

This appendix provides an up-to-date list of local atlases in the BNM survey area. Only publications that include maps of all species relevant to the area have been included. In addition to these local atlases, many provisional atlases and lists of butterflies have been published. Chalmers-Hunt (1989) provides a thorough review of local lists.

ENGLAND

Avon

Avon Butterfly Project. (in press). *The butterflies of the Bristol region.*

Bedfordshire

Arnold, V. W., Baker, C. R. B., Manning, D. V., and Woiwod, I. P. (1997). *The butterflies and moths of Bedfordshire.* Bedfordshire Natural History Society, Bedford.

Berkshire, Buckinghamshire, and Oxfordshire

Asher, J. (1994). *The butterflies of Berkshire, Buckinghamshire and Oxfordshire.* Pisces Publications, Oxford.

Baker, B. R. (1994). *The butterflies and moths of Berkshire.* Hedera Press, Uffington.

Knight, R. and Campbell, J. M. (1986). *An atlas of Oxfordshire butterflies*, Oxfordshire County Council Occasional Paper, No. 10. Oxfordshire County Council, Woodstock.

Steel, C. and Steel, D. (1985). *Butterflies of Berkshire, Buckinghamshire and Oxfordshire.* Pisces Publications, Oxford.

Cheshire

Shaw, B. T. (1999). *The butterflies of Cheshire.* National Museums and Galleries on Merseyside, Liverpool.

Cornwall and the Isles of Scilly

Frost, M. P. and Madge, S. C. (1991). *Butterflies in south-east Cornwall.* The Caradon Field and Natural History Club.

Worth, J. and Spalding, A. (ed.). (in press). *Butterfly atlas of Cornwall.*

Devon

Bristow, C. W., Mitchell, S. H., and Bolton, D. E. (1993). *Devon butterflies.* Devon Books, Tiverton.

Dorset

Thomas, J. and Webb, N. (1984). *The butterflies of Dorset*. Dorset Natural History and Archaeological Society, Dorset County Museum, Dorchester.

Thomas, J., Surry, R., Shreeves, B., and Steele, C. (1998). *New atlas of Dorset butterflies*. The Dorset Natural History and Archaeological Society, Dorset County Museum, Dorchester.

Essex

Corke, D. (1997). *The butterflies of Essex*. Lopinga books, Wimbish.

Emmet, A. M., Pyman, G. A., and Corke, D. (1985). *The larger moths and butterflies of Essex*. The Essex Field Club, London.

Gloucestershire

Meredith, G. H. J. (1989). Butterfly distribution in Gloucestershire 1975–1988. *The Gloucester Naturalist No. 4*.

Greater London

Plant, C. W. (1988). *The butterflies of the London area*. London Natural History Society, London.

Greater Manchester

Hardy, P. B. (1998). *Butterflies of Greater Manchester*. PGL Enterprises, Sale.

Hampshire

Taverner, J. H., Oates, M. R., and Green, D. G. (2000). *Butterflies of Hampshire*. Pisces Publications, Newbury.

Hertfordshire

Sawford, B. (1987). *Butterflies of Hertfordshire*. Castlemead Publications, Ware.

Kent

Philp, E. G. (1993). *The butterflies of Kent, an atlas of their distribution*. The Kent Field Club, Sittingbourne.

Leicestershire

Morris, R. (1990). *The butterflies of the Hinkley District*, Occasional Publications Series, No. 6. Leicestershire Entomological Society, Leicester.

Lincolnshire

Johnson, R. (1997). *The butterflies and moths of Lincolnshire*. Lincolnshire Naturalists' Society, Lincoln.

Norfolk

Hall, M. R. (1991). *An atlas of Norfolk butterflies*. Butterfly Conservation Norfolk Branch, Hunstanton.

Northumberland and Durham

Cook, N. J. (1990). *An atlas of butterflies of Northumberland and Durham.* Northumberland Biological Records Centre, Hancock Museum, Newcastle.

Dunn, T. C. and Parrack, J. D. (1986). *The moths and butterflies of Northumberland and Durham part 1: Macrolepidoptera.* Northern Naturalists' Union, Durham.

Shropshire

Riley, A. M. (1991). *A natural history of the butterflies and moths of Shropshire.* Swan Hill Press, Shrewsbury.

Somerset

Somerset Butterfly Group (1998). *Butterflies of Somerset—A provisional atlas.* Somerset Butterfly Group.

Staffordshire

Warren, R. G. (1984). *Atlas of the Lepidoptera of Staffordshire—Part 1: Butterflies.* Staffordshire Biological Recording Scheme, Stoke-on-Trent City Museum and Art Gallery, Hanley.

Suffolk

Mendel, H. and Piotrowski, S. H. (1986). *Butterflies of Suffolk—An atlas and history.* Suffolk Naturalists' Society, Ipswich.

Stewart, R. (in press). *Butterflies of Suffolk.*

Surrey

Collins, G. A. (1995). *Butterflies of Surrey.* Surrey Wildlife Trust, Woking.

Sussex

Gay, J. and Gay, P. (1996). *Atlas of Sussex butterflies.* Butterfly Conservation Sussex Branch, Henfield.

Wiltshire

Fuller, M. (1995). *The butterflies of Wiltshire.* Pisces Publications, Oxford.

Yorkshire

Barnham, M. and Foggitt, G. T. (1987). *Butterflies in the Harrogate District.* Barnham and Foggitt, Harrogate.

Garland, S. P. (1981). *Butterflies of the Sheffield area,* Sorby Record Special Series, No. 5. Sorby Natural History Society.

WALES

Whalley, P. (1996). *Butterflies of Gwynedd.* North Wales Wildlife Trust, Llanberis.

Fowles, A. (1984). The butterflies of Ceredigion. *Nature In Wales,* **3**, 25–43.

Morgan, I. K. (1989). *A provisional review of the butterflies of Carmarthenshire.* Nature Conservancy Council, Aberystwyth.

SCOTLAND

Smout, A. M. and Kinnear, P. K. (1994). *Butterflies of Fife: A provisional atlas*. Fife Nature, Fife.

Stewart, J., Barbour, D., and Moran, S. (1998). *Highland butterflies a provisional atlas*. Highland Biological Recording Group / Inverness Museum and Art Gallery, Inverness.

Thomson, G. (1980). *The butterflies of Scotland*. Croom Helm, London.

IRELAND

Hickin, N. (1992). *The butterflies of Ireland. A field guide*. Robert Rinehart, Schull.

Appendix 4

This appendix reproduces the instructions for recorders circulated by the BNM project to local co-ordinators, partner organisations, and recorders.

Butterflies for the New Millennium
Instructions for recorders

Butterflies for the New Millennium is a major project to map the distributions of butterflies across Britain and Ireland over the last 5 years of the 1990s. The data will provide essential information for conservation work and form the basis of a new Atlas of Butterflies. The following are instructions for recorders participating in the project.

A *record* is made up of a number of separate pieces of information, many of which can be given at different levels of accuracy. The essential information, which is needed to fulfil the aims of Butterflies for the New Millennium, is explained at some length on the special recording forms and in the 'Recording forms' section below. One of the most important pieces of information is the location of the sighting. This is recorded as a grid reference by referring to a map.

Sites and colonies

The main aim of the millennium project is to record and map butterfly *colonies*, which are generally correlated with *sites*, which may range from a roadside verge to a forest. It is only at *site* level that conservation can take place. It is convenient to use a grid of squares (10 km, 2 km, or 1 km) to define areas to be searched for colonies. For populous areas, such as much of southern England, a grid of 2 km squares is a convenient way to target recording; in upland/highland areas a 10 km grid is more practical.

Aim to visit *sites* in as many squares as possible, always identifying records from individual sites with at least 1 km square accuracy and preferably 100 m (6-figure), so that the data can be used to support conservation work. It may well be that in many 2 km squares, three out of four 1 km squares may have no useful habitat and records may come from only one 1 km square. Record the grid reference of the site or at least the 1 km square and *not* the 2 km square reference. Large sites, such as major forests, should be subdivided into 1 km squares or sub-sites for recording.

Timing of visits

You should try to visit each site at least four times in the season, preferably in early May, mid-June, mid-July and the second half of August, to catch the flight periods of all possible species.

Warm sunny weather between mid-morning and mid-afternoon is ideal, although some species may fly earlier and later in hot conditions; in cool overcast conditions you are unlikely to see butterflies in flight, but they may be found resting. Revisits to sites are also valuable to record species whose populations vary widely from year to year.

Identification

It is important that we collect reliable data for this project. There are several good field guides available to help with identification of species. If you are in any doubt about an identification, please indicate so on your recording form, so that it can be checked by a local expert.

Habitats

Try to find examples of different habitats to visit in your target area—for example, woodland, grassland, farmland and, where appropriate, heathland. These different habitats will support a different range of butterfly species.

In intensively agricultural areas, search for unploughed headlands, hedgerows, relics of woodland, tracks, and footpaths. Set-aside areas and some roadside verges often provide suitable habitat for many species. For urban areas, parks, gardens, churchyards, and 'waste' ground may provide suitable habitats for some species.

Access to sites

The 1:25 000 OS Pathfinder maps display public footpaths and rights of way very clearly; you should stick to these paths wherever possible. If you need to enter private property whilst carrying out fieldwork, please seek permission from the landowner first—this project does not convey any rights of access to private land. It is also important to respect the Country Code at all times.

Recording forms

For repeated visits to a particular site, use the Site Recording Form. The Casual Record Form is used to make a note of records from scattered locations.

Please enter on the recording form your name and address (or that of the recorder), the site name (or the name of a local geographical feature) and the grid reference of the site or locality. Please indicate also the type of habitat, using the codes listed on the form, and note the date of the visit. For each species seen, record an estimate of the numbers seen at that locality. Full instructions are given on the forms.

Where unusual or threatened species are recorded, it is useful to give as much detail as possible of the locality so that follow-up visits can be made to assess conservation needs. If you regard the sighting as confidential, please mark it as such on the form. Space for comments is provided on the forms.

Send completed forms promptly at the end of each recording season for computer entry to your local co-ordinator (usually printed on the forms) or to Butterfly Conservation at the address below.

Recording, mapping, and grid references

There is a clear distinction between the coverage for *recording* and the scale of *mapping*. For recording to provide effective data for local purposes, such as planning and conservation work or national analysis of how particular species are faring, records have to be related to sites. Therefore, the grid reference for any records of butterflies should be to at least 1 km square accuracy (four-figure grid reference), and preferably six-figure references which pinpoint a sighting to the nearest 100 m.

When it comes to mapping butterfly records, data are often plotted in a summary form on distribution maps e.g. as 2 km squares for local maps, or 10 km squares for the national scale. Generally, therefore, atlas organizers seek to ensure that some recording is carried out within each 2 km or 10 km square, to give a good level of *mapping* coverage.

How to work out a grid reference

Please use only the National Grid references from Ordnance Survey maps—not road maps, which may have non-standard grids.

A grid of light blue lines marks 1 km squares on 1:50 000 series (Landranger) OS maps. The grid reference of a 1 km square consists of the 100 km square code (e.g. NN), which will be marked somewhere on the map, followed by four numbers, which mark the bottom left-hand corner of the square (see example below). The first two numbers refer to the horizontal (east-west) scale and the second two the vertical (north-south). It is very important to put these in the right order.

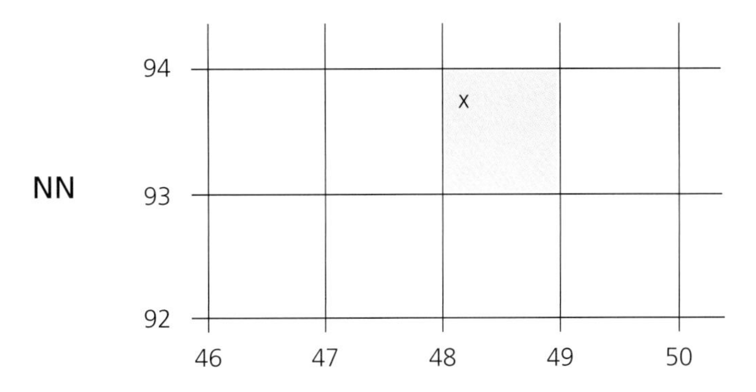

100 km square code: NN
Shaded 1 km square: NN4893

A 6-figure (100 m square) grid reference is derived by dividing the 1 km square into tenths, from the left (west) and the bottom (south). For example, the map reference of the cross in the shaded square above is given by combining the east-west reference, 482, with the north-south reference, 937, giving a full reference of: NN482937.

When you move into another 1 km square, record sightings there separately, so that important sites can be identified correctly.

If you do not have access to suitable maps or are not familiar with grid references, please do not be put off from recording the butterflies that you see. Contact your local co-ordinator and they may be able to help.

Appendix 5

Standard recording forms

This appendix reproduces the two standard recording forms designed and circulated by the BNM project (see p. 33). These are available from local co-ordinators or Butterfly Conservation.

Butterfly Casual Record Sheet

Britain and Ireland

Name:

Address/Tel:

BUTTERFLY CONSERVATION

in association with
Biological Records Centre

Year:	site info: OS Map reference / Site/Locality name							
Habitat type(s)								
Day/Month								
Chequered Skipper								
Small Skipper								
Essex Skipper								
Lulworth Skipper								
Silver-spotted Skipper								
Large Skipper								
Dingy Skipper								
Grizzled Skipper								
Swallowtail								
Wood White								
Clouded Yellow								
Brimstone								
Large White								
Small White								
Green-veined White								
Orange-tip								
Green Hairstreak								
Brown Hairstreak								
Purple Hairstreak								
White-letter Hairstreak								
Black Hairstreak								
Small Copper								
Small Blue								
Silver-studded Blue								
Brown Argus								
Northern Brown Argus								
Common Blue								
Chalkhill Blue								
Adonis Blue								
Holly Blue								
Duke of Burgundy								
White Admiral								
Purple Emperor								
Red Admiral								
Painted Lady								
Small Tortoiseshell								
Peacock								
Comma								
Small Pearl-bordered Fritillary								
Pearl-bordered Fritillary								
High Brown Fritillary								
Dark Green Fritillary								
Silver-washed Fritillary								
Marsh Fritillary								
Glanville Fritillary								
Heath Fritillary								
Speckled Wood								
Wall								
Mountain Ringlet								
Scotch Argus								
Marbled White								
Grayling								
Gatekeeper								
Meadow Brown								
Ringlet								
Small Heath								
Large Heath								
Other (specify)								

Recording Instructions

The Casual Record Sheet is for general recording of butterflies on a variety of sites, for example, when atlas recording in different areas.

Use one column to record sightings from each site on each separate day. An example of a completed column entry is shown alongside. Please enter information on this sheet as follows:

Name: Enter your name

Address/Tel: along with your address and telephone number.

Year: Enter the year during which the records are made. Please do not put records for more than one year on each sheet.

Site Information: Record here information about the site or locality:

OS Map Reference: Write the Ordnance Survey map reference of the site in the left-hand space. This should preferably be 6-figure for the centre-point of a small site (e.g. SP609104) or at least 4-figure (i.e. 1 km square, e.g. SU4796) for the locality. If the site is large, divide it into sub-sites and record each 1 km square separately.

Site name / Locality: Give the site name or name of a feature marked on the map, to identify the locality being recorded, in the right-hand space.

Habitat type(s): Enter the code(s) from the list of habitat types below that apply to the site/locality where the butterflies were actually seen.

Day / Month: Give the day and month of the record (as numbers).

SP617111 | Shabbington Wood
43
14 | 7
A
B
A
B
C, O
L
A

Species recorded

Please record all butterfly species seen at the named site/locality, marking the numbers seen on that date using the following codes:

Numbers seen: **A**: 1 **B**: 2-9 **C**: 10-29 **D**: 30-99 **E**: 100 or more

Please record also any eggs, larvae, pupae or mating seen, using codes: **O** (eggs), **L** (larvae), **P** (pupae) and **M** (mating), respectively.

Use the **Notes** box below to add further notes on any of the records. Thanks for your help!

Habitat codes

Code	Habitat	Code	Habitat
1	Sea shore / cliffs / salt marsh / dunes etc	54	Marshes / fens
2	Freshwater edges (lakes / rivers / canals)	6	Rocky inland habitats / screes etc
31	Heath / scrub	81	Fertilized / improved / reseeded grassland
34	Calcareous grassland	82	Crops
35	Acid grassland	83	Orchards / plantations / commercial forestry
38	Meadow (unimproved)	84	Tree lines / hedges / small woods
41	Broad-leaved deciduous woodland	85	Parks / gardens / churchyards
42	Natural coniferous woodland	86	Urban areas / industrial estates
43	Mixed woodland	87	Fallow / waste / disturbed land
51	Raised / lowland Bog	89	Quarries / pits
52	Blanket / upland Bog	90	Road / rail verges, cuttings etc

Please return completed forms to:

...or to
Butterfly Conservation, PO Box 444
Wareham, Dorset BH20 5YA

Notes

The information supplied here is completed and sent to Butterfly Conservation, the Biological Records Centre, and their partners on the understanding that the data provided by the recorder will be entered into a computerised database and will be used for nature conservation, research, education, and public information. The information remains the property of the recorder at all times.

Butterfly Site Recording Form

Britain and Ireland

	Name:	Year:
BUTTERFLY CONSERVATION *in association with* Biological Records Centre	Address & telephone number:	*list date(s) of visit(s) overleaf*

Site information:

Site name and status:	OS Grid ref:
Nearest town:	County:

Habitat types:

Please tick box(es) that apply to the area visited and give any other details about habitat / land use

Habitat	Code	
Sea shore / cliffs / salt marsh / dunes etc	1	
Freshwater edges (lakes / rivers / canals)	2	
Heath / scrub	31	
Calcareous grassland	34	
Acid grassland	35	
Meadow (unimproved)	38	
Broad-leaved deciduous woodland	41	
Natural coniferous woodland	42	
Mixed woodland	43	
Raised / lowland Bog	51	
Blanket / upland Bog	52	
Marshes / fens	54	
Rocky inland habitats / screes etc	6	

Habitat	Code	
Fertilized / improved / reseeded grassland	81	
Crops	82	
Orchards / plantations / commercial forestry	83	
Tree lines / hedges / small woods	84	
Parks / gardens / churchyards	85	
Urban areas / industrial estates	86	
Fallow / waste / disturbed land	87	
Quarries / pits	89	
Road / rail verges, cuttings etc	90	

Other details

If possible, please add a sketch map of the site and area visited:

Species seen *(overleaf)*

Enter the date and length (approx. minutes) of each visit and indicate weather conditions overleaf. Record the number of each species seen, using the following codes: ***A****: only 1 seen,* ***B****: 2-9,* ***C****: 10-29,* ***D****: 30-99,* ***E****: 100+.*
If early stages (ova, larvae or pupae) or mating of a particular species are seen, use codes ***O****,* ***L****,* ***P*** *or* ***M****, respectively.*

Please return completed forms to:

...or to
Butterfly Conservation
PO Box 444
Wareham
Dorset BH20 5YA

Recording Date																							
Day																							
Month																							
Length of visit (mins)																							
Weather conditions (**P**oor/**M**oderate/**I**deal)																							
Chequered Skipper																							
Small Skipper																							
Essex Skipper																							
Lulworth Skipper																							
Silver-spotted Skipper																							
Large Skipper																							
Dingy Skipper																							
Grizzled Skipper																							
Swallowtail																							
Wood White																							
Clouded Yellow																							
Brimstone																							
Large White																							
Small White																							
Green-veined White																							
Orange-tip																							
Green Hairstreak																							
Brown Hairstreak																							
Purple Hairstreak																							
White-letter Hairstreak																							
Black Hairstreak																							
Small Copper																							
Small Blue																							
Silver-studded Blue																							
Brown Argus																							
Northern Brown Argus																							
Common Blue																							
Chalkhill Blue																							
Adonis Blue																							
Holly Blue																							
Duke of Burgundy																							
White Admiral																							
Purple Emperor																							
Red Admiral																							
Painted Lady																							
Small Tortoiseshell																							
Peacock																							
Comma																							
Small Pearl-bordered Fritillary																							
Pearl-bordered Fritillary																							
High Brown Fritillary																							
Dark Green Fritillary																							
Silver-washed Fritillary																							
Marsh Fritillary																							
Glanville Fritillary																							
Heath Fritillary																							
Speckled Wood																							
Wall																							
Mountain Ringlet																							
Scotch Argus																							
Marbled White																							
Grayling																							
Gatekeeper																							
Meadow Brown																							
Ringlet																							
Small Heath																							
Large Heath																							
Other (specify)																							

Comments:

Appendix 6

UK monthly climate summaries (1995-9)

For each year of the BNM survey, monthly climate summary data are presented as the deviation from the 1951–80 average. Data kindly provided by Dr Mike Hulme, Climate Research Unit, University of East Anglia, Norwich on www.cru.uea.ac.uk/~mikeh/datasets/uk/month_uk

1995

MONTH	DAYTIME TEMP (°C)	RAINFALL (%)	SUNSHINE (%)
January	+0.5	+41	0
February	+2.0	+55	+7
March	0.0	+17	+23
April	+0.5	0	+1
May	+0.8	−19	+9
June	+0.5	−53	+7
July	+2.8	−33	+25
August	+4.1	−81	+53
September	+0.3	+48	+4
October	+2.1	+10	+16
November	+1.1	−17	+10
December	−2.5	−26	+1
Annual	**+1.0**	**−5**	**+15**

1996

MONTH	DAYTIME TEMP (°C)	RAINFALL (%)	SUNSHINE (%)
January	0.0	−31	−52
February	−1.0	+28	+23
March	−1.6	−21	−36
April	+0.4	0	−10
May	−1.8	0	−1
June	+0.8	−50	+19
July	+0.8	−31	+12
August	+1.1	+8	+7
September	+0.4	−60	+19
October	+0.7	+21	+5
November	−1.2	+37	+36
December	−1.9	−28	+12
Annual	**−0.1**	**−8**	**+3**

1997

MONTH	DAYTIME TEMP (°C)	RAINFALL (%)	SUNSHINE (%)
January	−1.1	−70	−1
February	+2.2	+74	+3
March	+2.6	−31	+5
April	+1.3	−41	−3
May	+0.2	+25	+17
June	−0.6	+65	-26
July	+1.2	−26	+17
August	+2.8	−4	+16
September	+0.8	−48	+24
October	+0.3	−21	+27
November	+1.6	+25	−18
December	+0.7	+20	−7
Annual	**+1.0**	**−4**	**+5**

1998

MONTH	DAYTIME TEMP (°C)	RAINFALL (%)	SUNSHINE (%)
January	+1.0	+27	+7
February	+3.5	−33	+2
March	+1.7	+35	−24
April	−0.5	+102	−9
May	+1.5	−47	+5
June	−0.5	+74	−16
July	−0.7	+24	−24
August	+0.2	−29	+10
September	+0.6	+7	−3
October	−0.8	+68	+8
November	−0.5	+2	+37
December	+0.3	0	+4
Annual	**+0.5**	**+4**	**+2**

1999

MONTH	DAYTIME TEMP (°C)	RAINFALL (%)	SUNSHINE (%)
January	+1.1	+52	+15
February	+0.7	+10	+42
March	+1.5	+16	+14
April	+0.9	+49	+1
May	+1.2	0	−5
June	−0.6	+37	−4
July	+1.3	−36	+14
August	+0.4	+30	+6
September	+1.7	+38	+9
October	+0.2	−9	+18
November	+0.8	−22	+15
December	-0.9	+58	+35
Annual	**+0.7**	**+17**	**+8**

Appendix 7

Useful contact addresses

Butterfly Conservation
Manor Yard
East Lulworth
Wareham
Dorset
BH20 5QP
01929 400 209
www.butterfly-conservation.org

Butterfly Conservation Scotland Office
Balallan House
Allan Park
Stirling
FK8 2QG

Butterfly Monitoring Scheme
Centre for Ecology and Hydrology
Monks Wood
Abbots Ripton
Huntingdon
PE28 2LS
01487 772 400
www.bms.ceh.ac.uk

Biological Records Centre
Centre for Ecology and Hydrology
Monks Wood
Abbots Ripton
Huntingdon
PE28 2LS
01487 772 400
www.brc.ac.uk

Dublin Naturalists' Field Club
35 Nutley Park
Donnybrook
Dublin 4
Republic of Ireland
+353 (0)1269 7469

Statutory conservation agencies

Joint Nature Conservation Committee
Monkstone House
City Road
Peterborough
Cambridgeshire
PE1 1JY
01733 562 626
www.jncc.gov.uk

Countryside Council for Wales
Plas Penrhos
Ffordd Penrhos
Bangor
Gwynedd
LL57 2JA
01248 370 444
www.ccw.gov.uk

English Nature
Northminster House
Peterborough
Cambridgeshire
PE1 1UA
01733 455 100/1/2
www.english-nature.org.uk

Environment and Heritage Service Northern Ireland
Commonwealth House
35 Castle Street
Belfast
BT1 1GU
028 9025 1477
www.ehsni.gov.uk

Scottish Natural Heritage
12 Hope Terrace
Edinburgh
EH9 2AS
0131 447 4784
www.snh.org.uk

Dúchas, The Heritage Service, National Parks and Wildlife
51 St Stephen's Green
Dublin 2
Republic of Ireland
+353 (0)1661 3111

Manx National Heritage
Douglas
Isle of Man
IM1 3LY
01624 648 000
www.gov.im/mnh

Jersey Environmental Services
South Hill
St Helier
Jersey
JE2 4US
01534 725 511
www.gov.je/esu

States of Guernsey—Board of Administration
PO Box 43
Sir Charles Frossard House
La Charroterie
St Peter Port
Guernsey
GY1 1FH
01481 717 000
www.gov.gg/boa

Other government agencies

The Countryside Agency
John Dower House
Crescent Place
Cheltenham
Gloucestershire
GL50 3RA
01242 521 381
www.countryside.gov.uk

Forestry Commission
231 Corstorphine Road
Edinburgh
EH12 7AT
0131 334 0303
www.forestry.gov.uk

Ministry of Agriculture, Fisheries and Food
Nobel House
17 Smith Square
London
SW1P 3JR
020 7238 3000
www.maff.gov.uk

The Heritage Council
Kilkenny
Republic of Ireland
+353 (0)567 0777
www.heritagecouncil.ie

Non-governmental organisations

Grazing Animals Project (GAP) Office and Forum for the Application of Conservation Techniques (FACT)
7A Friars Quay
Norwich
Norfolk
NR3 1ES
01603 756 070

The National Trust
36 Queen Anne's Gate
London
SW1H 9AS
020 7222 9251
www.nationaltrust.org.uk

The National Trust for Scotland
28 Charlotte Square
Edinburgh
EH2 4ET
0131 243 9300
www.nts.org.uk

The National Trust of Guernsey
26 Cornet Street
St Peter Port
Guernsey
01481 721 915

Royal Society for the Protection of Birds
The Lodge
Sandy
Bedfordshire
SG19 2DL
01767 680 551
www.rspb.org.uk

The Wildlife Trusts
The Kiln
Waterside
Mather Road
Newark
NG24 1WT
01636 677 711
www.wildlifetrust.org.uk

Biological recording contacts

National Biodiversity Network
c/o The Wildlife Trusts
The Kiln
Waterside
Mather Road
Newark
NG24 1WT
01636 670 090
www.nbn.org.uk

National Federation for Biological Recording
c/o Centre for Environmental Data and Recording
Ulster Museum
Botanic Gardens
Belfast
BT9 5AB
028 9038 3154

Biological Recording in Scotland
c/o Chesterhill
Shore Road
Anstruther
Fife
KY10 3DZ
01333 310 330

Appendix 8

Butterfly phenograms

The following phenograms show the number of butterflies recorded on a given day over the period 1995–9 in Britain and the Isle of Man plotted against latitude, expressed as northings on the Ordnance Survey National Grid. The background to these is described on p. 47. Those phenograms included within species accounts are not repeated here. Not all species are featured—either because the species is very rare and highly localized, or because the phenogram does not show any strong pattern.

A number of features are apparent from the phenograms:

- Species that are single-brooded show a single vertical bar. Double-brooded species have two vertical bars. A partial third brood can be seen in the south for a few double-brooded species.
- Where there is a shift in flight time from south to north, this shows as a tilt on the vertical bar(s). Where the flight time is shorter in the north, the bar can be seen as tapered.
- Where early and unseasonally warm weather brings butterflies out of hibernation, such as happened in mid-February 1996, this can be seen as a thin vertical bar.

Key to the phenograms

On each plot, the horizontal scale represents months of the year. The vertical scale represents grid lines at 100 km intervals in a northerly direction, which correspond to the horizontal lines on the main species maps. The numbering corresponds to the 100 km grid northings of the Ordnance Survey National Grid. Colour is used to indicate, on a logarithmic scale, the total number of butterflies seen on each day at each latitude.

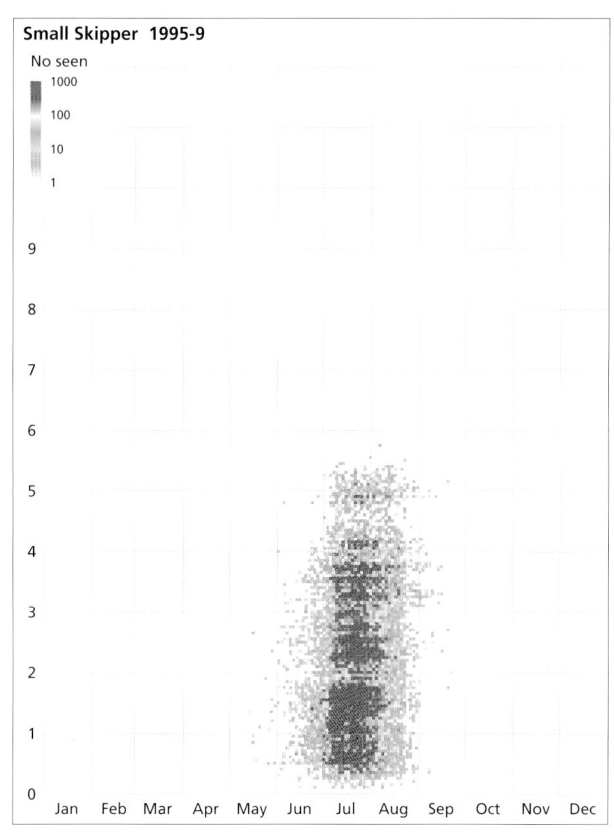
Small Skipper 1995-9
No seen
1000
100
10
1
9
8
7
6
5
4
3
2
1
0
Jan
Feb
Mar
Apr
May
Jun
Jul
Aug
Sep
Oct
Nov
Dec

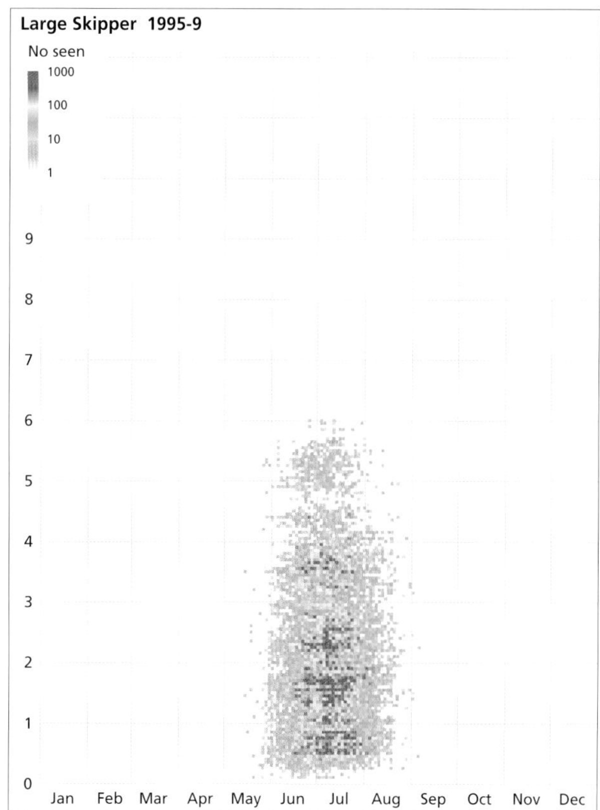
Large Skipper 1995-9
No seen
1000
100
10
1
9
8
7
6
5
4
3
2
1
0
Jan
Feb
Mar
Apr
May
Jun
Jul
Aug
Sep
Oct
Nov
Dec

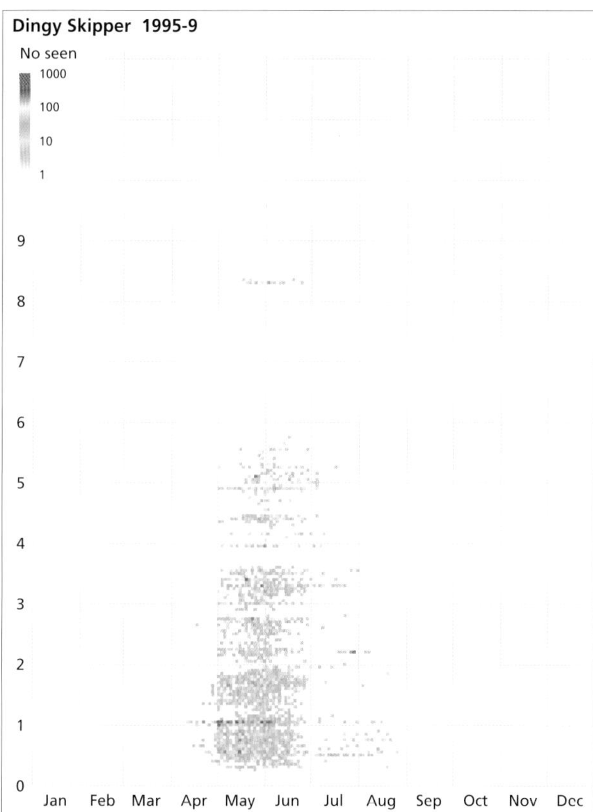
Dingy Skipper 1995-9
No seen
1000
100
10
1
9
8
7
6
5
4
3
2
1
0
Jan
Feb
Mar
Apr
May
Jun
Jul
Aug
Sep
Oct
Nov
Dec

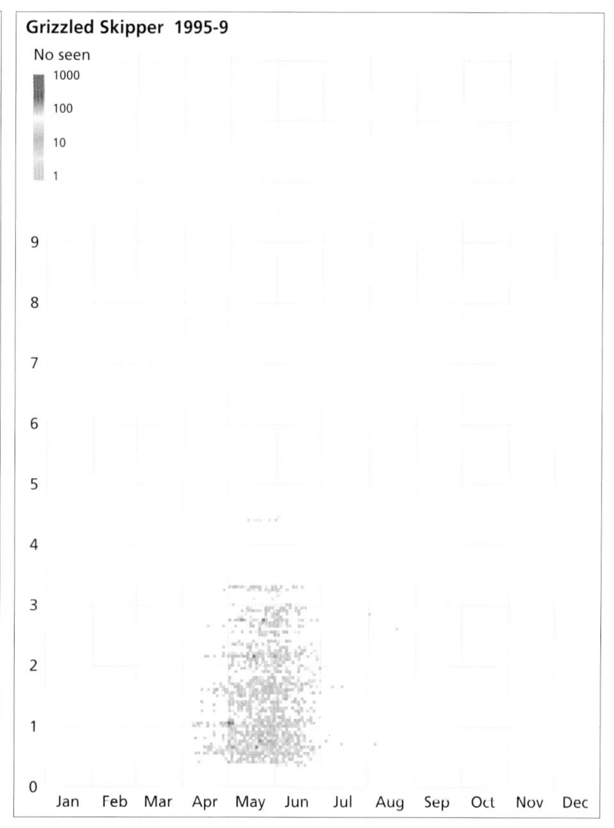
Grizzled Skipper 1995-9
No seen
1000
100
10
1
9
8
7
6
5
4
3
2
1
0
Jan
Feb
Mar
Apr
May
Jun
Jul
Aug
Sep
Oct
Nov
Dec

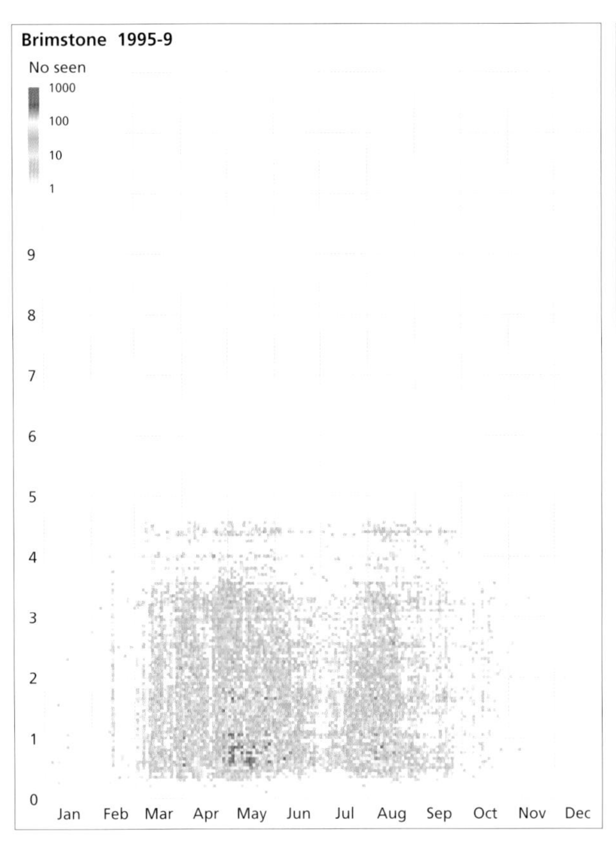
Brimstone 1995-9
No seen
1000
100
10
1
9
8
7
6
5
4
3
2
1
0
Jan
Feb
Mar
Apr
May
Jun
Jul
Aug
Sep
Oct
Nov
Dec

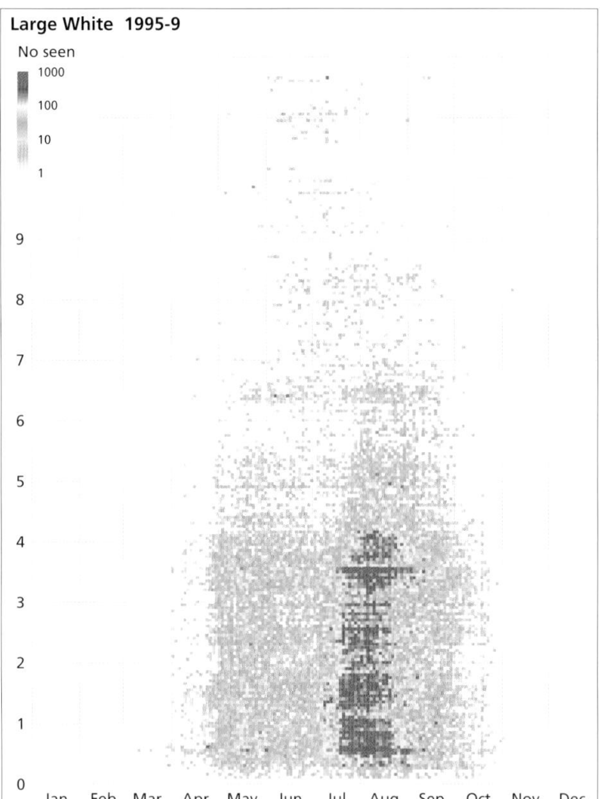
Large White 1995-9
No seen
1000
100
10
1
9
8
7
6
5
4
3
2
1
0
Jan
Feb
Mar
Apr
May
Jun
Jul
Aug
Sep
Oct
Nov
Dec

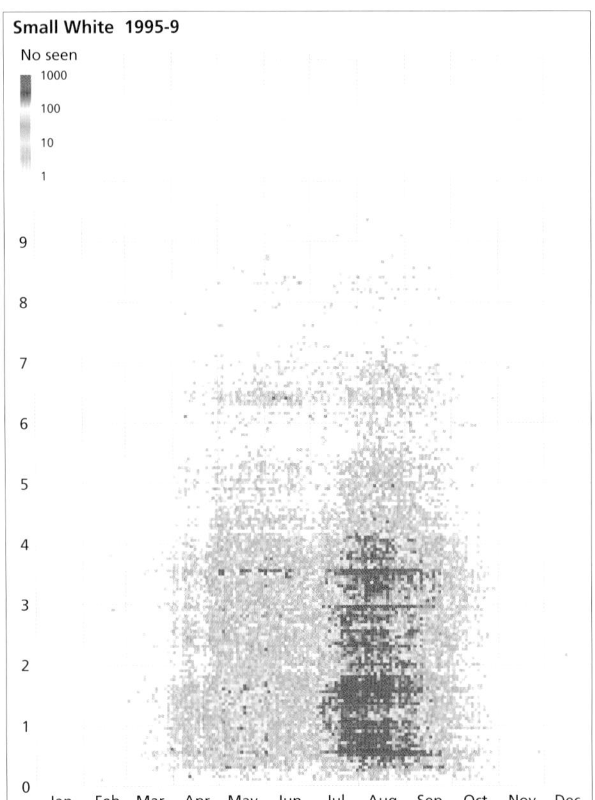
Small White 1995-9
No seen
1000
100
10
1
9
8
7
6
5
4
3
2
1
0
Jan
Feb
Mar
Apr
May
Jun
Jul
Aug
Sep
Oct
Nov
Dec

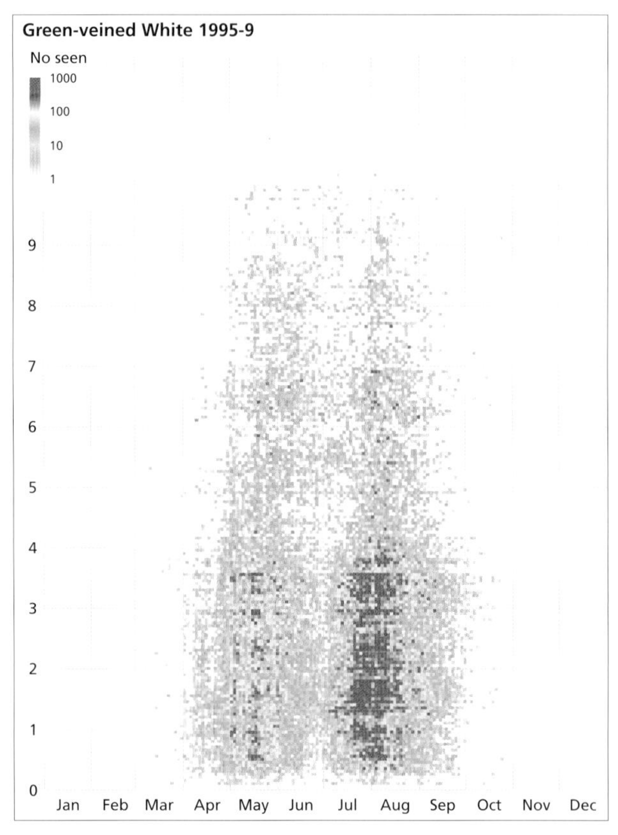
Green-veined White 1995-9
No seen
1000
100
10
1
9
8
7
6
5
4
3
2
1
0
Jan
Feb
Mar
Apr
May
Jun
Jul
Aug
Sep
Oct
Nov
Dec

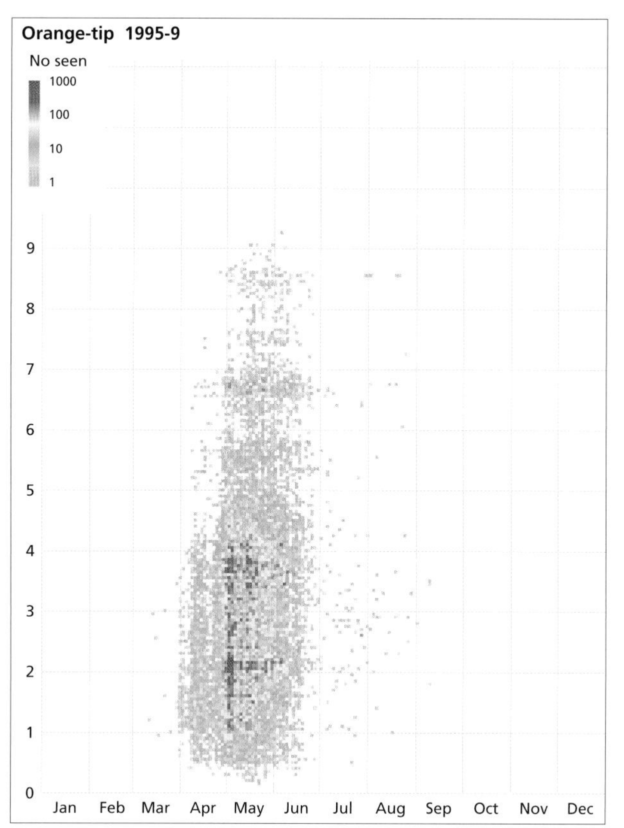
Orange-tip 1995-9
No seen
1000
100
10
1
9
8
7
6
5
4
3
2
1
0
Jan Feb Mar Apr May Jun Jul Aug Sep Oct Nov Dec

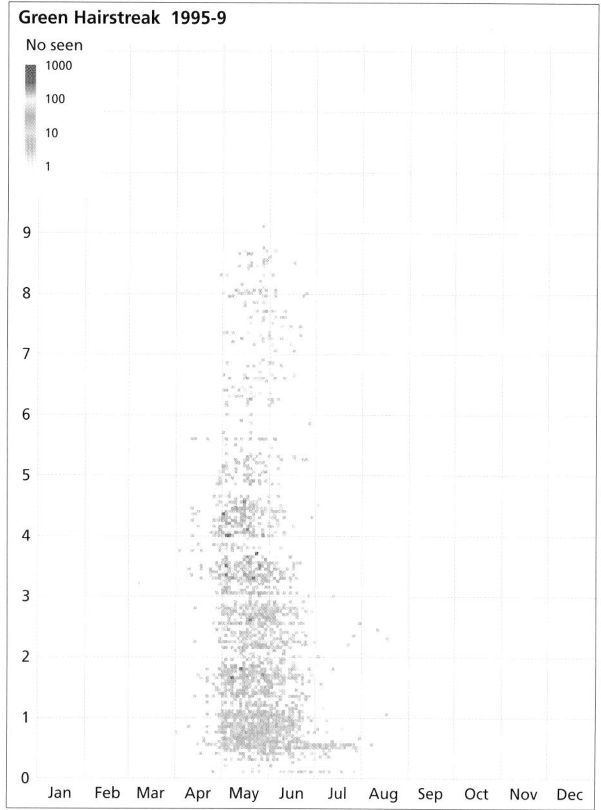
Green Hairstreak 1995-9
No seen
1000
100
10
1
9
8
7
6
5
4
3
2
1
0
Jan Feb Mar Apr May Jun Jul Aug Sep Oct Nov Dec

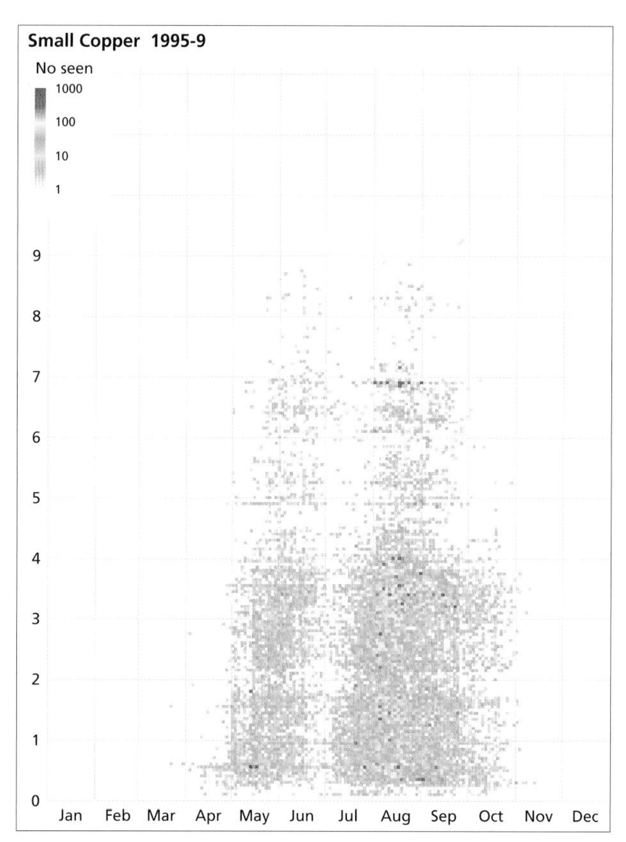
Small Copper 1995-9
No seen
1000
100
10
1
9
8
7
6
5
4
3
2
1
0
Jan Feb Mar Apr May Jun Jul Aug Sep Oct Nov Dec

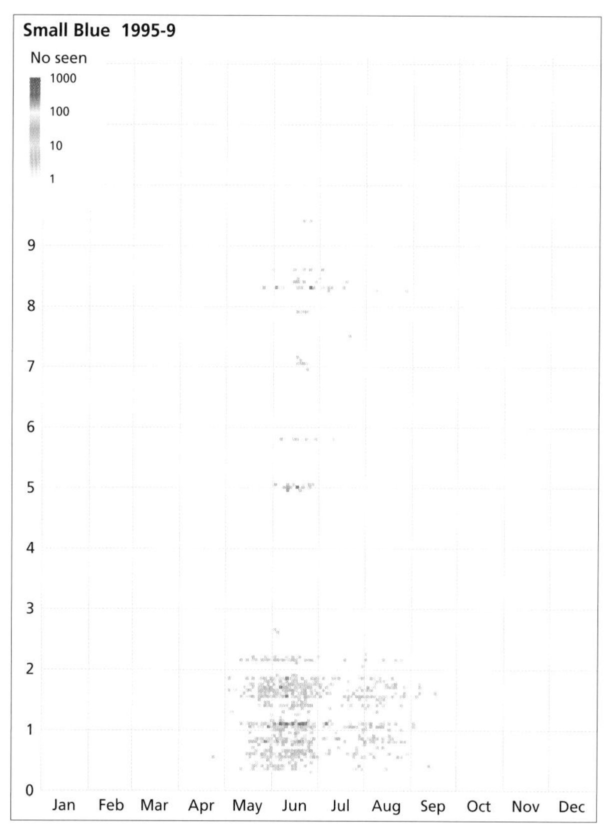
Small Blue 1995-9
No seen
1000
100
10
1
9
8
7
6
5
4
3
2
1
0
Jan Feb Mar Apr May Jun Jul Aug Sep Oct Nov Dec

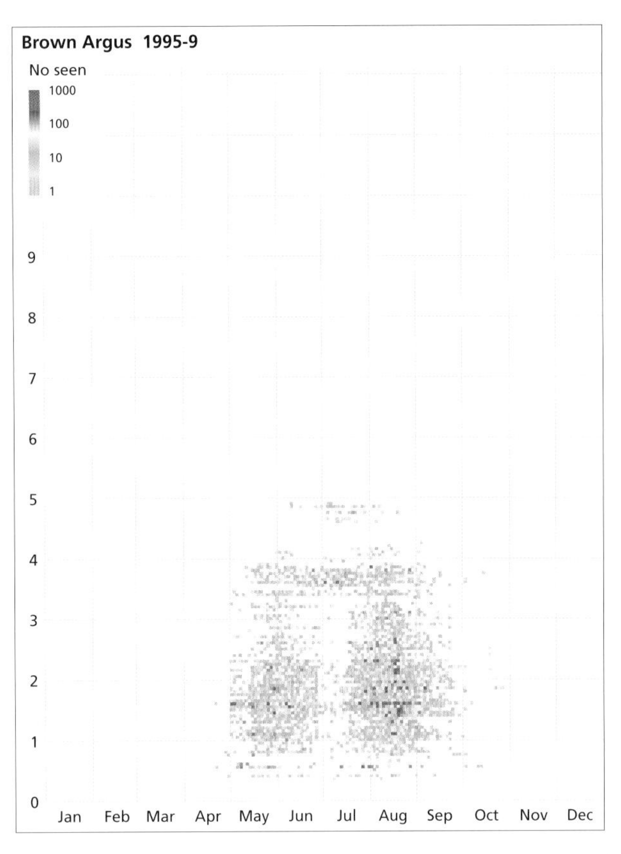
Brown Argus 1995-9
No seen
1000
100
10
1
9
8
7
6
5
4
3
2
1
0
Jan Feb Mar Apr May Jun Jul Aug Sep Oct Nov Dec

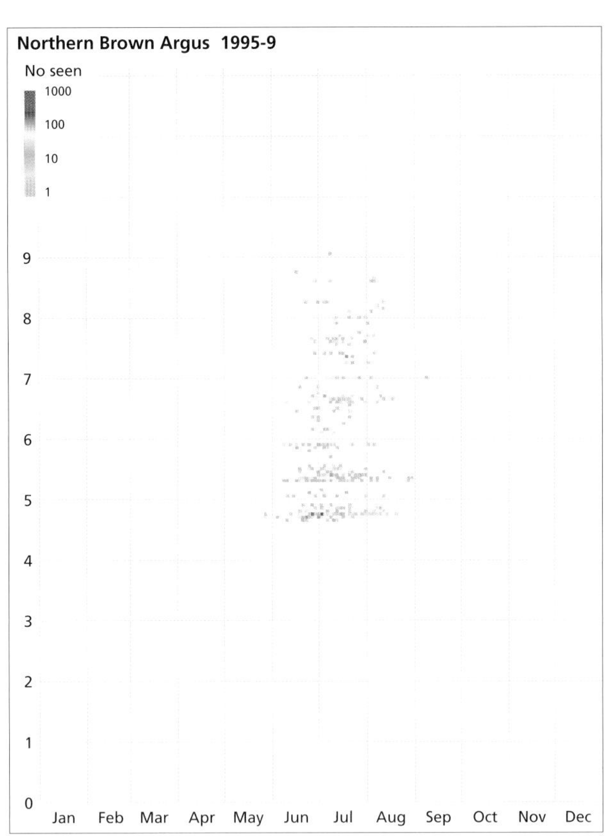
Northern Brown Argus 1995-9
No seen
1000
100
10
1
9
8
7
6
5
4
3
2
1
0
Jan Feb Mar Apr May Jun Jul Aug Sep Oct Nov Dec

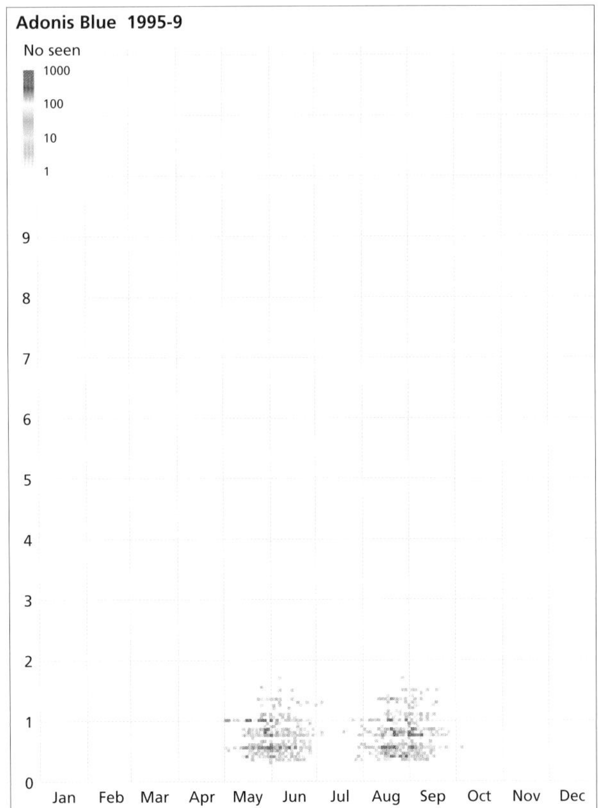
Adonis Blue 1995-9
No seen
1000
100
10
1
9
8
7
6
5
4
3
2
1
0
Jan Feb Mar Apr May Jun Jul Aug Sep Oct Nov Dec

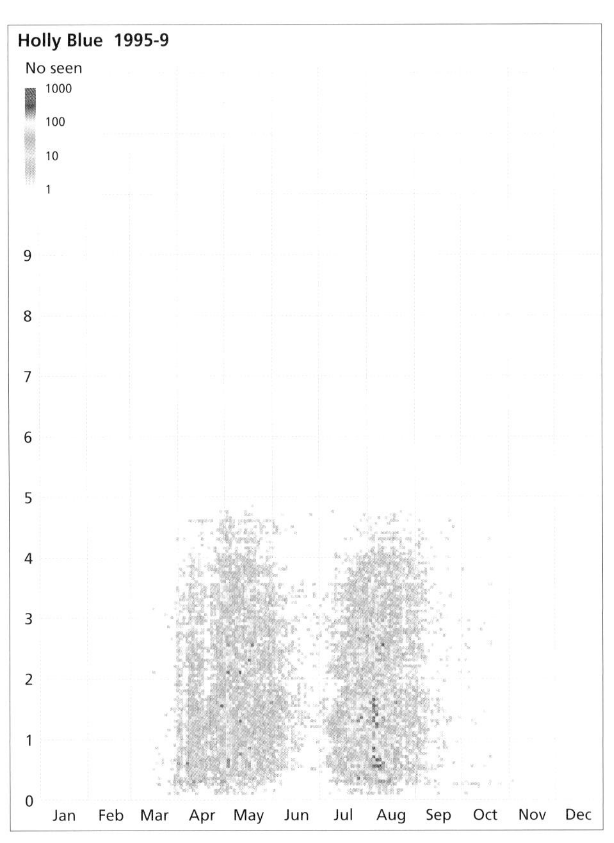
Holly Blue 1995-9
No seen
1000
100
10
1
9
8
7
6
5
4
3
2
1
0
Jan Feb Mar Apr May Jun Jul Aug Sep Oct Nov Dec

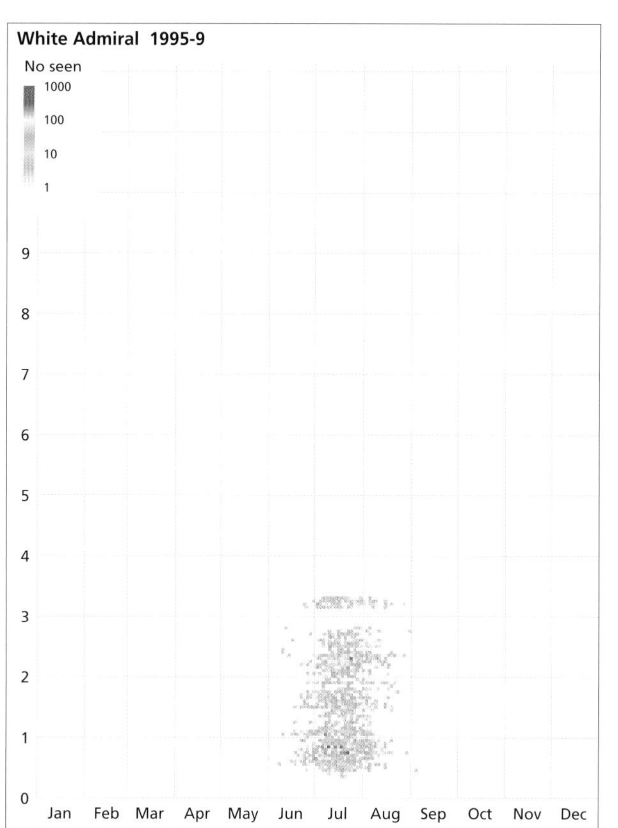
White Admiral 1995-9
No seen
1000
100
10
1
9
8
7
6
5
4
3
2
1
0
Jan Feb Mar Apr May Jun Jul Aug Sep Oct Nov Dec

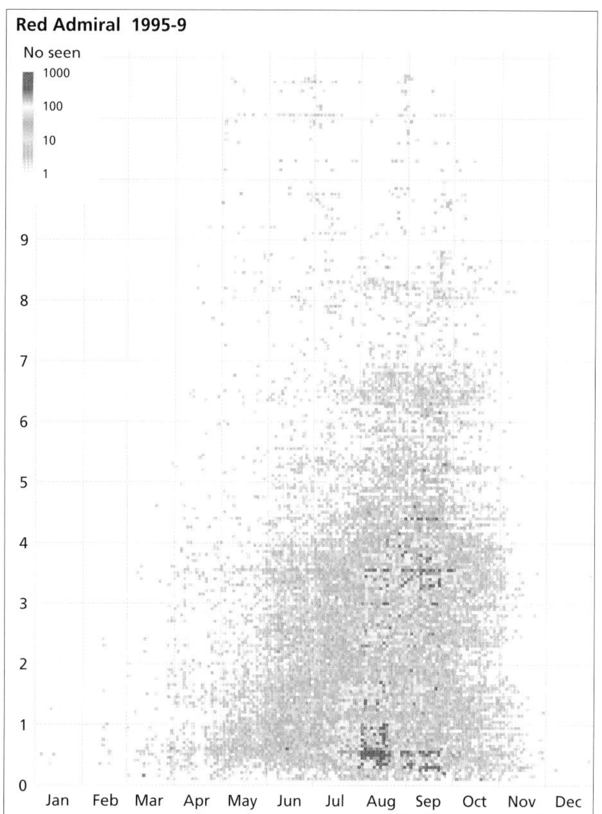
Red Admiral 1995-9
No seen
1000
100
10
1
9
8
7
6
5
4
3
2
1
0
Jan Feb Mar Apr May Jun Jul Aug Sep Oct Nov Dec

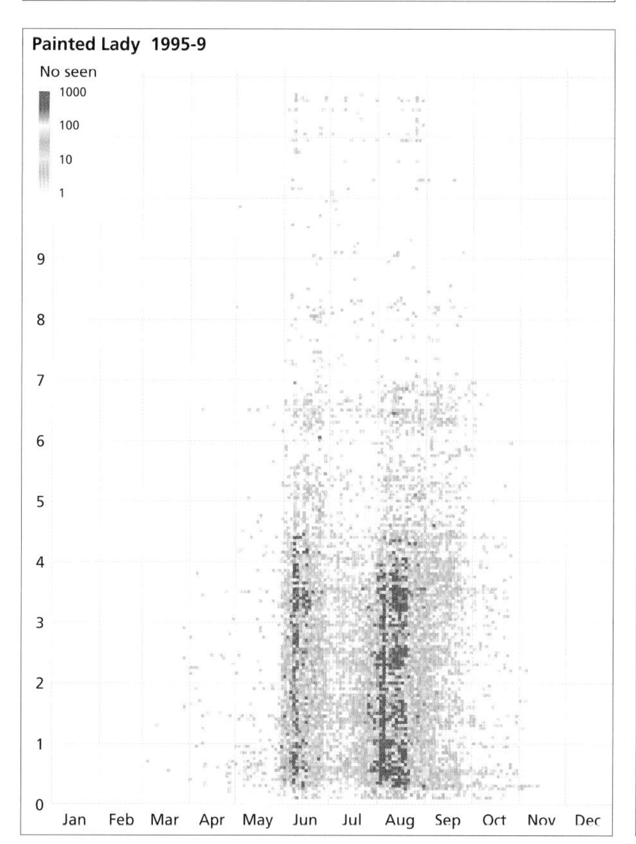
Painted Lady 1995-9
No seen
1000
100
10
1
9
8
7
6
5
4
3
2
1
0
Jan Feb Mar Apr May Jun Jul Aug Sep Oct Nov Dec

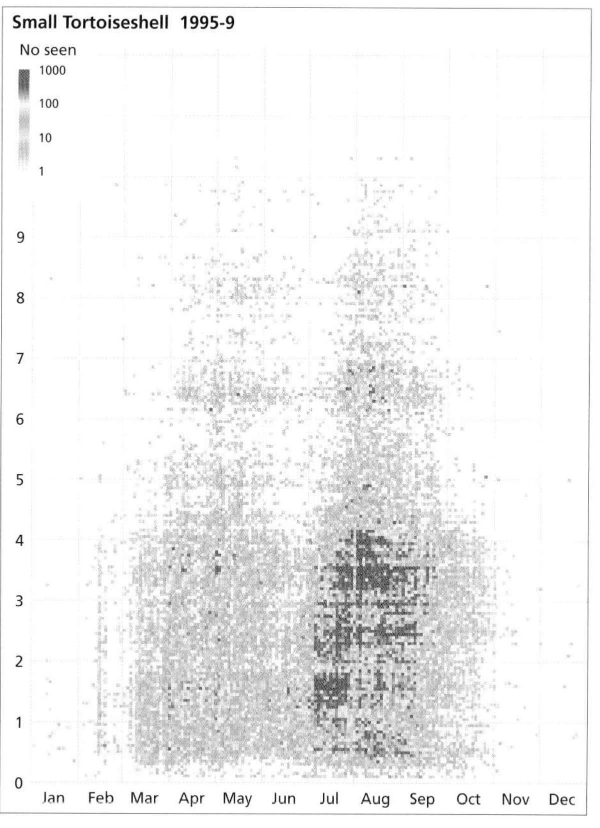
Small Tortoiseshell 1995-9
No seen
1000
100
10
1
9
8
7
6
5
4
3
2
1
0
Jan Feb Mar Apr May Jun Jul Aug Sep Oct Nov Dec

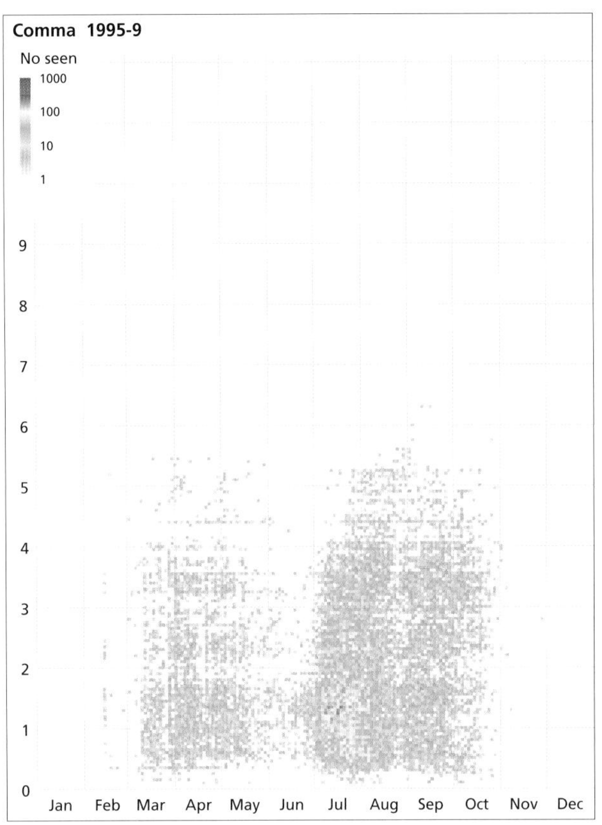
Comma 1995-9
No seen
1000
100
10
1
9
8
7
6
5
4
3
2
1
0
Jan
Feb
Mar
Apr
May
Jun
Jul
Aug
Sep
Oct
Nov
Dec

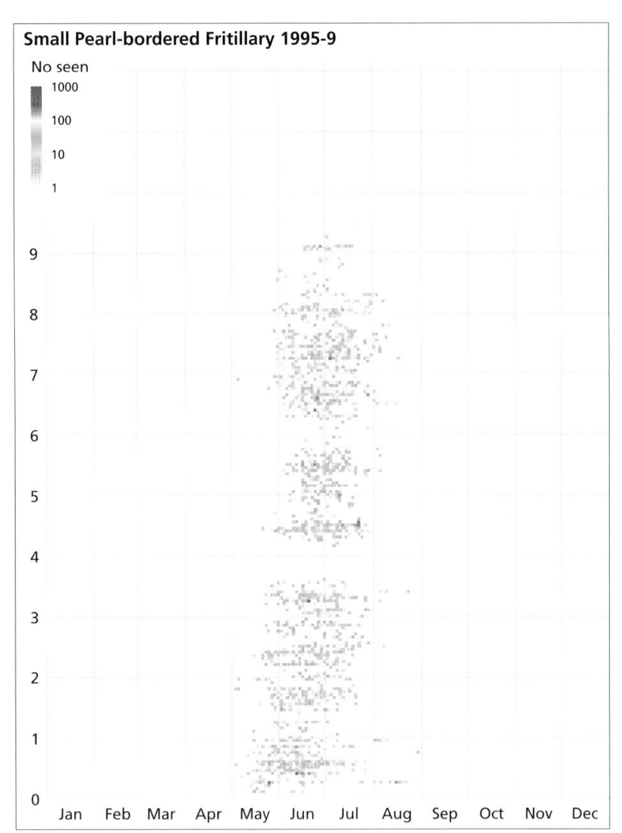
Small Pearl-bordered Fritillary 1995-9
No seen
1000
100
10
1
9
8
7
6
5
4
3
2
1
0
Jan
Feb
Mar
Apr
May
Jun
Jul
Aug
Sep
Oct
Nov
Dec

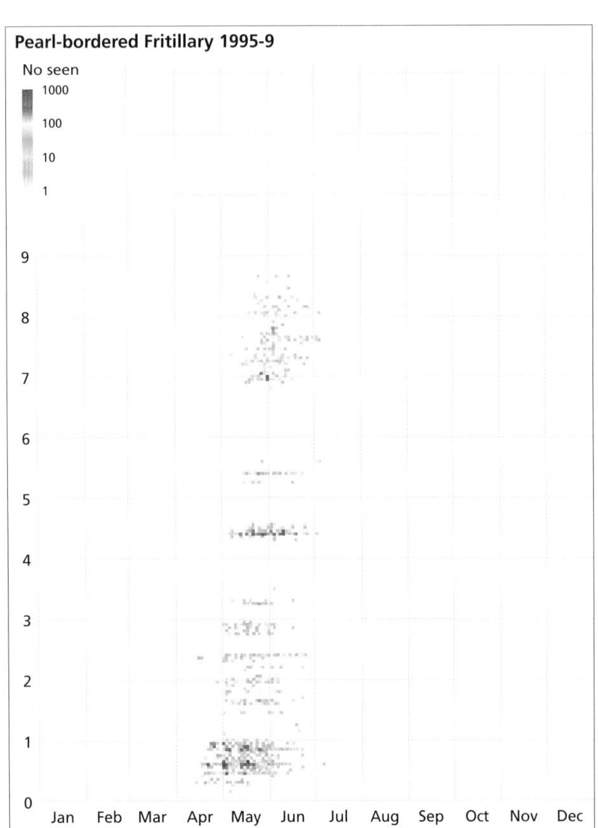
Pearl-bordered Fritillary 1995-9
No seen
1000
100
10
1
9
8
7
6
5
4
3
2
1
0
Jan
Feb
Mar
Apr
May
Jun
Jul
Aug
Sep
Oct
Nov
Dec

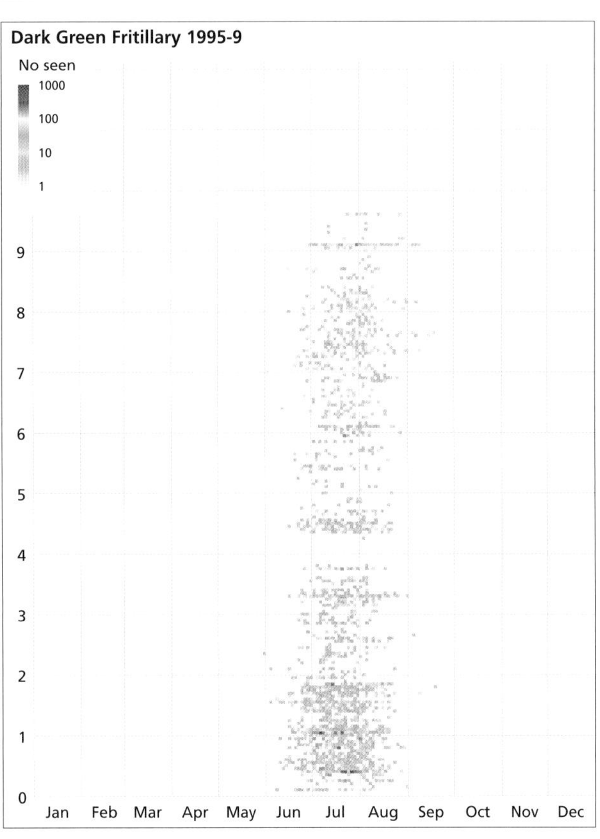
Dark Green Fritillary 1995-9
No seen
1000
100
10
1
9
8
7
6
5
4
3
2
1
0
Jan
Feb
Mar
Apr
May
Jun
Jul
Aug
Sep
Oct
Nov
Dec

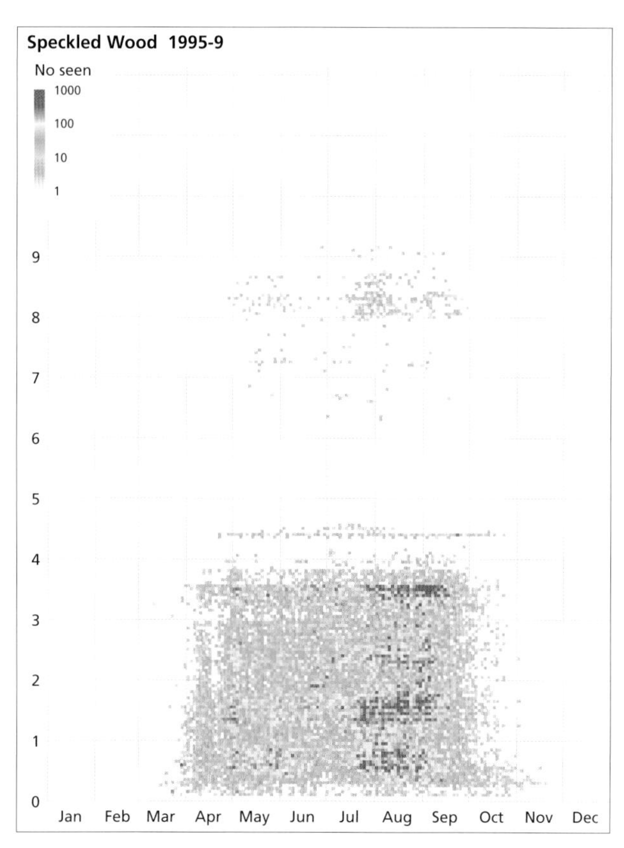

Speckled Wood 1995-9
No seen
1000
100
10
1
9
8
7
6
5
4
3
2
1
0
Jan
Feb
Mar
Apr
May
Jun
Jul
Aug
Sep
Oct
Nov
Dec

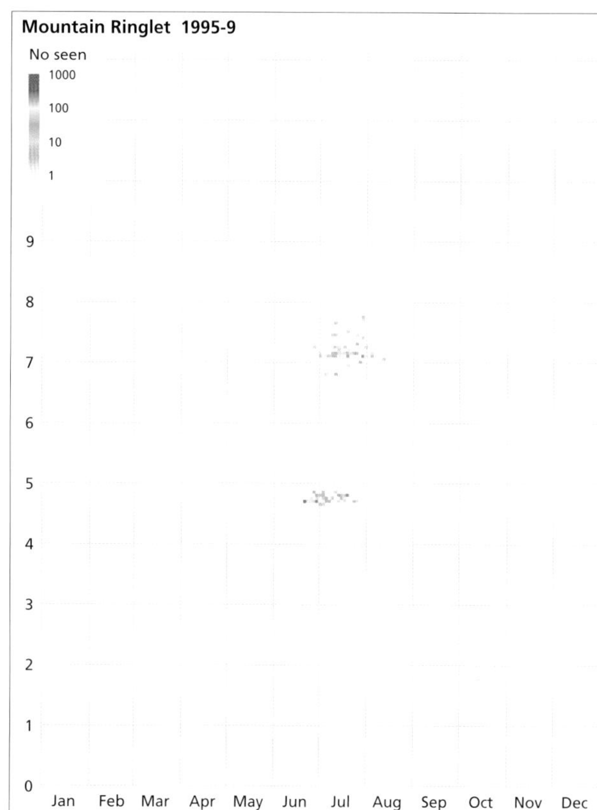

Mountain Ringlet 1995-9
No seen
1000
100
10
1
9
8
7
6
5
4
3
2
1
0
Jan
Feb
Mar
Apr
May
Jun
Jul
Aug
Sep
Oct
Nov
Dec

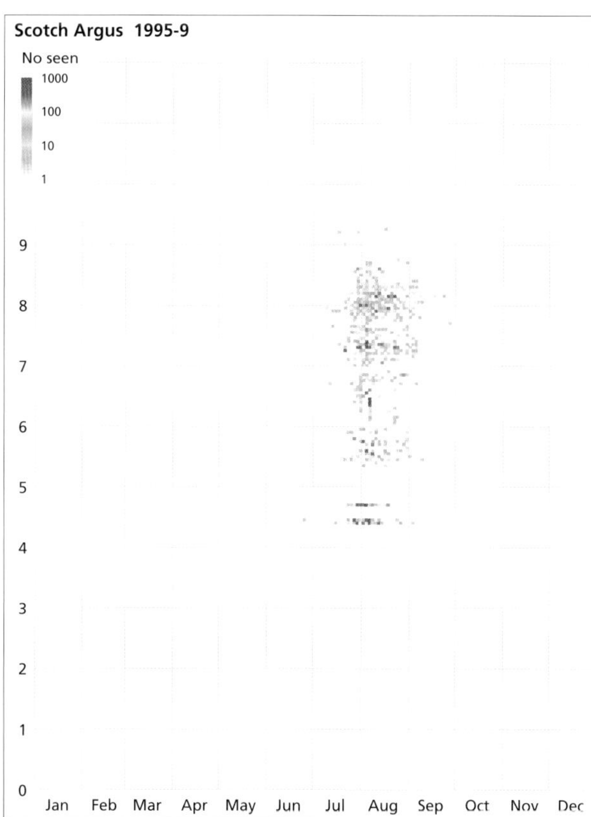

Scotch Argus 1995-9
No seen
1000
100
10
1
9
8
7
6
5
4
3
2
1
0
Jan
Feb
Mar
Apr
May
Jun
Jul
Aug
Sep
Oct
Nov
Dec

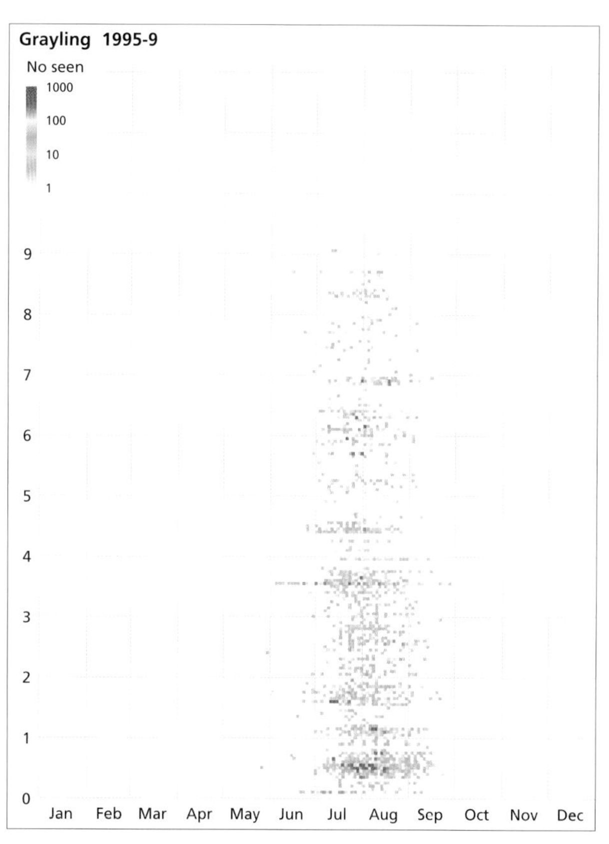

Grayling 1995-9
No seen
1000
100
10
1
9
8
7
6
5
4
3
2
1
0
Jan
Feb
Mar
Apr
May
Jun
Jul
Aug
Sep
Oct
Nov
Dec

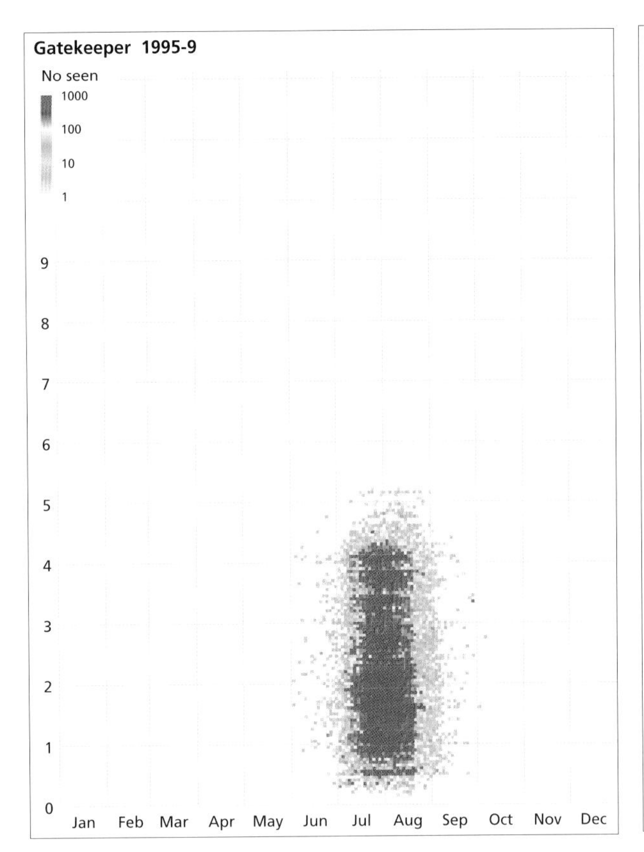
Gatekeeper 1995-9
No seen
1000
100
10
1
9
8
7
6
5
4
3
2
1
0
Jan Feb Mar Apr May Jun Jul Aug Sep Oct Nov Dec

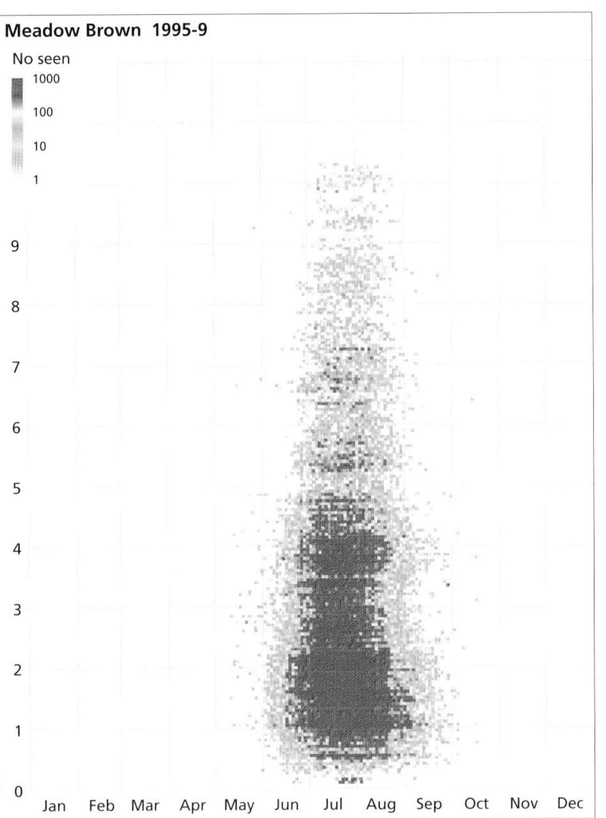
Meadow Brown 1995-9
No seen
1000
100
10
1
9
8
7
6
5
4
3
2
1
0
Jan Feb Mar Apr May Jun Jul Aug Sep Oct Nov Dec

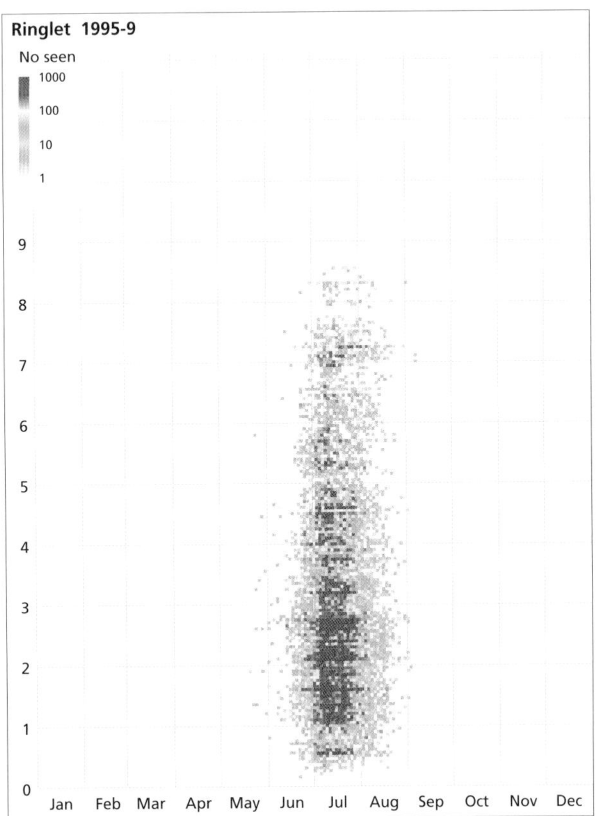
Ringlet 1995-9
No seen
1000
100
10
1
9
8
7
6
5
4
3
2
1
0
Jan Feb Mar Apr May Jun Jul Aug Sep Oct Nov Dec

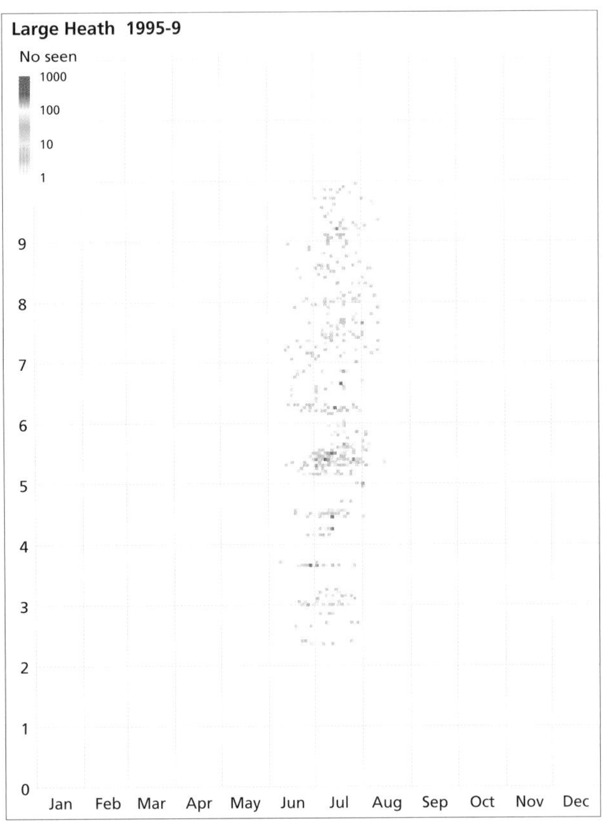
Large Heath 1995-9
No seen
1000
100
10
1
9
8
7
6
5
4
3
2
1
0
Jan Feb Mar Apr May Jun Jul Aug Sep Oct Nov Dec

Appendix 9

Vice-counties of Britain and Ireland

ENGLAND AND WALES

1. West Cornwall (with Scilly)
2. East Cornwall
3. South Devon
4. North Devon
5. South Somerset
6. North Somerset
7. North Wiltshire
8. South Wiltshire
9. Dorset
10. Isle of Wight
11. South Hampshire
12. North Hamsphire
13. West Sussex
14. East Sussex
15. East Kent
16. West Kent
17. Surrey
18. South Essex
19. North Essex
20. Hertfordshire
21. Middlesex
22. Berkshire
23. Oxfordshire
24. Buckinghamshire
25. East Suffolk
26. West Suffolk
27. East Norfolk
28. West Norfolk
29. Cambridgeshire
30. Bedfordshire
31. Huntingdonshire
32. Northamptonshire
33. East Gloucestershire
34. West Gloucestershire
35. Monmouthshire
36. Herefordshire
37. Worcestershire
38. Warwickshire
39. Staffordshire
40. Shropshire (Salop)
41. Glamorgan
42. Breconshire
43. Radnorshire
44. Carmarthenshire
45. Pembrokeshire
46. Cardiganshire
47. Montogomeryshire
48. Merionethshire
49. Caernarvonshire
50. Denbyshire
51. Flintshire
52. Anglesey
53. South Lincolnshire
54. North Lincolnshire
55. Leicestershire (with Rutland)
56. Nottinghamshire
57. Derbyshire
58. Cheshire
59. South Lancashire
60. West Lancashire
61. South-east Yorkshire
62. North-east Yorkshire
63. South-west Yorkshire
64. Mid-west Yorkshire
65. North-west Yorkshire
66. Durham
67. South Northumberland
68. North Northumberland (Cheviot)
69. Westmorland with North Lancashire
70. Cumberland
71. Isle of Man

113. Channel Islands

SCOTLAND

72. Dumfrieshire
73. Kirkcudbrightshire
74. Wigtownshire
75. Ayrshire
76. Renfrewshire
77. Lanarkshire
78. Peebleshire
79. Selkirkshire
80. Roxburghshire
81. Berwickshire
82. East Lothian (Haddington)
83. Midlothian (Edinburgh)
84. West Lothian (Linlithgow)
85. Fifeshire (with Kinross)
86. Stirlingshire
87. West Perthshire (with Clackmannan)
88. Mid Perthshire
89. East Perthshire
90. Angus (Forfar)
91. Kincardineshire
92. South Aberdeenshire
93. North Aberdeenshire
94. Banffshire
95. Moray (Elgin)
96. East Inverness-shire (with Nairn)
97. West Inverness-shire
98. Argyll Main
99. Dunbartonshire
100. Clyde Isles
101. Kintyre
102. South Ebudes
103. Mid Ebudes
104. North Ebudes
105. West Ross
106. East Ross
107. East Sutherland
108. West Sutherland
109. Caithness
110. Outer Hebrides
111. Orkney Islands
112. Shetland Islands (Zetland)

IRELAND

1. South Kerry
2. North Kerry
3. West Cork
4. Mid Cork
5. East Cork
6. Waterford
7. South Tipperary
8. Limerick
9. Clare
10. North Tipperary
11. Kilkenny
12. Wexford
13. Carlow
14. Laois
15. South-east Galway
16. West Galway
17. North-east Galway
18. Offaly
19. Kildare
20. Wicklow
21. Dublin
22. Meath
23. Westmeath
24. Longford
25. Roscommon
26. East Mayo
27. West Mayo
28. Sligo
29. Leitrim
30. Cavan
31. Louth
32. Monaghan
33. Fermanagh
34. East Donegal
35. West Donegal
36. Tyrone
37. Armagh
38. Down
39. Antrim
40. Londonderry

References

Alcamo, J. and Kreilemann, E. (1996). Emission scenarios and global climate prediction. *Global Environment Change*, **6**, 305–34.

Anon (1999). *Agenda 2000: implementing the CAP reforms 1999*. Working papers from the staff of the wildlife and countryside agencies in England and Wales. Countryside Agency, English Nature, Countryside Council for Wales.

Asher, J. (1992). *A programme for the co-ordination of butterfly recording in Britain and Ireland*. Unpublished report to Butterfly Conservation and Biological Records Centre.

Asher, J. (1995). Butterflies for the New Millennium. *Butterfly Conservation News*, **59**, 15.

Aspinall, R. J. (1987). The geography of the Small Copper butterfly *Lyceana phlaeas* (L.) in Northumberland. *Entomologist's Gazette*, **38**, 159–65.

Baker, M. (1992). The butterflies of Jersey—monitoring and rare species conservation (with reference to the Large Chequered Skipper *Heteropterus morpheus*). Unpublished M.Sc. thesis. University College London.

Baker, R. R. (1970). Bird predation as a selective pressure on the immature stages of the cabbage butterflies *Pieris rapae* and *P. brassicae*. *Journal of Zoology*, **162**, 43–9.

Baker, R. R. (1972*a*) The geographical origin of the British spring individuals of the butterflies *Vanessa atalanta* (L.) and *Cynthia cardui* (L.). *Journal of Entomology A*, **46,** 185–96.

Baker, R. R. (1972*b*). Territorial behaviour of the nymphalid butterflies, *Aglais urticae* and *Inachis io*. *Journal of Animal Ecology*, **41**, 453–69.

Baker, R. R. (1978). *The evolutionary ecology of animal migration*. Hodder and Stoughton, London.

Baldock, D. and Mitchell, K. (1998). *Farming with wildlife: implementation of the agri-environmental regulation in the UK*. WWF-UK, Godalming.

Barbour, D. A. (1986). Why are there so few butterflies in Liverpool? An answer. *Antenna*, **10**, 72–5.

Barker, S., Warren, M., Williams, M., and Davis, J. (1998). *Hedgerows for hairstreaks: hedgerow and woodland management for the Brown Hairstreak butterfly*. Leaflet from Butterfly Conservation, Wareham.

Barnett, L. K. (1995). Action for butterflies. *Butterfly Conservation News*, **59**, 22–3.

Barnett, L. K. and Warren, M. S. (1995*a*). *Species action plan: Silver-spotted Skipper Hesperia comma*. Butterfly Conservation, Wareham.

Barnett, L. K. and Warren, M. S. (1995*b*). *Species action plan: the Swallowtail Papilio machaon*. Butterfly Conservation, Wareham.

Barnett, L. K. and Warren, M. S. (1995*c*). *Species action plan: Large Copper Lycaena dispar*. Butterfly Conservation, Wareham.

Barnett, L. K. and Warren, M. S. (1995*d*). *Species action plan: Small Pearl-bordered Fritillary Boloria selene*. Butterfly Conservation, Wareham.

Barnett, L.K. and Warren, M. S. (1995*e*). *Species action plan: Pearl-bordered Fritillary Boloria euphrosyne*. Butterfly Conservation, Wareham.

Barnett, L. K. and Warren, M. S. (1995*f*). *Species action plan: High Brown Fritillary Argynnis adippe*. Butterfly Conservation, Wareham.

Barnett, L. K. and Warren, M. S. (1995*g*). *Species action plan: Marsh Fritillary Eurodryas (Euphydryas) aurinia*. Butterfly Conservation, Wareham.

Barnett, L. K. and Warren, M. S. (1995*h*). *Species action plan: Heath Fritillary Mellicta athalia*. Butterfly Conservation, Wareham.

Barnett, L. K. and Warren, M. S. (1995*i*). *Species action plan: Large Blue Maculinea arion*. Butterfly Conservation, Wareham.

Barr, C. J., Bunce, R. G. H., Clarke, R. T., Fuller, R. M., Furse, M. T., Gillespie, M. K., Groom, G. B., Hallam, C. J., Hornung, M., Howard, D. C., and Ness, M. J. (1993). *Countryside Survey 1990: main report, Volume 2*. HMSO, London.

Beaumont, S. (1995). The ecology of larvae of the Scotch Argus butterfly *Erebia ethiops* at

Arnside Knott SSSI, Cumbria. Unpublished M.Sc. thesis. University of London (Wye College).

Bink, F. A. and Siepel, H. (1996). Nitrogen and phosphorus in *Molinia caerulea (Gramineae)* and its impact on the larval development in the butterfly species *Lasiommata megera (Lepidoptera: Satyridae)*. *Entomologia Generalis*, **20**, 271–80.

Blackstock, T. H., Rimes, C. A., Stevens, D. P., Jefferson, R. G., Robertson, H. J., Mackintosh, J., and Hopkins, J. J. (1999). The extent of semi-natural grassland communities in lowland England and Wales: a review of conservation surveys 1978–96. *Grass and Forage Science*, **54**, 1–18.

Bobbink, R. and Willems, J. H. (1987). Increasing dominance of *Brachypodium pinnatum* (L.) Beauv. in chalk grasslands: a threat to a species-rich ecosystem. *Biological Conservation*, **40**, 301–14.

Bourn, N. A. D. and Thomas, J. A. (1993). The ecology and conservation of the Brown Argus butterfly *Aricia agestis* in Britain. *Biological Conservation*, **63**, 67–74.

Bourn, N. A. D. and Warren, M. S. (1996). *The impact of land enhancement schemes on the Marsh Fritillary, Eurodryas aurinia: a preliminary review in England*. Butterfly Conservation, Wareham.

Bourn, N. A. D. and Warren, M. S. (1997*a*). *Species action plan: Lulworth Skipper Thymelicus acteon*. Butterfly Conservation, Wareham.

Bourn, N. A. D. and Warren, M. S. (1997*b*). *Species action plan: Glanville Fritillary Melitaea cinxia*. Butterfly Conservation, Wareham.

Bourn, N. A. D. and Warren, M. S. (1997*c*). *Species action plan: Large Heath Coenonympha tullia*. Butterfly Conservation, Wareham.

Bourn, N. A. D. and Warren, M. S. (1997*d*) The road from Rio: conservation action for the UKs priority butterflies. *Butterfly Conservation News*, **65**, 7–9.

Bourn, N. A. D. and Warren, M. S. (1998*a*). *Species action plan: Brown Hairstreak Thecla betulae*. Butterfly Conservation, Wareham.

Bourn, N. A. D. and Warren, M. S. (1998*b*). *Species action plan: Black Hairstreak Satyrium pruni*. Butterfly Conservation, Wareham.

Bourn, N. A. D. and Warren, M. S. (1998*c*). *Species action plan: Adonis Blue Lysandra bellargus*. Butterfly Conservation, Wareham.

Bourn, N. A. D. and Warren, M. S. (1998*d*). *Species action plan: Duke of Burgundy Hamearis lucina*. Butterfly Conservation, Wareham.

Bourn, N. A. D. and Warren, M. S. (2000*a*). *Species action plan: Small Blue Cupido minimus*. Butterfly Conservation, Wareham.

Bourn, N. A. D. and Warren, M. S. (2000*b*). *Species action plan: Purple Emperor Apatura iris*. Butterfly Conservation, Wareham.

Bourn, N. A. D., Warren, M. S., and Kirkland, P. (1996). *Action for butterflies. Butterfly Conservation's guidelines for producing regional action plans*. Butterfly Conservation, Wareham.

Bourn, N. A. D., Whitfield, K. E. J., Pearman, G. S., and Roberts, E. (1999). *Site dossier and status of the Adonis Blue Polyommatus (Lysandra) bellargus in Dorset between 1997–1999*. Butterfly Conservation, Wareham.

Bourn, N. A. D., Jeffcoate, G. E., and Warren, M. S. (2000*a*). *Species action plan: Dingy Skipper Erynnis tages*. Butterfly Conservation, Wareham.

Bourn, N. A. D., Pearman, G. S., Goodger, B., Thomas, J. A., and Warren, M. S. (2000*b*). Changes in the status of two endangered butterflies over two decades and the influence of grazing management. In *Grazing Management: the principles and practice of grazing, for profit and environmental gain, within temperate grassland systems*, (ed. R. D. Rook), pp. 141–6. British Grassland Society Occasional Symposium No.34.

Boyd-Wallis, W. (1994). *Survey of Creag Meagaidh NNR for colonies of the Small Mountain Ringlet butterfly*. Unpublished report to Scottish Natural Heritage, Edinburgh.

Brakefield, P. M. (1982*a*). Ecological studies on the butterfly *Maniola jurtina* in Britain. I. Adult behaviour, microdistribution and dispersal. *Journal of Animal Ecology*, **51**, 713–26.

Brakefield, P. M. (1982*b*). Ecological studies on the butterfly *Maniola jurtina* in Britain. II. Population dynamics: the present position. *Journal of Animal Ecology*, **51**, 727–38.

Brakefield, P. M. (1987). Geographical variability in, and temperature effects on, the phenology of *Maniola jurtina* and *Pyronia tithonus* (*Lepidoptera, Satyrinae*) in England and Wales. *Ecological Entomology*, **12**, 139–48.

Brakefield, P. M. (1991). Genetics and the conservation of invertebrates. In *The scientific management of temperate communities for conservation*, (ed. I. F. Spellerberg, F. B. Goldsmith, and M. G. Morris), pp. 45–80. Blackwell Scientific Publications, Oxford.

Brasier, C. (1996). *Report on Forest Research*. Forestry Commission, Edinburgh.

Brereton, T. M. (1997). Ecology and conservation of the butterfly *Pyrgus malvae* (Grizzled Skipper) in south-east England. Unpublished Ph.D. thesis. University of East London.

Brereton, T. M. (1998). Pearl-bordered Fritillary national survey 1997. *Butterfly Conservation News*, **68**, 11–13.

Brereton, T. M. and Goodhand, A. (2000). Monitoring butterfly populations. *Butterfly Conservation News*, **74**, 6–9.

Brereton, T. M. and Warren, M. S. (1999). Ecology of the Pearl-bordered Fritillary butterfly in Scotland and possible threats from bracken eradication measures in woodland grant schemes. In *Bracken perceptions and bracken control in the British uplands*, (ed. J. Taylor), pp. 62–73. International Bracken Group, special publication No. 3.

Brereton, T. M., Bourn, N. A. D., and Warren, M. S. (1998*a*). *Species action plan: Grizzled Skipper Pyrgus malvae*. Butterfly Conservation, Wareham.

Brereton, T. M., Warren, M. S., and Roberts, E. (1998*b*). *Action for the Heath Fritillary: status, monitoring and conservation progress 1996 and 1997*. Butterfly Conservation (Confidential report), Wareham.

Brereton, T. M., Roberts, E., and Warren, M. S. (1999). *Action for the Pearl-bordered Fritillary: progress 1997/98; biodiversity action plan update*. Butterfly Conservation, Wareham.

Bretherton, R. F. (1951*a*). Our lost butterflies and moths. *Entomologist's Gazette*, **2**, 211–40.

Bretherton, R. F. (1951*b*). The early history of the Swallowtail butterfly (*Papilio machaon* L.) in England. *Entomologist's Record and Journal of Variation*, **63**, 206–11.

Bretherton, R. F. and Chalmers-Hunt, J. M. (1981). The immigration of Lepidoptera to the British Isles in 1980, with an account of the invasion of the Painted Lady *Cynthia cardui*. *Entomologist's Record and Journal of Variation*, **93**, 103–11.

Bretherton, R. F. and Chalmers-Hunt, J. M. (1985). The immigration of Lepidoptera to the British Isles in 1981, 1982, 1983: a supplementary note. *Entomologist's Record and Journal of Variation*, **97**, 76–84.

Brower, L. P. (1995). Understanding the migration of the Monarch butterfly (*Nymphalidae*) in North America: 1857–1995. *Journal of the Lepidopterists' Society*, **49**, 304–85.

Bryant, S. R., Thomas, C. D., and Bale, J. S. (1997). Nettle-feeding nymphalid butterflies: temperature, development and distribution. *Ecological Entomology*, **22**, 390–8.

Bryant, S. R., Bale, J. S., and Thomas, C. D. (1998) Modification of the triangle method of degree-day accumulation to allow for behavioural thermoregulation in insects. *Journal of Applied Ecology*, **35**, 921–7.

Bryant, S. R., Thomas, C. D., and Bale, J. S. (2000). Thermal ecology of gregarious and solitary nettle-feeding nymphalid butterfly larvae. *Oecologia*, **122**, 1–10.

Buckley, P. (1998). Legislation and protected areas for the conservation of biological diversity. In *Irish heritage and environment directory, 1999*, (ed. M. B. Deevy), pp. 13–20. Archaeology Ireland and The Heritage Council, Kilkenny.

Burghardt, F. and Fiedler, K. (1996). The influence of diet on growth and secretion behaviour of myrmecophilous *Polyommatus icarus* caterpillars (Lepidoptera: Lycaenidae). *Ecological Entomology*, **21**, 1–8.

Butterflies Under Threat Team (BUTT). (1986). *The management of chalk downland for butterflies*, Focus on Nature Conservation Series, No. 17. Nature Conservancy Council, Peterborough.

Butterfly Conservation. (1995). Lepidoptera restoration: Butterfly Conservation's policy, code of practice and guidelines for action. *Butterfly Conservation News*, **60**, 20–1.

Cabot, D. (1999). *Ireland: a natural history*. HarperCollins, London.

Chalmers-Hunt, J. M. (1977). The 1976 Invasion of the Camberwell Beauty (*Nymphalis antiopa* L.). *Entomologist's Record and Journal of Variation*, **89**, 89–105.

Chalmers-Hunt, J. M. (1989). *Local lists of Lepidoptera*. Hedera Press, Uffington.

Clark, G. C. and Dickson, C. G. C. (1971). *Life histories of the South African Lycaenid butterflies*. Purnell, Cape Town.

Clarke, H. (1993). *The butterflies of Jersey*, Discussion Paper in Conservation, No. 58. University College London.

Cleary, D. F. R., Aubert, J., Descimon, H., and Menken, S. B. J. (1995). Genetic differentiation and gene flow within and between *Colias alfacariensis* (Verity) and *Colias hyale* L. (Lepidoptera: Pieridae: Coliadinae). In *Proceedings of the section experimental and applied entomology of the Netherlands Entomological Society*, *Vol. 6*, (ed. M. J. Sommeijer and P. J. Francke), pp. 99–105. Netherlands Entomological Society, Amsterdam.

Clunas, A. (1986). The biology and habitat requirements of *Arica artaxerxes* Fabricius (Lepidoptera: Lycaenidae). Unpublished M.Sc. thesis. University of Edinburgh.

Collier, R. V. (1986). *The conservation of the Chequered Skipper in Britain*. Focus on Nature Conservation, No. 16. Nature Conservancy Council, Peterborough.

Coombes, S. and Tucker, V. (1996). The Monarch invasion of Great Britain 1995. *Butterfly Conservation News*, **63**, 30–5.

Corke, D. (1999). Are honeydew/sap-feeding butterflies (*Lepidoptera: Rhopalocera*) affected by particulate air-pollution? *Journal of Insect Conservation*, **3**, 5–14.

Corke, D. and Harding, P. T. (1990). A decade of county butterfly lists. *BBCS News*, **45**, 33–7.

Coulthard, N. (1982). An investigation of the habitat requirements and behavioural ecology of the Small Blue butterfly *Cupido minimus*, in relation to its distribution and abundance. Unpublished M.Sc. thesis. University of Aberdeen.

Council for the Protection of Rural England. (1999). *Hedging your bets: is hedgerow legislation gambling with our heritage?* CPRE, London.

Courtney, S. P. (1986). The ecology of Pierid butterflies; dynamics and interactions. *Advances in Ecological Research*, **15**, 51–131.

Courtney, S. P. and Duggan, A. E. (1983). The population biology of the Orange Tip butterfly *Anthocharis cardamines* in Britain. *Ecological Entomology*, **8**, 271–81.

Cowley, M. J. R., Thomas, C. D., Thomas, J. A., and Warren, M. S. (1999). Flight areas of British butterflies: assessing species status and decline. *Proceedings of the Royal Society of London B*, **266**, 1587–92.

Cribb, P. W. (1975). The long emergence period of the Meadow Brown. *Proceedings and Transactions of The British Entomological and Natural History Society*, **7**, 96.

Dandy, J. E. (1969). *Watsonian vice-counties of Great Britain*. The Ray Society, London.

Dannreuther, T. (1946). Records of the Bath White butterfly (*Pontia daplidice*) observed in the British Isles during 1945. *Journal of the Society for British Entomology*, **3**, 1–7.

Davey, P. (2000). Backtrack for the 1999 Monarch *Danaus plexippus* (Linn.) influx. *Atropos*, **9**, 17–18.

Davies, M. (1985). *White-letter Hairstreak project annual report 1984*. Butterfly Conservation West Midlands Branch.

Davies, M. (1986). *White-letter Hairstreak project annual report 1985*. Butterfly Conservation West Midlands Branch.

Davies, M. (1992). *The White-letter Hairstreak butterfly*. Butterfly Conservation, Colchester.

Davis, B. N. K., Lakhani, K. H., and Yates, T. J. (1991). The hazards of insecticides to butterflies of field margins. *Agriculture, Ecosystems and Environment*, **36**, 151–61.

Delmas, S. and Maechler, J. (1999). *Catalogue permanent de l'entomofaune. 2 Lepidoptera Rhopalocera*. Union de l'Entomologie Française, Dijon.

Dempster, J. P. (1967*a*). The control of *Pieris rapae* with DDT. 1. The natural mortality of the young stages of *Pieris*. *Journal of Applied Ecology*, **4**, 485–500.

Dempster, J. P. (1967*b*). The control of *Pieris rapae* with DDT. 2. Survival of the young stages of *Pieris* after spraying. *Journal of Applied Ecology*, **5**, 451–62.

Dempster, J. P. (1971). Some observations on a population of the Small Copper butterfly *Lycaena phlaeas* (L.) (Lep., Lycaenidae). *Entomologist's Gazette*, **22**, 199–204.

Dempster, J. P. (1992). Evidence of an oviposition-deterring pheromone in the Orange-tip butterfly, *Anthocharis cardamines* (L). *Ecological Entomology*, **17**, 83–5.

Dempster, J. P. (1995). The ecology and conservation of *Papilio machaon* in Britain. In *The ecology and conservation of butterflies*, (ed. A. S. Pullin), pp.137–49. Chapman & Hall, London.

Dempster, J. P. (1997). The role of larval food resources and adult movement in the population dynamics of the Orange-tip butterfly (*Anthocharis cardamines*). *Oecologia*, **111**, 549–56.

Dempster, J. P., King, M. L., and Lakhani, K. H. (1976). The status of the Swallowtail butterfly in Britain. *Ecological Entomology*, **1**, 71–84.

Dennis, R. L. H. (1984). Egg-laying sites of the Common Blue butterfly, *Polyommatus icarus* (Rottemburg) (Lepidoptera: Lycaenidae): the edge effect and beyond the edge. *Entomologist's Gazette*, **35**, 85–93.

Dennis, R. L. H. (1985*a*). Choice of egg-laying sites in the Green-veined White butterfly (*Artogeia napi*) (L.) (Lep. Pieridae). *Bulletin of the Amateur Entomologists' Society*, **44**, 199–219.

Dennis R. L. H. (1985*b*). *Polyommatus icarus* (Rottemburg) (Lepidoptera: Lycaenidae) on Brereton Heath in Cheshire; voltinism and switches in resource exploitation. *Entomologist's Gazette*, **36**, 175–9.

Dennis, R. L. H. (ed.) (1992*a*). *The ecology of butterflies in Britain.* Oxford University Press, Oxford.

Dennis, R. L. H. (1992*b*). Islands, regions, ranges and gradients. In *The ecology of butterflies in Britain*, (ed. R. L. H. Dennis), pp. 1–21. Oxford University Press, Oxford.

Dennis, R. L. H. (1993). *Butterflies and climate change.* Manchester University Press, Manchester.

Dennis, R. L. H. (2000). Progressive bias in species status is symptomatic of fine-grained mapping unit subject to repeated sampling. *Biodiversity and Conservation.*

Dennis, R. L. H. and Bardell, P. (1996). The impact of extreme weather events on local populations of *Hipparchia semele* (L.) (Nymphalidae) and *Plebejus argus* (L.) (Lycaenidae): hindsight, inferences and lost opportunities. *Entomologist's Gazette*, **47**, 211–25.

Dennis, R. L. H. and Bramley, M. J. (1985). The influence of man and climate on dispersion patterns within a population of adult *Lasiommata megera* (L.) (Satyridae) at Brereton Heath, Cheshire (U.K.). *Nota Lepidopterologica*, **8**, 309–24.

Dennis, R. L. H. and Eales, H. T. (1997). Patch occupancy in *Coenonympha tullia* (Muller, 1764) (Lepidoptera: Satyrinae): habitat quality matters as much as patch size and isolation. *Journal of Insect Conservation*, **1**, 167–76.

Dennis, R. L. H. and Eales, H. T. (1999). Probability of site occupancy in the large heath butterfly *Coenonympha tullia* determined from geographical and ecological data. *Biological Conservation*, **87**, 295–301.

Dennis, R. L. H. and Hardy, P. B. (1999). Targeting squares for survey: predicting species richness and incidence of species for a butterfly atlas. *Global Ecology and Biogeography Letters*, **8**, 443–54.

Dennis, R. L. H. and Thomas, C. D. (2000). Bias in butterfly distribution maps: the influence of hot spots and access. *Journal of Insect Conservation*, **4**, 73–7.

Dennis, R. L. H. and Williams, W. R. (1986). Butterfly 'diversity'—regressing and a little latitude. *Antenna*, **10**, 108–12.

Dennis, R. L. H. and Williams, W. R. (1987). Mate location behaviour of the Large Skipper butterfly *Ochlodes venata*: flexible strategies and spatial components. *Journal of the Lepidopterists' Society*, **41**, 45–64.

Dennis, R. L. H., Shreeve, T. G., and Sparks, T. H. (1998*a*). The effects of island area, isolation and source population size on the presence of the Grayling butterfly (*Hipparchia semele* (L.) (Lepidoptera: Satyrinae) on British and Irish offshore islands. *Biodiversity and Conservation*, **7**, 765–76.

Dennis, R. L. H., Sparks, T. H., and Shreeve, T. G. (1998*b*). Geographical factors influencing the probability of *Hipparchia semele* (L.) (Lepidoptera: Satyrinae) occurring on British and Irish off-shore islands. *Global Ecology and Biogeography Letters*, **7**, 205–14.

Dennis, R. L. H., Sparks, T. H., and Hardy, P. B. (1999). Bias in butterfly distribution maps: the effects of sampling effort. *Journal of Insect Conservation*, **3**, 33–42.

Department of Arts, Heritage, Gaeltacht and the Islands. (1998). *National report Ireland: first national report on the implementation of the Convention on Biological Diversity by Ireland*. Department of Arts, Heritage, Gaeltacht and the Islands, Dublin.

Department of the Environment. (1994). *Biodiversity, the UK action plan*. HMSO, London.

Department of the Environment. (1995). *Biodiversity: the UK steering group report*. HMSO, London.

Department of the Environment, Transport and the Regions. (1998). *Sites of Special Scientific Interest: better protection and management*. DETR, London.

Department of the Environment, Transport and the Regions. (1999*a*). *A better quality of life: a strategy for sustainable development in the United Kingdom*. DETR, London.

Department of the Environment, Transport and the Regions. (1999*b*). *Quality of life counts: indicators for a strategy for sustainable development for the United Kingdom: a baseline assessment*. DETR, London.

DNFC. (1986). *Recollections and reflections: 100 years of the Dublin Naturalists' Field Club*. Dublin Naturalists' Field Club, Dublin.

Dover, J. W. (1996). Factors affecting the distribution of butterflies on arable farmland. *Journal of Applied Ecology*, **33**, 723–34.

Dowdeswell, W. H. (1981). *The Life of the Meadow Brown*. Heinemann, London.

Dowdeswell, W. H., Fisher, R. A., and Ford, E. B. (1940). The quantitative study of populations in the Lepidoptera. 1. *Polyommatus icarus* Rott. *Annals of Eugenics*, **10**, 123–6.

Duffey, E. (1968). Ecological studies on the Large Copper butterfly *Lycaena dispar batavus* (Obth.) on Woodwalton Fen National Nature Reserve, Cambridgeshire, England, 1969–73. *Biological Conservation*, **12**, 143–58.

Duffey, E. (1977). The re-establishment of the Large Copper butterfly *Lycaena dispar batavus* (Obth.) on Woodwalton Fen National Nature Reserve, Huntingdon. *Journal of Animal Ecology*, **5**, 69–96.

Dunbar, D. (ed.) (1993). *Saving butterflies*. Butterfly Conservation, Dedham.

Dunn, T. C. and Parrack, J. D. (1986). *The moths and butterflies of Northumberland and Durham part 1: Macrolepidoptera*. Northern Naturalists' Union, Durham.

Dyck, H. van, Matthysen, E., and Dhondt, A. A. (1997). The effect of wing colour on male behavioural strategies in the Speckled Wood butterfly. *Animal Behaviour*, **53**, 39–51.

Eales, H. T. (1999*a*). Postponed emergence: a possible survival tactic in the Orange-tip butterfly *Anthocharis cardamines* (Lep.: Pieridae). *Entomologist's Record and Journal of Variation*, **111**, 241.

Eales, H. T. (1999*b*). The current status of the Large Heath butterfly *Coenonympha tullia* Muller (Lep.: Satyridae). *Entomologist's Record and Journal of Variation*, **111**, 5–9.

Ellis, H. A. (1994). The status of the Wall Brown butterfly, *Lasiommata megera*, in Northumberland, 1965–91, in relation to local weather. *Transactions of the Natural History Society of Northumbria*, **57**, 135–52.

Ellis, H. A. (1998).The current status and history of the Large Skipper *Ochlodes venata* (Bremer and Grey) in Northumberland. *The Vasculum*, **83**, 41–67.

Ellis, H. A. (1999*a*). The Small Skipper *Thymelicus sylvestris* Poda (Lep.: Hesperiidae) in north-east England: history and current status. *Entomologist's Record and Journal of Variation*, **111**, 223–5.

Ellis, H. A. (1999*b*). Return of the Comma *Polygonia c-album* L. (Lep.: Nymphalidae) to Northumberland: historical review and current status. *Entomologist's Record and Journal of Variation*, **111**, 227–31.

Ellis, S. (1995). Ecological studies of the butterflies of magnesian limestone grassland. Unpublished Ph.D. thesis. University of Sunderland.

Ellis, S. (1999*a*). Metapopulation structure and grazing management of the Durham argus butterfly *Aricia artaxerxes ssp. salmacis* in north-east England. Unpublished report to English Nature, Peterborough.

Ellis, S. (1999*b*). *North East Regional Action Plan*. Butterfly Conservation, Wareham.

Elmes, G. W. and Free, A. (1994). *Climate change and rare species in Britain*. HMSO, London.

Emmet, A. M. and Heath, J. (ed.) (1989). *The moths and butterflies of Great Britain and Ireland, 7, part 1, Hesperiidae-Nymphalidae, the butterflies*. Harley Books, Colchester.

Emmet, A. M. and Heath, J. (ed) (1990). *The moths and butterflies of Great Britain and Ireland, 7, part 1, Hesperiidae-Nymphalidae, the butterflies*. Harley Books, Colchester. (Paperback edition revised with minor corrections).

English Nature. (1993). *Roads and nature conservation: guidance on impacts, mitigation and enhancement*. English Nature, Peterborough.

English Nature. (1999*a*). House of Commons Agriculture Committee: inquiry into Agreement on Agenda 2000: CAP reform. Memorandum by English Nature, Peterborough.

English Nature. (1999*b*). *Natural Areas: helping to set the regional agenda for nature*. Series of eight reports. English Nature, Peterborough.

English Nature. (1999*c*). *Position statement on sustainable development*. English Nature, Peterborough.

Fearn, G. M. (1973). Biological flora of the British Isles: *Hippocrepis comosa* L. *Journal of Ecology*, **61**, 915–26.

Feber, R. E. and Smith, H. (1995) Butterfly conservation on arable farmland. In *The ecology and conservation of butterflies*, (ed. A. S. Pullin), pp. 84–97. Chapman & Hall, London.

Feber, R. E., Smith, H., and Macdonald, D. W. (1996). The effects on butterfly abundance of the management of uncropped edges of arable fields. *Journal of Applied Ecology*, **33**, 1191–205.

Feber, R. E., Smith, H., and Macdonald, D. W. (1999). The importance of spatially variable field margin management for two butterfly species. In *Aspects of applied biology 54, field margins and buffer zones: ecology, management and policy*, (ed. N. D. Boatman, B. N. K. Davies, K. Chaney, R. Feber, G. R. de Snoo, and T. H. Sparks), pp. 155–62. The Association of Applied Biologists, Wellesbourne.

Feber, R. E., Bourn, N., Brereton, T., and Warren, M. (2000). *Site dossier and status of the Heath Fritillary (Mellicta athalia) on Exmoor in 1999*. Butterfly Conservation, Wareham.

Feltwell, J. (1982). *Large White butterfly—the biology, biochemistry and physiology of Pieris brassicae (Linnaeus)*, Series Entomologica, **18**. W. Junk, The Hague.

Feltwell, J. (1995). *The conservation of butterflies in Britain past and present*. Wildlife Matters, Battle.

Fiedler, K., Krug, E., and Proksch, P. (1993). Complete elimination of hostplant quinolizidine alkaloids by larvae of a polyphagous lycaenid butterfly, *Callophrys rubi*. *Oecologia*, **94**, 441–5.

Findlay, R., Young, M. R., and Findlay, J. A. (1983). Orientation behaviour in the Grayling butterfly: thermoregulation or crypsis? *Ecological Entomology*, **8**, 145–53.

Firbank, L. G., Smart, S. M., van de Poll, H. M., Bunce, R. G. H., Hill, M. O., Howard, D. C., Watkins, J. W., and Stark, G. J. (1999). *Causes of change in British vegetation. ECOFACT Volume 3*. Institute of Terrestrial Ecology, Grange Over Sands.

Ford, E. B. (1945). *Butterflies*. Collins, London.

Forestry Authority. (1996). *Woodland Improvement Grant—Project 3: woodland biodiversity: coppice for butterflies*. Leaflet from Forestry Authority, Cambridge.

Fox, R. (ed.) (1997*a*). *Butterflies for the New Millennium proceedings of 3rd annual meeting and Welsh regional meeting*. Butterfly Conservation, Wareham.

Fox, R. (ed.) (1997*b*). *Butterflies for the New Millennium proceedings of Scottish regional meeting*. Butterfly Conservation, Wareham.

Fox, R. (1998*a*). Butterflies for the New Millennium—Ireland. *Butterfly Conservation News*, **69**, 3.

Fox, R. (1998*b*). Butterflies for the New Millennium—a new atlas of butterflies in Britain and Ireland. *British Wildlife*, **9**, 176–9.

Fox, R. (1999*a*). *Butterflies for the New Millennium proceedings of 4th national meeting*. Butterfly Conservation, Wareham.

Fox, R. (1999*b*). A million for the millennium. *Butterfly Conservation News*, **71**, 6–8.

Fox, R. (1999*c*). Butterflies for the New Millennium—a new atlas of butterflies in Britain and Ireland. *Sanctuary*, **28**, 10–11.
Fox, R. (1999*d*). Butterflies for the New Millennium. *British Birds*, **92**, 193.
Fox, R. (2000). *Butterflies for the New Millennium proceedings of 5th national meeting*. Butterfly Conservation, Wareham.
Fox, R. (in press). Butterflies and moths. In *Changing wildlife of Great Britain and Ireland*, (ed. D. L. Hawksworth). Taylor and Francis, London.
Fric, Z. and Konvička, M. (2000). Adult population structure and behaviour of two seasonal generations of the European Map Butterfly, *Araschnia levana*, species with seasonal polyphenism (Nymphalidae). *Nota Lepidopterologica*, **23**, 2-25.
Friends of the Earth. (1994). *Losing interest*. Friends of the Earth, London.
Fry, G. L. A. and Cooke, A. S. (1984). *Acid deposition and its implications for nature conservation in Britain*, Focus on Nature Conservation Series, No. 7. Nature Conservancy Council, Peterborough.
Fry, R. and Lonsdale, D. (1991). *Habitat conservation for insects—a neglected green issue*. Amateur Entomologists' Society, Middlesex.
Fuller, R. J. and Warren, M. S. (1991). Conservation management in ancient and modern woodlands: responses of fauna to edges and rotations. In *The scientific management of temperate communities for conservation*, (ed. I. F. Spellerburg, F. B. Goldsmith, and M. G. Morris), pp. 445–72. Blackwell Scientific, Oxford.
Fuller, R. J. and Warren, M. S. (1993). *Coppiced woodlands—their management for wildlife* (*second edition*). JNCC, Peterborough.
Fust, H. J. (1868). On the distribution of Lepidoptera in Great Britain and Ireland. *Transactions of the Royal Entomological Society of London*, **4**, 417–517.
Gaston, K. J. and Kunin, W. E. (1997). Rare-common differences: an overview. In *The biology of rarity: Causes and consequences of rare–common differences*, (ed. W. E. Kunin and K. J. Gaston), pp. 12–29. Chapman & Hall, London.
Geiger, H., Descimon, H., and Scholl, A. (1988). Evidence for speciation within nominal *Pontia daplidice* in southern Europe. *Nota Lepidopterologica*, **11**, 7–20.
Gibbons, D. W., Reid, J. B., and Chapman, R. A. (1993). *The new atlas of breeding birds in Britain and Ireland: 1988–1991*. T. & A. D. Poyser, London.
Gibbs, J., Brasier, C., and Webber, J. (1994). *Dutch Elm Disease in Britain*, Forestry Commission Research Information Note, No. 252. Forestry Commission, Edinburgh.
Gibson, C. W. D. (1995). *Creating chalk grasslands on former arable land: a review*. Bioscan, Oxford.
Gilbert, O. L. and Anderson, P. (1998). *Habitat creation and repair*. Oxford University Press, Oxford.
Gilpin, M. and Hanski, I. (ed.). (1991). *Metapopulation dynamics: empirical and theoretical investigations*. Academic Press, London.
Goulson, D., Ollerton, J., and Sluman, C. (1997*a*). Foraging strategies in the small skipper butterfly, *Thymelicus flavus*: when to switch ? *Animal Behaviour*, **53**, 1009–16.
Goulson, D., Stout, J. C., and Hawson, S. A. (1997*b*). Can flower constancy in nectaring butterflies be explained by Darwin's interference hypothesis ? *Oecologia*, **112**, 225–31.
Greatorex-Davies, J. N. (1997). The recent spread of the Brown Argus butterfly and some likely causes. In *Butterfly Monitoring Scheme report to recorders 1997*, (ed. J. N. Greatorex-Davies and E. Pollard), pp. 18–27. Institute of Terrestrial Ecology Monks Wood, Huntingdon.
Greatorex-Davies, J. N. and Roy, D. B. (2000). *Butterfly Monitoring Scheme report to recorders 1999*. Institute of Terrestrial Ecology Monks Wood, Huntingdon.
Gutiérrez, D. and Thomas, C. D. (2000). Marginal range expansion in a host-limited butterfly species *Gonepteryx rhamni*. *Ecological Entomology*, **25**, 165–70.
Gutiérrez, D., Thomas, C. D., and León-Cortés, J. L. (1999). Dispersal, distribution, patch-network and metapopulation dynamics of the Dingy Skipper butterfly (*Erynnis tages*). *Oecologia*, **121**, 506–17.
Hall, M. L. (1981). Butterfly Monitoring Scheme. *Instructions for independent recorders*. Institute of Terrestrial Ecology, Cambridge.

Hanski, I. (1998). Metapopulation dynamics. *Nature*, **396,** 41–9.

Hanski, I. (1999). *Metapopulation ecology*. Oxford University Press, Oxford.

Hanski, I. and Gilpin, M. (1991). Metapopulation dynamics: brief history and conceptual domain. *Biological Journal of the Linnean Society*, **42**, 3–16.

Hanski, I., Kussaari, M., and Nieminen, M. (1994). Metapopulation structure and migration in the butterfly *Melitaea cinxia*. *Ecology*, **75**, 747–62.

Harding, P. T. (ed.) (1995). *Butterflies for the New Millennium proceedings of a meeting held at ITE, Monks Wood*. Butterfly Conservation, Wareham.

Harding, P. T. (ed.) (1996). *Butterflies for the New Millennium 2nd annual meeting*. Butterfly Conservation, Wareham.

Harding, P. T. and Green, S. V. (1991). *Recent surveys and research on butterflies in Britain and Ireland: a species index and bibliography*. Biological Records Centre, Huntingdon.

Harding, P. T. and Greene, D. M. (1984). Butterflies in the British Isles: a new data base. *Annual report of the Institute of Terrestrial Ecology 1983*, 48–9. Institute of Terrestrial Ecology, Cambridge.

Harding, P. T., Asher, J., and Yates, T. J. (1995). Butterfly monitoring 1—recording the changes. In *Ecology and conservation of butterflies*, (ed. A. S. Pullin), pp. 3–22. Chapman & Hall, London.

Hardy, P. B. (1998). *Butterflies of Greater Manchester*. PGL Enterprises, Sale.

Hardy, P. B. and Dennis, R. L. H. (1999). The impact of urban development on butterflies within a city region. *Biodiversity and Conservation*, **8**, 1261–79.

Hardy, P. B., Hind, S. H., and Dennis, R. L. H. (1993). Range extension and distribution-infilling among selected butterfly species in north-west England: evidence for inter-habitat movements. *Entomologist's Gazette*, **44**, 247–55.

Harvey, G. (1997). *The killing of the countryside*. Jonathan Cape, London.

Heal, H. (1965). The Wood White *Leptidea sinapis* and the railways. *Irish Naturalists' Journal*, **15**, 8–13.

Heath, J. (1974). A century of change in the Lepidoptera. In *The changing flora and fauna of Britain*, (ed. D. L. Hawksworth), pp. 275–92. Academic Press, London.

Heath, J. (1981). *Threatened Rhopalocera (butterflies) in Europe*, Nature and Environment, No. 23. Council of Europe, Strasbourg.

Heath, J., Pollard, E., and Thomas, J. A. (1984). *Atlas of butterflies in Britain and Ireland*. Viking, Harmondsworth.

Heslop, I. R. P., Hyde, G. E., and Stockley, R. E. (1964). *Notes and views on the Purple Emperor*. Southern Publishing Company, Brighton.

Hickin, N. (1992). *The butterflies of Ireland. A field guide*. Roberts Rinehart, Schull.

Hill, J. K., Thomas, C. D., and Lewis, O. T. (1996). Effects of habitat patch size and isolation on dispersal by *Hesperia comma* butterflies: implications for metapopulation structure. *Journal of Animal Ecology*, **65**, 725–35.

Hill, J. K., Thomas, C. D., and Huntley, B. (1999). Climate and habitat availability determine 20th century changes in a butterfly's range margin. *Proceedings of the Royal Society of London B*, **266**, 1197–206.

Hill, J. K., Thomas, C. D., Fox, R., Moss, D., and Huntley, B. (in press). Analysis and modelling range changes in UK butterflies. In *Insect movement: mechanisms and consequences*, (ed. I. P. Woiwod, D. R. Reynolds, and C. D. Thomas). CAB International, Wallingford.

Hillis, J. P. (1975). New locations for the Purple Hairstreak, *Quercusia quercus* (L.) and Small Blue, *Cupido minimus*. *Irish Naturalists' Journal*, **18**, 193.

Hoare, A. J. (1999). An evaluation of downland turf recreation using invertebrates as indicators. Unpublished Ph.D. thesis. University of Bournemouth.

Hobson, R. (2000*a*). *Managing chalk grassland for the Marsh Fritillary butterfly*. Technical advice note from Butterfly Conservation, Wareham.

Hobson, R. (2000*b*). *Managing damp grassland for the Marsh Fritillary butterfly*. Technical advice note from Butterfly Conservation, Wareham.

Holloway, J. (1998). Geranium Bronze *Cacyreus marshalli*. *Atropos*, **4**, 3–6.

Holloway, J. (1999). Geranium Bronze *Cacyreus marshalli* update. *Atropos*, **6**, 73.

Holloway, J. D. (1980). A mass movement of *Quercusia quercus* (L.) (Lepidoptera: Lycaenidae) in 1976. *Entomologist's Gazette*, **31**, 150.

Holloway, S. (1996). *The historical atlas of breeding birds in Britain and Ireland: 1875–1900.* T. & A.D. Poyser, London.

Honey, M. R. (1993). *Cacyreus marshalli*, a recent addition to the European fauna and details of its spread. *Butterfly Conservation News*, **53**, 18–9.

Hoole, J. C., Joyce, D. A., and Pullin, A. S. (1999). Estimates of gene flow between populations of the Swallowtail butterfly, *Papilio machaon* in Broadland, U.K. and implications for conservation. *Biological Conservation*, **89**, 293–9.

Hulme, M. and Jenkins, G. (1998). *Climate change scenarios for the United Kingdom: summary report.* Climate Research Unit, University of East Anglia, Norwich.

IUCN. (1994). *IUCN red list categories.* IUCN, Gland.

Jarvis, F. V. L. (1959). Biological notes on *Aricia agestis* (D. & S.) in Britain. *Entomologist's Record and Journal of Variation*, **71**, 169–78.

Jarvis, F. V. L. (1966). The genus *Aricia* (Lep: Rhopalocera) in Britain. *Proceedings and transactions of the South London Entomological and Natural History Society*, **1966**, 37–60.

Jeffcoate, S. (1999). Genetically modified crops: Butterfly Conservation's position. *Butterfly Conservation News*, **72**, 22–3.

Joy, J. (1995). *Heathland management for the Silver-studded Blue.* Leaflet from English Nature, Peterborough.

Joy, J. (1997). *West Midlands Regional Action Plan.* Butterfly Conservation, Wareham.

Joy, J. (1998*a*). *Bracken for butterflies.* Leaflet from Butterfly Conservation, Wareham.

Joy, J. (1998*b*). (ed. N. A. D. Bourn). *National Action Plan for Wales.* Butterfly Conservation, Wareham.

Joy, J. and Pullin, A. S. (1997). The effects of flooding on the survival and behaviour of overwintering Large Heath butterfly *Coenonympha tullia* larvae. *Biological Conservation*, **82**, 61–6.

Joy, J. and Pullin, A. S. (1999). Field studies on flooding and survival of overwintering large heath butterfly *Coenonympha tullia* larvae on Fenn's and Whixall Mosses in Shropshire and Wrexham, UK. *Ecological Entomology*, **24**, 426–431.

Joy, J., Dennis, R. L. H., Miles, A., and Hinde, J. (1999). Atypical habitat choice by White Admiral butterflies, *Ladoga camilla* (L.) (Lepidoptera: Nymphalidae), at the edge of their range. *Entomologist's Gazette*, **50**, 169–79.

Karsholt, O. and Razowski, J. (ed.) (1996). *The Lepidoptera of Europe.* Apollo Books, Stenstrup.

Kirby, K. (1992). *Habitat management for invertebrates: a practical handbook.* RSPB, Sandy.

Kirkland, P. (1995). The beleaguered Marsh Fritillary. *Butterfly Conservation News*, **59**, 6–7.

Kirkland, P. (1996). *A review of the distribution, ecology and behaviour of the Scotch Argus (Erebia aethiops Esper, 1777).* Unpublished report to the British Ecological Society, London.

Kuussaari, M. (1998). The biology of the Glanville Fritillary butterfly (*Melitaea cinxia*). Unpublished Ph.D. thesis. University of Helsinki.

Kuussaari, M., Saccheri, I., Camara, M., and Hanski, I. (1998). Allee effect and population dynamics in the Glanville Fritillary butterfly. *Oikos*, **82**, 384–92.

Lavery, T. A. (1993). A review of the distribution, ecology and status of the Marsh Fritillary *Euphydryas aurinia* Rottemburg, 1775 (Lepidoptera: Nymphalidae) in Ireland. *Irish Naturalist's Journal*, **24**, 192–9.

Lees, E. (1980). A theory to account for the spread of the Orange Tip butterfly in northern England. *Entomologist's Record and Journal of Variation*, **92**, 122–3.

León-Cortés, J. L., Cowley, M. J. R., and Thomas, C. D. (1999). Detecting decline in a formerly widespread species: how common is the Common Blue butterfly *Polyommatus icarus*? *Ecography*, **22**, 643–50.

León-Cortés, J. L., Cowley, M. J. R., and Thomas, C. D. (2000). The distribution and decline of a widespread butterfly *Lycaena phlaeas* in a pastoral landscape. *Ecological Entomology*, **25**, 285–94.

Le Quesne, W. (1946). The butterflies of Jersey. *Entomologists Monthly Magazine*, **82**, 22–3.
Lewington, R. (1999). *How to identify butterflies*. HarperCollins, London.
Lewis, O. T., Thomas, C. D., Hill, J. K., Brookes, M. I., Crane, T. P. R., Graneau, Y. A., Mallet, J. L. B., and Rose, O. C. (1997). Three ways of assessing metapopulation structure in the butterfly *Plebejus argus*. *Ecological Entomology*, **22**, 283–93.
Lisney, A. A. (1960). *A bibliography of British Lepidoptera 1608–1799*. Chiswick Press, London.
Loertscher, M. (1991). Population biology of two satyrine butterflies, *Erebia meolans* and *Erebia aethiops* (Lepidoptera: Satyridae). *Nota Lepidopterologica*, suppl. **2**, 22–31.
Loertscher, M., Erhardt, A., and Zettel, J. (1997). Local movement patterns of three common grassland butterflies in a traditionally managed landscape. *Mitteilungen der Schweizerischen Entomologischen Gesellschaft*, **70**, 43–55.
Long, R. (1970). Rhopalocera of the Channel Islands. *Entomologist's Gazette*, **21**, 241–51.
Long, R. (1977). Entomological section report. *Annual Bulletin Société Jersiaise*, 1976.
Mackey, E. C., Shewry, M. C., and Tudor, G. J. (1998). *Land cover change: Scotland from the 1940s to the 1980s*. The Stationery Office, Edinburgh.
Maes, D. and Swaay, C. A. M. van. (1997). A new methodology for compiling national Red Lists applied to butterflies (Lepidoptera, Rhopalocera) in Flanders (N-Belgium) and the Netherlands. *Journal of Insect Conservation*, **1**, 113–24.
Marren, P. (1998). A short history of butterfly-collecting in Britain. *British Wildlife*, **9**, 362–70.
Marrs, R. H., Frost, A. J., and Plant, R. A. (1989). A preliminary report on the impact of herbicide drift on plant species of conservation interest. *Brighton Crop Protection Conference—Weeds 1989*, **2**, 795–802.
Martin, J. and Gurrea, P. (1988). Establishment of a population of *Danaus plexippus* (Linnaeus, 1758) (Lep.: Danaidae) in southwest Europe. *Entomologist's Record and Journal of Variation*, **100**, 163–8.
Mavi, G. S. (1992). A critical review on the distribution and host-range of Pea Blue butterfly, *Lampides boeticus* (Linn.). *Journal of Insect Science*, **5**, 115–19.
McGowan, G. (1997). The distribution of the Small Mountain Ringlet, *Erebia epiphron mnemon* (Haworth, 1812), in Scotland. *Entomologists' Gazette*, **48**, 135–45.
McKay, H. V. (1991). Egg-laying requirements of woodland butterflies; Brimstones (*Gonepteryx rhamni*) and Alder Buckthorn (*Frangula alnus*). *Journal of Applied Ecology*, **28**, 731–43.
Melling, P. M. (1987). The ecology and population structure of a butterly cline. Unpublished Ph.D. thesis. University of Newcastle.
Mendel, H. (1995). It was migration—the exceptional abundance of the Large White butterfly *Pieris brassicae* (L.) in 1992. *Entomologist's Record and Journal of Variation*, **107**, 293–4.
Mendel, H. and Piotrowski, S. H. (1986). *Butterflies of Suffolk—an atlas and history*. Suffolk Naturalists' Society, Ipswich.
Moriarty, F. (1969*a*). Butterflies and insecticides. *Entomologist's Record and Journal of Variation*, **81**, 276–8.
Moriarty, F. (1969*b*). The sublethal effects of synthetic insecticides on insects. *Biological Review*, **44**, 321–57.
Morton, A. C. G. (1985). The population biology of an insect with a restricted distribution: *Cupido minimus* Fuessly (Lepidoptera: Lycaenidae). Unpublished Ph.D. thesis. University of Southampton.
Moss, D. and Pollard, E. (1993). Calculation of collated indices of abundance of butterflies based on monitored sites. *Ecological Entomology*, **18**, 77–83.
Murray, J. B. and Souter, R. (1999). *Hertfordshire and Middlesex butterfly and moth report for 1998*. Butterfly Conservation Hertfordshire and Middlesex Branch.
Murray, J. B. and Souter, R. (2000). *Hertfordshire and Middlesex butterfly and moth report for 1999*. Butterfly Conservation Hertfordshire and Middlesex Branch.
Nature Conservancy Council. (1984). *Nature Conservation in Great Britain*. Nature Conservancy Council, Shrewsbury.

Nelson, B. (1998). *Northern Ireland Action Plan*. Northern Ireland Branch of Butterfly Conservation, Belfast.

Nelson, J. (1996). The Monarch *Danaus plexippus* influx into Britain and Ireland in October 1995. *Atropos*, **1**, 5–10.

New, T. R. (1997). *Butterfly conservation (second edition)*. Oxford University Press, Oxford.

Nylin, S. (1992). Seasonal plasticity in life history traits: growth and development in *Polygonia c-album*. *Biological Journal of the Linnean Society*, **47**, 301–23.

Nylin, S., Janz, N., and Wedell, N. (1996). Oviposition plant preferences and offspring performance in the Comma butterfly: correlations and conflicts. *Entomologia Experimentalis et Applicata*, **80**, 141–4.

Oates, M. R. (1986). *The ecology and conservation of the Duke of Burgundy Fritillary*. Unpublished report to the World Wide Fund for Nature, Godalming.

Oates, M. R. (1993). The management of southern limestone grasslands. *British Wildlife*, **5**, 73–82.

Oates, M. R. (2000). The Duke of Burgundy—conserving the intractable. *British Wildlife*, **11**, 250-7.

Oates, M. R. and Warren, M. S. (1990). *A review of butterfly introductions in Britain and Ireland*. World Wide Fund for Nature (WWF-UK), Godalming.

Pain, D. J. and Pienkowski, M. W. (ed.) (1997). *Farming and birds in Europe: the Common Agricultural Policy and its implications for bird conservation*. Academic Press, London.

Parmesan, C., Ryrholm, N., Stefanescu, C., Hill, J. K., Thomas, C. D., Descimon, H., Huntley, B., Kaila, L., Kullberg, J., Tammaru, T., Tennant, J., Thomas, J. A., and Warren, M. S. (1999). Poleward shifts in geographical ranges of butterfly species associated with regional warming. *Nature*, **399**, 579–83.

Pearman, G. S., Goodger, B., Bourn, N. A. D., and Warren, M. S. (1998). *The changing status of the Lulworth Skipper (Thymelicus acteon) and Adonis Blue (Lysandra bellargus) in south-east Dorset over two decades*. Butterfly Conservation, Wareham.

Perring, F. H. and Sell, P. D. (ed.) (1968). *Critical supplement to the atlas of the British flora*. Nelson, London.

Perring, F. H. and Walters, S. M. (ed.) (1962). *Atlas of the British flora*. Nelson, London.

Peterken, G. F. (1981). *Woodland conservation and management*. Chapman & Hall, London.

Peterken, G. F. and Allison, H. (1989). *Woods, trees and hedges: a review of changes in the British countryside*, Focus on Nature Conservation, No. 22. Nature Conservancy Council, Peterborough.

Pivnick, K. A. and McNeil, J. N. (1985). Mate location and mating behaviour of *Thymelicus lineola*. *Annals of the Entomological Society of America*, **78**, 651–6.

Pollard, E. (1979). Population ecology and change in range of the White Admiral *Ladoga camilla* in England. *Ecological Entomology*, **4**, 61–74.

Pollard, E. (1981). Aspects of the ecology of the Meadow Brown butterfly, *Maniola jurtina* (L.) (Lepidoptera: Satyridae). *Entomologist's Gazette*, **32**, 67–74.

Pollard, E. (1985). Larvae of *Celastrina argiolus* (L.) (Lepidoptera: Lycaenidae) on male Holly bushes. *Entomologist's Gazette*, **36**, 3.

Pollard, E. (1988). Temperature, rainfall and butterfly numbers. *Journal of Applied Ecology*, **25**, 819–28.

Pollard, E. (1991). Changes in the flight period of the Hedge Brown butterfly *Pyronia tithonus* during range expansion. *Journal of Animal Ecology*, **60**, 737–48.

Pollard, E. (1994). Was it a migration—the exceptional abundance of the large white butterfly *Pieris brassicae* (L.) in 1992. *The Entomologist*, **133**, 211–6.

Pollard, E. and Cooke, A. S. (1994). Impact of Muntjac deer *Muntiacus reevesi*, on egg-laying sites of the White Admiral *Ladoga camilla* in a Cambridgeshire wood. *Biological Conservation*, **70**, 189–91.

Pollard, E. and Eversham, B. C. (1995). Butterfly monitoring 2—interpreting the changes. In *Ecology and conservation of butterflies*, (ed. A. S. Pullin), pp. 23–36. Chapman & Hall, London.

Pollard, E. and Greatorex-Davies, J. N. (1997). Flight-periods of the Small Heath butterfly, *Coenonympha pamphilus* (Linnaeus) (Lepidoptera: Nymphalidae, Satyrinae) on chalk downs and in woodland in southern England. *Entomologist's Gazette*, **48**, 3–7.

Pollard, E. and Greatorex-Davies, J. N. (1998). Increased abundance of the Red Admiral butterfly *Vanessa atalanta* in Britain: the roles of immigration, overwintering and breeding within the country. *Ecology Letters*, **1**, 77-81.

Pollard, E. and Hall, M. L. (1990). Possible movement of *Gonepteryx rhamni* (L.) (Lepidoptera: Pieridae) between hibernating and breeding areas. *Entomologist's Gazette*, **31**, 217–20.

Pollard, E. and Moss, D. (1995). Historical records of the occurrence of butterflies in Britain: examples showing associations between annual number of records and weather. *Global Change Biology*, **1**, 107–13.

Pollard, E. and Yates, T. J. (1993*a*). *Monitoring butterflies for ecology and conservation*. Chapman & Hall, London.

Pollard, E. and Yates, T. J. (1993*b*). Population fluctuations of the Holly Blue butterfly, *Celastrina argiolus* (L.) (Lepidoptera: Lycaenidae). *Entomologist's Gazette*, **44**, 3–9.

Pollard, E., Hall, M. L., and Bibby, T. J. (1984). The Clouded Yellow butterfly migration of 1983. *Entomologist's Gazette*, **35**, 227–31.

Pollard, E., Hall, M. L., and Bibby, T. J. (1986). *Monitoring the abundance of butterflies 1976–85*. Nature Conservancy Council, Peterborough.

Pollard, E., Moss, D., and Yates, T. J. (1995). Population trends of common British butterflies at monitored sites. *Journal of Applied Ecology*, **32**, 9–16.

Pollard, E., Greatorex-Davies, J. N., and Thomas, J. A. (1997). Drought reduces breeding success of the butterfly *Aglais urticae*. *Ecological Entomology*, **22**, 315–8.

Pollard, E., Swaay, C. A. M. van, Stefanescu, C., Lundsten, K. E., Maes, D., and Greatorex-Davies, J. N. (1998*a*). Migration of the Painted Lady butterfly *Cynthia cardui* in Europe: evidence from monitoring. *Diversity and Distributions*, **4**, 243–53.

Pollard, E., Woiwod, I. P., Greatorex-Davies, J. N., Yates, T. J., and Welch, R. C. (1998*b*). The spread of coarse grasses and changes in numbers of Lepidoptera in a woodland nature reserve. *Biological Conservation*, **84**, 17–24.

Pope, C. R. (1988). The status of the Glanville Fritillary on the Isle of Wight. *Proceedings of the Isle of Wight Natural History and Archaeology Society*, **8**, 33–42.

Porter, A. H., Wenger, R., Geiger, H., Scholl, A., and Shapiro, A. M. (1997). The *Pontia daplidice-edusa* hybrid zone in northwestern Italy. *Evolution*, **51**, 1561–73.

Porter, J. (1997). *The colour identification guide to caterpillars of the British Isles*. Viking, Harmondsworth.

Porter, K. (1982). Basking behaviour in larvae of the butterfly *Euphydryas aurinia*. *Oikos*, **38**, 308–12.

Praeger, R. L. (1901). Irish topographical botany. *Proceedings of the Royal Irish Academy* (3rd series), **7**, i-clxxxvii and 1–410.

Pratt, C. (1983). A modern review of the demise of *Aporia crataegi* : the Black-veined White. *Entomologist's Record and Journal of Variation*, **95**, 45–52, 161–6, and 232–7.

Pratt, C. (1986). A history and investigation into the fluctuations of *Polygonia c-album*: the Comma butterfly. *Entomologist's Record and Journal of Variation*, **98**, 197–203 and 244–50.

Pratt, C. (1987). A history and investigation into the fluctuations of *Polygonia c-album*: the Comma butterfly. *Entomologist's Record and Journal of Variation*, **99**, 21–7 and 69–80.

Preston, C. D., Pearman, D. A., and Dines, T. D. (in press). *New atlas of the British and Irish flora*. Oxford University Press, Oxford.

Pullin, A. S. (1986). Effect of photoperiod and temperature on the life cycle of different populations of the Peacock butterfly *Inachis io*. *Entomologia Experimentalis et Applicata*, **41**, 237–42.

Pullin, A. S. (1988). Environmental cues and variable voltinism patterns in *Aglais urticae*. *Entomologist's Gazette*, **39**, 101–12.

Pullin, A. S. (ed.) (1995). *Ecology and conservation of butterflies*. Chapman & Hall, London.

Pullin, A. S. (1997). Habitat requirements of *Lycaena dispar batavus* and implications for re-establishment in England. *Journal of Insect Conservation*, **1**, 177–85.

Pullin, A. S. and Bale, J. S. (1989). Effects of low temperature on diapausing *Aglais urticae* and *Inachis io*: cold hardiness and overwintering survival. *Journal of Insect Physiology*, **35**, 277–81.

Pullin, A. S., McLean, I. F. G., and Webb, M. R. (1995). Ecology and conservation of *Lycaena dispar*: British and European perspectives. In *The ecology and conservation of butterflies*, (ed. A. S. Pullin), pp. 150–64. Chapman & Hall, London.

Pullin, A. S., Cameron, J. C., and James, C. H. (1997). Use of molecular genetics in the formulation of insect conservation strategies. In *The role of genetics in conserving small populations*, (ed. T. E. Tew, T. J. Crawford, J. W. Spencer, D. P. Stevens, M. B. Usher, and J. Warren), pp. 80–6. JNCC, Peterborough.

Pullin, A. S., Balint, Z., Balletto, E., Buszko, J., Coutsis, J. G., Kulfan, M., Lhonore, J. E., Settele, J., and Made, J. G. van der. (1998). The status, ecology and conservation of *Lycaena dispar* (Lycaenidae: Lycaenini) in Europe. *Nota Lepidopterologica*, **21**, 94–100.

Quinn, R. M., Gaston, K. J., Blackburn, P. M., and Eversham, B. C. (1997). Abundance-range size relationships of macrolepidoptera in Britain: the effects of taxonomy and life history variables. *Ecological Entomology*, **22**, 453–61.

Rackham, O. (1986). *The history of the countryside*. Dent, London.

Raemakers, I. P. and Made, J. G. van der. (1991). Het Spiegeldikkopje, habitat en beheer. *De Levende Natuur*, **4**, 117–22.

Rafe, R. W. and Jefferson, R. G. (1983). The status of *Melanargia galathea* (Lepidoptera: Satyridae) on the Yorkshire Wolds. *Naturalist*, **108**, 3–7.

Ravenscroft, N. O. M. (1990). The ecology and conservation of the Silver-studded Blue (*Plebejus argus* L.) on the Sandlings of East Anglia. *Biological Conservation*, **53**, 21–36.

Ravenscroft, N. O. M. (1994*a*). The ecology of the Chequered Skipper butterfly *Carterocephalus palaemon* Pallas in Scotland. I. Microhabitat. *Journal of Applied Ecology*, **31**, 613–22.

Ravenscroft, N. O. M. (1994*b*). The ecology of the Chequered Skipper butterfly *Carterocephalus palaemon* Pallas in Scotland. II. Foodplant quality and population range. *Journal of Applied Ecology*, **31**, 623–30.

Ravenscroft, N. O. M. (1994*c*). The feeding behaviour of *Carterocephalus palaemon* (Lepidoptera: Hesperiidae) caterpillars: does it avoid host defences or maximize nutrient intake? *Ecological Entomology*, **19**, 26–30.

Ravenscroft, N. O. M. and Warren, M. S. (1996*a*). *Species action plan: Chequered Skipper Carterocephalus palaemon*. Butterfly Conservation, Wareham.

Ravenscroft, N. O. M. and Warren, M. S. (1996*b*). *Species action plan: Silver-studded Blue Plebejus argus*. Butterfly Conservation, Wareham.

Ravenscroft, N. O. M. and Warren, M. S. (1996*c*). *Species action plan: Northern Brown Argus Aricia artaxerxes*. Butterfly Conservation, Wareham.

Ravenscroft, N. O. M. and Warren, M. S. (1996*d*). *Species action plan: the Mountain Ringlet Erebia epiphron*. Butterfly Conservation, Wareham.

Redway, D. B. (1981). Some comments on the reported occurrence of *Erebia epiphron* (Knoch) (Lepidoptera: Satyridae) in Ireland during the nineteenth century. *Entomologist's Gazette*, **32**, 157–9.

Revels, R. (1994). The rise and fall of the Holly Blue butterfly. *British Wildlife*, **5**, 236–9.

Rich, T. (1998). Squaring the circles—bias in distribution maps. *British Wildlife*, **9**, 213–9.

Roberts, R. E., Bourn, N. A. D., and Thomas, J. A. (1999). *Black Hairstreak site dossier, 1999*. Butterfly Conservation (confidential report), Wareham.

Robertson, P. A., Clarke, S. A., and Warren, M. S. (1995). Woodland management and butterfly diversity. In *The ecology and conservation of butterflies*, (ed. A. S. Pullin), pp. 113–22. Chapman & Hall, London.

Rodwell, J. S. (ed.) (1991*a*). *British plant communities, vol. 1: woodlands and scrub*. Cambridge University Press, Cambridge.

Rodwell, J. S. (ed.) (1991*b*). *British plant communities, vol. 2: mires and heaths*. Cambridge University Press, Cambridge.

Rodwell, J. S. (ed.) (1992). *British plant communities, vol. 3: grassland and montane communities*. Cambridge University Press, Cambridge.

Rodwell, J. S. (ed.) (1995). *British plant communities, vol. 4: aquatic communities, swamps, and tall-herb fens*. Cambridge University Press, Cambridge.

Rodwell, J. S. (ed.) (1999). *British plant communities, vol. 5: maritime communities and vegetation of open habitats*. Cambridge University Press, Cambridge.

Rothschild, M. and Marren, P. (1997). *Rothschild's reserves: time and fragile nature*. Harley Books, Colchester and Balaban Publishers, Rehovot.

Roy, D. B. and Sparks, T. H. (2000). Phenology of British butterflies and climate change. *Global Change Biology*, **6**, 407-16.

Roy, D. B., Rothery, P., Pollard, E., Moss, D., and Thomas, J. A. (in press). Butterfly numbers and weather: predicting historical trends in abundance and the future effects of climate change. *Journal of Animal Ecology*.

Royal Society for the Protection of Birds. (1995). *Environmental land management schemes*. RSPB, Sandy.

Royal Society for the Protection of Birds. (1998*a*). *Land for life: a future for the UK's special places for wildlife*. RSPB, Sandy.

Royal Society for the Protection of Birds. (1998*b*). *Agenda 2000: CAP and rural policy reform*. RSPB, Sandy.

Royal Society for the Protection of Birds. (1999). *The future of livestock farming in the UK*. RSPB, Sandy.

Saccheri, I., Kuussaari, M., Kankare, M., Vikman, P., Fortelius, W., and Hanski, I. (1998). Inbreeding and extinction in a butterfly metapopulation. *Nature*, **392**, 491–4.

Shannon, S. (1995). A study of the Small Mountain Ringlet *Erebia epiphron* Knoch. in south west Cumbria. Unpublished M.Sc. thesis. University of Leicester.

Sharrock, J. T. R. (1976). *The atlas of breeding birds in Britain and Ireland*. T. & A.D. Poyser, Calton.

Shaw, B. T. (1999). *The butterflies of Cheshire*. National Museums & Galleries on Merseyside, Liverpool.

Shaw, M. R. and Bland, K. P. (1994). Foodplant and parasitoids (Hymenoptera: Braconidae and Diptera: Tachinidae) of *Callophrys rubi* (L.) (Lepidoptera: Lycaenidae) on two wet heaths in Argyll, Scotland. *Entomologist's Gazette*, **45**, 27–8.

Sheail, J. (1998). *Nature conservation in Britain: the formative years*. Stationery Office, London.

Sheldrick, R. D. (ed.) (1997). *Grassland management in Environmentally Sensitive Areas*. British Grassland Society.

Shirt, D. B. (ed.). (1987). *British red data books, number 2 insects*. Nature Conservancy Council, Peterborough.

Shreeve, T. G. (1986*a*). Egg-laying by the Speckled Wood butterfly (*Pararge aegeria*): the role of female behaviour, host plant abundance and temperature. *Ecological Entomology*, **11**, 229–36.

Shreeve, T. G. (1986*b*). The effect of weather on the life cycle of the Speckled Wood butterfly *Pararge aegeria*. *Ecological Entomology*, **11**, 325–32.

Sinha, S. N., Lakhani, K. H., and Davies, B. N. K. (1990). Studies on the toxicity of insecticidal drift to the first instar larvae of the Large White butterfly *Pieris brassicae* (Lepidoptera: Pieridae). *Annals of Applied Biology*, **116**, 27–41.

Siriwardena, G. M., Baillie, S. R., Buckland, S. T., Fewster, R. M., Marchant, J. H., and Wilson, J. D. (1998). Trends in the abundance of farmland birds: a quantitative comparison of smoothed Common Bird Census indices. *Journal of Applied Ecology*, **35**, 24–43.

Skelton, M. (1999). Successful overwintering by Clouded Yellow *Colias croceus* in southern England. *Atropos*, **8**, 3–6.

Skelton, M. (2000). UK overwintering of Clouded Yellow. *Butterfly Conservation News*, **73**, 26.

Skinner, B. and Parsons, M. (1998). The immigration of Lepidoptera to the British Isles in 1995. *Entomologist's Record and Journal of Variation*, **110**, 197–227.

Smyllie, W. J. (1992). The Brown Argus butterfly in Britain—a range of *Aricia* hybrids. *Entomologist*, **111**, 27–37.

Smyllie, W. J. (1998). Similarities between British and north-west European *Aricia* 'subspecies' (Lepidoptera: Lycaenidae). *Nachr. Entomologischer Verein Apollo N. F.*, **19**, 69–88.

South, R. (1906). *The butterflies of the British Isles*. Warne, London.

Sparks, T. H. (1997). First leafing dates of trees in Surrey between 1947 and 1996. *The London Naturalist*, **76**, 15–20.

Sparks, T. H. and Yates, T. J. (1997). The effect of spring temperature on the appearance dates of British butterflies 1883–1993. *Ecography*, **20**, 368–374.

Sparks, T. H., Porter, K., Greatorex-Davies, J. N., Hall, M. L., and Marrs, R. H. (1994). The choice of oviposition sites in woodland by the Duke of Burgundy butterfly *Hamearis lucina* in England. *Biological Conservation*, **70**, 257–64.

Stace, C. (1997). *New flora of the British Isles (second edition)*. Cambridge University Press, Cambridge.

Stamp, L. D. (1955). *Man and the land*. Collins New Naturalist No. 31, London.

Stewart, J., Barbour, D., and Moran, S. (1998). *Highland butterflies: a provisional atlas*. Highland Biological Recording Group / Inverness Museum and Art Gallery, Inverness.

Sutcliffe, O. L. and Thomas, C. D. (1996). Open corridors appear to facilitate dispersal by Ringlet butterflies (*Aphantopus hyperantus*) between woodland clearings. *Conservation Biology*, **10**, 1359–65.

Sutcliffe, O. L., Thomas, C. D., and Djunijanti, P. (1997*a*). Area-dependent migration by Ringlet butterflies generates a mixture of patchy population and metapopulation attributes. *Oecologia*, **109**, 229–34.

Sutcliffe, O. L., Thomas, C. D., Yates, T. J., and Greatorex-Davies, J. N. (1997*b*). Correlated extinctions, colonizations and population fluctuations in a highly connected Ringlet butterfly metapopulation. *Oecologia*, **109**, 235–41.

Sutcliffe, R. (1994). The Clouded Yellow invasion of Scotland, 1992. *Glasgow Naturalist*, **22**, 389–96.

Sutcliffe, R. and Kirkland, P. (1998). Scottish Orange Tip survey 1997. *Butterfly Conservation News*, **68**, 20.

Sutton, S. L. and Beaumont, H. E. (ed.). (1989). *Butterflies and moths of Yorkshire: distribution and conservation*. Yorkshire Naturalists' Union, Doncaster.

Swaay, C. A. M. van. (1995). Measuring changes in butterfly abundance in the Netherlands. In *The ecology and conservation of butterflies*, (ed. A. S. Pullin), pp. 230–47. Chapman & Hall, London.

Swaay, C. A. M. van. (1998). *Mogelijkheden voor het Spiegeldikkopje in Noord-Brabant—eindrapport* (Possibilities for the Large Chequered Skipper in Noord-Brabant—final report). De Vlinderstichting, Wageningen.

Swaay, C. van and Warren, M. S. (1999). *Red data book of European butterflies* (*Rhopalocera*), Nature and Environment, No. 99. Council of Europe, Strasbourg.

Swash, G. and Belding, R. (1999). *Arable reversion in Environmentally Sensitive Areas and Countryside Stewardship Schemes*: *guidelines for project officers*. Farming and Rural Conservation Agency, London.

Tansley, A. G. (1939). *The British Islands and their vegetation*. Cambridge University Press, Cambridge.

Thomas, C. D. (1985). The status and conservation of the butterfly *Plebejus argus* L. (Lepidoptera: Lycaenidae) in north-west Britain. *Biological Conservation*, **33**, 29–51.

Thomas, C. D. (1994). Local extinctions, colonizations, and distributions: habitat tracking by British butterflies. In *Individuals*, *populations and patterns in ecology*, (ed. S. R. Leather, A. D. Watt, N. J. Mills, and K. F. A. Walters), pp. 319–36. Intercept, Andover.

Thomas, C. D. (1995). Ecology and conservation of butterfly metapopulations in the fragmented British landscape. In *The ecology and conservation of butterflies*, (ed. A. S. Pullin), pp. 46–63. Chapman & Hall, London.

Thomas, C. D. and Abery, J. C. G. (1995). Estimating rates of butterfly decline from distribution maps: the effect of scale. *Biological Conservation*, **73**, 59–65.

Thomas, C. D. and Jones, P. M. (1993). Partial recovery of a skipper butterfly (*Hesperia comma*) from population refuges: lessons for conservation in fragmented landscapes. *Journal of Animal Ecology*, **62**, 472–81.

Thomas, C. D. and Kunin, W. E. (1999). The spatial structure of populations. *Journal of Animal Ecology*, **68**, 647–57.
Thomas, C. D., Thomas, J. A., and Warren, M. S. (1992). Distribution of vacant and occupied habitats in fragmented landscapes. *Oecologia*, **92**, 563–7.
Thomas, C. D., Hill, J. K., and Lewis, O. T. (1998). Evolutionary consequences of habitat fragmentation in a localized butterfly. *Journal of Animal Ecology*, **67**, 485–97.
Thomas, C. D., Glen, S. W. T., Lewis, O. T., Hill, J. K., and Blakeley, D. S. (1999). Population differentiation and conservation of endemic races: the butterfly, *Plebejus argus*. *Animal Conservation*, **2**, 15–21.
Thomas, J. A. (1974). Factors influencing the numbers and distribution of the Brown Hairstreak, *Thecla betulae*, and Black Hairstreak, *Strymonidia pruni*. Unpublished Ph.D. thesis. University of Leicester.
Thomas, J. A. (1975*a*). *The Black Hairstreak: conservation report.* Unpublished report by the Institute of Terrestrial Ecology and the Nature Conservancy Council. ITE, Huntingdon.
Thomas J. A. (1975*b*). Some observations on the earlier stages of the Purple Hairstreak butterfly, *Quercusia quercus* (Linnaeus) (Lep., Lycaenidae). *Entomologist's Gazette*, **26**, 224–6.
Thomas, J. A. (1980). Why did the Large Blue become extinct in Britain? *Oryx*, **15**, 243–7.
Thomas, J. A. (1983*a*). A quick method for estimating butterfly numbers during surveys. *Biological Conservation*, **27**, 195–211.
Thomas, J. A. (1983*b*). The ecology and status of *Thymelicus acteon* (Lepidoptera: Hesperiidae) in Britain. *Ecological Entomology*, **8**, 427–35.
Thomas, J. A. (1983*c*). The ecology and conservation of *Lysandra bellargus* (Lepidoptera Lycaenidae) in Britain. *Journal of Applied Ecology*, **20**, 59–83.
Thomas, J. A. (1984). The conservation of butterflies in temperate countries: past efforts and lessons for the future. In *The biology of butterflies*, (ed. R. I. Vane-Wright and P. R. Ackery), pp. 333–53. Academic Press, London.
Thomas, J. A. (1989). *Hamlyn guide to the butterflies of the British Isles*. Hamlyn, London
Thomas, J. A. (1990*a*). The conservation of Adonis Blue and Lulworth Skipper butterflies—two sides of the same coin. In *Calcareous grasslands—ecology and management*, (ed. S. H. Hillier, D. W. H. Walton, and D. A. Wells), pp. 112–7. Bluntisham Books, Huntingdon.
Thomas, J. A. (1990*b*). The return of the Large Blue butterfly. *British Wildlife*, **1**, 2–13.
Thomas, J. A. (1991). Rare species conservation: case studies of European butterflies. In *The scientific management of temperate communities for conservation*, (ed. I. F. Spellerburg, F. B. Goldsmith, and M. G. Morris), pp. 149–98. Blackwell Scientific, Oxford.
Thomas, J. A. (1993). Holocene climate changes and warm man-made refugia may explain why a sixth of British butterflies possess unnatural early-successional habitats. *Ecography*, **16**, 278–84.
Thomas, J. A. (1995). The ecology and conservation of *Maculinea arion* and other European species of butterfly. In: *The ecology and conservation of butterflies*, (ed. A. S. Pullin), pp. 180–97. Chapman & Hall, London.
Thomas, J. A. (1999*a*). The Large Blue butterfly—a decade of progress. *British Wildlife*, **11**, 22–7.
Thomas, J. A. (1999*b*). Return of the Large Blue. *Butterfly Conservation News*, **71**, 18–21.
Thomas, J. A. and Lewington, R. (1991). *The butterflies of Britain and Ireland.* Dorling Kindersley, London.
Thomas, J. A. and Simcox, D. (1982). A quick method for estimating larval populations of *Melitaea cinxia* L. during surveys. *Biological Conservation*, **22**, 315–22.
Thomas, J. A. and Snazell, R. (1989). Declining fritillaries: the next challenge in the conservation of Britain's butterflies. *Annual report of the Institute of Terrestrial Ecology*, 54–6.
Thomas, J. A., Thomas, C. D., Simcox, D. J., and Clarke, R. T. (1986). The ecology and declining status of the Silver-spotted Skipper butterfly (*Hesperia comma*), in Britain. *Journal of Applied Ecology*, **23**, 365–80.
Thomas, J. A., Moss, D., and Pollard, E. (1994). Increased fluctuations of butterfly populations towards the northern edges of species' ranges. *Ecography*, **17**, 215–20.
Thomas, J. A., Simcox, D. J., Wardlaw, J. C., Elmes, G. W., Hochberg, M. E., and Clarke,

R. T. (1998). Effects of latitude, altitude and climate on the habitat and conservation of the endangered butterfly *Maculinea arion* and its *Myrmica* ant hosts. *Journal of Insect Conservation*, **2**, 39–46.

Thomson, G. (1980). *The butterflies of Scotland*. Croom Helm, London.

Tilzey, M. (1998). *Sustainable development and agriculture*, English Nature Research Report, No. 278. English Nature, Peterborough.

Tinbergen, N. (1972). The courtship of the Grayling *Eumenis (Satyrus) semele* L. (Lep. Satyridae). In *The animal and its world: explorations of an ethologist. Volume 1: field studies 1932–1972*, (ed. N. Tinbergen), pp. 197–249. Allen and Unwin, London.

Tolman, T. and Lewington, R. (1997). *Butterflies of Britain and Europe*. HarperCollins, London.

Tucker, J. M. (1991). *Problems posed by the hibernation and migration habits of the Red Admiral (Vanessa atalanta)*. Butterfly Conservation Occasional Paper, No. 1. Butterfly Conservation, Colchester.

Tucker, J. M. (1997). *The Red Admiral butterfly*. Butterfly Conservation, Colchester.

Tunmore, M. (1996). The 1995 Camberwell Beauty *Nymphalis antiopa* (L.) influx. *Atropos*, **1**, 2–5.

Tunmore, M. (2000). The 1999 Monarch *Danaus plexippus* (Linn.) influx into the British Isles. *Atropos*, **9**, 4–16.

Turner, J. R. G., Gatehouse, C. M., and Corey, C. A. (1987). Does solar energy control organic diversity? Butterflies, moths and the British climate. *Oikos*, **48**, 195–205.

UK Biodiversity Group. (1998*a*). *Tranche 2 Action Plans. Volume I—vertebrates and vascular plants*. English Nature, Peterborough.

UK Biodiversity Group. (1998*b*). *Tranche 2 Action Plans. Volume II—terrestrial and freshwater habitats*. English Nature, Peterborough.

UK Biodiversity Group. (1999*a*). *Tranche 2 Action Plans. Volume III – plants and fungi*. English Nature, Peterborough.

UK Biodiversity Group. (1999*b*). *Tranche 2 Action Plans. Volume IV – invertebrates*. English Nature, Peterborough.

UK Biodiversity Group. (1999*c*). *Tranche 2 Action Plans. Volume VI – terrestrial and freshwater species and habitats*. English Nature, Peterborough.

Vanholder, B. (1996). The migration of *Danaus plexippus* (Linnaeus, 1758) during October 1995 in the UK (Lepidoptera: Nymphalidae, Danainae). *Nota Lepidopterologica*, **19**, 129–40.

Vickery, M. L. (1995). Gardens: the neglected habitat. In *Butterfly conservation and ecology*, (ed. A. S. Pullin), pp.123–34. Chapman & Hall, London.

Vos, R. de and Rutten, A. L. M. (1996). Migrating Lepidoptera in 1995. *Entomologische Berichten*, **56**, 177–91.

Wacher, J. (1998). Successful overwintering of Painted Lady *Cynthia cardui* in the UK. *Atropos*, **5**, 19–20.

Wallis DeVries, M. F. and Raemakers, I. (1998). Grazing impact on butterflies in coastal dunes. *Annales de Zootechnie*, **47**, 514.

Warren, M. S. (1984). The biology and status of the Wood White butterfly, *Leptidea sinapis* (L.) (Lepidoptera: Pieridae), in the British Isles. *Entomologist's Gazette*, **35**, 207–23.

Warren, M. S. (1985). The influence of shade on butterfly numbers in woodland rides, with special reference to the Wood White, *Leptidea sinapis*. *Biological Conservation*, **33**, 147–64.

Warren, M. S. (1987). The ecology and conservation of the Heath Fritillary *Mellicta athalia*. *Journal of Applied Ecology*, **24**, 467–513.

Warren, M. S. (1990). *The Chequered Skipper Carterocephalus palaemon in northern Europe*. Butterfly Conservation Occasional Paper, No. 2. Butterfly Conservation, Colchester.

Warren. M. S. (1991). The successful conservation of an endangered species, the Heath Fritillary *Mellicta athalia* Rott. in Britain. *Biological Conservation*, **55**, 37–56.

Warren, M. S. (1992*a*). The conservation of British butterflies. In *The ecology of butterflies in Britain*, (ed. R. L. H. Dennis), pp. 246–74. Oxford University Press, Oxford.

Warren, M. S. (1992*b*). Britain's vanishing fritillaries. *British Wildlife*, **3**, 282–96.

Warren, M. S. (1993*a*). A review of butterfly conservation in central southern Britain: I. Protection, evaluation and extinction on prime sites. *Biological Conservation*, **64**, 25–35.

Warren, M. S. (1993*b*). A review of butterfly conservation in central southern Britain: II. Site management and habitat selection of key species. *Biological Conservation*, **64**, 37–49.

Warren, M. S. (1994*a*). Autecology and conservation needs of the High Brown Fritillary: annual report for 1993/94. Butterfly Conservation, Wareham.

Warren, M. S. (1994*b*). The UK status and suspected metapopulation structure of a threatened European butterfly, the Marsh Fritillary *Eurodryas aurinia*. *Biological Conservation*, **67**, 239–49.

Warren, M. S. (1995*a*). Managing local micro-climates for the High Brown Fritillary. In *The ecology and conservation of butterflies*, (ed. A. S. Pullin), pp. 198–210. Chapman & Hall, London.

Warren, M. S. (1995*b*). Habitats directive—list of candidate sites announced. *Butterfly Conservation News*, **60**, 7.

Warren, M. S. (1996). Giant step to save woodland butterflies: the coppice for butterflies challenge. *Butterfly Conservation News*, **63**, 18–20.

Warren, M. S. and Bourn, N. A. D. (1997). The impact of grassland management on threatened butterflies in ESA's. In *Grassland management in Environmentally Sensitive Areas*, (ed. R. D. Sheldrick), pp. 138–43. British Grassland Society.

Warren, M. S. and Bourn, N. A. D. (1998). *Species action plan: Wood White Leptidea sinapis*. Butterfly Conservation, Wareham.

Warren, M. S. and Fuller, R. J. (1993). *Woodland rides and glades—their management for wildlife (second edition)*. JNCC, Peterborough.

Warren, M. S. and Key, R. S. (1991). Woodlands: past, present and potential for insects. In *The conservation of insects and their habitats*, (ed. N. M. Collins and J. A. Thomas), pp. 155–212. Academic Press, London.

Warren, M. S. and Oates, M. R. (1995). The importance of Bracken habitats to fritillary butterflies and their management for conservation. In *Bracken: an environmental issue*, (ed. R.T. Smith and J. A.Taylor), pp. 178–81. The International Bracken Group, Leeds.

Warren, M. S. and Thomas, J. A. (1992). Butterfly responses to coppicing. In *The ecological effects of coppicing*, (ed. G.P. Buckley), pp. 249–70. Chapman & Hall, London.

Warren, M. S. and Thomas, J. A. (1993). Conserving the Silver-Spotted Skipper in practice. *Butterfly Conservation News*, **54**, 21–6.

Warren, M. S., Thomas, C. D., and Thomas, J. A. (1984). The status of the Heath Fritillary butterfly *Mellicta athalia*, in Britain. *Biological Conservation*, **29**, 287–305.

Warren, M. S., Roberts, G., Roper, P., and Barnett, L. K. (1994). *New life for old woods—the Land Rover woodlands campaign information pack*. Butterfly Conservation, Dedham.

Warren, M. S., Barnett, L. K., Gibbons, D. W., and Avery, M. I. (1997). Assessing national conservation priorities: an improved red list of British butterflies. *Biological Conservation*, **82**, 317–28.

Warren, M. S., Thomas, J. A., and Wilson, R. J. (1999). *Management options for the Silver-spotted Skipper butterfly: a study of the timing of grazing at Beacon Hill NNR, Hampshire, 1986–98*. Butterfly Conservation, Wareham.

Warrington, S. and Brayford, J. P. (1995). Some aspects of the population ecology and dispersal of the Small Skipper butterfly *Thymelicus sylvestris* in a series of linked grasslands. *The Entomologist*, **114**, 201–9.

Webb, D. A. (1980). The biological vice-counties of Ireland. *Proceedings of the Royal Irish Academy*, **80**, 179–96.

Wedell, N., Nylin, S., and Janz, N. (1997). Effects of larval host plant and sex on the propensity to enter diapause in the Comma butterfly. *Oikos*, **78**, 569–75.

Wells, T. C. E., Bell, S., and Frost, A. (1981). *Creating attractive grasslands using native plant species*. Nature Conservancy Council, Shrewsbury.

Wells, T. C. E., Bell, S., and Frost, A. (1986). *Wild flower grasslands from crop-grown seed and hay-bales*, Focus on Nature Conservation Series, No. 15. Nature Conservancy Council, Peterborough.

West, I. (1993). *The Swallowtail butterfly*. Butterfly Conservation, Colchester.

Whalley, P. and Lewington, R. (1996). *Butterflies (revised edition)*. Mitchell Beazley Publishers, London.

White, J. and Doyle, G. (1982). The vegetation of Ireland: a catalogue raisonne. *Royal Dublin Society Journal of Life Sciences*, **3**, 289–368.

Whitfield, K. E. J. (1999). An investigation of the status and recolonisation patterns of Adonis Blue butterfly colonies in the South Wessex Downs. Unpublished M.Sc. thesis. University of Warwick.

Wickman, P. O. (1988). Dynamics of mate-searching behaviour in a hilltopping butterfly, *Lasiommata megera* (L.): the effects of weather and male density. *Zoological Journal of the Linnean Society*, **93**, 357–77.

Wickman, P. O. (1992). Mating systems of *Coenonympha* butterflies in relation to longevity. *Animal Behaviour*, **44**, 141–8.

Wickman, P. O., Garcia-Barros, E., and Rappe-George, C. (1994). The location of landmark leks in the Small Heath butterfly, *Coenonympha pamphilus*: evidence against the hot-spot model. *Behavioural Ecology*, **6**, 39–45.

Wiklund, C. and Åhrberg, C. (1978). Host plants, nectar source plants, and habitat selection of males and females of *Anthocharis cardamines* (Lepidoptera). *Oikos*, **31**, 169–83.

Williams, C. B. (1958). *Insect migration*. Collins, London.

Willmott, K. J. (1978). The history of *Argynnis lathonia* in Britain. *Bulletin of the Amateur Entomologists' Society*, **37**, 72–80.

Willmott, K. J. (1990). *The Purple Emperor butterfly*. Butterfly Conservation, Colchester.

Willmott, K. J. (1994). Locating and conserving the elusive Purple Emperor. *British Wildlife*, **5**, 288–95.

Willmott, K. J. (1999). *The Holly Blue butterfly*. Butterfly Conservation, Colchester.

Wilson, J. (1997). The management of limestone grassland and scrub for butterflies: a case study for Warton Crag RSPB reserve. In *RSPB conservation review 11*, (ed. C. J. Cadbury and S. Niemann), pp. 80–7. RSPB, Sandy.

Wilson, R. (1998). The Queen of Spain Fritillary *Argynnis lathonia*—a new British breeding species? *Atropos*, **5**, 3–7.

Wilson, R. J. and Bourn, N. A. D. (1998). *The Silver-spotted Skipper butterfly Hesperia comma: review of trends on monitored sites*. Unpublished report to Butterfly Conservation, Wareham.

Windig, J. J. and Nylin, S. (1999). Adaptive wing asymmetry in males of the Speckled Wood butterfly (*Pararge aegeria*). *Proceedings of the Royal Society of London B*, **266**, 1413–18.

Wynne, G., Avery, M., Campbell, L., Gubbay, S., Hawkswell, S., Juniper, T., King, M., Newbery, P., Smart, J., Steel, C., Stones, T., Stubbs, A., Taylor, J., Tydeman, C., and Wynde, R. (1995). *Biodiversity challenge (second edition)*. RSPB, Sandy.

Glossary and abbreviations

agri-environment schemes Government-funded schemes providing grants to farmers and land managers to improve or maintain the natural beauty and wildlife of their land. Examples include Countryside Stewardship and Environmentally Sensitive Areas in England, Rural Environment Protection Scheme in the Republic of Ireland, Tir Gofal in Wales, and the Scottish Rural Stewardship Scheme.

annual index The annual sum of the weekly counts of a species at a site recorded using the butterfly transect method.

ASSI Area of Special Scientific Interest (designated under Northern Ireland legislation).

biodiversity The variety of life on Earth or any given part of it.

Biodiversity Action Plan (BAP) The UK Government's plan for the protection and sustainable use of biodiversity.

biotope Distinctive vegetation type or land use.

bivoltine Having two generations or broods each year.

BMS Butterfly Monitoring Scheme.

BNM Butterflies for the New Millennium.

BRC Biological Records Centre.

BSE Bovine Spongiform Encephalopathy ('mad cow disease').

Butterfly*Net* The network of local co-ordinators set up by the Butterflies for the New Millennium project. Co-ordinators receive, check, and computerize records before transferring them to the central database.

calcareous grassland Species-rich, often specialized, plant communities that occur on soils derived from chalk and limestone.

carr A woodland habitat, characterized by willows or Alder, and often associated with fens and other wetlands.

CEH Centre for Ecology and Hydrology (incorporates the former ITE).

cocoon A silky, protective envelope formed by a larva and in which the pupa develops.

collated index Shows trends in relative abundance of a species, calculated by combining annual indices from sites monitored by the butterfly transect method.

colony A group of individuals occurring in a discrete patch of habitat apparently separated from other groups of the same species.

Common Agricultural Policy (CAP) An agreement between member states of the European Union to support farm incomes by maintaining agricultural prices.

conservation headlands Arable field margins with reduced chemical applications and other forms of management to encourage wildlife.

coppice management A traditional method of management of broad-leaved trees, producing a supply of poles by cutting just above the base of the trunk on a regular cycle.

CORINE A biotope classification system largely based on vegetation composition, developed as a framework for comparing habitats across Europe. Now superceded by EUNIS.

Culm grassland A mixture of marshy grassland, rush pasture, wet heathland, and scrub characteristic of north Devon and Cornwall, named after the underlying geology (the Culm Measures).

DNFC The Dublin Naturalists' Field Club.

diapause A form of dormancy, characteristic of a particular stage of insect life cycles, during which metabolic activity is greatly decreased during a certain (inclement) period of the year.

endemic A species or other taxon restricted to a particular region or country, and which, as far as is known, evolved or is relic there.

ESA Environmentally Sensitive Area.

eskers Long, winding ridges of glacial material, often with base-rich soils, deposited by melt-water streams flowing under glaciers.

fenland Wetland habitats with peat soils, which receive water and nutrients from the soil, rock and ground water as well as from rainfall. Fens may have high or low nutrient status, the former being fed by mineral-enriched waters and occurring mainly in the lowlands.

flight period The length of the adult (flying) stage of a butterfly's life cycle.

flush A patch of wet ground, usually on a slope, where water flows diffusely rather than in a channel.

foodplant The plant species on which butterfly larvae feed.

genetically modified organism Living organism transformed by genes artificially introduced from another species.

habitat A place in which a particular plant or animal lives, characterized by its biological and physical attributes. Often used in a wider sense, referring to major assemblages of plants and animals found together.

hibernaculum The shelter of an overwintering (hibernating) larva, usually formed from a leaf.

honeydew A sugary substance excreted by aphids upon which some butterflies feed. Also refers to secretions produced by some butterfly larvae.

immigrant A species that has established a breeding population in a region outside its historically known native range.

instar The stage of development and growth between successive moults in insect larvae. The first instar describes the larva on hatching from the egg until the first moult, after which it becomes a second instar and so on.

introduction The intentional or accidental release of an organism to a place outside its historically known native range.

ITE Institute of Terrestrial Ecology (now CEH).

JNCC Joint Nature Conservation Committee.

lek A gathering of male animals that females visit only to mate.

Levana Computer software developed for butterfly recording as part of the Butterflies for the New Millennium project.

machair Sandy, often lime-rich, grassland above the high-water mark usually inland from a beach: found in northern Scotland (mainland and islands) and western Ireland and used for extensive grazing and rotational hay cropping.

MAFF Ministry of Agriculture, Fisheries and Food.

metapopulation Several discrete populations in close proximity that are interconnected by movement of individuals.

microclimate Climatic conditions (e.g. sunshine, temperature, and humidity) that occur on a very small scale, for example at the base of a clump of grass, or a depression in the ground.

migrant (cf. vagrant) A butterfly that moves instinctively from one geographical region to another.

NNR National Nature Reserve.

osmaterium Tubular, forked process possessed by the larvae of some swallowtail species. When a larva is alarmed the osmaterium is everted and produces a strong odour.

parasite An organism that is intimately associated with and metabolically dependant upon another living organsim (the host) for the completion of its life cycle, and is typically detrimental to the host.

parasitoid An animal, usually an insect, that is parasitic during the larval stage of its life cycle but becomes free living when adult.

phenology The seasonal occurrence of life cycle events such as the separate life stages of a species.

reintroduction The intentional release of an organism into a part of its native range from which it has become extinct.

(**habitat**) **restoration** The improvement of degraded habitats; usually achieved by altering the management regime.

Rhos pasture Species-rich, wet grassland and heath in Wales, which is used as pasture and forms a key habitat for the Marsh Fritillary.

rides Pathways cut to provide access through woodland, often benefiting butterflies by creating open sunny areas where larval foodplants and nectar sources can grow.

RSPB Royal Society for the Protection of Birds.

ruderal Inhabiting disturbed sites.

scent-brands Scales on the upperside of the forewings of male butterflies which are modified for distributing pheromones to attract females.

semi-natural habitat Habitat comprising mainly native plants, that has been modified to a limited extent by human activity.

set-aside Arable land that is taken out of production temporarily, as part of a CAP scheme to reduce surpluses. A subsidy is paid to the farmer for such land.

SSSI Site of Special Scientific Interest (designated under British legislation).

succession The progressive change in the plant and animal species occupying an area or habitat over a long period of time, from the initial colonization of an area to mature ecosystem.

sustainable development Development that meets the needs of the present with the intention of not compromising the ability of future generations to meet their own needs.

transect A fixed route, walked regularly through a site, used to monitor changes in the abundance of butterflies.

unimproved grassland Grassland comprising mainly native species, which has not been ploughed, drained, cultivated, or intensively fertilized.

univoltine Having one brood or generation each year.

vagrant (cf. migrant) A butterfly outside its normal geographical range.

voltinism The number of generations of a butterfly species each year.

Index

Notes

- For all butterfly species, the **Species Accounts** (in Chapter 5) are indexed under the *vernacular name* in English with cross-reference to the *scientific name*. Their equivalents are listed in taxonomic order on pages 366–7.
- The pages covered by the individual **Species Accounts**, usually shown in **bold**, the first of which is the main distribution map.
- These accounts contain detailed sections (outlined on pages 50–1) including some supplementary distribution maps, photographs, life cycle descriptions, trends in distribution and abundance, conservation, and management. Most of the information pertaining to each species is located within these species accounts and with few exceptions (e.g. 'ant-butterfly relationships'), topics within them are not separately indexed.
- Use of larval foodplants is also described in the individual Species Accounts and is not separately indexed, except where of particular conservation importance. A full list of both vernacular and scientific names of foodplants is given in Appendix 2 (pages 368–71).
- Page references to *figures* are shown in *italics*.

abundance 333–6; *see also* flight areas
addresses, useful contact 387–9
Adonis Blue (*Polyommatus bellargus*) **168–71**
  causes of decline 337, 339, 341
  changes in distribution 326, 329, 332, 333–4
  habitat *19*, *167*, 339, 360
  phenogram *394*
  rabbit grazing 352
afforestation *351*
Aglais urticae, *see* Small Tortoiseshell
agriculture *342*, 342–3, 351–2, 361
  Common Agricultural Policy (CAP) 340, 355, 358, 364
  intensification 338–9, *342*, 351
  pattern of *18*
  traditional 336
  reform proposals 364
agri-environment schemes *see* Countryside Stewardship, Environmentally Sensitive Areas, Tir Gofal
air pollution, *see also* atmospheric pollution
  acid rain 342–3
  nitrogen deposition 342–3, 351
altitude, butterfly distribution and *14*, *26*, *254*
American Painted Lady (*Cynthia virginiensis*) 198
Anthocharis cardamines, *see* Orange-tip
ancient woodland 22–3, 338
Annual Index, *see* Butterfly Monitoring Scheme
ant-butterfly relationships, *see also Myrmica*, *Lasius*
  Adonis Blue 169–70
  Brown Argus 153–4
  Brown Hairstreak 122
  Chalkhill Blue 165
  Common Blue 162
  Green Hairstreak 118
  Holly Blue 174
  Large Blue 177–9
  Northern Brown Argus 157
  Purple Hairstreak 126
  Silver-studded Blue 149–50, 349
  Small Blue 145–6
Apatura iris, *see* Purple Emperor
Aphantopus hyperantus, *see* Ringlet
Aporia crataegi, *see* Black-veined White
arable fields and crops *24*, *25*, *342*
Araschnia levana, *see* Map
Areas of Special Scientific Interest, Northern Ireland (ASSI) 9, 355, 364
Argynnis adippe, *see* High Brown Fritillary
Argynnis aglaja, *see* Dark Green Fritillary
Argynnis lathonia, *see* Queen of Spain Fritillary
Argynnis paphia, *see* Silver-washed Fritillary
Aricia agestis, *see* Brown Argus
Aricia artaxerxes, *see* Northern Brown Argus
atmospheric pollution, effects of 351
  on Gatekeeper 271
  on Purple Hairstreak 126–7
  on Ringlet 278

Bath White (*Pontia daplidice*) **296–7**
Berger's Clouded Yellow (*Colias alfacariensis*) **292–3**
Bern Convention 50
bias, *see* recording
Biodiversity Action Plan, UK (BAP) 7, 9, 50, 347, 357, 364–5
  BAP Status 50, 348–9
  Habitat Action Plans 347, 349
  Priority Species 347–9
  Regional Action Plans (BC) 350, 359
  Species Action Plans 7, 347, 349–50, 359
  Species of Conservation Concern 347–9
Biodiversity Action Plan (Ireland) 347
Biological Records Centre (BRC) 1, 3–5, 365, 387; *see also* Centre for Ecology and Hydrology
  and Butterflies for the New Millennium Project 1, 8–9
Black Hairstreak (*Satyrium pruni*) **132–5**
  causes of decline, 337
  changes in distribution 326, 328, 334
  habitat *22*, *24*
  in historic data sets 49
  reintroduction 341
  on reserves 360
Black-veined White (*Aporia crataegi*) **294–5**
  comparison with historic data sets *295*
  extinction 333–4, 337, 346
Blackthorn 22–24, *134*
Bloxworth Blue, *see* Short-tailed Blue
bog habitats *28*, 338, 351
Boloria euphrosyne, *see* Pearl-bordered Fritillary
Boloria selene, *see* Small Pearl-bordered Fritillary
bracken habitats *20*, 337, 340, 351–2
Brimstone (*Gonepteryx rhamni*) **96–9**
  changes in distribution 327, 330, 332, 334
  climate change and 343
  habitat *22*, *24*, *27*
  phenogram *392*
British Butterfly Conservation Society 1; *see also* Butterfly Conservation
Brown Argus (*Aricia agestis*) **152–5**
  changes in distribution *155*, 327, 329, 331–2, 334
  climate change and 343–4
  flight area loss 335
  habitat *19*, *29*
  phenogram *394*
  set-aside fields 344–5
  as wider countryside species 325, 329

Brown Hairstreak (*Thecla betulae*) **120–3**
causes of decline 337
changes in distribution 326, 330–1, 334
egg surveys *5, 121–2*
habitat *24*
hedgerow management 351
buckthorns *22–4, 27*, 97–9, *98*
Butterflies for the New Millennium (BNM) project 1, 7–10
bias and constraints 37–8, 325
comparison with BMS 39
comparison with historic data sets 41, 48–9, 325–335
co-ordinators 32–3
data collation, validation and verification 36–8
improving coverage 34–6, 329
instructions for recorders 377–84
interpreting the data 40–9
mapping the data 40–5
numbers (of butterflies) seen 32, 41–3
numbers (of records) *40*
postcard surveys *35*, 36
recording and data collection 30–9, 377–84
recording effort (intensity) *42–3*, 44, 329
targeted recording *35*, 325
verification of records 36
weather during survey period 385–6
butterfly atlases (local) 6, 373–6
Butterfly Conservation (BC) 1, 365, 387
Action for Butterflies Project 347
in Biodiversity Challenge Group 347
and Butterflies for the New Millennium Project 8–9, 365
Code on Species Restoration 362
foundation (1968) 346
Regional Action Plans 350, 359
Species Action Plans 7, 347, 349–50, 359, 364
Ten-point Plan for the conservation of butterflies 364
UK BAP partner 347
butterfly habitats 11–29; *see also* habitats
Butterfly Monitoring Scheme (BMS) 5–7, 38–9, 335–6, 387
annual index 39, 46, 422
climate and weather 343
collated index 45–6, 51; *see also* individual species accounts
comparison with BNM project 39
interpretation of data 45–7
site locations *6*, 46
transect walk 6–7, 38–9
Butterfly*Net* 6, 32–3, 36–8

Cacyreus marshalli, *see* Geranium Bronze
calcareous grassland *19*, 337–41
Callophrys rubi, *see* Green Hairstreak
Camberwell Beauty (*Nymphalis antiopa*) **310–13**
migrations *312–13*
phenogram 313
CAP, *see* Common Agricultural Policy
carr woodland 27
Carterocephalus palaemon, *see* Chequered Skipper
CEH *see* Centre for Ecology and Hydrology
Celastrina argiolus, *see* Holly Blue
Centre for Ecology and Hydrology (CEH) 1, 6, 387
Centre for Environmental Data and Recording, Belfast 33
chalk grassland *19*, 337–41; *see also* geology; grassland
Chalkhill Blue (*Polyommatus coridon*), **164–7**
causes of decline 337
changes in distribution 326, 334
habitat *19, 167*
chemicals, agricultural 342, 351–2
Chequered Skipper (*Carterocephalus palaemon*) **52–5**
causes of decline 337
changes in distribution 326, 333–4
extinction in England 54
habitat *23*
reintroduction 53
targeted recording 44
churchyards 28
climate 11, *17*, 343
and BNM survey 37–8, 385–6
and BMS 343
climate change 47, 343–4, 352–5, *353–4*; *see also* individual species accounts
Clouded Yellow (*Colias croceus*) **92–5**; *see also* Berger's Clouded Yellow, Pale Clouded Yellow
changes in distribution 327, 330
climate change and 345
as migrant 2
Coenonympha pamphilus, *see* Small Heath
Coenonympha tullia, *see* Large Heath
Colias alfacariensis, *see* Berger's Clouded Yellow
Colias croceus, *see* Clouded Yellow
Colias hyale, *see* Pale Clouded Yellow
collated index, *see* Butterfly Monitoring Scheme
collecting 343, 352
Comma (Polygonia c-album) **208–11**
in BMS 46
changes in distribution *210*, 327, 331–2, 344
habitat *23*
phenogram *396*
Committee for the Protection of British Lepidoptera (1925) 346
Common Agricultural Policy (CAP) 340, 355, 358, 364
Common Birds's-foot-trefoil 19–20, 23, 29, 49
Common Blue (*Polyommatus icarus*) **160–3**
causes of decline 337
changes in distribution 327, 330
flight area loss 334
habitats *19–21, 23, 25, 29*
pesticides and 342
phenogram *47*, 48
Common Rock-rose, distribution in Northern Britain *159*
conifer plantation *23, 351*
conservation in the new millennium 346–60
conservation legislation 8–9, 50
conservation status (of butterfly species) 50
coppice management (of woodland) *22*, 337–8, 340–1, *361*
CORINE biotope codes 32
Countryside Council for Wales (CCW) 9, 33, 387
Countryside Stewardship Scheme (England) 71, 357–8
Cupido argiades, *see* Short-tailed Blue
Cupido minimus, *see* Small Blue
Cyaniris semiargus, *see* Mazarine Blue
Cynthia cardui, *see* Painted Lady

Danaus plexippus, *see* Monarch
Dark Green Fritillary (Argynnis aglaja) **224–7**
causes of decline 337
changes in distribution 326, 330, 334
flight area loss 335
habitats *19–24, 26*
phenogram *396*
data collation 36–8
declines of butterflies, causes 336–43; *see also* individual species accounts
agricultural intensification 342–3, 351–2
airborne pollution 342–3
collecting 343, 352
development, urban 352
habitat fragmentation and isolation 336–7, 340–2, 350–1
habitat loss 336–9
habitat management change 339–40, 350–2
improvement of semi-natural grassland 337–8
deer
browsing 22, 26, 135, 187
overpopulation 135
derelict land *29*
development 12
industrial 352
sustainable 364
urban 352
Dingy Skipper (*Erynnis tages*) **76–9**
bias in recording 43
changes in distribution 326, 329–31, 334
climate change and 345
habitats *19, 23, 25, 29*
loss of flight area 335
phenogram *391*
threats 337
distribution maps 40–9; *see also* individual species accounts
improving coverage *34*

The Dublin Naturalists' Field Club (DNFC) 1, 9, 33, 365, 387
Duchas: The Heritage Service (Republic of Ireland) 9, 388
Duke of Burgundy (*Hamearis lucina*) **180–3**
causes of decline 337, 340
changes in distribution 326, 329, 331, 334
habitats *19, 22, 24, 183*, 340
local extinction 335

ecological change (succession) 13, 352
elms 22–4, *131*
Dutch Elm Disease 129–131, *130–1*, 309, 331
English Nature 9, 33, 387
Natural Areas 359
Environment and Heritage Service (Northern Ireland) 9, 387
Environmentally Sensitive Area(s) (ESA) 71, *357*, 358, 361
Erebia aethiops, *see* Scotch Argus
Erebia epiphron, *see* Mountain Ringlet
Erynnis tages, *see* Dingy Skipper
Essex Skipper (Thymelicus lineola) **60–3**,
changes in distribution *62*, 327, 329, 331–2
habitat *20, 25*
introduction to Cornwall 62
separation from Small Skipper 2, 62
Euphydryas aurinia, *see* Marsh Fritillary
European Union
Common Agricultural Policy 340, 355, 358, 364
Habitats and Species Directive 347, 363
Everes argiades, *see* Short-tailed Blue
extinction, of butterfly species 333–5, 337, 343, 346
Black-veined White 294–5
Large Blue 177–8
Large Copper 141–3
Large Tortoiseshell 307–8
Mazarine Blue 302–3

farming, *see* agriculture
fenland *27*, 85–7, 141–3, 336–8
fertilizers 342
field margins *24*, 339, 342, 364
flight area(s) 334–5, 341
flight period 47–8; *see also* phenogram, phenology, and individual species accounts
forestry, *see* woodland
Forestry Commission
Coppice for Butterflies Challenge 359
Forest Design Plans 361
Forest Enterprise 388
grants for woodland management 360
Forum for the Application of Conservation Techniques (FACT) 360, 389
future action (recommendations for) 362–5

garden butterfly survey *7*, 33
garden habitats *28*
Gatekeeper [Hedge Brown] (*Pyronia tithonus*) **268–71**
changes in distribution *270*, 327, 330–2, 334–5
climate change and 344
at edge of range *3, 270*
flight area loss 335
habitat *24–5*
pesticides and 342
phenogram *398*
genetic diversity 349–50
genetically modified (GM) crop cultivation 351
geology 11, *14*
chalk and limestone geology of southern England *167, 170*
Geranium Bronze (*Cacyreus marshalli*) **304–5**
introduction into Britain 2, 304–5
Glanville Fritillary (*Melitaea cinxia*) **236–9**
causes of decline 337, 340
changes in distribution 326, 328, 334, 346
habitats *21*
global warming 47; *see also* climate change
Gonepteryx rhamni, *see* Brimstone
government agencies 387–8
grasses *19, 29*
grassland 19–21, 336–41
acid *20*, 341
arable *18*
calcareous *19*, 337–41
chalk *19*
coastal (and dunes) *21*
damp *20*
hay meadows *20*
improved 13, *21*, 337–8
limestone 19
lowland 338–9
management 339–40, 360, *362*
restoration 361–2
semi-natural 13, *19*, 336–41, *340*
sward height and structure 19, 20
upland *26*
Grayling (*Hipparchia semele*) **264–7**
causes of decline 337, 339
changes in distribution 326, 330, 334,
flight area loss 334–5
genetic diversity 349
habitat(s) *20–21, 27, 29, 266*
phenogram *397*
grazing
abandonment of 178, 337, 339–40, *340*
deer 22, 26, 135
livestock 12, *20*, 26–8, 33, 360–2, *361–2*
over-grazing 119, 234, 340, *350*
rabbit 19, 22, 71, 178, 339, 352
Grazing Animals Project (GAP) 360, 389
Green Hairstreak (*Callophrys rubi*) **116–19**
bias in recording 43, 325, 328
causes of decline 337
changes in distribution 326, 328–30, 334
extinction, local 335
habitats *19, 23–7*
phenogram *393*
Green-veined White (*Pieris napi*) **108–11**
changes in distribution 327, 330, 334
habitats *20, 24–6, 28*
in historic data sets 49
improved coverage *34*
pesticides and 342
phenogram *392*
grid reference 30–1, 379
Grizzled Skipper (Pyrgus malvae) **80–3**
bias in recording 37, 43
causes of decline 337
changes in distribution 326, 334
climate change and 343
habitats *19, 23, 25, 29*, 352
phenogram *391*
priority 349
threats 337

habitat(s) *19–29*; *see also* specific habitats such as grassland, heathland, woodland
bog(s) *28*, 338, 351
bracken *20*, 337, 340, 351–2
churchyards 28
CORINE biotope codes 32
destruction/loss 336–9
fenland *27*, 85–7, 141–3, 336–8
fragmentation and isolation 336–7, 340–2, 350–1
geology and soils *15–16*
management 19–29, 339–40, 350, 360–2
montane *26*, 253, *255*
post-industrial *29*, 352
restoration 55, 361–2, 364
semi-natural 12–13, *19*, 336–41, *340*, 346–7
set-aside 25, 344–5
upland *26*
urban *29*
wetland *27–8*
Habitat Action Plans 347, 349; *see also* Biodiversity Action Pan
habitat specialist species 3, *46*, 324–31
causes of decline 327–9, 336–9
changes in distribution in Britain 326–8
changes in distribution in Ireland 330–1
in historic data sets 48–9
and urban development 352
Hamearis lucina, *see* Duke of Burgundy
Hare's-tail Cottongrass, *Eriophorum vaginatum*
distribution of *286*
habitat *28, 287*
headlands, conservation, *see* field margins
Heath Fritillary (*Melitaea athalia*) **240–3**
causes of decline 337, 340
changes in distribution 326, 330, 334, 346
climate change and 344
habitat 2, *22, 242*
on reserves 360
heathland *27*, 149–51, 265–7, *266*, 337, *342*

Hedge Brown, *see* Gatekeeper
hedge(row)s *24*, 324, 337–9, 364
  arable field margins *24*
  loss 121–3, 337–9, 345
  management 24, *339*, 351
  restoration 364
herbicides 25, 339, 342, 351
The Heritage Council (Republic of Ireland) 9, 33, 388
Hesperia comma, *see* Silver-spotted Skipper
Heteropterus morpheus, *see* Large Chequered Skipper
High Brown Fritillary (*Argynnis adippe*) **220–3**
  causes of decline 335, 337, 340
  changes in distribution 326–8, 334
  in ESAs 358
  extinction, local 335
  habitat *20*, *22*,
  on reserves 360
Hipparchia semele, *see* Grayling
history of butterfly recording 3–6, *48*, *49*
Holly 22–4, 28
Holly Blue (*Celastrina argiolus*) **172–5**
  abundance cycles *174*, *175*, 333
  changes in distribution 326–7, 330–3
  habitat *24*, *28*
  phenogram *394*
honeydew 12, 126
Horseshoe Vetch
  distribution map *166*
  habitat *19*
hot-spots (or 'honeypot' areas) 43, 45, *358–9*, 359

immigration of butterflies, *see* migration
improved grassland, *see* grassland
Inachis io, *see* Peacock
insecticides, *see* pesticides
Institute of Terrestrial Ecology (ITE) *see* Biological Records Centre, Centre for Ecology and Hydrology
International Union for the Conservation of Nature (IUCN) 347–8
instructions to BNM recorders 377–9
introduction(s) *see* reintroductions
island species, *see* habitat specialist species
Issoria lathonia, *see* Queen of Spain Fritillary

Joint Nature Conservation Committee (JNCC) 1, 6, 9, 36, 387

Kidney Vetch 19, 21, 29; *see also* Small Blue

Ladoga camilla, *see* White Admiral
Lampides boeticus, *see* Long-tailed Blue
land use *17*, 12–18, 344–5
landscape (management and restoration) 355–9, 364
Large Blue (*Maculinea arion*) **176–9**
  collecting 343
  extinction 326, 334, 337, 339, 346
  habitat *179*
  reintroduction 177–9, 334, 341, 362
Large Chequered Skipper (*Heteropterus morpheus*) **288–9**
  in Channel Islands 2
Large Copper (*Lycaena dispar*) **140–3**
  causes of decline 336–8
  collecting 343
  extinction 333–4, 346
  fenland loss 336–8
  habitat *27*
  races *dispar*, *batavus* and *rutilus* 141–2
  reintroduction 141–3
Large Heath (*Coenonympha tullia*) **284–7**
  causes of decline 337–8, 351
  changes in distribution 326, 330, 334
  habitat *26*, *28*
  races *davus*, *scotica* and *polydama* 285
Large Skipper (*Ochlodes venata*) **72–5**
  changes in distribution 327, 331–2
  habitat *23–5*, *29*
  loss of flight area 335
  phenogram *391*
Large Tortoiseshell (*Nymphalis polychloros*) **306–9**
  extinction 326, 333–4, 337
Large White (*Pieris brassicae*) **100–3**
  changes in distribution 327, 330, 334
  habitat *24*, *25*, *28*
  pesticides and 342
  phenogram *392*
Lasiommata megera, *see* Wall
Lasius species of ant 174
  alienus 150, 154, 170
  niger 150
latitude 47, 390; *see also* phenogram
leaf litter 22
legislation 8–9, 50, 362, 364
Leptidea sinapis, *see* Wood White
Levana butterfly recording system 36
Limenitis camilla, *see* White Admiral
limestone grassland *19*
local (county, regional and national) butterfly atlases 6, 373–6
Long-tailed Blue (*Lampides boeticus*) **298–9**
Lulworth Skipper (*Thymelicus acteon*) **64–7**
  changes in distribution 326, 328, 334,
  European status 349
  grazing and 340
  habitat *19*
  in historic data sets 49
Lycaena dispar, *see* Large Copper
Lycaena phlaeas, *see* Small Copper
Lysandra bellargus, *see* Adonis Blue
Lysandra coridon, *see* Chalkhill Blue
Maculinea arion, *see* Large Blue
Maniola jurtina, *see* Meadow Brown
Manx National Heritage 388
Map (*Araschnia levana*) **314–15**
  introduction(s) 362
mapping the data 40–5; *see also* Butterflies for the New Millennium project
Marbled White (*Melanargia galathea*) **260–3**
  changes in distribution *262*, 327, 331–2
  grazing and 340
  habitat *19*, *25*, *29*
  as wider countryside species 325
Marsh Fritillary (*Euphydryas aurinia*) **232–5**
  causes of decline 337, 339–42
  changes in distribution 326, 328–31, 333–4, 339–40
  EC Habitats Directive 363
  In ESA and 358
  habitat *19–20*, 341–2
matrix species, *see* wider countryside species
Mazarine Blue (*Polyommatus semiargus*) **302–3**
  extinction 333, 337, 346
Meadow Brown (*Maniola jurtina*) **272–5**
  causes of decline 337
  changes in distribution 327, 330, 334
  flight area loss 335–6
  habitats *19*, *21*, *28–9*
  in historic data sets 49
  phenogram *398*
Melanargia galathea, *see* Marbled White
Melitaea cinxia, *see* Glanville Fritillary
Melitaea athalia, *see* Heath Fritillary
Mellicta athalia, *see* Heath Fritillary
metapopulation (dynamics) 341
migration(s) 46
  Bath White 296–7
  Berger's Clouded Yellow 292
  Camberwell Beauty 311–13
  climate change and 345
  Clouded Yellow 2, 93–5, 327, 330, 345
  Large Tortoiseshell 307
  Large White 102
  Long-tailed Blue 298–9
  Map 314–15
  Monarch 321–3
  overwintering of migrants 93–5, 193–5, 290, 318, 345
  Painted Lady 2, 198–9, 327, 330, 345
  Pale Clouded Yellow 290–1
  Queen of Spain Fritillary 317–8, 345
  Red Admiral 2, 193–5, 327, 330
  Short-tailed Blue 300–1
  reverse (southerly) migration 195, 198
  Small White 106
Ministry of Agriculture, Fisheries and Food (MAFF) 388
Ministry of Defence (MOD) 33
Monarch (*Danaus plexippus*) **312–18**
montane habitats *26*, 253, *255*
moorland *26*

Mountain Ringlet (*Erebia epiphron*) **252–5**
bias in recording 329
changes in distribution 326, 329, 334–5
habitat *26*, *255*
phenogram *397*
postcard survey 36
threat from climate change 345
Myrmica (species of ant) 174
ruginodis 126
sabuleti 118, 154, 170, 177–8
scabrinodis 177
myxomatosis 71, 339, 352

National Biodiversity Network 1, 365, 389
National Grid *31*, 47, *379*
National Trust, (NT) 33, 179, 243, 254, 389
National Trust for Scotland 389
Natura 2000 sites 363–4
Natural Areas initiative 359
Natural Heritage Zones (Scotland) 359
nature reserves 9, 336, 346, 352, 359–61
Neozephyrus quercus, *see* Purple Hairstreak
nettles 24, 28, *194*, 324, 345
non-adult stages 32, 50–1; see also individual species accounts
Northern Brown Argus (Aricia artaxerxes) **156–9**
causes of decline 337
changes in distribution 326, 334
distinction from Brown Argus in historic data set 327
habitats *19*, *29*
phenogram *394*
numbers (of butterflies) seen
codes for 32
instructions to recorders 377–84
interpretation 41–3
Nymphalis antiopa, *see* Camberwell Beauty
Nymphalis polychloros, *see* Large Tortoiseshell

Ochlodes venata, *see* Large Skipper
Orange-tip (*Anthocharis cardamines*) **112–15**
abundance in BMS 46
changes in distribution *115*, 327, 330, 332, 344
habitats *20*, *24–5*, *28*
phenogram *393*
postcard survey *35*
Ordnance Survey National Grid(s) *31*, *379*

Painted Lady (*Vanessa cardui*) **196–9**
changes in distribution 327, 330
climate change and 345
as migrant 2, 327, 330
phenogram *395*
Pale Clouded Yellow (*Colias hyale*) **290–1**
Papilio machaon, *see* Swallowtail
Pararge aegeria, *see* Speckled Wood
parks 28
Peacock (*Inachis io*) **204–7**
changes in distribution *206*, 327, 330–2
climate change and 344
habitats *24–5*, *28*
phenogram *206*
Pearl-bordered Fritillary (*Boloria euphrosyne*) **216–9**
abundance in BMS 46
causes of decline 337, 340, 351–2
changes in distribution 46, 326, 328–31, 334
discovery in Ireland (in 1922) 2
in ESAs 358
extinction, local 335
habitat *20*, *22–3*, 351–2, 358
Pearl-bordered Fritillary survey 33
phenogram *396*
targeted recording 33, 44
Perennial Rye-grass 13, 78, 338
pesticides 342
phenogram 47–8, 51, 390, *391–8*
Camberwell Beauty *313*
Common Blue *47*
Small Heath *282*
Peacock *206*
Purple Hairstreak *127*
Wall *250*
phenology 36, 46–7
Pieris brassicae, *see* Large White
Pieris napi, *see* Green-veined White
Pieris rapae, *see* Small White
Plebeius argus, *see* Silver-studded Blue
Polygonia c-album, *see* Comma
Polyommatus bellargus, *see* Adonis Blue
Polyommatus coridon, *see* Chalkhill Blue
Polyommatus icarus, *see* Common Blue
Polyommatus semiargus, *see* Mazarine Blue
Pontia daplidice, *see* Bath White
postcard surveys 35–6, 114
priority species 347–9, *358*
Purple Emperor (*Apatura iris*) **188–91**
causes of decline 337
changes in distribution *191*, 326, 329, 331–2, 334
climate change and 345
habitat *22*
Purple Hairstreak (*Neozephyrus quercus*) **124–7**
changes in distribution *126*, 327, 329–331
habitat *22*, *28*
phenogram *127*
recording of 325, 331–2
as wider countryside species 325
Pyrgus malvae, *see* Grizzled Skipper
Pyronia tithonus, *see* Gatekeeper

quarries (disused) *29*
Queen of Spain Fritillary (*Issoria lathonia*) **316–18**
climate change and 345
Quercusia quercus, *see* Purple Hairstreak

rabbit
grazing 19, 22, 71, 178, 339
myxomatosis 71, 339, 352
viral haemorrhagic disease 352
race(s) 2, 349–50
of Bath White 296
of Large Copper 141–2
of Large Heath 285
of Northern Brown Argus 157
of Swallowtail 85–6
railway line/tracks *25*
Wood White habitat 89–91
rainfall *17*, *354*, 385–6; *see also* climate change
RECORDER, biological data management program, 36
Red Admiral (*Vanessa atalanta*) **192–5**
changes in distribution 327, 330
habitat *28*,
migration 2, 193–5
phenogram *395*
reverse (southerly) migration 195
Red Data Book
Britain 347
European 50, 347
re-establishment *see* reintroduction
Regional Action Plans (BC) 350, 359
reintroduction(s) (and introductions) 341, 361–2
Adonis Blue 171, 362
Black Hairstreak 134, 341
Black-veined White 294–5
Chequered Skipper 53–5
Grizzled Skipper 82
Heath Fritillary 242–3, 362
Large Blue *176*, 177–9, 341, 362
Large Copper 141–3
Marbled White 262
Marsh Fritillary 234–5
ride management *see* woodland
Ringlet (*Aphantopus hyperantus*) **276–9**
changes in distribution *278*, 327, 330–2
climate change and 344
flight area loss 335
habitats *20*, *23–5*, 342
phenogram *398*
roadside verges *25*, 90, 113–5, 261–2
Royal Society for the Protection of Birds (RSPB) 33, 389

Satyrium pruni, *see* Black Hairstreak
Satyrium w-album, *see* White-letter Hairstreak
Scotch Argus (*Erebia aethiops*) **256–9**
causes of decline 337
changes in distribution 327, 334
habitat *23*, *26*, *258*
phenogram *397*
postcard suvey 36
as wider countryside species 325
Scottish Natural Heritage (SNH) 9, 33, 388
Natural Heritage Zones 359
scrub 13, 23, *24*, 29

semi-natural habitat 12–13, 18, *19*, 336–41, *340*, 346–7
set-aside fields 25, *155*, 345
Short-tailed Blue (*Cupido argiades*) **300–1**
Silver-spotted skipper (*Hesperia comma*) **68–71**
causes of decline 337, 341
changes in distribution 326, 328–9, 332, 334
conservation needs 339, 341
habitat *19*, 339
rabbit grazing 71, 339, 352
Silver-studded Blue (*Plebeius argus*) **148–51**
causes of decline 337, 339
changes in distribution 326, 328–9, 334–5
flight area loss 335
genetic diversity 349
habitat *27*, *342*
Silver-washed Fritillary (*Argynnis paphia*) **228–31**
causes of decline 337
changes in distribution *231*, 325, 329, 332, 334
climate change and 345
extinction, local 335
habitat *22*
Site(s) of Special Scientific Interest (SSSI) 8–9, 39, 355, 359, 364
Small Blue (*Cupido minimus*) **144–7**
causes of decline 337
changes in distribution 326, 328, 330–1, 334
extinction, local 335
habitats *19*, *22*, *25*, *29*
phenogram *393*
Small Copper (*Lycaena phlaeas*) **136–9**
causes of decline 337
changes in distribution 327, 330, 334
flight area loss 335
habitats *20*, *27–8*
phenogram *393*
Small Heath (*Coenonympha pamphilus*) **280–3**
causes of decline 337
changes in distribution 327, 330, *336*, 337
flight area loss 335–6
habitats *19–21*, *26–7*, *29*
in historic data sets 49
phenogram *282*
Small Pearl-bordered Fritillary (*Boloria selene*) **212–5**
causes of decline 337, 340, 351
changes in distribution 326, 329, 334
extinction, local 335
habitats *20*, *22–3*, *26*, *28*
Small Skipper (*Thymelicus silvestris*) **56–9**
changes in distribution 327, 331–2, 337
flight area loss 334
habitats *19–20*, *24–5*, *29*
phenogram *391*
Small Tortoiseshell (*Aglais urticae*), **200–3**
changes in distribution 327, 330, 334
habitats *24–5*, *28–9*
pesticides and 342
Small White ((Pieris rapae) **104–7**
changes in distribution 327, 330, 334
habitats *24*, *28*
phenogram *392*
soils 11, *16*
Special Area of Conservation (Ireland) 347
species
of Conservation Concern 347–9, *359*
extinct species 294–5, 302–3, 306–7, 333–5, 337, 343, 346
habitat specialists 3, *46*, 324–31; *see also* habitat specialist species
legal status 50, 348, 363
migrant 2, 50; *see also* migration(s)
priority (BAP) 347–9, *358*
rare migrants 290–3, 296–301, 309–18
restoration *see* reintroduction
richness *44–5*
sub-species *see* race(s)
of the wider countryside 3, 324–32, 355, 357; *see also* wider countryside species
Species Action Plan(s) 7, 347, 350, 359, 364
Speckled Wood (*Pararge aegeria*) **244–7**
abundance in BMS 39, 46
changes in distribution *246*, 327, 330–2, 344
climate change and 344
flight area loss 335
as habitat specialist 324
habitats *22*, *28*, 342
phenogram *397*
statutory conservation organisations 9, 387–8
sub-species *see* race(s)
Swallowtail (*Papilio machaon*) **84–7**
causes of decline 336–8
changes in distribution 326, 334,
fenland loss 336–8
habitat *27*, 346
races *britannicus* and *gorganus* 85–6

temperature *17*, *353*, 385–6; *see also* climate, climate change
Thecla betulae, *see* Brown Hairstreak
Thymelicus acteon, *see* Lulworth Skipper
Thymelicus lineola, *see* Essex Skipper
Thymelicus sylvestris, *see* Small Skipper
Tir Gofal *357*, 358
transect (walk) 6–7, 38–9; *see also* Butterfly Monitoring Scheme

upland(s) *26*; *see also* montane habitat
urban habitats 28, *29*, 352

vagrant(s) 2, 41; *see also* migration(s), and individual species accounts
validation (of data) 36
Vanessa atalanta, *see* Red Admiral
Vanessa cardui, *see* Painted Lady
verge(s), roadside *25*, 90, 113–5, 261–2
verification (of data) 36–7
Vice-county(ies) (Watsonian) 3–4, *399*, 400
viral haemorrhagic disease 71, 352

Wall (*Lasiommata megera*) **248–51**
abundance in BMS 46
changes in distribution 327, 329–334, 336, 344
decline, causes of 337
flight area loss 335
habitats *19–21*, *25*, *29*
phenogram *250*
weather, and recording for BNM 37–38, 385–6; *see also* climate
wetland habitats *27*, *28*
White Admiral (*Limenitis camilla*) **184–7**
changes in distribution *186*, 326, 329, 332, 334
climate change and 343
habitat *22*
phenogram *395*
White-letter Hairstreak (*Satyrium w-album*) **128–31**
causes of decline 337
changes in distribution 327, 329, 331–2, 334
habitat *22*, *24*
recording of 331
as wider countryside species 325
wider countryside species 3, 324–32, 355, 357
changes in distribution in Britain 327, *329*
changes in distribution in Ireland 330
flight areas 335
hedgerow loss 338–9
in historic data sets 48–9, 325–32
Wildlife and Countryside Act 50, 348
Wood White (*Leptidea sinapis*) **88–91**
causes of decline 335, 337, 340
changes in distribution 326, 330–1, 333–4, 346
habitat *23*, *25*
woodland 12–13, 22–3
afforestation *351*
ancient 22–3, 337, 344
broad-leaved 13, *22*, 338, 340–1
carr 27
clearings *22*, 324
coniferous 13, *23*, 341
coppice(ing) *22*, 337, 340–1
management 22–3, 338–41, 344, 360–1
rides and glades 22, *23*, 342–3
woodland edge 22–3, *134*, 352

Message to recorders: we still need your help

The distribution maps presented in this Atlas are based on the most comprehensive survey of butterflies ever undertaken in Britain and Ireland. The results show that many species are undergoing rapid changes and continued recording and monitoring are vital. Moreover, some areas remain under-recorded (e.g. parts of Ireland, northern England, Scotland, and Wales) and there are still colonies of rare species to be discovered. Further records are needed to help Butterfly Conservation and other organizations to conserve our butterflies effectively in the future.

Anyone can help record butterflies and contribute to the various ongoing recording schemes. These are designed to suit all levels of experience and whatever amount of time individuals have available. For full details please contact your local co-ordinator or:

Butterfly Conservation
Manor Yard
East Lulworth
Wareham
Dorset BH20 5QP

Tel: 01929 400209

Or visit our website:
www.butterfly-conservation.org

Recorders in the Republic of Ireland should contact:

The Dublin Naturalists' Field Club
35 Nutley Park
Donnybrook
Dublin 4

Photo and artwork credits

Cover/spine : Marsh Fritillary. Alan Barnes
Back cover: Chalk grassland. English Nature (Peter Wakeley)

Frontispiece: Wall. Richard Lewington
Page 4: EB Ford. John Haywood
Page 4: John Heath. Biological Records Centre
Page 5: Ernie Pollard. Biological Records Centre
Page 5: Jeremy Thomas. Chris Newbold
Page 19: Chalk grassland. English Nature (Peter Wakeley)
Page 20: Wet meadow. Robert Thompson.
Page 20: Acidic grassland. Martin Warren
Page 21: Coastal grassland. English Nature (Peter Wakeley)
Page 21: Silage harvesting. English Nature (Peter Wakeley)
Page 22: Broad-leaved woodland. English Nature (Peter Wakeley)
Page 22: Coppice woodland. English Nature (Peter Wakeley)
Page 23: Woodland ride. English Nature (Peter Wakeley)
Page 23: Conifer plantation. Alan Barnes
Page 24: Scrub habitat. Robert Thompson
Page 24: Arable field margin. Alan Barnes
Page 25: Arable fields. Gail Jeffcoate
Page 25: Roadside verge. English Nature (Peter Wakeley)
Page 25: Disused railway line. English Nature (Peter Wakeley)
Page 26: Mountains and upland grassland. Alan Barnes
Page 26: Moorland. Alan Barnes
Page 27: Lowland heathland. English Nature (Peter Wakeley)
Page 27: Fenland. English Nature (Peter Wakeley)
Page 28: Bog habitat. Robert Thompson
Page 28: Garden habitat. Stephen Jeffcoate
Page 29: Disused quarry. English Nature (Peter Wakeley)
Page 29: Urban habitat. Butterfly Conservation
Page 35: Orange-tip. F. Rodway
Page 53: Chequered Skipper. Alan Barnes
Page 54: Chequered Skipper larva. Martin Warren
Page 57: Small Skipper. Robert Thompson
Page 61: Essex Skipper. Alan Barnes
Page 65: Lulworth Skipper. Alan Barnes
Page 66: Lulworth Skipper larva. Martin Warren
Page 69: Silver-spotted Skipper. Alan Barnes
Page 69: Silver-spotted Skipper eggs. Ken Willmott
Page 73: Large Skipper. Alan Barnes
Page 74: Large Skipper eggs. Ken Willmott
Page 77: Dingy Skipper. Alan Barnes
Page 78: Dingy Skipper eggs. Ken Willmott
Page 81: Grizzled Skipper. Alan Barnes
Page 82: Grizzled Skipper larva. Martin Warren
Page 85: Swallowtail. Robert Thompson
Page 86: Swallowtail larva. Jim Asher
Page 89: Wood White. Martin Warren
Page 90: Wood White larva. Ken Willmott
Page 93: Clouded Yellow. Martin Warren
Page 97: Brimstone. Robert Thompson
Page 97: Brimstone larva. Ken Willmott
Page 101: Large White. Robert Thompson
Page 101: Large White eggs. Jim Asher
Page 102: Large White larvae. Martin Warren
Page 102: Large White pupa. Ken Willmott
Page 105: Small White. Ken Willmott
Page 105: Small White eggs. Martin Warren
Page 106: Small White larva. Martin Warren
Page 106: Small White pupa. Jim Asher
Page 109: Green-veined White. Robert Thompson
Page 109: Green-veined White eggs. Martin Warren
Page 113: Orange-tip. Alan Barnes
Page 114: Orange-tip egg. Jim Asher
Page 117: Green Hairstreak. Robert Thompson
Page 118: Green Hairstreak larva. Ken Willmott
Page 121: Brown Hairstreak. Alan Barnes
Page 122: Brown Hairstreak egg. Jim Asher
Page 125: Purple Hairstreak. Alan Barnes
Page 125: Purple Hairstreak larva. Alan Barnes
Page 129: White-letter Hairstreak. Alan Barnes
Page 130: Larval feeding galleries. Roger Key
Page 131: Die-back of hedgerow elm. English Nature (Peter Wakeley)
Page 133: Black Hairstreak. Robert Thompson
Page 133: Black Hairstreak pupa. Jim Asher
Page 134: Blackthorn habitat. Martin Warren
Page 137: Small Copper. Alan Barnes
Page 137: Small Copper larva. Ken Willmott
Page 141: Large Copper. Robert Thompson
Page 142: Water Dock. Martin Warren
Page 145: Small Blue. Butterfly Conservation
Page 146: Small Blue larva. Jim Asher
Page 149: Silver-studded Blue. Alan Barnes
Page 150: Silver-studded Blue larva. Ken Willmott
Page 153: Brown Argus. Alan Barnes
Page 155: Set-aside fields. English Nature (Peter Wakeley)
Page 157: Northern Brown Argus. Ken Willmott
Page 157: Northern Brown Argus eggs. Martin Warren
Page 158: Northern Brown Argus larva. Jim Asher
Page 161: Common Blue. Robert Thompson
Page 161: Common Blue larva. Jim Asher
Page 165: Chalkhill Blue. Alan Barnes
Page 169: Adonis Blue. Ken Willmott
Page 169: Adonis Blue larva. Ken Willmott
Page 173: Holly Blue. Alan Barnes
Page 173: Holly Blue larva. Ken Willmott
Page 177: Large Blue. Martin Warren
Page 178: Large Blue larva. Jeremy Thomas
Page 179: Former Large Blue site. Jeremy Thomas
Page 181: Duke of Burgundy. Alan Barnes
Page 182: Duke of Burgundy larva. Martin Warren

Page 183: Duke of Burgundy habitat. Ken Willmott
Page 185: White Admiral. Alan Barnes
Page 185: White Admiral larva. Martin Warren
Page 189: Purple Emperor. Robert Thompson
Page 190: Purple Emperor larva. Tony Hoare
Page 193: Red Admiral. Alan Barnes
Page 194: Red Admiral larval tent. Jim Asher
Page 197: Painted Lady. Martin Warren
Page 198: Painted Lady larva. Jim Asher
Page 201: Small Tortoiseshell. Alan Barnes
Page 201: Small Tortoiseshell larvae. Martin Warren
Page 205: Peacock. Robert Thompson
Page 206: Peacock larvae. Jim Asher
Page 209: Comma. Robert Thompson
Page 213: Small Pearl-bordered Fritillary. Robert Thompson
Page 214: Small Pearl-bordered Fritillary larva. Martin Warren
Page 217: Pearl-bordered Fritillary. Robert Thompson
Page 218: Pearl-bordered Fritillary larva. Martin Warren
Page 221: High Brown Fritillary. Robert Thompson
Page 222: High Brown Fritillary larva. Martin Warren
Page 225: Dark Green Fritillary. Robert Thompson
Page 226: Dark Green Fritillary larva. Martin Warren
Page 229: Silver-washed Fritillary. Alan Barnes
Page 230: Silver-washed Fritillary larva. Martin Warren
Page 233: Marsh Fritillary. Alan Barnes
Page 234: Marsh Fritillary larvae. Robert Thompson
Page 237: Glanville Fritillary. Robert Thompson
Page 238: Glanville Fritillary larval nest. Ken Willmott
Page 241: Heath Fritillary. Alan Barnes
Page 242: Heath Fritillary larva. Martin Warren
Page 243: Heath Fritillary habitat. Martin Warren
Page 245: Speckled Wood. Alan Barnes
Page 249: Wall. Robert Thompson
Page 253: Mountain Ringlet. Ken Willmott
Page 255: Mountain Ringlet habitat. Alan Barnes
Page 257: Scotch Argus. Robert Thompson
Page 258: Scotch Argus habitat. Alan Barnes
Page 261: Marbled White. Robert Thompson
Page 261: Marbled White larva. Jim Asher
Page 265: Grayling. Alan Barnes
Page 266: Grayling habitat. Ken Willmott
Page 269: Gatekeeper. Alan Barnes
Page 269: Gatekeeper larva. Ken Willmott
Page 273: Meadow Brown. Robert Thompson
Page 273: Meadow Brown larva. Jim Asher
Page 277: Ringlet. Alan Barnes
Page 277: Ringlet larva. Jim Asher
Page 281: Small Heath. Robert Thompson
Page 281: Small Heath larva. Ken Willmott
Page 285: Large Heath. Robert Thompson
Page 285: Large Heath larva. Ken Willmott
Page 287: Large Heath habitat. Martin Warren
Page 288: Large Chequered Skipper. Kars Veling (Dutch Butterfly Conservation)
Page 289: Large Chequered Skipper larva. Martin Warren
Page 290: Pale Clouded Yellow. Kars Veling (Dutch Butterfly Conservation)
Page 292: Berger's Clouded Yellow. Robert Thompson
Page 294: Black-veined White. Alan Barnes
Page 296: Bath White. Robert Thompson
Page 298: Long-tailed Blue. Robert Thompson
Page 300: Short-tailed Blue. Jim Asher
Page 302: Mazarine Blue. Martin Warren
Page 304: Geranium Bronze. Jim Asher
Page 305: Geranium Bronze larva. John Holloway
Page 307: Large Tortoiseshell. Alan Barnes
Page 308: Large Tortoiseshell larva. Ken Willmott
Page 311: Camberwell Beauty. Alan Barnes
Page 311: Camberwell Beauty larva. Ken Willmott
Page 314: Map. Alan Barnes
Page 315: Summer brood Map. Alan Barnes
Page 317: Queen of Spain Fritillary. Alan Barnes
Page 318: Queen of Spain Fritillary. Alan Barnes
Page 321: Monarch. Robert Thompson
Page 339: Flailed hedgerow. Ken Willmott
Page 340: Coastal grassland. Martin Warren
Page 342: Intensive cereal farm. English Nature (Peter Wakeley)
Page 344: Power station. English Nature (Peter Wakeley)
Page 350: Overgrazing by sheep. Stephen Jeffcoate
Page 351: Conifer plantation. English Nature (Paul Glendell)
Page 360: Coppice woodland. Martin Warren
Page 361: Grazing with traditional livestock. English Nature (Peter Wakeley)
Page 361: Extensive cattle grazing. Martin Warren